GEOGRAPHY AND GEOGRAPHERS

*Anglo-American
Human Geography
since 1945*

Fifth Edition

R. J. JOHNSTON

Professor of Geography, University of Bristol

ARNOLD

A member of the Hodder Headline Group
LONDON • NEW YORK • SYDNEY • AUCKLAND

First published in Great Britain in 1979
Second edition 1983
Third edition 1987
Fourth edition 1991
Fifth edition published in 1997 by
Arnold, a member of the Hodder Headline Group
338 Euston Road, London NW1 3BH

Copublished in the US, Central and South America by John Wiley & Sons, Inc.
605 Third Avenue, New York, NY 10158-0012

Whilst the advice and information in this book is believed to be true and accurate
at the time of going to press, neither the author nor the publisher can accept any
legal responsibility or liability for any errors or omissions that may be made.

British Library Cataloguing in Publication Data
A catalogue entry for this book is available from the British Library

Library of Congress Cataloging-in-Publication Data
A catalog record for this book is available from the Library of Congress

ISBN 0 340 65263 2
ISBN 0 470 23652 3 (Wiley)

Typeset in 10/12pt Sabon by Phoenix Photosetting, Chatham, Kent
Printed and bound in Great Britain by
J W Arrowsmith Ltd, Bristol

To Rita

Contents

Preface

To the first edition

Most students reading for a degree or similar qualification are required to pay some attention to their chosen discipline's academic history. The history presented to them commonly ends some time before the present. This has advantages for the historian, because the past is often better interpreted from the detachment of a little distance and there is less chance of hurting scholars still alive. But there is a major disadvantage for the students. In virtually every other component of their courses they will be dealing with the discipline's current literature, and so if the history ends some decades ago, they are presented with the contemporary substance but not with the contemporary framework, except where this is very clearly derivative of the historical context.

This state of affairs is unfortunate. Students need a conspectus of the current practice in their chosen discipline and should encounter a relevant overview which describes, and perhaps explains too, what scholars believe the philosophy and methodology of the discipline presently are and should be. Such an overview will allow the substantive courses comprising the rest of the degree to be placed in context and appreciated as examples of the disciplinary belief system as well as ends in themselves.

As a discipline expands, so the need for a course in its 'contemporary history' will grow too. In the last few decades, for example, most disciplines, and certainly human geography, have expanded greatly, with expansion measured by the number of active participants and their volume of published work. And the more active members there are, almost certainly the greater the variety of work undertaken, making it difficult for individual students to provide their own conspectus of the discipline from their own reading. Hence the need for a 'contemporary history' course at the present time.

The present book is the outcome of teaching such a course for several years, and is offered as a guide for others, both teachers and students. As with all texts, it has many idiosyncrasies. The course on which it is based is taught to final-year students reading for honours degrees in geography, and is probably best used by people at that level since it assumes familiarity with the concepts and language of human geography. Further, my view is that students probably benefit most from a framework after experiencing some of its contents; this book provides the matrix for organizing the individual parts rather than a series of slots into which parts can be placed later, which would be the case if the course were taught early in the degree.

A series of constraints circumscribes the contents and approach of this contemporary history. First, it deals only with human geography, for several reasons. The most important is that I find the links between physical and human geography tenuous, as those disciplines are currently practised. The major link between them is a sharing of techniques and research procedures, but these are shared with other disciplines too, and are insufficient foundation for a unified discipline. (What price a department of factorial ecology?) Further, my own competence, work and interests lie wholly within human geography and although I have been trained with and by and have worked among physical geographers, and have obtained stimuli from this, I am incompetent to write about physical geography. And finally, much of the human geography discussed here is of North American origin, and many geographers there, especially in the United States, encounter no physical geography as it is understood in British universities. To a considerable extent, therefore, human and physical are separate, if not independent, disciplines. Throughout the book, I use the terms 'geography' and 'human geography' interchangeably.

The second constraint is cultural; the subject matter of the book is Anglo- American human geography. Most of the work discussed emanates from either the United States or the United Kingdom: there are some contributions from workers in Australia, Canada, and New Zealand, but the efforts of geographers in the rest of the world are largely ignored. (A partial exception is Sweden, which has major academic links with the Anglo-American tradition; much Swedish geographical work is published in English.) Such academic parochialism in part reflects personal linguistic deficiencies, but it is not entirely an idiosyncratic decision. Contacts between Anglo-American human geography, on the one hand and, say, that of France and of Germany, on the other, have been few in recent decades, so to concentrate on the former is not to commit a major error in separating a part from an integrated whole.

The final constraint is temporal, for the book is concerned with Anglo-American human geography during the decades since the Second World War only. Again, this is in part a reflection of personal competence, for I have been personally involved in academic geography for the last twenty years. But the Second World War was a major watershed in so many aspects of history, not least academic practice, and much of the methodology and philosophy currently taught in human geography has been initiated since then.

This book, then, is a history of Anglo-American human geography since 1945. It does not purport to be an objective history, for such an enterprise is impossible. The material included represents subjective judgements: the stress on certain topics is subjective, too, and so is the organization. But although not objective, nor intended to be, the book is neutral. My own opinions are not stated, and nor intendedly implied in anything that has been written (though some of them may be identifiable). There is no commentary, only a presentation of what I perceive to be the salient features.

Arising out of this intended neutrality is a second characteristic of the book, its dependence on the written statements of others. There are few lengthy quotations, but many short ones. Most arguments have, of necessity, been précised or paraphrased, in which case there may be unintentional distortion of the emphasis in the original. My aim has been to report what others have written, and the contents of Chapters 2–6 depend entirely on the published record, for no use has been made of personal memoirs (however valuable these might be, if a representative set could be collected).[1] A consequence of this orientation is a large bibliography. Some authors are commended by reviewers for the utility of their bibliographies. Mine indicates the material on which the book is based.

Within the terms of reference set, I have not attempted simply to provide a chronology of arguments. As well as describing the changing contents of Anglo-American human geography I have tried to account for them, to suggest why certain things were said and done, by a particular person in a particular place. Such an attempted 'explanation' of events cannot be neutral, and so it has been handled by writing a book within a book. The outer volume is Chapters 1 and 7; the inner is 2–6: the former presents the subjective account, phrased in terms of the models developed by historians of science, and the latter contains the neutral description. The inner book can be read independently of the outer; the contents of Chapter 7, however, are derived from all of those preceding it.

The account provided in the outer book is not idealist (using that term in the same way as in Chapter 5), for in attempting to explain changes in attitudes to philosophical and methodological topics I rely on my own modelling of scientific progress rather than on the views of the change-agents themselves. The modelling is set in the context of other studies of the history of academic disciplines, notably the physical sciences. By the end of Chapter 7, this model turns out to be something of a straw man. I am not the first to use it within human geography, however, and so my

1 The first three editions of the book had seven chapters: the 'inner book' was Chapters 2–6 and the 'outer book' was Chapters 1 and 7. The fourth edition had nine chapters, making the 'inner' and 'outer books' 2–8 and 1 and 9 respectively; the present edition has added a further chapter to the 'inner book'.

presentation, criticism and then replacement of it represent a contribution to the general enterprise of writing the history of geography (and perhaps of other social sciences).

One problem with writing 'contemporary history' is knowing when to stop. There is a constant flow of new material which can be employed in the task and there is always the probability (sometimes the certainty, using publishers' advertising) of something very useful appearing tomorrow. Eventually, a halt must be called. For this book, it was in mid-1978, although some later material is referenced (because I was privileged to read it before it went to press). By the time it appears, therefore, this book will of necessity be a period piece, although the arguments in Chapter 7 suggest that a few years will elapse before a major revision is needed.

It was Malcolm Lewis who designed the course of which the material in this book became a part; Stan Gregory was responsible for my teaching it. Both are in no way responsible for the outcome. I am grateful to them for the opportunity to undertake what turned out to be, for me, an enjoyable and stimulating task. The comments (direct and indirect) of the various student audiences have helped to redefine the contents in various ways over the years, and I am grateful for these.

Preparation of this book has been helped materially by several people. My wife Rita has, as always, been of great assistance in a variety of ways, not least in reading the whole manuscript twice and giving much advice on the presentation. Secondly, Walter Freeman has read the entire manuscript and given much of his time in discussing both its contents and its context; I am deeply grateful to him for his interest, his kindness, and his continued friendship. Alan Hay, too, read the complete first draft and commented freely and very helpfully on many aspects of the work. Both he and Walter Freeman felt that there is not enough of me and my opinions in the book: I hope that they understand why. The diagrams were produced by Stephen Frampton and Sheila Ottewell, and Joan Dunn has yet again created an excellent typescript out of a messy manuscript. My thanks to all.

I make it clear in Chapter 1 that the progress of any individual's academic career depends considerably on the actions (and sometimes inactions) of others. No academic is an island. Many people have helped me in the last twenty years, but I should like to express special thanks to Percy Crowe, Walter Freeman, Basil Johnson, Murray Wilson, Barry Johnston, Michael Wise, Stan Gregory and Ron Waters.

Autumn 1978

To the second edition

The generally favourable reaction to this book, and the continued demand for it, suggest that a successful formula was identified in its writing, and that no basic changes are needed. Nevertheless, it was felt that a second edition was desirable, rather than a reprint. The opportunity has been taken to correct errors, to update (hence expanding the bibliography substantially) the 'inner volume' (to mid 1982; the first edition was written in 1978), and to respond to some of the criticisms, in particular with regard to the contents of the 'outer volume', on which my ideas have changed somewhat.

The preface to a second edition should not be used simply as a response to particular reviews of the first. But the reviews of this book have raised a number of general issues that are worthy of brief discussion here – and which have been taken into account when preparing the revision.

The first point concerns the nature of the entire enterprise. In the original Preface, I stated that although the book was not objective history – 'for such an enterprise is impossible' – it was neutral, in that it contained no personal commentary. In retrospect, this was an over-statement. The book is basically a reconstruction, from published material, of philosophical and methodological debates within human geography; it is more concerned with writing about human geography than with writings that are human geography (as one reviewer so clearly pointed out), and is something of a political history of human geography. In effect, it is an exercise using the philosophy of idealism outlined in Chapter 5 (despite my statement to the contrary in the original Preface). The literature that I review represents the theories (or ideologies) of its authors regarding the nature of human geography. The framework into which I set that literature represents my theory of the history of human geography, which has been constructed – and is continually reconstructed to ensure that the two cohere (see Chapter 7). It is, then, a personal statement, set in the context of my own socialization as a human geographer.

This orientation has not been changed. I still rely entirely on published evidence for my history – while accepting the value of 'oral history' of various kinds, which is now providing valuable published materials (Browning, 1982; Buttimer, 1983; Buttimer and Hagerstrand, 1980). But the revision has allowed me to remedy some important omissions, and to reorganize certain sections to emphasize more fully the contributions of particular individuals. It is their contributions to philosophical and methodological debates that are the focus of the book. Substantive contributions are widely reviewed in a variety of places – most notably the excellent journal *Progress in Human Geography*. It is not a major purpose of this book to summarize what human geographers have done, but rather to concentrate on how they have done it, and why.

Some criticism has been directed at the book for its separation of human from physical geography; indeed I have been 'accused' of doing the discipline a political disservice by advancing this separation. I stand by my statement in the original Preface, however – 'I find the links between physical and human geography tenuous, as they are currently practised'. During the 'spatial science' era (Chapters 3 and 4) there was a common interest among physical and human geographers in methodological questions, especially of a statistical and mathematical nature, and the interactions that took place are recognized here. And certain methodological frameworks – notably systems analysis (Chapter 4) – seemed to provide common bonds. These bonds remain, and are presented here. But as each of the two fields has moved away from quantitative description to studies of, first, how (process studies to physical geographers) and, then, why (process studies to human geographers), so the differences between the two have been magnified. There is very little basis for bonds with physical geography in the approaches to human geography discussed in Chapters 5 and 6 of this book. Of course, the subject matter of human geography is closely concerned with that of physical geography, but there is no apparent need (and certainly no published evidence of its realization) for an integration of the two fields as they are currently practised. Some argue that such integration is (or should be) achieved through the study of resources, but a review of the relevant literature (Johnston, 1983a) provides no evidence to support that claim. Human and physical geography overlap. There are benefits to be achieved from that, especially in a pedagogical context, and I would not strongly advocate their institutional separation. But they are separate fields, each overlapping more with other fields than with each other.

Internal to human geography, it has been suggested that my treatment is unbalanced. In particular, cultural, historical and regional geography are ignored, relative to economic, social and political geography. To the extent that this is so, it reflects relative contributions to the literature being reviewed. The focus of this book is philosophical and methodological debate, not substantive contribution.

One of the most fundamental criticisms concerns the underlying rationale for the book, that there is a separate history of human geography worthy of investigation. By presenting such a history, am I not drawing artificial boundaries? To this, I would respond that the boundaries already existed. Human geography is an institutionalized discipline in the countries studied here; at the school level in Britain it is clearly defined by the syllabuses for public examinations. The boundaries around human geography are indeed artificial, because the subject matter of the social sciences should not be compartmentalized. But a desire to break down those boundaries in some way does not detract from the need for a history which, among other things, should demonstrate to students how that desire has developed within the academic community of human geographers in recent decades.

Despite the institutionalization of human geography, there is nevertheless considerable movement through its boundaries, involving contributions by human geographers to other disciplines and vice versa. This may be understated in the book. It takes a variety of forms. The basic flows are of ideas from other fields into human geography. These ideas are assimilated into the recipient discipline, and in return human geographers may seek to demonstrate the results of that assimilation to the source discipline. This is usually done by way of publications in the journals of the latter; there has been more of it in some source disciplines (agricultural history, for example) than others (such as political science; Laponce, 1980). Some individuals have transferred their disciplinary allegiance as a result: there are several British professors of sociology who were trained as geographers, for example. But are they still human geographers, whose work should be reviewed here?

Certainly, inter-disciplinary links are substantial, even if not always apparent, and should not be ignored. Developments in human geography in recent decades have been indelibly influenced by these links – many of them completely informal, because they involved 'World Three' (p. 280) only. But such links do not deny the existence of a discipline called human geography into which many students are socialized each year and whose history is of intrinsic interest.

Finally, why confine the treatment to the 'English-speaking world', basically to Britain and North America? My answer is the same as it was in the Preface to the first edition: competence. The need for a more thorough international survey is clear, however, and small steps have been taken to meet it (Johnston and Claval, 1984). There are other things we are ignorant about. None of the arguments presented here is in any way backed by solid quantitative evidence, where that would be relevant. Gatrell (1982) has pointed to one possible direction, but the routes are wide open. Indeed, the tasks ahead of us are many

> The 1980s promise to be challenging, existing and fraught years, not least to the academic community. For geographers, the uncertainty is a double one: both in what to study and how to study it (Robson, 1982, p.1).

In continuing the work on this subject, and in the formulation of the second edition, my debts from the original book remain. I am particularly indebted to my publishers for constant encouragement, and also for stimulating me to provide a companion volume – *Philosophy and Human Geography* – that outlines the philosophies the debates over which are reviewed here. The contributors to *The Dictionary of Human Geography* stimulated me considerably, as did the valuable comments of the kind reviewers of the first edition. To them all, I am extremely grateful. As always, I am deeply indebted to Rita Johnston and Joan Dunn.

Summer 1982

To the third edition

When the second edition of this book was prepared in 1982, it seemed unlikely that another would be needed only four years later. Events have

proved different, however, for three reasons. First, there has been a continued flowering of geographical scholarship, in all three of the major research
paradigms identified here; to retain the contemporary perspective of the
book it has been necessary to incorporate these, so that several sections
have been substantially rewritten and extended. Secondly, the political
response to economic recession in the late 1970s and, so far, the whole of
the 1980s has put substantial pressures on researchers to contribute to economic and social change in particular ways, and the reactions of geographers to such pressures have been incorporated in the present narrative.
Finally, there has been increased interest in the history of human geography,
bringing new insights to the processes of change. These, too, have been
incorporated here. The basic thesis of the book has not been altered, however; indeed, re-evaluation of the work of Kuhn (as suggested by Mair's,
1986, critique) has led to greater confidence being expressed in the concepts that he introduced.

The Preface to the Second Edition provides an extended defence and
amplification of the approach taken in the book, and no further discussion is intended here. I am grateful to those who continue to comment favourably upon it and consider it a valuable introduction for
student use. To them, as again to Rita Johnston and Joan Dunn, my
thanks.

December 1986

To the fourth edition

The previous three editions of this book have each been written at approximately four year intervals, so the preparation of this fourth version has
increased the turnover time somewhat (by about six months, or 12.5 per
cent). The reason for the haste is in large part the continued production of
substantial volumes of high quality scholarship by Anglo-American human
geographers, and especially the continued vitality of the debate over competing views of the discipline. More than sufficient new material has been
produced since the third edition was written 3½ years ago to require the
rewriting of several sections of the book and the inclusion of new sections
and sub-sections. Within a short period of years, the production of a large
number of edited volumes with titles such as *Geography in America*,
Horizons in Human Geography, *Remaking Human Geography*,
Remodelling Geography, and *The Power of Geography* all testified to the
discipline's vitality and called for a reworking of a book that sets out to
provide students of geography with a conspectus of its 'contemporary history'.

The basic organizational framework of the book has not been altered;
the contents have simply been updated. But one significant change has

been introduced, however. The chapter structure increasingly seemed anachronistic as the debates over the discipline continued, and it was decided to increase the number in order that separately identifiable approaches should be individually treated, and clearly so. Thus the book now has two more chapters than it had in the previous editions, which it is hoped provides a clearer structure within which the debates can be situated.

The Preface to the Second Edition was used to answer some of the criticisms of the book – what it covers and what it excludes; how this material is structured – and the Preface to the Third Edition made brief reference to continuing debates over the material used, and especially its interpretation. Similar criticisms have been made since (notably in Stoddart, 1987) but no detailed response is offered here; the Preface to my book on *Environmental Problems* (Johnston, 1989b) continued that debate, however.

In producing yet another edition of this book I am grateful to those who have reviewed it and made valuable comments (especially Robin Flowerdew), those who continue to cite it and to recommend it to students, and to those who buy it. The four editions have each been commissioned by different editors, and I am grateful to all with whom I have been involved at Arnold for their continued support. As always, my greatest debt is to Rita Johnston, not least for convincing me that I should learn word-processing.

April 1990

To the current edition

Readers who compare this edition of *Geography and Geographers* with its predecessor will rapidly identify why a new version was needed: the six years that have elapsed since the previous edition was completed have seen a continued flowering of work within human geography, associated with stimulating debate over the discipline's nature and practice. Indeed, such has been the volume of work published during the current decade, and the new lines that it has promoted, that a further chapter (Chapter 8) has been added to the book, with some consequent restructuring of the others.

Given these extensive changes, a case could have been made for reducing the coverage in some chapters, on the grounds that much of what was emphasized in the late 1970s (when the book was first written) is much less relevant to students nearly two decades on. That has been resisted, however: the goal remains to interpret Anglo-American human geography since 1945, and much that is practised as geography today builds on the substan tial foundations erected in the 1950s and 1960s. Others may focus on the favoured approaches at any one time, but an overall synthesis is the purpose here.

The desirability of such an enterprise has recently been questioned by Clive Barnett (1995, p. 417), who opened his provocative piece by indicating: 'doubts about the value and relevance of expending energy studying the history of geography as a means of throwing light on the state of the discipline today'. Claims 'about the necessity of attending anew to the origins of modern geography in order to understand the contemporary nature of the discipline' have 'rather a hollow-sounding ring' for him, and he thought that the links 'are more often loudly asserted than convincingly demonstrated'. He identified the main reason why academics are interested in their discipline's history as political rather than intellectual: it is part of establishing and sustaining a professional identity. Much of what was undertaken as geography in the past is 'simply now redundant' (p. 418), but is too often used 'not because of the unquestionable need to understand geography's past as it necessarily still bears upon the present' but rather as an entrée to certain types of study – such as those investigating geographers' roles in nineteenth- and twentieth-century western imperialism.

Only at the end of his critique does Barnett suggest that (p. 419):

> what needs to be most urgently addressed by critical human geographers is not the distant past of geography but a set of questions about what all this theory is doing in geography, how it came to be here and what we can hope to do with it.

However, so much depends on one's interpretation of 'distant past'! For the present book, some of the material considered in Chapters 3 and 6 might be represented as redundant since it has little direct relevance to what is currently practised as geography (which is, of course, defined more widely here than the 'critical human geography' which Barnett is addressing). But without appreciating the trends covered in Chapter 3 the current situation in 'geography as spatial science' could only be partially understood; similarly, although very few human geographers now subscribe to the philosophies of idealism and phenomenology, an evaluation of the debates outlined in Chapter 6 is central to appreciating the current popularity of much (non-marxist and non-realist) social theory among human geographers. Barnett concludes that:

> if you want to understand the institution of academic geography as it is currently constituted, then maybe the best place to start is by actually examining the discipline as it exists in the here and now. When you have done that, the bits of geography's past which perhaps deserve further attention, and the bits which do not, might become a little clearer.

But even this is conceding too much. What I really want to believe is that we would be better served if we simply let the dead bury their dead ... might it not be possible actively to forget about the past and to act instead with no regard at all for what has gone before?

In response, Livingstone (1995, p. 420) argued that:

What makes geography a tradition – a contested tradition – ... is precisely that it has a story, a history ... I would go so far as to claim that it is only when we take the tradition seriously that we can appreciate how incoherences become evident, how new questions emerge and how older practices fail to provide the resources to deal with new issues.

This book adopts the same stance. It rejects Barnett's position that by excluding the past from our studies we 'might find that we invent something which is a surprise, something whose form and content we cannot now fully anticipate and which, therefore, would be something genuinely new' (p. 419). History constrains us in so many ways, but it also enables us – without appreciating it, the prospect of us independently coming up with something 'genuinely new' is remote, because we would have no map on which to chart a voyage of discovery. Understanding our discipline's past and present does not allow us to anticipate its future form and content, of course, for geography is reproduced by 'knowing individuals'; but only by appreciating what they know (or knew) can we understand why they chose to follow certain routes and reject others.

Knowing what people know (or knew) is an interpretative exercise, so there can be no 'objective history' of a discipline: the interpreter selects from the available evidence what appears to be relevant and structures the material accordingly. Others may approach the task in a different way – from a different position. My own context is clear: I have been an academic geographer for much of the period studied in this book (I became an undergraduate in 1959) and have contributed slightly to what has occurred. I am part of the past I am writing about, and so it is not a 'foreign country' to me; I have written about the world of academic geography as I have experienced it, and have drawn on others' writings in that light. In Livingstone's words (1995, p. 421), therefore, critics must appreciate the 'situatedness of [my] knowledge claims'.

'Situatedness' is an increasingly used term in studies of the history of geography: context is not all, but it is crucial. It has certainly been crucial in the preparation of this book, as the Prefaces to the earlier editions indicate; this volume does not present 'the history' of human geography over the last half-century, therefore, but only 'a history' – indeed, 'my history', because although it is not presented autobiographically I have been part of most of it and my participation has coloured my interpretation in a great variety of ways.[2]

In preparing this edition, I am indebted to Laura McKelvie for her patience and encouragement, and to my new colleagues at the University of Bristol for welcoming me to their stimulating collegial environment: Peter Dicken, Derek Gregory, Charles Pattie, Dave Rossiter and Pete Taylor did much to sustain me and my geographical work during the period between the production of the fourth and fifth editions, and I am delighted to have this opportunity to thank them publicly. My greatest debt is to Rita Johnston, for being a much-loved part of the contented environment within which it is possible to undertake a project of this size.

May 1996

---

2 I have, however, relied almost entirely on the published record in constructing 'my history', while recognising Anne Buttimer's comment (in a review of a biography of Vidal de la Blache: 1995, p. 406) that 'Geography based on books is mediocre. With maps one can do a little better. But the only way to do it well is in the field'. By analogy, a history of geographical thought which relies on published work will be incomplete: to do it well one needs 'field work . . . into the lived world, personality and broader contexts in which [a] scholar's field unfolded'.

1

The nature of an academic discipline

This book is a study of an academic discipline, particularly of changes in its content. Content cannot be appreciated fully without understanding context, however, which this chapter provides. To study an academic discipline is to study a miniature society, which has a stratification system, a set of rewards and sanctions, and a series of bureaucracies, not to mention a large number of interpersonal conflicts (some academic, some not). An outsider may perceive academic work as objective, but many subjective decisions must be taken: what to study and how; whether to publish the results; where to publish them and in what form; what to teach; whether to question the work of others publicly; and so on. As with all human decisions, they are made within the constraints set by the wider society.

Studying an academic discipline involves studying a society within a society; both set the constraints to individual and group activity. Two questions are focused on here; 'how is academic life organized?'; and 'how does that academic work, basically its research, proceed?' Use of the term 'society within a society' strongly implies that academic life does not proceed independently in its own closed system but rather is open to the influences and commands of the encompassing wider society. A third necessary question, therefore, is 'what is the nature of the society which provides the environment for the academic discipline being studied, and how do the two interact?'

Academic life: the occupational structure

Pursuit of an academic discipline in modern society is part of a career, undertaken for financial and other gains; most of its practitioners see their career as a profession, complete with entry rules and behavioural norms. In the initial development of almost all current academic disciplines, some of the innovators were amateurs, perhaps financing their activities from individual wealth. There are virtually no such amateurs now; very few of the research publications in human geography are by other than either a professional academic trained in the discipline or a member of a related discipline. The profession is not geography, however. Rather, geography is the discipline professed by individuals who probably state their profession as university teacher or some similar term. Indeed, the great majority of academic geographers are teachers in universities or comparable institutions of higher education. (Recent years have seen the invention of new types of higher education institution; 'university' is used here as a generic term.) Academic geographers are distinguished from other professional geographers (many of whom are also teachers) by their commitment to all three of the basic canons of a university; to propagate, preserve and advance knowledge. The advancement of knowledge – the conduct of fundamental research and the publication of its original findings – identifies an academic discipline; the nature of its teaching follows from the nature of its research.

The academic career structure

University staff members – hereafter termed academics – follow their chosen occupation within a well-defined career structure, of which two variants are relevant to this discussion: the *British model* and the *American model* (Figure 1.1). Entry to both is by the same route. With rare exceptions, the individual must have been a successful student as an undergraduate and, more especially, as a postgraduate. The latter involves pursuing original research, guided by one or more supervisors expert in the relevant specialized field. The research results are almost invariably presented as a thesis for a research degree (usually the Ph.D.), which is examined by relevant experts; possession of a Ph.D. is an almost obligatory entrance ticket for both models.

While pursuing the research degree, most postgraduate students obtain experience in teaching undergraduates, particularly in practical and tutorial classes: some universities, notably those operating the American model, finance many research students through such teaching activities. The individual may then proceed to further research experience, or may gain appointment to a limited-tenure university teaching post, offering an 'apprenticeship' in the teaching aspects of the profession whilst either the research degree is completed or the individual's research expertise is consolidated.

Beyond these limited-tenure positions lie the permanent teaching posts, which is where the two models deviate (Figure 1.1). In the *British model*, the first level of the career structure is the lectureship. For the first years of that appointment (usually three) the lecturer is on probation: training in teaching and related activities is provided, annual reports are made on

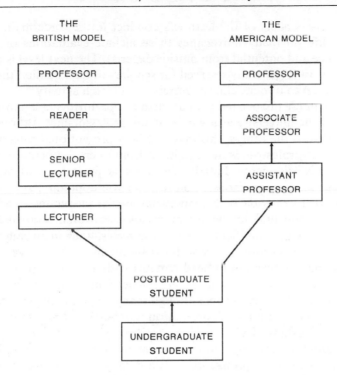

Fig 1.1 The academic career ladder. Note that whereas in the American model it is usual to progress through the various stages in an orderly sequence, it is quite common to miss some of them in the British model. (Some British universities have slightly different nomenclature – the universities created after 1992 have principal as well as senior lectureships; Oxford and Cambridge have no senior lectureships.)

progress as teacher, researcher and administrator, and advice is offered on adaptation to the demands of the profession. At the end of the probationary years, the appointment is either terminated or confirmed.

Appointment in this model is almost always to a department, which is probably named for the discipline which its staff members profess. The prescribed duties involve undertaking research and such teaching and administration duties as directed by the head of the department. The lecturers are on a salary scale and receive an annual increment; accelerated promotion may be possible. In the United Kingdom there is an 'efficiency bar' after a certain number of years' service; 'crossing the bar' involves a promotion, and is determined by an assessment of the lecturer's research, teaching and administrative activities.

Beyond the lectureship are further grades into which the individual can be promoted. First is the senior lectureship for which there is no allocation to departments but competition across the entire institution; entry is based again on assessments of the lecturer's conduct in the three main areas of academic life (in many universities these include evaluations of research performance and potential from outside experts). The next level is the readership, a position generally reserved for scholars of high quality; the criteria for promotion to it focus almost entirely on research activity.

The final grade (associated with academic departments as against administration of the university as a whole) is the professorship. Although superior in status to the others, this may not be a promotional category: until recently most professors were appointed from open competition preceded by public advertisements. Initially the posts of professor and of head of department were synonymous, so a professor was appointed as an administrative head, to provide both managerial and academic leadership. With growing departmental size in recent decades, however, and the development of specialized sub-fields within disciplines, each requiring separate leadership, it has become common for departments to have several professors. In some, the position of head remains with a single appointee; in the majority, the headship is a position independent of the professoriat to which other grades may aspire, although many heads are also professors.

Finally in the British model, promotion is possible to non-advertised, personal professorships. These positions are becoming increasingly common, as the possibilities for appointments to established 'chairs' are reduced. In most cases, a personal professorship is awarded for excellence in research and scholarship.

The system under the *American model* is simpler (Figure 1.1). There are three permanent staff categories: assistant professor, associate professor, and professor. The first contains two sub-categories – those staff on probation who do not have 'tenure' and those with security of tenure. Each category has its own salary scale but not automatic annual increments. (In the United States, these scales vary from university to university – according to prestige, state, etc.) Salary adjustments result from personal bargaining, on the basis of academic activity in the three areas already listed, and salary scales for the categories may overlap within a department. Movement from one category to another is promotional, as a recognition of academic excellence. Professorships are simply the highest promotional grade and do not carry obligatory major administrative tasks. Heads of departments in the American model are separately appointed, usually for a limited period only, and they need not be professors (usually referred to as 'full professors').

Status, rewards and promotion prospects

The rewards for academics are the salary levels, the profession's general status, the particular status of its individual promotional categories, and the relative independence and flexibility of the working conditions. The tangible rewards – the payments – are determined by the academic's position within the career structure. Promotion must be 'earned'.

Academic work has three main components – research, teaching and administration. Entrants to the profession have little or no experience of academic administration and their only teaching has probably been as assistants to others. Thus it is very largely on proven research ability that an aspirant's potential for an academic career is judged. Research ability can be partly equated with probable teaching and administrative competence, since all require the same personal qualities – enthusiasm, orderliness, incisive thinking and ability to communicate orally and in writing. To a considerable extent, however, academics gain their first position on faith in their potential, hence the usual probationary periods before tenure is granted.

Once admitted to the profession, staff undertake all three types of work, so that promotional prospects can be more widely assessed. Teaching and administration have traditionally carried less weight with those responsible for promotions than research. This is partly because of perceived difficulties in assessing performance in the first two, and partly because of a general academic ethos which gives prime place to research activity in peer evaluation.

Although not necessarily progressive in terms of increasing level of difficulty, administrative tasks tend to be more complex and demanding of political and personal judgement and skills as the academic becomes more senior. A person's ability to undertake a certain task can often only be fully assessed after promotion to the relevant position (this has been called the 'Peter principle'. Peter and Hull, 1969), so promotion must frequently be based on perceived potential; although it is possible to point to somebody whose administrative skills are insubstantial, it is not always easy to assess who will be able to cope with the more demanding tasks.

It has long been argued that assessment of teaching ability is difficult (although clear inability is often very apparent). Student and peer evaluations are possible and are increasingly used, as is external scrutiny of students' work. But criteria for judging teaching performance at university level are ill-defined, and expectations of a lecturer, tutor or seminar leader often vary quite considerably within even a small group of students, as well as among external assessors. Thus the majority of teachers are usually accepted as competent and undistinguished, and their performance is neither help nor hindrance to their promotion prospects. (The 'inspections' of teaching sessions – most of them lectures – undertaken during the 1995 Quality Assessment of Teaching in English University Departments of Geography invariably rated the majority as 'satisfactory', with a substantial minority being deemed 'excellent' and very few – none in most cases – 'unsatisfactory': Johnston, 1996f.)

The main criterion for promotion is usually research performance, therefore, although a relatively undistinguished record in this area can be

compensated by excellence elsewhere. (Some argue that teaching competence is not only undervalued, but that as a consequence of the emphasis on
research in most universities it is receiving less attention: Jenkins, 1995.)
How is research ability judged? Details of how research is undertaken are
considered in the next section; here the concern is with its assessment rather
than with stimulus and substance.

Successful research involves making an original contribution to a field of
knowledge. It may involve the collection, presentation and analysis of new
information, within an accepted framework; it may be the development of
new ways of collecting, analysing and presenting facts; it may comprise
promoting a new way of ordering facts – a new theory or hypothesis; or it
may be some combination of all three. Its originality is judged through its
acceptance by those of proven expertise in the particular field. The generally accepted validation procedure is publication, hence adages such as
'unpublished research isn't research' and 'publish or perish'.

The main publishing outlets for research findings in most disciplines are
their scholarly journals, which operate fairly standard procedures for scru-
tinising submitted contributions. (It is widely believed that journals published by academic societies have higher standards than those published by
commercial companies, but the validity of this is difficult to assess.)
Manuscripts are submitted to the editor, who seeks the advice of qualified
academics on the merits of the contribution; these referees will recommend
either publication, rejection, or revision and resubmission. When accepted,
a manuscript will enter the publication queue, which may be up to two
years long.

Although widely accepted, this procedure is flawed, because it is operated by human decision-makers. The opinions of both editor (on the
manuscript, and on the choice of referees) and referees may be biased or
partial, so a paper can be rejected by one journal but accepted by
another, even without alteration. (On the canons of editorship see, for
example, Hart, 1990, and Taylor, 1990c; for a radical alternative, see
Symanski and Picard, 1996.) Most disciplines have an informal prestige
ranking of journals, and it is considered more desirable to publish in
some rather than others. (This prestige ranking is sometimes taken into
consideration by appointment and promotion committees.) The issue of
the standing of journals has become crucial in the UK since the mid-
1980s because of the use of publication data as a performance indicator
for the periodic Research Assessment Exercises which not only rate every
department but also determine its future funding (the better get more
money: see Johnston, 1993b).

Some research results are published in book form, rather than as journal
articles. Most academic books in geography are texts, however, published
by commercial companies whose main interest is marketability among the
large student population. The textbook may be innovative in the way that
it orders and presents material, and can be beneficial to its author's

reputation (and bank balance), but it is not usually a vehicle for demonstrating research ability. Many companies, including university presses, publish research monographs, however, presenting the results of major research projects to relatively small academic markets. Their decisions on whether to publish are made on academic as well as commercial grounds, usually with the aid of academic referees, and their output is validated through the journal book-review columns.

Processes of promotion and appointment: patronage

Whatever the weighting given by appointment and promotion committees to the three main academic activities (others are pursued by some, such as consulting for outside bodies and contributions to the work of learned societies), how they make assessments is frequently contentious. One of the main sets of 'objective' information which can be presented to them is lists of publications, most of which have been validated by academic journals. But how is such information evaluated?

Two modes of assessment are widely used and relied upon: the written opinion of a third party (either a referee, selected by the applicant, or an assessor, selected by the committee); and the interview, sometimes associated with presentation of a lecture or seminar. The British model places considerable weight on the former. An applicant must supply a list of referees who will provide an opinion on her or his suitability for the post: applicants are likely to ask people, especially senior people, they have worked with, and are believed to be favourably inclined towards them, to act in this capacity. As appointment committees tend to be swayed very much by these reports, especially when preparing a shortlist of candidates to be interviewed, the opinions of well-respected referees are often crucial. Thus certain leaders in a subject often find it easier to get their candidates appointed to posts than do others: there is a considerable element of patronage involved in obtaining a university post, especially a first university post.

Promotions under the British model are frequently strongly influenced by the head of the candidate's department, although he or she is expected to consult senior colleagues before making a recommendation. Reports must be made on each lecturer: annually during the probation period; at the confirmation stage after probation; on reaching the 'efficiency bar'; and for either accelerated promotion within the lecturer scale or promotion to senior lecturer. This also contains a strong element of patronage, although constraints are built into the committee system to promote fairness for all, including the right either to present one's own case or to appeal against a decision. Some universities use outside referees and assessors when considering promotions to senior lecturer; promotions to readerships and professorships invariably involve such consultation.

Referees, assessors and interviews are also used in the appointment of professors. Referees nominated by the candidates provide a confidential report on their potential for the job in question. Assessors nominated by the university evaluate the list of candidates, suggest other worthy candidates who might be approached, and in many cases attend the formal interviews: their potential patronage power is great.

Procedures are slightly different under the American model. More weight is usually given to an enlarged interview, with candidates giving seminars to the department and meeting with various groups of staff. (Such procedures are increasingly used in British universities.) The reference is still important, however, especially for aspirants to first teaching positions, and a letter from a respected leader in a field can be very influential in gaining an appointment for a former student. The individual candidate is more active in this system than in the British, however, perhaps canvassing for interviews during the annual conferences of scholarly societies, for example. For promotions and for salary rises there is considerable bargaining between individual, department chair and university administrators: external referees' opinions may be sought when tenure is being confirmed, or promotion to full professor proposed.

In all of these procedures the applicants depend to a considerable extent on the opinions of senior academic colleagues in evaluating their potential, performance and prospects. Some opinions carry more weight than others, and so it is important when developing a career to identify potentially influential individuals, to keep them informed of your work, and to enlist their support for your advancement. Because of the lack of truly objective criteria for measuring research, teaching and administrative ability, such patronage is crucial.

Other rewards and the sources of status

The tangible rewards of an academic career are the salary and the life style, plus the occasional extra earnings that are possible – for examining, writing, lecturing and consulting. In addition the hours are flexible; the possibilities for travel are considerable; and the constraints on when, where and how work is done are relatively few, compared to most other professions. Other, less tangible rewards include involvement with the intellectual development of students. There is also the charisma associated with recognized excellent teachers, and even more so for leading researchers, whose publications are widely read, whose invitations to give outside lectures are many, and whose opinions as examiners, referees and reviewers are widely canvassed. The conduct of research brings its own rewards, apart from the charisma; the satisfaction from having identified and solved a significant problem is often considerable.

Academic life also offers a reward common to many social systems: power. Patronage is power, as is work as examiner, referee or reviewer. Others' careers are being affected, and exercise of this power can bring with it the loyalty and respect of those who benefit. Because the academic system is so dependent on individuals' opinions, because some individuals' opinions are more valued than others', and because power is a 'commodity' widely desired in most societies, many academics seek positions of influence.

Among the most influential positions in an academic discipline are the administrative headships of departments. Their holders can instruct other staff members (usually after consultation) in their teaching and administrative duties; they are frequently used by staff and students as referees for job and other (such as research grant) applications; and their reports are crucial in the promotions procedure. The departmental organization of universities is a bureaucratic device which makes for relative ease in administering what may be a very large institution. (It also tends to fossilize disciplinary boundaries, as will be discussed later.) Departmental heads not only have power over members of their own discipline's staff, they also participate in their university's administration, and they treat within its committee system for departmental resources. As in all bureaucracies, there is a tendency for the status and power of the various departmental heads to be a function of the size of their 'empires' (Tullock, 1976; large departments in British universities certainly get the highest research ratings: Johnston, Jones and Gould, 1995). Heads of large departments, especially those which are also growing (growth being widely considered a 'good thing') and those attracting large numbers of students and research grants/contracts, are very often the most influential within a bureaucracy. They have the incentive to build up their departments, which usually means increasing student numbers, since universities tend to allocate resources to departments according to the numbers taught. This brings power and prestige, both within the university and beyond; it is an added reward for the academic bureaucrat, and the power over resources which it involves usually benefits the whole department.

Finally, the academic bureaucrat can gain power beyond the home university through, for example, positions on committees. These may be concerned with the subject through professional bodies, with the allocation of research moneys by public or private foundations, or with a wide range of public duties. Again, the status and power obtained can spill over to others, since patronage is important in all of those roles.

The academic working environment

The continuing goal of an academic discipline is the advancement of knowledge, pursued within its own particular areas of study. Its individual members contribute by conducting research and reporting their

findings, by integrating material into the disciplinary corpus, and by pedagogical activities aimed at informing about, promoting and reproducing
the discipline: they may also argue for the discipline's 'relevance' to society at large. But there is no fixed set of disciplines, nor any one correct
division of academia according to subject matter. As Livingstone (1992,
pp. 28–9) puts it:

> The idea that there is some eternal metaphysical core to geography
> independent of historical circumstances will simply have to go ...
> geography has meant different things to different people in different
> places and thus the 'nature' of geography is always negotiated. The
> task of geography's historians, at least in part, is thus to ascertain
> how and why particular practices and procedures come to be
> accounted geographically legitimate and hence normative at different
> moments in time and in different spatial settings.

Those disciplines currently in existence are contained within boundaries
established by earlier communities of scholars. The boundaries are porous,
so disciplines interact. Occasionally the boundaries are changed, usually
through the establishment of a new discipline that occupies an enclave
within the pre-existing division of academic space. (The boundaries are not
necessarily the same in all comparable institutions.)

Just as there is no immutable set of disciplinary boundaries so there is no
right way of undertaking research nor, in many cases, any exact criteria for
determining whether research findings, and even more their interpretations,
are correct. The 'right' and 'wrong' ways of doing research, the 'correct'
and 'incorrect' interpretations of research findings, and the 'proper' ways of
presenting knowledge and training students are all the product of decisions
by academics themselves. Thus, not surprisingly, there is considerable
debate within disciplines over these issues. At any one time, there may be
consensus within a disciplinary community regarding both its subject matter and its research methods. But controversy is just as likely, as academics
discuss the relevance of particular research findings, the validity of certain
research methodologies, and so on. Indeed, there is controversy over the
definition of knowledge itself.

The study of the controversies and consensuses that characterize academic disciplines is the function of historians of science. (Science is used here
as a very broad term encompassing the entire range of academic disciplines
in the natural and social sciences.) To the outsider, the workings of a
science are generally mysterious – especially regarding disciplines that
require a great degree of prior training before original research is undertaken and whose literature is virtually impenetrable to the untrained. It is
generally believed, however, that science is an objective activity undertaken
within very strict rules, involving the continuous excitement of the search
for new discoveries. Indeed, scientists frequently present themselves in this

light. Certain values are universally subscribed to within academia, it is argued, with the main five being (according to Mulkay, 1975, p. 510):

1 the norm of *originality*: academics strive to advance knowledge, and conduct original research designed to discover and account for aspects of the world as yet not fully understood;
2 the norm of *communality*: all information is shared within the academic community – it is transferred through accepted channels (notably the research journals) and its provenance is always recognized when it is being used;
3 the norm of *disinterestedness*: academics are devoted to their subject, and their main reward is the satisfaction of participating in the advancement of knowledge – a reward that may bring with it charisma and promotion;
4 the norm of *universalism*: judgements are made on entirely impartial criteria, which take into account the academic merits of work only and make no reference to the personalities of the researchers; and
5 the norm of *organized scepticism*: knowledge is furthered by a continuing process of constructive criticism, in which academics are always reconsidering both their own work and that of others.

According to these five norms, therefore, academic work is carried out neutrally; there is a complete lack of partiality, self-seeking, secrecy and intellectual prejudice. Objective assessment criteria are assumed, as are high levels of ability and humility on the part of members of the academic community.

Against this ideal view of science and scientists, which many of the latter promote, are the results of studies in the history of science (most of which refer to the physical sciences: Mulkay, 1975; Mulkay, Gilbert and Woolgar, 1975). These show that the procedures adopted are frequently neither objective nor neutral, and they present a picture of disciplines growing 'by gathering more detail in areas already investigated, and by stumbling across new sets of facts in areas of experience never previously investigated' (Barnes, 1974, p. 5). Science is a culture, with each discipline comprising one or more subculture(s). It has its own rules and procedures which are open to change and hence are the subject of debate, even conflict. Furthermore (as will be suggested in more detail below), scientific cultures are parts of wider cultures, and although to some extent scientists – because of their (self-imposed) expert status – can impose their own views of themselves on their host societies, they are subject to external influences (see Barnes, 1974).

Each scientific discipline is a separate academic community, therefore; many are groupings of several related communities. Their goal is the advancement of knowledge, but the definitions of advancement and of knowledge itself are influenced, if not determined, by the community's members. Thus study of a discipline's history is not simply a chronology of its successes. It is an investigation of the sociology of a community, of its debates, deliberations and decisions as well as of its findings. (See Watson, 1968, for an example of this; for a fictional account, see Cooper, 1952. Livingstone, 1992, p. 30, suggests that 'it might be helpful if we were to think of geography as a tradition that evolves like a species over time'.) Most investigations take one of two general forms: the empiricist approach portrays history as it happened, with each community treated separately; the other seeks generalizations on how sciences progress, which involves either an inductive approach – studying individual histories and identifying their common themes – or a deductive framework that provides a model of scientific progress against which events in a particular discipline are compared. Most of the models currently available have been developed for, and tested on, the physical sciences. Their relevance to human geography has been suggested, however (e.g. Haggett and Chorley, 1967; Harvey, 1973), and they are presented here as a backcloth for a later evaluation of the substantive material arranged in Chapters 2–9.

Kuhn, normal science and revolutions

A framework for studying the history of science that has received much attention (indeed, the book is one of the most frequently cited by social scientists) is provided by T. S. Kuhn's (1962, 1970a) *The Structure of Scientific Revolutions*. As Kuhn (1970b) and many commentators have pointed out, this is a study in the sociology of science: it is a positive interpretation – a presentation of what scientists do – rather than a normative programme – an argument for what scientists *ought* to do. It generalizes from its observations, identifying common elements in disciplinary histories: Kuhn's goal was to identify 'what science, scientific research as it is actually practised, is really like' (Barnes, 1982, p. 1).

Kuhn argued that scientists work in communities of researchers and teachers who share a common approach. They operate within an agreed philosophy, concur on the theoretical focus of their work and use accepted methodological procedures. They use those procedures to solve problems identified within the theoretical framework, thereby adding to knowledge (the store of problems solved) and extending the range of their theory. The framework, its procedures and its empirical substance are codified in their textbooks. What they share is termed a *paradigm*; 'a scientific community consists of men who share a paradigm' (Kuhn, 1970a, p. 176). As Popper (1959) expresses it, this means that once scientists are socialized into a research field they can proceed directly to its unsolved problems. The existing framework defines these problems, providing both a context and a methodology for tackling them. The researcher does not set out to tackle a problem *de novo*, therefore, but on the basis of what is already known in the chosen field. A paradigm is thus 'an accepted problem-solution'

(Barnes, 1982, p. xiv), which by its very nature poses the next problem. Progress in science is achieved by problem-solving.

To undertake research within a paradigm requires understanding and acceptance of its philosophy and methodology. This involves a period of training, during which the potential new researcher is socialized into the paradigmatic culture, that is, its way of thinking about its scientific problems. The key is provided by the paradigm's textbooks, the summaries of its literature which set out what is known and how more knowledge can be obtained. The training prepares individuals for work within an accepted mould:

> Scientific training is dogmatic and authoritarian . . . [it] does not generate or encourage traits such as creativity or logical rigour; rather it equips scientists so that it is possible for them to be creative, or rigorous, or whatever else, in the context of a specific form of culture (Barnes, 1982, pp. 1–17).

Having been integrated into the paradigm, the scientist joins a research community. Such communities, sometimes termed 'invisible colleges', operate through close interaction involving attendance at specialized conferences and the private circulation of research papers in pre-publication form, and reflected in citations to each other's work (Crane, 1972); increasingly, electronic communications media allow community members to be in frequent, almost constant, contact and collaborative work without regular meetings is now common. Individual members receive recognition for the value of their work and attract the patronage of leaders who may assist in their career advancement; some communities have only a few patrons, perhaps even one, and may be identified as a particular school of thought.

Research within a paradigm involves filling in the gaps; the researcher 'has to make the unknown into an instance of the known, into another routine case' (Barnes, 1982, p. 49). The paradigm provides the needed resources – guidance but not direction. Success, which brings the rewards of recognition, patronage and status, involves conforming to the paradigmatic norms (Mulkay, 1975, p. 515):

> It is clear that the quality or significance of a scientist's work is judged in relation to the existing set of scientific assumptions and expectations. Thus, whereas radical departures from a well defined framework are unlikely to be granted recognition early under normal circumstances, original contributions which conform to established preconceptions will be quickly rewarded.

Thus the dominant norm of academic life is not one of the five listed above (p. 11), but conformity. Science is not the constant search for novel discoveries but rather the careful application of agreed procedures to the solution of problems in order to extend existing well-structured bodies of knowledge. Judgements are being made all of the time, but within an academic environment carefully structured by the training process. Science progresses through filling the gaps in a pre-defined framework.

This operation of a paradigm is known as *normal science* (Kuhn, 1962, pp. 35–6):

> Perhaps the most striking feature of the normal research problems . . . is how little they aim to produce major novelties, conceptual or phenomenal. Sometimes . . . everything but the most esoteric detail of the result is known in advance . . . the range of anticipated, and thus assimilable, results is always small compared with the range that imaginations can conceive . . . the aim of normal science is not major substantive novelties . . . the results gained in normal research are significant because they add to the scope and precision with which the paradigm can be applied. . . . Though its outcome can be anticipated, often in detail so great that what remains to be known is itself uninteresting, the way to achieve that outcome remains very much in doubt. . . . The man who succeeds proves himself an expert problem-solver.

Within normal science, therefore, the researcher has available:

1 an accepted body of knowledge, ordered and interpreted in a particular way;
2 an indication of the puzzles that remain to be solved; and
3 a set of procedures for puzzle-solving.

Training within a paradigm places the researcher in a rut, and provides the tools for extending the paradigmatic body of knowledge (deepening the rut). The result is 'conventional, routinised practice' (Barnes, 1982, p. 11). This does not imply that normal scientific activity is one of 'long periods of dreary conformity' (p. 13), however, because extending and developing knowledge are not simply 'a matter of following instructions or rules. Rather, normal science is a test of ingenuity and imagination, with paradigms figuring largely among the cultural resources of the scientist' (p. 13). Solving problems is rarely easy; Kuhn (1970c, pp. 36–9) uses chess-playing as an example, arguing that much ingenuity as well as existing knowledge (practice) must be brought to bear if the problems posed in individual games are to be solved successfully.

Scientists are not omniscient. They do not understand everything – even within a particular paradigm – so their predictions occasionally prove inaccurate. While the process of normal science continues, therefore, slowly accumulating extra knowledge as problems are solved, it sometimes throws up anomalies, findings that are not in accord with the paradigm's assumptions. These must be accounted for (Barnes, 1982, p. 53):

> Puzzle-solving activity frequently attempts to show that what is *prima facie* anomalous is either the spurious product of bad equipment or

technique, or a familiar phenomenon in disguise. And most anomalies are successfully assimilated in this way.

Either the work was badly done or the researcher interpreted the results wrongly. Minor adaptation of the paradigm may be needed, but the general process of normal science continues.

Some anomalies cannot readily be accounted for, however, and they continue to worry a few scientists. Their persistence leads to work on alternative paradigms, potential new frameworks that structure knowledge so that there are no anomalies. This is *extraordinary research*, conducted outside the bounds of the accepted paradigm. When it is successful, a 'revolutionary episode' is in progress. Regarding such research, Kuhn (1962, pp. 89–90) notes that:

Almost always the men who achieve these fundamental inventions of a new paradigm have been either very young or very new to the field whose paradigm they change ... obviously these are the men who, being little committed by prior practice to the traditional rules of normal science, are particularly likely to see that those rules no longer define a playable game and to conceive another set that can replace them.

The result is an alternative paradigm. Adherents of the current normal science are then asked to choose between two competing views of their subject: either the accepted mode of working is to be maintained, despite the anomalies, or a new sub-culture is to be adopted. The invitation is to discard existing authorities and habits and to take up new ones, which are claimed to be superior, because they are better predictors of that aspect of the world being studied. If the need for change is accepted, then a revolution in scientific practice occurs; one paradigm is replaced by another.

That choice between competing paradigms is an extremely difficult one because of their incommensurability; there are no common standards against which both can be compared. As Kuhn (1970a, p.148) points out: 'the proponents of competing paradigms will often disagree about the list of problems that any candidate for paradigm must resolve. Their standards or their definitions of science are not the same.' The new paradigm will almost certainly use some of the language and procedures of the one it is seeking to replace, but in slightly different ways, giving rise to considerable misunderstanding in discussions between the rival paradigms' protagonists. More importantly, however, the two groups of scientists may be looking at the world in very different ways:

Both are looking at the world, and what they look at has not changed. But in some areas they see different things, and they see them in different relations one to the other. That is why a law that cannot even be demonstrated to one group of scientists may occasionally seem intuitively obvious to another (p. 150).

Thus any switch from one paradigm to another is not forced simply by logic. It is what Kuhn calls a 'gestalt switch', a decision to abandon one way of viewing the world and replace it by another on intuitive grounds, rather than through the application of strict scientific criteria; a subjective decision has to be made that one is better than the other. Not all scientists may come to the same intuitive decision, therefore, so that some continue to work in the context of a paradigm that others have discredited.

Kuhn represents scientific activity as researchers trained to employ a proven mode of looking at their subject matter using an accepted methodology for solving problems. They proceed in a steady, cumulative manner, adding to the store of knowledge: small modifications may be needed to accommodate minor anomalies encountered *en route*. Anomalies are occasionally detected which cannot either be explained away or accommodated. Some researchers may focus on these, developing new paradigms to account for them, as well as everything else that was already known. When this is achieved, the new paradigm is presented for approval. A revolution is invited, for the alternative paradigms are incommensurable and only one can be right: the community is asked to redirect its work. Science proceeds in a steady fashion along well-trodden paths, therefore, with occasional major breaks in its continuity marked by important changes in the organization of its material, in the definition of its problems, and in its techniques for problem-solving.

Criticisms of Kuhn's approach, and his response

Kuhn's work stimulated a great deal of debate among philosophers of science, because it challenged orthodox views of scientific progress and implied, especially in the concept of revolutions, that some scientific decision-making was 'irrational' (see Watkins, 1970, on Kuhn's analogy between scientific and religious communities). Indeed, the introduction and treatment of the concept of the revolutionary episode attracted much of the attention (see Stegmuller, 1976), because most of the views that Kuhn was challenging were normative rather than positive; they prescribed what science *should* be like whereas he described what it was *actually* like.

A major problem with Kuhn's original presentation for many commentators was the varying use of the term 'paradigm'. Masterman (1970) identified no less than 21 different usages, which 'makes paradigm elucidation genuinely difficult for the superficial reader' (p. 61). From these, she distilled three main groups of definitions:

1 the *metaparadigms* (or metaphysical paradigms), which can be equated with 'world views', or general organizing principles;
2 the *sociological paradigms*, which are the concrete scientific achievements of a community defining their working habits; and

3 the *artefact* or *construct paradigms*, the classic works that provide the
   tools for future work.

The second provides the structure within which individual scientists
work, whereas the third provides their means for puzzle-solving; the first
provides their overall view of the nature of science and its objects of
study.

Kuhn accepted Masterman's representation of his work, and later writings
clarified his views, focusing almost entirely on the second and third of
Masterman's definitions. He indicated that if he rewrote the original book
he would give primacy not to paradigms but to scientific communities (Kuhn,
1970a). These operate at various scales: the global community of natural
scientists, for example; the main professional groups (physicists, chemists
etc.); and intra-professional groups working on particular empirical problems (such as nuclear physicists). He focused on the last: 'Communities of
this sort are the units that this book has presented as the producers and val-
idators of scientific knowledge. Paradigms are something shared by members
of such groups' (p. 178). With regard to the use of the term 'paradigm' in
this context, Kuhn (1977, p. 460) suggested a two-part definition, the second
component fitting within the first: 'One sense of "paradigm" is global, embracing all the shared commitments of a scientific group; the other isolates a particularly important sort of commitment and is thus a subset of the first'. The
global, or sociological, definition is of a *disciplinary matrix*: 'what the members of a scientific community, and they alone, share' (p. 460). It comprises
(Kuhn, 1970c, pp. 152 ff): the accepted generalizations; shared commitments
to particular models, or guiding frameworks for the construction and validation of theories (elsewhere – Kuhn, 1977, p. 501 – he equates the disciplinary matrix with a theory); and shared values regarding procedures. It also
contains a fourth element that provides the second, and subsidiary, definition of paradigms, and which he claimed (Kuhn, 1977, p. xx) was the originally intended meaning. This is the set of *exemplars* (the construct or artefact
paradigms in Masterman's classification), which are:

> the concrete problem-solutions that students encounter from the start
> of their scientific education, whether in laboratories, in examinations,
> or at the ends of chapters in science texts. To these shared examples
> should be added at least some of the technical problem-solutions
> found in the periodical literature that scientists encounter during their
> post-educational research careers and that also show them by exam-
> ple how their job is to be done (p. 187).

Solving a problem involves recourse to these exemplars; scientists search
for analogies, classic works that will suggest how the current problem
should be approached. Students learn about the exemplars, and apply them
in their own training, in order to appreciate the empirical content of their
disciplinary matrix. This provides them with a way of working:

One of the fundamental techniques by which the members of a group, whether an entire culture or a specialists' sub-community within it, learn to see the same things when confronted with the same stimuli is by being shown examples of situations that their predecessors in the group have already learned to see as like each other and as different from other sorts of situations (pp. 193–4).

Revolutions can therefore involve:

1 the replacement of one exemplar by another;
2 modification of the existing set of exemplars to accommodate new material; or
3 the replacement of the disciplinary matrix.

The last, a revolution in the sociological paradigm, is likely to be a major event in the history of a science; the first and second can occur without affecting the fundamentals of the disciplinary matrix.

The usual interpretation of Kuhn's work is that scientific communities operate within one paradigm, according to the normal science model, whereas according to Masterman (1970) and others it is possible to have long periods of either non-paradigm, multi-paradigm or dual-paradigm activity. This possibility is addressed in Lakatos' (1978a) alternative view of the history of sciences. With Kuhn, Lakatos accepts that no theory (or paradigm) is falsified until a better one is available, but he posits long debates between competing theories. Central to those debates are what he calls *research programmes* (which are similar to Kuhn's paradigms as disciplinary matrices). Each programme has a hard core, an irrefutable central set of beliefs. Its methodological rules include a negative heuristic which directs attention away from the hard core, and it also includes a positive heuristic, directing researchers towards problems in the 'protective belt' that surrounds the core. Anomalies are solved in this protective belt, by the development of auxiliary hypotheses which can account for the observed deviations and so protect the programme's core.

Progress within a research programme depends on the ingenuity of the scientists in devising hypotheses that are both consistent with the core and can account for anomalies. Success 'hardens' the core, and is measured by the number of 'verifications' or successful predictions, not by failures ('refutations'). The core's content is never in doubt. (According to Lakatos, 1978b, it is conventionally accepted and 'irrefutable', defining the problems within a preconceived plan: p. 110.) Research programmes may falter, however. Their positive heuristic may fail to produce successful predictions, so that the programme enters a degenerative phase, when it is ripe for replacement by a new programme still in the progressive phase of expanding its content substantially. A programme shift occurs because the superiority of the new over the old can be demonstrated, in much the same way as Kuhn's paradigm shifts, except that the programme shift usually takes a considerable time. Change is not instantaneous, Lakatos claims (despite its presentation as such in 'scientific folklore': p. 85). Nor are there crucial experiments, the results of which convince a community that one programme is right and the other wrong. (Crucial experiments may be identified with hindsight, but these are not recognized as such at the time when they are reported. There is a danger, Lakatos and others claim, of rewriting history as a series of crucial experiments.) Thus much of the history of a discipline may involve two or more competing research programmes coexisting (Lakatos, 1978a, p. 69):

> The history of science has been and should be a history of competing research programmes (or, if you wish, 'paradigms'), but it has not been and must not become a succession of periods of normal science: the sooner competition starts the better for progress.

Lakatos not only recognized but also required theoretical pluralism (which led to criticism from Barnes, 1982, that his is a normative view of science, not a positive one like Kuhn's); within this pluralist situation (Lakatos, 1978a, p. 92): 'Criticism of a programme is a long and often frustrating process and one must treat budding programmes leniently'.

Although Lakatos denies Kuhn's concept of normal science as one of paradigm dominance in a discipline, he accepts that progressive research programmes comprise relatively routine extensions to knowledge through the testing of new hypotheses derived from the positive heuristic. Refutations are rare; the aim is verification and progress. Popper argued against this, claiming that 'science is essentially critical' (Popper, 1970, p. 55) and characterized not by normal science but by extraordinary research. To him, the normal scientist is really an applied scientist (Popper, 1970, pp. 52–3):

> 'Normal' science . . . is the activity of the non-revolutionary, or more precisely, the not-too-critical professional: of the science student who accepts the ruling dogma of the day; who does not wish to challenge it; and who accepts a new revolutionary theory only if almost everybody else is ready to accept it – if it becomes fashionable by a kind of bandwagon effect. To resist a new fashion needs perhaps as much courage as was needed to bring it about . . . The success of the 'normal' scientist consists, entirely, in showing that the ruling theory can be properly and satisfactorily applied in order to reach a solution of the puzzle in question.

For Popper, science consists of bold conjectures and the conduct of experiments designed to refute them. In his view of science (p. 77) hypotheses can never be verified, only falsified, so that disciplines are in constant revolution as researchers seek to prove that each other's theories are wrong.

Popper's concept of a 'revolution in permanence' has been criticized by Barnes (1982) as normative and not descriptive of science as actually practised. Kuhn's (1970, p. 243) response is that: '[Popper] and his group argue that the scientist should try at all times to be a critic and a proliferator of alternate theories. I urge the desirability of an alternate strategy which reserves such behaviour for special occasions.' According to this alternative strategy, most scientists are 'normal scientists' working routinely rather than critically. This view is supported by Eilon (1975), who classified management scientists into:

1 *chroniclers*, whose role is to describe reality within the constraints of a particular paradigm;
2 *dialecticians*, who stimulate debate and progress – a dialectician 'believes it is necessary to debate and argue issues in order to elicit the facts . . . challenging stated views or records, in order to uncover what otherwise may remain hidden from an innocent observer' (Eilon, 1975, p. 361);
3 *puzzle-solvers*, who advance the empirical content of the accepted paradigm;
4 *empiricists*, who are like chroniclers in their focus on description;
5 *classifiers*, who take information generated by empiricists and chroniclers and order it within the paradigm framework;
6 *iconoclasts*, who destroy cherished beliefs, the assumptions on which they are built, the deductions, conclusions and interpretations, and expose the incompatibilities between theory and practice, between predicted and observed worlds – such iconoclasts may be destructive only, or they may be constructive in their presentation of alternative theories, in which case they are the agents of paradigm change; and
7 *change-agents*, who are the applied scientists concerned not so much with the establishment of theories and facts but rather with the use of existing knowledge to create a better world.

Individuals may be in more than one of these archetypes (not necessarily contemporaneously). If Kuhn and Lakatos are correct, however, most will be chroniclers, puzzle-solvers, empiricists and classifiers (Mercer, 1977). The change-agents are conformists too, since they apply the accepted normal science, not challenge it. Thus Popper focuses on a few key scientists only, and seeks to characterize the whole of science as if they were typical. (Note, however, that in defending this view, Magee (1975, p. 41) claims that normal scientists' work can be studied in Popperian terms. They may 'take for granted, in order to solve problems at a lower level, theories which only a few of their colleagues are questioning', but they should be seeking intra-paradigm progress by using the methodology of conjecture and refutation. Barnes (1982) also contends that much 'normal science' research involves refuting conjectures.)

Popper's arguments were extended by Feyerabend (1975), who claimed that history is a series of accidents – which is how it should be. Science should therefore be allowed to evolve as a series of accidents, since this is the best way of ensuring progress. Most discoveries have been made by individuals who either deliberately or unwittingly flouted the rules, but modern scientific education tries to prevent this by constraining its practitioners into myopic paradigms. Feyerabend claims that the only rule in science should be 'anything goes', in a scientific anarchy whose hallmark, like a political anarchy's, is (Feyerabend, 1975, p. 187): 'opposition to the established order of things: to the state; its institutions, the ideologies that support and glorify those institutions'. Like Popper, however, Feyerabend presents a normative model, concerned with major scientific developments and not with their everyday extensions. Regarding the latter, if Lakatos rather than Kuhn is correct, revolutions take a long time so a discipline will be characterized by dissensus rather than consensus over much of its history. Is it likely that a body of scholars (especially a large body) will agree on ends and means and will not differ at all, except very occasionally and then only for short periods, on fundamental interpretations? From research on radio astronomy, Mulkay (1978, p. 11) concluded that 'Scientific consensus . . . in a given area of interest is seldom complete'. A discipline may contain several branches (Mulkay, 1975, p. 520):

> In science, new problem areas are regularly created and associated social networks formed . . . The onset of growth in a new area typically follows the perception, by scientists already at work in one or more existing areas, of unresolved problems, unexpected observations or unusual technical advances, the pursuit of which lies outside their present field. Thus the exploitation of a new area is usually set in motion by a process of scientific migration.

This does not involve a paradigm shift, nor even a degenerating research programme. Rather, it indicates the establishment of a new paradigm-exemplar (in Lakatos' terms, a shift to another area of the positive heuristic or protective aureole). A new branch may be identified either by its focus on methods and techniques, or by its attention to, as yet, relatively ignored subject matter.

The branching process suggested by Mulkay allows for debate and movement within a paradigm/research programme without substantial conflict, since the disciplinary matrix/hard core is not being attacked, let alone the 'world view' (see p. 16). Kuhn (1977, p. 462) accepts this, noting that: 'Individual scientists, particularly the ablest, will belong to several such [branches] . . . either simultaneously or in succession', and Suppe (1977b, p. 498) notes that inter-branch consensus is usual:

> to account for normal science all [one] has to assume is that scientists in a particular scientific community are in sufficient agreement on what theory to employ, what counts as good and bad science, what

the relevant questions are, what sort of work to take as exemplary, and so forth.

The disciplinary matrix/hard core is central to the paradigm/research programme, not the exemplars and the type of problem attacked.

Training in its disciplinary matrix allows researchers to move among branches of a scientific community, therefore. Some branches may eventually break away, establishing new communities – by 'quiet revolutions' (Johnston, 1978c). Such breaks are rarely favoured by the parent community on political grounds, for they are likely to be against the interests of the disciplinary bureaucracy; separate communities competing for students and research funds, yet covering similar subject-matter, are unpopular. Thus communities may be prepared to accommodate dissenting groups, even potential revolutionaries, rather than risk breakaway success. Such dissent is often contained, however, and may even be repressed (see Lichtenberger, 1984, for an example from within geography). Toulmin (1970, p. 45) notes that the conflict is frequently intergenerational: 'perhaps every new generation of scientists having any original ideas or "slant" of its own finds itself, at certain points and in certain respects, at cross-purposes with the immediately preceding generation'. Its seniors may decide to accommodate the differences, 'for the good of the subject'. But according to Mulkay (1978, p. 116): 'there are now several well documented studies of instances where demonstrably competent scientists have been excluded from a field of study as their ideas have come to diverge from those in the majority'. Thus the competition between paradigms/research programmes may be contained within a discipline. In others, it results in the establishment of separate disciplines, usually after a stage in which the nascent discipline operates as a separate community within the parent body.

Kuhn and the social sciences

These critiques of Kuhn's ideas cover most of the elements of his thesis. Lakatos questions whether revolutions are achieved in short periods, Popper doubts the existence of normal science, and Mulkay suggests that several paradigms (as exemplars) can coexist within a discipline which is not characterized by consensus. None, however, questions the basic (if implicit) world view that science is the study of an empirical world, in which subject and object can be separated, and that scientific progress is measured by the volume of successful predictions. This assumes a certain philosophy of knowledge, as made clear below (p. 33).

The assumption regarding predictive success as a measure of scientific progress raises doubts about the validity of Kuhn's thesis – plus those of Lakatos, Popper and Mulkay – to the social sciences. It has not prevented social scientists trying to fit Kuhn's model to their disciplines, however, an exercise which Kuhn himself thought non-viable (because of the great variety of incommensurable writings: Kuhn, 1970a, p. 165) and which has drawn scathing criticism (Barnes, 1982, p. 120):

> The popularity of debates about whether sociology has a paradigm, or whether there have been scientific revolutions in economics or in psychology, attests more to the prevalence of intellectual laziness than to the significance of Kuhn's thinking. He himself . . . has stressed that a case for their utility has only been made in the context of the history of the natural sciences.

Kuhn's model is not positivist, but his conception of science is, for he talks of paradigms (especially exemplars) being assessed according to their predictive ability (1970c, p. 185). This is not a valid approach according to some social scientists, as parts of this book will show. Particular paradigms/research programmes may be assessed on the criterion of predictive accuracy, with degenerative phases following poor predictions (as Blaug, 1975, suggests for economics), but how does one assess different world views which offer varying basic conceptions of the nature of science? (This is Masterman's 1970, first definition of a paradigm, which Kuhn did not respond to: see also Gutting, 1980, and Hacking, 1983.) A switch from one world view to another, from one conception of science to another, should fit Kuhn's model in that the two conceptions are incommensurable and the decision must involve an 'act of faith' rather than one compelled by defined criteria logically applied. But Kuhn's work does not address such a switch within a social science, since all of his examples come from the natural sciences (as Kuhn, 1977, himself stressed).

A potential framework for analysing a social science discipline outwith the particular world view of Kuhn and his critics is outlined in Foucault's (1972) *The Archaeology of Knowledge*. This offers a method for the 'pure description' of discourses, unified systems of statements (however expressed) that can only be understood within their context (Foucault, 1972, p. 97): 'One cannot say a sentence, one cannot transform it into a statement, unless a collateral space is brought into operation'. Anything said or written can only be understood by those privy to the context of the discourse; as with a language, one can only comprehend the words, and the sequence in which they are used, because one understands the rules governing their use.

According to this view, the history of ideas is the history of discourses, of systems (somewhat akin to languages) used for the discussion of subjects that are defined within the discourses. (There is, for example, no fixed definition of madness; each discourse that discusses it has its own definition and any discussion of madness is particular to that definition and the discourse in which it is set.) The nature of a discourse is deposited in an archive, and it is the function of archaeology to reconstitute the discourse from the archive, to describe what was being done in a particular configuration (Foucault, 1972, pp. 138–40, 157–65). It describes the nature of the discourse without imposing a prescribed framework. As Foucault (1972, pp. 33–4) describes nineteenth-century medical science:

> [it] was characterized not so much by its objects or concepts as by a certain *style*, a certain constant manner of statement ... medicine no longer consisted of a group of traditions, observations and heterogeneous practices, but of a corpus of knowledge that presupposed the same way of looking at things ... [in addition it was] a group of hypotheses about life and death, of ethical choices, of therapeutic decisions, of institutional regulations, of teaching models, as a group of descriptions ... if there is a unity, its principle is not therefore a determined form of statements; is it not rather the group of rules, which, simultaneously or in turn, have made possible purely perceptual descriptions? What one must characterize and individualize is the coexistence of these dispersed and heterogeneous statements; the system that governs their division, the degree to which they depend upon one another, the way in which they interlock or exclude one another, the transformation that they undergo, and the play of their location, arrangement, and replacement.

Discourses are therefore sets of mutually agreed rules which govern description and discussion among the members who agree those rules, and are similar to the scientific communities given central place in Kuhn's (1970a) later statements. They are not independent of wider conditions, however. Archaeology may identify separate discursive formations, but these must be related to what Foucault calls the *episteme*, defined as (Foucault, 1972, p. 19):

> something like a world-view, a slice of history common to all branches of knowledge, which imposes on each one the same norms and postulates, a general stage of reason, a certain structure of thought that the men *of* a particular period cannot escape.

This implies that a particular world view (macro-paradigm?) predominates during a period and influences (controls?) the contents of individual discourses. Presumably, too, there must be periods of competition between *epistemes* (which may involve a similar process to that outlined by Lakatos). Foucault does not outline how such change occurs, however. One is left with the concept of discourses set in a societal matrix, but with little indication of how the nature of the matrix changes, presumably in part through its interactions with those discourses.

Foucault's ideas set the study of scientific disciplines more firmly in the context of the wider environment in which (and for which) they are practised than is the case with the other approaches discussed here, all of which very largely abstract the study of science from its social milieu. (As Berdoulay, 1981, expresses it: 'little interest is paid to historical contexts or intellectual climates since the focus is placed on the internal evolution of each science': p. 9.) Such abstraction is especially unfortunate for the social sciences, the disciplines that investigate and interact with their milieux and whose contents, in the broadest sense if not in detail, are likely to be strongly influenced by that context. Just how social science and society interact remains to be mapped out in detail (as, of course, does the interaction between natural science and society because the contents of the former, too, are clearly influenced by the environmental constraints), and whether Foucault's proposal of a single dominating *episteme* at any time is valid is open to doubt. That human geography should be studied in its social context appears an irrefutable claim, however. The remainder of this chapter outlines that context.

The external environment

The discussion so far has suggested that scientific disciplines (and/or discourses) are communities, small societies which are microcosms of their containing social systems. As such, the proper study of how they operate is sociological, although philosophy may provide a normative framework.

Sociological studies of academic disciplines accept that such communities are not autonomous. Their members are not isolated, and they need the support of a wider society in order to exist: society employs academics to teach and research. Whereas scientists may be able to some extent to impose their own priorities over what type of work is done, they are strongly influenced by external factors (Barnes, 1982, pp. 102–3):

> Social, technical and economic determinants routinely affect the rate and direction of scientific growth. . . . It is true that much scientific change occurs despite, rather than because of, external direction or financial control. . . . Progress in the disinterested study [of certain] . . . areas has probably occurred just that bit more rapidly because of their relevance to other matters.

The study of a discipline must be set in its societal context. It must not be assumed that members of academic communities fully accept the social context and its directives and impulses. They may wish to counter it, and use their academic base as a focus for their discontent and promotion of alternatives. But the (potential) limits to that discontent are substantial. Most university academic communities are dependent (indirectly if not directly) for their existence on public funds disbursed by governments which may use their financial power to influence, if not direct, what is taught and researched. Some universities (notably in the USA) are dependent on private sources of finance, so they must convince their sponsors and students that their work is relevant to current societal concerns (as Taylor, 1985c, suggests).

In that context, World War II is more than a convenient period from which to commence this history of human geography; it marked a major watershed in the development of the societies which are the prime focus of the book – the United Kingdom and the United States. It cannot be considered in isolation, however. Just as important for the present discussion are the world-wide economic depression which preceded it and the Cold War, the economic boom, and then the recession and restructuring which followed.

For the first time, a major international conflict was not determined solely by the sacrifices of men in battles on land and sea, although there were many during World War II. And the extra dimensions of this war did not just involve the development of air space as a further arena for conflict. The war was fought not only between military forces with guns and bombs, but also between scientists, and victory was hastened, if not ensured, by the scientific superiority of the Allied Powers, most obviously at Hiroshima and Nagasaki. Science and technology had long been major elements in the developing industrialization of the western world, but their dominance was established between 1939 and 1945, and there was to be no retreat from the many technological advances made by the researchers involved in the military effort. Thus the war heralded the dominance of the machine.

Associated with this growth in scientific activity, and its consequent prestige (with governments and with society at large), was a parallel development of social engineering. The major economic depression of the 1930s, finally triggered by the Wall Street crash of 1929, had a massive impact on governments and stimulated many measures aimed at relieving poverty and deprivation, simultaneously assuaging the liberal conscience. In the United States, this was represented by the New Deal legislation of President Roosevelt's governments, which aimed at relief and encouragement to industry and, through the Social Security Act, public support for those who, by no fault of their own, were indigent. Similar measures were introduced by the National Government in Britain (such as the 'Butler' Education Act of 1944); more were foreshadowed by plans formulated during the war, such as those promoted in the Beveridge Report on social security, and the landslide victory of the Labour Party in 1945 heralded the introduction of many social democratic policies which gave government a much greater peacetime role in the organization of the economy and society than previously envisaged.

This development of social engineering was associated with a rising status for the social sciences, and a great expansion in their activities. Economics achieved prestige first, notably through the contributions of Keynes and others to solving problems of the depression, the organization of economies during wartime, and participation in the planning of a new world economic order after the war. Others followed. Social psychological research was widely used in the evaluation of personnel by the armed forces and after the war opinion surveys proved valuable to politicians and related groups while market research was increasingly used by industry and commerce (along with psychology in its advertising efforts). All of these fields adopted the 'scientific methods' of the more prestigious hard sciences, and their successes were envied by other disciplines, such as sociology, social administration and, later, geography. To be scientific was to be respectable and useful.

The war years saw the end of the economic deprivations of the depression, as manufacturing output was boosted to provide the machinery of war. During the Cold War period that followed, military production remained considerable and kept many people in work. Further, there were many years of doing without to be compensated for, and with full employment, government direction of, and increasing involvement in, economic affairs, plus the need to re-equip industries, the two decades following the war were characterized by an economic boom in the western world. Apart from the greater government involvement, this era in industrial development was marked by another major new characteristic, the development of the giant firm, including the multi-national corporation. The concentration and centralization of capital proceeded rapidly; the average size of firms and factories increased and the economy of the world became dominated by a relatively small number of concerns.

Rebuilding the ravaged war arenas placed new demands on societies, and the planning profession emerged from earlier obscurity to take on a major role in preparing the blueprint for a new social order. The need for such action was realized during the war in the United Kingdom with the preparation of a series of reports concerned with future land-use patterns, and with the spatial distribution of economic activity, at local and regional scales. Cities were to be rebuilt; New Towns were to be constructed; a more balanced inter-regional distribution of industry and employment was to be ensured; agricultural land was to be protected; and residential environments were to be improved: all of this made great demands on social scientists, as well as engineers. The greater degree of commitment to the private ownership of land in the United States led to a slightly slower acceptance of the need for spatial planning there, but its heyday came with the rapid growth of problems involved in catering for the upsurge of ownership and use of the automobile: transport planning and engineering soon became growth industries, allied with the automobile industry and the companies which constructed the major highways.

All of these tasks – economic growth and planning; spatial planning; social administration; technological change; management, etc. – generated a need for educated personnel, and the universities received unprecedented demands for their graduates to serve the new needs of society. Education expanded rapidly; existing universities and colleges grew and many more were founded. Science and social science departments expanded to meet student demand. Their extra staff were involved in research, which became a salient component of academic life and so increased the tempo of paradigm development and questioning. Rather than places where a small elite were educated and a few privileged individuals followed their research interests, the universities became centres of society's development – the 'white-hot technological revolution' which Harold Wilson promised the British in the early 1960s. Research projects became bigger, supported by large grants from outside bodies and carried out by specially employed graduates, and the rate and volume of publication increased exponentially (Stoddart, 1967a; Johnston, 1995a, 1995b, 1996a).

The years from 1945 to about 1965 were a period of scientific and technological dominance, therefore. It was argued that the problems of production had been solved, because enough goods and services could be provided to satisfy all. The problems of distribution were still being solved, for as yet there was inequality of provision on all spatial scales. But these could be handled, it was argued, and the prospect of a prosperous and healthy life for all was widely canvassed. Academic disciplines were contributing substantially to this problem-solving by their own scientific progress. Advances in the natural sciences and technologies were solving the problems of production – of food, housing and consumer goods – as well as of ill-health. And advances in the social sciences were aiding in the management of success. Investment in education was thus investment in social progress (as well as an investment in the life chances of the individuals involved).

The deprivations of the 1920s and 1930s produced political responses characterized by a determination not only to ensure no return to economic recession but also to provide permanent protection to those who suffered short- or long-term deprivation. The Beveridge Report in the UK identified five causes for concern: want, disease, ignorance, squalor and idleness. Policies aimed at their removal led to the creation of a major edifice known as the Welfare State, which offered basic minimal standards of living for all, through such mechanisms as: a National Health Service that was free at the point of demand; a universal education service for all aged between 5 and 14 (later 15 and then 16), plus subsidized further and higher education systems; subsidized state housing; universal child allowances; unemployment benefits; benefits and care for the sick and disabled; and universal pensions. Other countries created different structures, but the overall pattern in the 'developed world' was of a major state role in sustaining its citizens (see Johnston, 1992).

Growth of the 'interventionist state' involved a major change to the nature of polity as well as society. A much wider range of activities came within the government's orbits, and the public bureaucracies grew accordingly – with consequent benefits in both employment opportunities for graduates and outlets for applying scientific and social scientific research.

The decades after World War II saw the dominance of the welfare-corporate state in public affairs, with many professional groups (including scientists and social scientists) providing policy advisers and analysts.

Despite the successes of this initial post-war period, some doubted the desirability of such an all-embracing state. These concerns were growing during the 1960s, and by the early 1970s were having a major impact on the world scene. The seeds of the doubt were many. Initially they focused, especially in the United States, on the problems of nuclear weapon development and of war, particularly the increasingly unpopular conflict in Vietnam and surrounding countries where technology was not carrying all before it against strong popular resistance and the casualty rates were increasingly deplored. Beneath the humanitarian concerns were doubts about the inequalities that continued, both on a world scale and within the 'successful' capitalist societies. Poverty was not being alleviated; if anything the disparities between rich and poor were being extended, and absolute standards of living remained appallingly low in many parts of the world, condemning the majority of the population to short lives of continuous deprivation. The prospects for solving these problems were much less rosy than they had been a few years earlier.

Furthermore, it was increasingly realized that success was being bought at considerable cost. Scientific and technological advancement required the dominance of the machine and the large factory. Work for large proportions of the labour force was being made more repetitive and boring, as skilled tasks were taken over by automated production and assembly lines. Alienation of the individual from society was increasing. Particular groups suffered more than others, as the results of prejudice and discrimination. Ethnic minority groups and women were the main sufferers, and they were the focus of civil rights movements. Finally, interest was kindled in the growing degradation of the environment to fuel the production goals of advanced capitalist societies.

The problems of the dignity of the individual, the repression of minorities, the quality of life and the depletion of environmental resources were not new in the late 1960s. What was realized then, however, was that the form of social 'progress' advanced during the previous decades was in many cases exacerbating and not solving such problems. As the realization grew, so the proposed solutions varied. To some, the problems could be solved by greater state involvement, on an international and national scale. Human and civil rights must be protected; greater equality must be achieved through the redistribution of wealth; the environment must be conserved, and where necessary preserved. In the language of the previous section, the research programme is maintained, but major efforts are made within its positive heuristic to solve the many anomalies. To others, this solution was insufficient. It would simply generate new anomalies because, according to the developing critique of capitalist systems, these are necessary to social 'progress': capitalism, it was argued, necessarily survives on inequalities, on alienation and on the rape of the environment. Only a new research programme could change society for the better.

Universities were the focus of much of this developing concern. There were major confrontations between students and other elements of society in the late 1960s, for example, in particular at Berkeley, Chicago, London and Paris, and the student body was in the forefront of the anti-war movements. Some academics, notably in the social sciences, focused on the 'managerial' issues involved in producing a 'better, more equal' society, and these were considered more crucial than the 'production' issues covered by the natural sciences and technologies. A threefold division developed. In one, the need for social scientists to become more active in developing solutions to the problems of distribution and environmental depletion was advanced: social science must become more 'relevant', more 'policy-oriented', within the constraints of the proposed societal 'research programme'. A second argued for greater concentration on the problem of the alienation of the individual, who should be released from the overbearing dominance of the machine and the big organization and should be encouraged to take a much greater part in creating her/his own life. The individual was to be protected against the increasingly distrusted expert. Finally, a third group developed a critique of capitalist society, seeking to show that while specific problems may be soluble, this would merely lead to others, while the general problems would remain because they are endemic to that mode of social organization. The call was for major social reform – to some, revolution – as the only long-term solution to the problems of human dignity and inequality.

The force of these arguments can be identified in a variety of ways, not least the declining popularity of scientific and technological subjects among students and the growing demand for places in the social sciences and humanities in the 1960s and 1970s. A further major problem then arose, because the capitalist world spent much of the 1970s in economic recession, partly stimulated by the decision of the Arab countries during the 1973 Yom Kippur war to use oil as an economic weapon. The price of this vital raw material was increased many-fold in a few years, with major impacts on all economies. To many analysts, however, the recession had already set in – as illustrated by Britain's problems in the late 1960s – and the Arab decision was just a major additional stimulus. The recession reflected the failure of state policies of demand management following Keynesian principles, it was argued, and the search for an alternative saw a growing divergence between political parties. In Britain, for example, the relatively high degree of consensus over demand management broke down after 1970, with the Conservative party promoting a greater emphasis on free markets and a reduction in the role of the state, whereas the Labour party, shifting to the left, promoted the opposite. This growing polarization added to Britain's problems, some claim, because of the uncertainty it engendered about the future; a change of government would bring a major shift in policy.

To some social theorists, recession and its major impacts (notably unemployment) should accelerate the demands for reform and revolution. Both the United Kingdom and the United States at the end of the 1970s elected right-wing governments dedicated to radical programmes of economic regeneration by a reduction in public expenditure and a liberalization of those capitalist forces that produce the inequalities so widely condemned only a decade earlier. The role and size of the state were to be significantly reduced (Gamble, 1988); its task was to provide the infrastructure for a buoyant economy (particularly through sound management of public finances and the removal of restrictive practices – notably, though not only, by trades unions). Enterprise was to be encouraged by a system which rewarded successful initiatives and risk-taking, and the result would be a wealthier society from which all would benefit, while the state provided protection only for those in proven need (Johnston, 1992).

Education has been subject to significant policy changes since the 1970s because of this shift. For the first time for more than two decades, expenditure on higher education in Britain was cut in 1981, and the numbers of students undertaking undergraduate degrees and postgraduate training were reduced. The cuts were selective, with relative protection for science and technology – growth in which was seen as necessary for economic progress – and substantial reductions in the social sciences, believed by many politicians to be the homes of left-wing radicals and the fomenters of discontent. Research funds were similarly cut and redistributed. The reaction to these policies has been in part an attempt to defend academic freedom and independence and the need for a 'critical conscience' within society (which involves curricula designed to develop critical intellects as well as foster 'transferable skills') but there has also been a desire to re-orient work within disciplines to make them more 'relevant' to current societal concerns.

The educational system within which the research components of academic disciplines are located has been closely involved in economic, social and political change during the last forty years. It benefited from the boom years of the first two decades. Expansion was rapid and academic activity was considerable. It was then cut back, especially in the UK, as – to some decision-makers at least – expansion, especially expansion in the arts and social sciences (excluding business studies and management), was seen as an unaffordable luxury. Economic progress did not require large numbers of students and potential researchers being trained in disciplines with little relevance to perceived societal needs and working on topics that were critical of societal structure. Disciplines and scholars had to prove their relevance and sell their skills in the market-place. Academic freedom was not entirely removed, but was to be curtailed, simply by denying it resources if it was felt that freedom was being abused. The result is that after participating gleefully in the booms of the 1950s and 1960s, academia went into deep depression, suffering internal and external crises of confidence and subject to considerable political direction.

There was a reversal in political attitudes towards higher education at the end of the 1980s, if not also in the funding provided. In the United Kingdom, for example, there was a growing realization that national competition in the restructured global economy required developing human resources to their full potential, through a substantial increase in the participation rate of the 18–25-year-olds and an expansion of continuing education for older people. Thus despite a demographic downturn (which particularly affected the size of the teenage population in the socio-economic classes from which few traditionally moved into higher education), universities, polytechnics and colleges were expected to expand their provision and numbers. But, in line with the economic ideology of the governing party, they were expected to become more efficient in doing their teaching (i.e. to take more students without additional resources), and to obtain more income (especially for research and for continuing education) from sources other than the state. Further, there was an increased emphasis on the 'customer pays' principle (for long the norm in North America), with students meeting more of the costs of their education (in part, it was argued, because they would benefit from higher incomes as a consequence of their qualifications and expertise) and so being more concerned with getting 'value for money'; courses with a clear vocational orientation became more popular.

The UK government also became increasingly concerned over accountability for the funds provided to the universities for research. These were thought to be too widely spread and their use subject to insufficient evaluation. From the late 1980s on, therefore, funding for research in each discipline was to be concentrated on those institutions where work was evaluated as of high quality internationally, and a series of regular Research Assessment Exercises was instituted which provide gradings of all departments; the higher the grade achieved, the more money provided and the greater the amount of time and resources available for research. At the same time, the Research Councils, which provide funding for individual projects, were restructured to promote 'user interests' and were directed to focus their funding on areas of national economic importance identified through a regular Technology Foresight exercise. Financial pressure was thus used to promote a particular orientation to research (and teaching) within universities, one that was sympathetic to the ideology of the 'enterprise economy' (Johnston 1995a, 1995b).

Departments of geography and their individual members had to respond to these political, institutional and potential market changes, and to restructure their course and research offerings accordingly. The nature of the science that they practised and taught was necessarily strongly influenced by the 'culture of the times' (which does not mean that all conformed to it). As Livingstone (1992, p. 347) expresses it, this is because: 'geography changes as society changes, and ... the best way to understand the tradition to which geographers belong is to get a handle on the different social and intellectual environments within which geography has been practised'.

Three types of science

Little has been said so far about various conceptions of the nature of scientific activity, except for the material on different paradigms as 'world views'. It has been implicit that geographers share a world view, and that any differences in what they have done, and still do, reflect adherence to either or both of separate sociological paradigms and different exemplars (see p. 16): changes in the relative importance of such paradigms and exemplars reflect responses by geographers, both individually and collectively, to changes in their external environment. But is this the case? Has there been just one 'world view' which all human geographers have shared during the period under review, or are there different conceptions of the discipline?

In outline form, there are three separate conceptions of the nature of a scientific discipline such as human geography. Each has its own world view, its beliefs regarding both the nature of knowledge (its epistemology, which answers such questions as 'What can be known?' and 'How can we know it?') and the means of obtaining knowledge, plus its beliefs on the uses to which knowledge can and/or should be put. The remainder of this book illustrates the relative importance of those three conceptions within Anglo-American human geography since 1945 and the debates among the protagonists of the various positions. The present section provides a brief introductory outline of the three world views. (The classification is drawn directly from Habermas, 1972, and draws on the presentation in Johnston, 1986a and 1989a.)

The empirical (or analytic) sciences draw on an empiricist world view; knowledge comes from direct experience and is based on the senses, especially visual observation. Their methodology calls for accurate observation and reportage. Some identify this as a neutral, value-free position, in which the 'facts speak for themselves', a view countered by those who argue that all observation is theoretically based – what is recorded as present in a place reflects what is being sought and what is deemed to be important, and its classification reflects a prior selection of categories. Thus empiricist work is not the presentation of unordered material, but the recording of information within an agreed and approved conceptual framework; as illustrated below (p. 42), for geographers in the decades prior to 1945 this involved the collection and recording of material within a framework which identified the physical environment as the major determinant of the pattern of human activity on the earth's surface.

A particular form of empirical science generally known as *positivism* is the approach most frequently assumed to be characteristic of all science. Its goal is not only to describe (in a geographical context, to show what is where), but also to explain (to say why it is there). Such explanation is provided by presenting individual occurrences as examples of general laws (of the form 'if A then B': if B is present in a place, it is because A is there also). The goal of positivist science is to identify laws, thereby providing not only an explanatory procedure (the distribution of B can be accounted for by the distribution of A in the above example) but also a predictive device – the presence of further occurrences of A can be used to predict future occurrences of B.

Empirical science involves the collection and reporting of information; positivist empirical science involve its use to produce a particular form of explanation. The individual event is an example of the operation of one or more general laws. The predictive content of those laws can be used in processes of technical control; the presence of B in a place can be ensured (if it is desired) by putting A there, or its absence (if wished) can be guaranteed by preventing the occurrence of A there. Successful positivist physical science can be used to manipulate and control the environment through the application of known physical laws; successful positive social science can be used to manipulate and control society through the application of known social laws.

The hermeneutic sciences deny the existence of a separate empirical world outside the individual observing it. No observation and description can be neutral, it is argued, but involve interpretation of the world as it is perceived through a system of meanings, which are human constructs developed by individuals through a continuous process of socialization and resocialization in contact with others. Thus as a human I am more than a combination of living cells; I have both powers of reason and emotional traits. I share many characteristics with others, some (such as age and sex) based on biological criteria, but others (such as religious beliefs and class position) based on human constructs that are far from universal – my interpretation of my class position may differ from that of others, who may also disagree among themselves. All of these characteristics influence how I act, because I draw on them, and my interpretation of what they mean (what I think a person in 'my class' should do, for example), to guide my thinking and acting: thus the only way to understand what I do is to understand me. I observe the world and ascribe meanings to what I see, and I then act in accordance with those meanings, which may differ from the meanings that an empirical scientist observing me may choose to use. According to a hermeneutic scientist, the meanings that matter are mine, since they are the foundations for my behaviour.

In the hermeneutic sciences, therefore, general laws of human behaviour are impossible, because humans, with their powers of memory and reason, cannot be treated as equivalent to machines that always respond in the same way to an identical stimulus (which is what the positivist sciences proclaim). Hermeneutic science does not offer explanations, but rather understandings. Its goal is to appreciate what people believe, how those beliefs develop within societies, and how they are drawn upon as the bases for actions. Such appreciation helps one to understand the past and the present, and may provide a guide to the future, but it is in no way predictive: it cannot say, 'if A, then B'.

Empirical sciences, especially positivist sciences, can be applied in strategies aiming at control of environments and societies. Hermeneutic sciences cannot. This does not mean that they cannot be applied within society, however; far from it. The understanding that is gained from hermeneutic appreciation can be used to promote mutual appreciation, thereby enriching societies by making people better aware both of others and of themselves.

Critical sciences differ from both of the other two, for they accept neither the implicit determinism of the positivist nor the voluntarism of the hermeneutic: the former implies that people ultimately have no control over their lives whereas the latter implies that they have complete control. According to critical sciences, people live within societies which are complex organizations created to ensure both individual, day-to-day and collective, inter-generational survival. Those organizations involve rules which must be operated if the society is to continue. At the most fundamental level, they must ensure sufficient food for all, for example, but in different types of society that is done in different ways: in capitalist societies, for example, food is only produced if it can be sold for a profit; in socialist societies it is produced according to collectively agreed plans to meet collectively determined needs.

People are free to interpret a society's rules in a variety of ways, as long as they do not transgress its boundaries between the acceptable and the unacceptable; in capitalist societies, for example, the rules require the production of food for sale at a profit but do not determine what foods will and will not be profitable – that is decided by individuals, both separately and collectively, and over time the types of food produced may change. Hermeneutic processes are involved, therefore, because the operation of a society depends on how people interpret its rules; further, conditions change (as a result of environmental variations, both temporary and secular, for example) and these changes must be interpreted and responded to. In the critical sciences, therefore, it is necessary to appreciate the basic rules by which a society operates in order to achieve a fundamental understanding; to appreciate what happens in particular circumstances, it is necessary to appreciate how the people operating the rules interpret them. (All sports have basic rules. Participants interpret those rules, and plan courses of action within them. To understand the sport, you must know the rules; to appreciate a particular game, you must understand how the participants have decided to operate within those rules.)

Critical science is also applicable, but in a different way. Its goal is to ensure that people understand the rules by which a society operates (even though they may be hidden and unwritten, and can only be determined through abstraction from the many different interpretations of the rules that lead to the empirical world of appearances). Once people understand the rules, then they understand the fundamentals of the society – in technical terms, they are emancipated. They are then freed from constraints to their understanding, and can, if they wish, become involved in the transformation of society, changing the rules to a set which they find more acceptable.

These, then, are three very different conceptions of science, of what it is, how it is done, and what its purpose is. If people disagree over the relative merits of the different views, then there will be debate between adherents of the various causes (and perhaps variants within each) promoting their own views. Any one discipline may therefore incorporate competition over the relative merits of each approach, as well as competition within each as to the proper way for that particular form of science to be practised. There can be a hierarchy of debates within a discipline such as human geography, therefore: at the most fundamental level, there is debate over the relevant world view; lower down, there can be debate within each world view over the conduct of research; and at the lowest there can be debate over procedural details, as set out by different exemplars.

Human geography is not alone among the social sciences in having experienced all three levels of debate in recent decades. In general terms, the initial debates were within the empirical/analytic world view, and were followed by the introduction, in turn, of cases for each of the other world views, as well as debates within each regarding procedures and exemplars. Thus the task of charting the history of Anglo-American human geography over the last five decades involves identifying the major features in these various debates. That is the role of the next eight chapters; they are followed by one which evaluates those debates, not by trying to reconcile the various positions, but rather by setting them in context and seeking to appreciate why and when they occurred and were settled (to the extent that they were).

Conclusions

The thesis of this chapter is that the history of an academic discipline must be set in a context comprising three elements: the occupational structure; the organizational framework for research; and the societal environment. These three interact in a variety of ways. The occupational structure is very much constrained by the societal environment, for example, as indicated by the negligible opportunities for promotion in British universities in the late 1970s and early 1980s because of the cuts in educational funding. Similarly, the framework for research, although established by and for academics, is subject to societal support. Some frameworks are much more acceptable than others, and so are more likely to receive the needed public finance.

In the following chapters, the content of human geography in Britain and North America since World War II is reviewed, within the context set by the discussion here. The emphasis is on 'extraordinary research' rather than the cumulative achievements of 'normal science', stressing the debates over how research should be done in human geography. No attempt is made to test the models outlined in the section above on 'the academic working environment', although the ideas outlined there have clearly influenced the organization of the book. The main purpose is to present a reasoned summary of debates in human geography (the discipline being defined as comprising that which is claimed as human geography), based entirely on the published record. The relationship between this summary and the models of scientific progress only resurfaces in the final chapter.

|2|

Foundations

Although this book is about human geography since 1945, any discussion of that period requires a brief outline of the nature of the discipline in preceding decades, for a variety of reasons. The first is that although 1945 was a watershed year in many aspects of the social, economic and intellectual life of the countries being considered, it was not a significant divide in the views on geographical philosophy and methodology. Not surprisingly, the war years were not a major period of intra-disciplinary academic debate. Most academics spent the time either on active service or in associated intelligence activities (some of those involved in the latter retained their teaching commitments); the everyday activities of teaching, pure research and administration were replaced by commitment to the war effort. It took a few years for academic life to return to something like normality, to assimilate the large numbers of new staff needed to replace the losses of the war years, to teach the backlog of students, and to react to the new social and economic environments.

A second reason for surveying the period preceding that being studied relates to the processes of change in academic work. New paradigms are responses to the perceived failings of those currently in favour, not inventions produced in an intellectual vacuum. The post-Second World War changes were reactions to the philosophies and methodologies developed and taught in the preceding decades; reactions to them cannot be studied without some knowledge of what went before.

Finally, change is not instantaneous in academic life. A new research programme usually takes years to mature, while experimentation with alternatives takes place, the programmatic statements are written, and converts are won over by the prophets of the new approach. Meanwhile the current paradigm prevails (or paradigms, if several have considerable support). Its adherents continue in their accepted ways, researching, publishing, and teaching generations of undergraduates according to the conventional wisdom. Even when a new paradigm has been crystallized, it may have to coexist with its predecessors for several years, while competing for support among academics. This may be especially characteristic of the social sciences, in which interpretations of data are frequently more subjective than in the physical sciences; it is quite feasible for two or more separate world views to find adherents at the same time, quite possibly in the same academic department.

Geography in the modern period

The hallmark of an academic discipline, according to one of geography's chroniclers (James, 1972), is an educational organization which provides specialist training in the subject. James dates the beginning of such an organization for geography around 1874, when the first university geography departments were established in Germany (see also Taylor, 1985c): Britain and the United States followed a little later, with the main developments coming in the twentieth century. Before 1874, geography was investigated either by amateurs or by scientists trained in other fields. With the inauguration of its own specialized institutional training, geography left its classical age and entered what James terms its modern period, which lasted for about eighty years, being superseded after 1945 by what James calls its contemporary period.

James's modern period is virtually co-extensive with the decades surveyed in Freeman's (1961) *A Hundred Years of Geography* which, with James's book, is one of the few attempts to provide a history of the discipline (see also Freeman, 1980a; Stoddart, 1986; Gaile and Willmott, 1989; Livingstone, 1992: Martin and James, 1992). Freeman identified six main trends in the geographical literature.

1 *The encyclopaedic trend,* associated with the collection of new information about the world, particularly areas little known to Western Europeans and North Americans. Although the great age of discovery was over, and by the late nineteenth century much of the world had been visited by Anglo-Saxon explorers, vast tracts remained, notably in Africa, which if not *terra incognita* were almost empty on contemporary maps. Indeed, at the beginning of geography's modern period much of the North American continent itself remained to be settled by permanent farms.
2 *The educational trend,* which, as James stresses, characterizes an academic discipline needing to propagate its knowledge, establish its relevance, and ensure its reproduction. Much work was undertaken to achieve a solid foundation of geographical work in schools, colleges and universities, involving both proselytizers and the architects of curricula (Freeman, 1980a, 1980b; Stoddart, 1986).

3 *The colonial trend* reflects a major environmental preoccupation during the early decades of the modern period, especially in Britain whose empire was being consolidated and developed into a spatial division of labour based on its metropolitan hub and covering a considerable proportion of the earth's surface (see the essays in Bell, Butlin and Heffernan, 1995). Organization of the commercial world required a great deal of information about the various countries concerned, the provision of which became a major task of geographical research whilst its propagation was the keystone of geographical education.

4 *The generalizing trend* describes the use to which data collected in the encyclopaedic and colonial traditions were increasingly employed. Academic study involved more than collecting and collating facts: these had to be interpreted, and the methods and aims of such interpretation defined the early paradigms of the discipline's development.

5 *The political trend* was reflected in the contemporary uses made of geographical expertise. Isaiah Bowman was a chief adviser to Woodrow Wilson at the conferences which re-drew the map of the world after the First World War (on which see Martin, 1980, and Smith, 1994), for example, and the work of geopoliticians such as Haushofer was influential on the *Lebensraum* ideology of Nazi Germany (James and Martin, 1981; Kost, 1989; Parker, 1985).

6 *The specialization trend* was a reaction to the growth of knowledge and the inability of any one individual to master it all, even within the single discipline of geography. Prior to the modern period, scientists and other academics could be extremely catholic in their interests and expertise, but as the volume of research literature increased and the techniques of investigation demanded longer and more rigorous training so it became necessary for the individual to specialize, first as a geographer and then within geography, focusing either on one substantive area or on a particular region of the earth's surface.

Some of these trends represent philosophies, some methodologies, and some ideologies with regard to the purpose of academic geography. From them it is possible to identify three paradigms (as disciplinary matrices) which characterized the modern period in human geography as a whole and of its component parts, such as urban geography.

Exploration

The first approach was carried over into the modern period from the classical, for exploration was the major activity recognized as geography through most of the nineteenth century. The collection and classification of information about 'unknown' parts of the earth (unknown, that is, to Western Europeans and North Americans) were undertaken by geographers. Many of their expeditions were financed through the geographical societies which were founded then (Freeman, 1961, 1980b; Bell, Butlin and Heffernan, 1995); these, in turn, obtained money from commercial as well as philanthropical sources, for the information gathered was of great value to mercantile interests.

As well as supporting and sponsoring exploratory expeditions, the geographical societies also undertook major educational roles. Their lecture meetings provided opportunities for the general public to see and hear of the new discoveries, and their officers worked hard to establish the teaching of geography in schools and universities: the Royal Geographical Society of London (RGS), for example, was involved in discussions which led to the establishment of geography teaching at England's two oldest universities, Oxford and Cambridge (Stoddart, 1975a; Freeman, 1980b; Cameron, 1980).

The importance of exploration declined as geography matured in its new academic-discipline status, although in 1899 Halford Mackinder felt it necessary to establish his credentials as a geographer by becoming the first recorded person to climb Mt. Kenya. Much *terra incognita* remained, however, and the geographical societies maintained their interest in and sponsorship of expeditions; the RGS still sponsors scientific expeditions and *The Geographical Journal* reports their findings. The work undertaken is very different from that of a century ago in most cases, reflecting developments in scientific technology and the available store of knowledge, but basic activities such as accurate map-making remain crucial to many successful expeditions. Many of the Society's meetings involve word and picture presentations of the results of expeditions to all parts of the world. In 1995 the RGS merged with the Institute of British Geographers, established in 1933 by geographers in universities to promote academic research through conferences and publications, and the joint society now has a much wider remit.

The American Geographical Society (AGS) in New York is the American counterpart of the RGS. It too has broadened from the exploration role on which it focused early in the modern period, and it now sponsors research in many other areas of geography while still supporting investigations of relatively unstudied parts of the globe, such as the Arctic. The exploration tradition is maintained in the United States by the National Geographical Society and its popular journal, the *National Geographical Magazine*. (A journal which it launched in 1984 – *National Geographic Research* – seeks to bridge the gap between academic research and a wider audience.) Other societies have taken over some of the professional roles: there are separate academic bodies in both the United States and the United Kingdom (the Association of American Geographers and the Institute of British Geographers – though see above regarding the merger of the IBG and the RGS) as well as those which concentrate on geographical education (the National Council for Geographical Education and the Geographical Association respectively).

Although most of it was not strictly exploration, the work summarized by Freeman under the colonial trend can be included here, since its aims were the collection, collation and dissemination of information. Much of the material was about commercial activities and infrastructure, as in volumes such as Chisholm's *Handbook of Commercial Geography* (first edition, 1899) and *Gazetteer of the World* (1895), which were aimed at the world of commerce, with companion volumes for schools (Wise, 1975). Their content comprised statistics and descriptions of production and trade, and a training in this type of geography was boring to many with its focus on the assimilation of large bodies of factual knowledge ('capes and bays' geography). But the need for such geographical expertise was widely recognized, and was called on when geographers were made responsible for the preparation of intelligence reports about areas in which Allied troops were likely to be engaged, work characterized by the set of British Admiralty Handbooks.

Environmental determinism and possibilism

These two competing approaches represent the first attempts at generalization by geographers during the modern period. Instead of merely presenting information in an organized manner, either topically or by area, geographers sought explanations for the patterns of human occupation of the earth's surface. Their major initial source for explanations was the physical environment, and a theoretical position was established around the belief that the nature of human activity was controlled by the parameters of the physical world within which it was set.

The origins of this environmental determinism lie in the work of Charles Darwin, whose seminal book *On the Origin of Species* (first published in 1859) influenced many scientists (see Livingstone, 1992). His notions regarding evolution were taken up by an American geographer, William Morris Davis, in his cycle-of-erosion model of landform development. Ideas of natural selection and adaptation formed the basis of statements regarding environmental determinism, including Davis's (1906) programmatic paper identifying the core of geography as the relationship between the physical environment, as the control, and human behaviour as the response (Stoddart, 1966; Martin, 1981; see also Campbell and Livingstone, 1983, and Livingstone, 1984 on the influence of Lamarckism in the development of geography, and Peet, 1985a, on a similar theme).

Chief among the early nineteenth-century environmental determinists was a German geographer, Ratzel, whose American disciple Ellen Churchill Semple opened her book *Influences of Geographic Environment* (1911) with the statement that 'Man is the product of the earth's surface'. In some hands, the environmental influences adduced were gross, and with hindsight it is hard to believe that they were taken seriously; Tatham (1953), for example, illustrates the extent to which authors were prepared to credit all aspects of human behaviour with an environmental cause.

Reaction to the extreme generalizations of the environmental determinists led to a counter-thesis, that of possibilism, which presented the individual as an active rather than a passive agent. Led by French geographers, followers of the *Annales* school historian Lucien Febvre, possibilists presented a model of people perceiving the range of alternative uses to which they could put an environment and selecting that which best fitted their cultural dispositions. Taken to extremes, this approach could be as ludicrous as that which it opposed, but possibilists generally recognized the limits to action which environments set, and avoided the great generalizations which characterized their antagonists.

Debate over environmental determinism and possibilism continued into the 1960s (Lewthwaite, 1966; Spate, 1957, for example, proposed a middle ground with the concept of 'probabilism'). The determinist cause was continued in the period between the world wars by writers such as Ellsworth Huntington, who advanced theories relating the course of civilization to climate and climatic change. Perhaps the most doughty advocate was the Australian Griffith Taylor, whose views so angered politicians interested in the settlement of outback Australia that he was virtually hounded out of his homeland (Powell, 1980a). He argued that possibilists had developed their ideas in temperate environments such as northwestern Europe, which offer several viable alternative forms of human occupance. But such environments are rare: in most of the world – as in Australia – the environment is much more extreme and its control over human activity accordingly that much greater. He coined the phrase 'stop-and-go' determinism to describe his views. In the short term, people might attempt whatever they wished with regard to their environment, but in the long term, nature's plan would ensure that the environment won the battle and forced a compromise out of its human occupants.

Many debates begin as two opposing, extreme views, and end with a compromise accepted by all but the most fervent devotees of each polar position. Thus the lengthy discussion among geographers about whether people are free agents in their use of the earth or whether there is a 'nature's plan' slowly dissolved as the antagonists realized the merits in each case. (And some geographers studied people–environment interactions outside the confines of these debates: see Fleure, 1919.) But while environmental determinism was strongly promoted by geographers, respect for the discipline declined somewhat in the eyes of the academic community at large, which rejected the approach. As a consequence, geography's next focus, which nevertheless had strong roots in environmental determinism, was very much an introspective and conservative one.

The region and regional geography

This third approach dominated British and American geography for much of the first half of the present century. Like environmental determinism, it was an attempt at generalization, but it lacked structured explanation and so was of a very different type from the increasingly discredited law-making attempts of the previous writings (although much of it had strong implicit connections with environmental determinism). Much early development took place in Britain, involving work on two scales (Freeman, 1961, p. 84; Johnston, 1984d). On the large scale were efforts, such as Herbertson's (1905), to divide the earth into major natural regions, usually on the basis of climatic parameters and thus having some links with the earlier determinism. On the smaller scale, the aim was to identify individual areas with particular characteristics:

> The fundamental idea was that the small area would legitimately be expected to show some distinct individuality, if not necessarily entire homogeneity, through a study of all its geographical features – structure, climate, soils, vegetation, agriculture, mineral and industrial resources, communications, settlement and distribution of population. All these, it has often been said, are united in the visible landscape, linked into one whole and dependent one on another. And more, every area, save those few never occupied by man, has been influenced, developed and altered by human activity, and therefore the landscape is an end-product, moulded to its present aspect by successive generations of people. The practice has therefore been to take an evolutionary view and . . . to attempt to reconstruct the landscape as it was a hundred, or a thousand years ago (Freeman, 1961, p. 85).

Some of this work, exemplified by Herbertson's (1905), was the precursor of the ecosystem concept.

Hartshorne and the American view

The ideas and methods of regional geography were taken up a little later in the United States. In the late 1930s, two non-geographers published a major survey of American regionalism (Odum and Moore, 1938) and in 1939 the Association of American Geographers published a monograph – Richard Hartshorne's *The Nature of Geography: A Critical Survey of Current Thought in the Light of the Past* – which was rapidly established as the definitive statement of the current orthodoxy (see Stoddart, 1990). As Hartshorne (1948, 1979) later made clear, there was much debate among American geographers during the 1930s (most of it apparently unpublished) about the nature of their discipline. He was concerned about both the tone and the content of that debate (in particular by Leighley, 1937), and in 1938 he submitted a paper to the *Annals of the Association of American Geographers*, as a contribution to the philosophical discussions. He then proceeded to Europe for fieldwork on boundary problems, as part of his ongoing research into political geography. This work was frustrated by the political situation, and he spent his time reading European, mainly German, work on the nature of geography. He used this to extend his 1938 paper, adding the crucial sub-title; the result was a 'paper' of 491 pages (some 230,000 words) which became the major philosophical and method-ological contribution to the literature of geography in English then available.

A synopsis of Hartshorne's book, and his interpretations of others' works, notably Hettner's, are not possible in a few paragraphs, and only the main conclusions can be highlighted here. It should be stressed that Hartshorne's statements were positive ones – of what geography is: they were only normative in the sense of him saying that geography should be what others (notably Hettner) have said that it is. Thus Livingstone (1992, p. 306) describes his project as seeking 'to determine the nature of geography from scrutinizing its history', and Lukermann (1990, p. 58) claimed that the *Nature* was 'a search for authority to validate the conclusions drawn from selected premises – largely formulated by Hettner, who had philosophical associations and leanings rather than historical associates'. Others argue that he was selective in his use of Hettner's material (Butzer, 1990: Gregory, 1994, p. 51, claims that 'Hartshorne's views were developed through a highly selective exegesis of a German intellectual tradition. His approval of Hettner (in particular) was unrestrained, but the regional geography that he constructed was purged of both the physico-ecological and the cultural-historical implications that were indelibly present in Hettner').

Hartshorne argued forcefully that the focus of geography is areal differentiation, the mosaic of separate landscapes on the earth's surface (see Agnew, 1990, on the representation of Hartshorne's focus as 'arcal variation' rather than 'areal differentiation'). Thus geography is:

a science that interprets the realities of areal differentiation of the world as they are found, not only in terms of the differences in certain things from place to place, but also in terms of the total combination of phenomena in each place, different from those at every other place (p. 462).

Also geography is concerned to provide accurate, orderly and rational descriptions and interpretations of the variable character of the earth's surface (p. 21) and

seeks to acquire a complete knowledge of the areal differentiation of the world, and therefore discriminates among the phenomena that vary in different parts of the world only in terms of their geographic significance – i.e. their relation to the total differentiation of areas. Phenomena significant to areal differentiation have areal expression – not necessarily in terms of physical extent over the ground, but as a characteristic of an area of more or less definite extent (p. 463).

The principal purpose of geographical scholarship is thus synthesis, the integration of relevant characteristics to provide a total description of a place – or region – which is identifiable by its peculiar combination of those characteristics. There is then, according to Hartshorne, a close analogy between geography and history; the latter provides a synthesis for 'temporal sections of reality' whereas the former performs a similar task for 'spatial sections of the earth's surface' (p. 460).

Hartshorne also indicated the methodology to be used for this integrating science – 'the ultimate purpose of geography, the study of areal differentiation of the world, is most clearly expressed in regional geography' – and accepted procedures were necessary for regional identification. Regions are characterized by their homogeneity on prescribed characteristics, selected for their salience in highlighting areal differences. Two types were identified: the *formal (or uniform) region,* in which the whole area is homogeneous with regard to the phenomenon or phenomena under review, and the *nodal (or functional) region* in which the unity is imparted by organization around a common node, which may be the core area of a state or a town at the centre of a trade area. Identification of such regions 'depends first and fundamentally on the comparison of maps depicting the areal expression of individual phenomena, or of interrelated phenomena ... geography is represented in the world of knowledge primarily by its technique of map use' (pp. 462–4). Hartshorne emphasized map use. Although it is valuable for geographers to know something about the preparation and construction of maps, the sciences of surveying and map projections are of only secondary interest to them; their prime task is map interpretation. Much of the information to be interpreted may have been placed on the maps by geographers during their fieldwork, and the role and nature of fieldwork were of considerable interest to American geographers during the period when Hartshorne was developing his ideas.

Preparation of a regional synthesis required materials both from other sciences specializing in certain phenomena (though usually not their areal patterning) and from the topical systematic specialisms within geography which complemented, but were eventually subsidiary to, regional geography. Physical, economic, historical and political were the main systematic sub-divisions recognized within geography at the time Hartshorne wrote, although a later survey, set firmly within the regional paradigm, identified many other 'adjectival geographies', including population, settlement, urban, resources, marketing, recreation, agricultural, mineral production, manufacturing, transportation, soils, plant, animal, medical, and military, plus climatology and geomorphology (James and Jones, 1954). A number were of only minor importance, however, so that despite the apparent diversity of interests among geographers of the time, the 'classic' regional study usually followed a sequence comprising physical features, climate, vegetation, agriculture, industries, population and the like (Freeman, 1961, p. 142) and was summarized by a synthesis of the individual maps to produce a set of formal regions.

To most geographers of the period spanning World War II, including those who contributed to the survey edited by James and Jones (1954), which may have been why they were invited, regional geography was at the forefront of their discipline's scholarship and systematic studies were the providers of information for that enterprise: thus to James, 'Regional geography in the traditional sense seeks to bring together in an areal setting various matters which are treated separately in topical geography' (1954, p 9). Urban geographers studied towns because they 'constitute distinctive areas' (Mayer, 1954, p. 143), in line with the regional concept; political geographers studied the functions and structures of an area 'as a region homogeneous in political organization, heterogeneous in other respects' (Hartshorne, 1954a, p. 174); and in defining the 'new' field of social geography, Watson (1953, p. 482) saw it 'as the identification of different regions of the earth's surface according to associations of social phenomena related to the total environment' (see also Johnston, 1993a). Each topical specialism produced its own regionalization (as in agricultural geography, especially O. E. Baker's papers in *Economic Geography* during the 1920s and 1930s outlining the agricultural regions of various parts of the world), and each had links with the relevant systematic sciences – social geography with sociology, for example. The key differentiating factor between the two was the geographer's focus on the region, both the specialist's single-attribute region and the synthesiser's multi-attribute region.

Given this focus on the region, the literature unsurprisingly contained many contributions discussing their nature and delimitation, for virtually every region was in effect a generalization, complete homogeneity being very rare over more than a small area. Much effort was expended developing methods for defining multi-attribute regions; in agricultural geography, for example, it culminated in the statistical procedure developed by Weaver (1954). But on the small scale it was widely accepted that regional delimitation should be based on personal interpretation of landscape assemblages. For this, the model was the work by the French geographer Paul Vidal de la Blache and his followers on the *pays* of their homeland, small regional units with distinct physical characteristics, notably in soils and drainage, and associated agricultural specialisms (Buttimer, 1971, 1978a).

One systematic specialism which stood slightly apart from the others was historical geography, whose study was based on the argument that investigations of the past were needed in order to comprehend the regional patterns of the present. Two approaches to historical geography can be recognized from the 1920s on. The first, often thought of as the British approach and closely associated with the work of H. C. Darby (Perry, 1969), involved the detailed study of past geographies using a series of cross-sections whose locations in time were almost always determined by the available source material, such as the Domesday Book of *c*. 1086 which was analysed in great depth by Darby and his associates (culminating in Darby, 1977; Perry, 1979). These cross-sectional analyses, complete with regionalizations in many cases, were linked by a narrative outlining the changes between the periods studied: most emphasis was placed on the cross-sections, however, for which data allowed analysis rather than interpretation (see Darby, 1973, 1983a).

The second approach was largely American in its provenance, centred on the works of Carl Sauer and his associates. (On the differences between Hartshorne and Sauer, see Lukermann, 1990, and Butzer, 1990.) Their focus was the study of processes leading to landscape change up to the present, beginning at the prehuman stage of occupance (Mikesell, 1969): most of the work was conducted either outside the United States (particularly in Latin America) or in that country's less industrialized parts. Sauer's (1925) first methodological statement constrained geographical endeavour closely to the generic study of landscapes, emphasizing their cultural features (although work was also done on the borderlands between geography and botany); there was no glorification of the region, however. In his later 'sermons' – as he called his methodological and philosophical statements – Sauer (1941, 1956a) encouraged research over a much wider field, but emphasized the study of cultural landscapes, and the links which he had forged with anthropology, to produce a creative art-form whose hallmark was that it was not prescribed by pattern or method: the human geographer is obliged 'to make cultural processes the base of his thinking and observation' (Sauer, 1941, p. 24). The work involved neither detailed reconstruction of past geographies nor close consideration of regional boundaries: instead it led to a catholic historical geography whose rationale (Clark, 1954, p. 95) was that: 'through its study we may be able to find more complete and better answers to the problems of interpretation of the world both as it is now and as it has been at different times in the past'. Not all American historical geographers followed this lead – Brown (1943), for example, worked on detailed reconstructions of past periods (Meinig, 1989) – but the 'Berkeley school' which Sauer founded and led for almost five decades had many followers and a particular point of view, focused on a single iconoclast (Hooson, 1981). Sauer's influence was continued by his students, notably Leighley Parsons and Clark (Bushong, 1981).

The major statement of the approach engendered by Sauer was the symposium on 'Man's role in changing the face of the earth' which was conceived by Thomas (1956: see Glacken, 1983) and resulted in a substantial publication (52 chapters plus discussions: 1193 pages in all) that had a very substantial impact. The range of material included was vast. Its theme was identified by Sauer (1956b, p. 49) as

> the capacity of man to alter his natural environment, the manner of his so doing, and the virtue of his actions. It is concerned with historically cumulative effects, with the physical and biologic processes that man sets in motion, inhibits, or deflects, and with the differences in cultural conduct that distinguish one human group from another.

It presented no grand methodology or set of general findings – indeed, in his closing statement Sauer (1956c) criticized the tendency of American authors who 'have an inclination to universalize ourselves' (p. 1133). The volume stressed both diversity in human response to environments and its impacts on them, with cultural differences providing the basis for that diversity. Mumford's (1956, p. 237) conclusion to the symposium is very similar to that advanced in the 1980s by adherents of structuration theory (see p. 237): 'the future is not a blank page; and neither is it an open book' (p. 1142).

A contributor to Thomas's symposium, and also a member of the Berkeley school, was Glacken (1956), whose later *magnum opus* focused on the conceptions of nature that have been current in western thought at various times and places. *Traces on the Rhodian Shore* (Glacken, 1967) is a major survey of interpretations of nature, in which he demonstrates 'how all-pervading teleology has been in the history of Western interpretation of nature' (Glacken, 1983, p. 32). Like the Thomas symposium before it, this book is widely recognized as a classic on society–nature inter-relationships. But it was published at a time when that topic was receiving a rapidly diminishing amount of attention, and its impact was consequently less than might otherwise have occurred.

The British view

British geographers were less concerned with philosophical and methodological debate than their American counterparts during the 1920s, 1930s and 1940s (though see the exchange in the *Scottish Geographical Magazine* during the late 1930s, initiated by Crowe, 1938). They were apparently more pragmatic, less prone to contemplate the nature of their subject and more prepared, perhaps, to adopt the well-used adage that 'Geography is what geographers do'. But they accepted that the *raison d'être* of geography was synthesis, the integration of the findings of various systematic studies, with a strong emphasis on genesis, as in the studies of geomorphology and historical geography (Darby, 1953). According to Wooldridge and East (1958): 'geography . . . fuses the results, if not the methods, of a host of other subjects . . . [it] is not a science but merely an aggregate of sciences (p. 14). [I]ts *raison d'être* and intellectual attraction arise in large part from the shortcomings of the uncoordinated intellectual world bequeathed us by the specialists' (pp. 25–6) and 'in its simplest essence the geographical problem is how and why does one part of the earth's surface differ from another' (p. 28).

All of these statements indicate a strong trans-Atlantic common body of opinion (see Stoddart, 1990, regarding Hartshorne's influence on Woodridge) although, despite a statement that 'The purpose of regional geography is simply the better understanding of a complex whole by the study of its constituent parts' (p. 159), the British writers did not elevate the regional doctrine as much as did their American counterparts. (Nor were they carried to excesses of environmental determinism in earlier decades.) Nevertheless, Wooldridge (1956, p. 53) wrote in 1951 that:

> the aim of regional geography . . . is to gather up the disparate strands of the systematic studies, the geographical aspects of other disciplines, into a coherent and focused unity, to see nature and nurture, physique and personality as closely related and interdependent elements in specific regions.

He argued that in any department of geography each staff member should be committed to the study of a major region (p. 64).

A major difference between British and American geography by the 1950s was in attitude to physical geography, the study of the land surface, the atmosphere and the oceans, and their faunal and floral inhabitants. Both countries had strong traditions of work on these topics, and many geographers had academic roots in the associated field of geology. But this tradition had slowly dissolved in North America (the United States much more than Canada) and interest in the physical environment, and particularly its understanding as against its description, waned (see Leighley, 1955). This may have been a consequence of the excesses of environmental determinism, with a subsequent desire to remove all traces of that connection and to see society as the formative agent of landscape patterns and change: associated with this was an attempt to redefine geography in the 1920s as the study of human ecology, presenting people as simultaneously reacting and adjusting to environments while at the same time attempting to adjust the environment to their own needs (Barrows, 1923). Thus with regard to geomorphology – the science of landform genesis – Peltier (1954, p. 375) wrote:

> the geographer needs precise, factual information about particular places. What landforms actually exist in a given area? How do they differ? Where are they? What are their distribution patterns? The geomorphologist may concern himself with questions of structure,

process, and stage, but the geographer wants specific answers to the questions: what? where? and how much?

What geographers were interested in, according to this view, was the geography of landforms: geomorphology, the genetic study of landforms, was a part of geology and, unlike historical geography, was deemed irrelevant to the geographical enterprise. Similar reactions saw the wholesale removal of climatology and biogeography from American geographical curricula, and their replacement by introductory courses in physical geography which described landforms, climates, and plant assemblages – usually in a regional context – but paid little or no attention to their origins. There has been a substantial revival of physical geography in the USA in recent decades, however (see Marcus, 1979 and the essays in Gaile and Willmott, 1989), as a reaction to its perceived relevance to the attack on environmental problems (see p. 338) and the technological advances in remote sensing and associated technologies.

This American trend was not repeated in Britain where, according to Wooldridge and East (1958, p. 47):

> To treat geography too literally as an affair of the 'quasistatic present' is to make both it and its students seem foolish and superficial. It is true that our primary aim is to describe the present landscape; but it is also to interpret it. . . . Our study has therefore always to be evolutionary. . . . It is unscholarly to take either landforms or human societies as 'given' and static facts, though we must not let temporal sequences obscure spatial patterns.

Geography students at British universities in the 1950s rarely specialized in either physical or human geography, except perhaps in the final year of their course. Both were considered essential parts of a geographical education, as contributions to the genetic study of regional landscapes which was the integrating focus of geographical scholarship (see Johnston and Gregory, 1984; Cosgrove, 1989a). As researchers, most British geographers specialized in either physical or human geography (though rarely exclusively so), but almost all had a regional specialism as well, in which they 'integrated' studies from 'both sides' of their subject, as widely illustrated in the regional textbooks of the period; the 'dogma of regional synthesis' (Darby, 1983b, p. 25) was being softened, however, and geographers were increasingly turning their attention from regions to problems.

Conclusions

This chapter has presented an extremely brief outline of geography during its 'modern period', since the focus of the book is on the ensuing 'contemporary period'. Three approaches have been identified, although deeper analysis may well indicate more coexisting during any one time (see Taylor, 1937) . All three lasted into the contemporary period, although one, the regional, dominated in the years before and just after World War II. Its main focus was on areal differentiation, on the varying character of the earth's surface, and its picture of that variation was built up from parallel topical studies of different aspects of the physical and human patterns observed. By the 1950s, initially in America and then in Britain too, there was growing disillusionment with the empiricist philosophy of regional geography. The topical specialisms slowly came to dominate disciplinary practice and the regional synthesis was increasingly ignored.

|3|

Growth of systematic studies and the adoption of 'scientific method'

Dating the origin of a change in the orientation of a discipline, or even a part of it, is difficult. Several pieces which contain the kernel of the new ideas can usually be found in its literature, but often these are derivative of the earlier teachings of others, whose views may never have been published; others may promote that which is later adopted, but have no impact on those developments, for a variety of reasons (see Johnston, 1993a, 1996b). Change can also emanate contemporaneously from several separate though usually not entirely independent nodes, as various iconoclasts introduce stimuli. An attempt to locate the first stirrings against regional geography would be a futile exercise, therefore. Instead, the present chapter isolates the most influential statements published by geographers, and traces their impact on the geographical community.

Change within a discipline involves both dissatisfaction with existing approaches and the promotion of an acceptable alternative disciplinary matrix (if not world view: see p. 16). The 1950s saw widespread dissatisfaction, as indicated by Freeman (1961), who noted that 'disappointment with the work of regional geographers has led many to wonder if the regional approach can ever be academically satisfying and to turn to specialization or some systematic branch of the subject' (p. 141). He suggested three reasons for such disappointment. First, much regional classification was naive, particularly on the large scale, where generalizations such as Herbertson's world climatic regions were found on detailed investigation to contain too many discrepancies. The second, and perhaps most important to many people, was the 'weary succession' of physical and human activity 'facts' which characterized so much regional writing (though not all, as exemplified by James's, 1942, *Latin America*): 'The trouble has perhaps been that many regional geographers have tried to include too much' (p. 143). Third, he claimed that the model of regional writing, derived from work on the French *pays*, suggested that the whole of the earth's surface could be divided into such clear regions, each with its own character: that this proved false was reflected by many pedestrian studies of areas lacking such personality. (A satirical essay highlighting the poverty of much regional geography was published anonymously in an early issue of the *IBG Newsletter*: Anon, 1968. The author's name was given as Llwynog Llwyd, 'an exiled Celt ... [working in] an Antipodean University' – it was Keith Buchanan.)

Whereas Freeman focused on the failings of regional geography as practised, a case was set out in the United States that insistence on the primacy of regional geography was undermining the associated systematic studies. This was put forcefully by Ackerman (1945) in a report on his experience of working in the wartime intelligence services. He identified two major failings of professional geographers there: their inability to handle foreign languages; and the weakness of their topical specialisms. Regarding the latter, he criticized much geographical work during the preceding quarter of a century as conducted by scholars who were 'more or less amateurs in the subject on which they published' (p. 124), so that when called upon to provide intelligence material for wartime interpretation what the geographers produced was extremely thin in its content. Regional geographers could provide only superficial analyses, and the division of labour within the discipline whereby people specialized in different areas of the earth was both inefficient and ineffective. (Gould, 1979, p. 140, calls the geography of the fifty years prior to 1950 'bumbling amateurism and antiquarianism'.)

Ackerman suggested that rectifying this major deficiency in geographical work required much more research and training in the systematic specialisms: this need not necessarily be contrary to the philosophy of the subject which gave primacy to regional synthesis, he claimed, since more detailed systematic studies would lead to greater depth in regional interpretations. There is little evidence that his paper had an immediate impact, however, and publications over the next few years, as illustrated by the abstracts of papers presented at the Association of American Geographers' annual conferences (published in the *Annals* each year) indicated no major shift in the orientation of academic work with the return to post-war 'normality': two of the few exceptions are the abstracts presented by Garrison and McCarty at Cleveland in 1953, which were clearly based on a different methodology to that widely used (see below, p. 63). The systematic fields had undoubtedly been gaining in importance prior to Ackerman's statement, and continued to do so, as indicated by the extent of their treatment in the review volume edited by James and Jones (1954). But it was not until the mid-1950s that this volumetric change in the substance of geographical research was matched by any widespread changes in its methodology and philosophy.

Schaefer's paper and the response

As Hartshorne published his major statement of the regional paradigm in the United States, and as it was there, rather than in Britain, that philosophy and methodology were apparently debated most earnestly, it is perhaps not surprising that the revolution against the paradigm originated on that side of Atlantic. The first major shot was a – posthumously published – paper by Schaefer (1953) which is often referred to by those seeking the origins of the 'quantitative and theoretical revolutions' (on its impact, see both Cox, 1995 and Getis, 1993). Schaefer was originally an economist: he joined the group of geographers teaching in the economics department at the University of Iowa after his escape from Nazi Germany (Bunge, 1979).

Schaefer claimed that his paper was the first to challenge Hartshorne's interpretation of the works of Hettner and others. He criticized Hartshorne's exceptionalist claims for regional geography, and presented an alternative case for geography adopting the philosophy and methods of the positivist school of science (see Martin, 1990, p. 72). He first outlined the nature of a science and then defined the peculiar characteristics of geography as a social science. He argued that a claim for geography as the integrating science which put together the findings of the individual systematic sciences was arrogant, and that in any case its products were somewhat lacking in 'startlingly new and deeper insights' (p. 227). A science is characterized by its explanations, and explanations require laws: 'To explain the phenomena one has described means always to recognize them as instances of laws.' In geography, Schaefer argued, the major regularities which are described refer to spatial patterns: 'Hence geography has to be conceived as the science concerned with the formulation of the laws governing the spatial distribution of certain features on the surface of the earth' (p. 227) and these spatial arrangements of phenomena, not the phenomena themselves, should be the subject of geographers' search for lawlike statements. Geographical procedures would then not differ from those employed in the other sciences, both natural and social: observation would lead to a hypothesis – about the inter-relationship between two spatial patterns, for example – and this would be tested against large numbers of cases, to provide the material for a law if it were thereby verified.

The argument against this definition of geography as the science of spatial arrangements was termed *exceptionalist*. It claims that geography does not share the methodology of other sciences because of the peculiar nature of its subject matter – the study of unique places, or regions (and compares geography with history, which studies unique periods of time). Using analogies from physics and economics, Schaefer argued that geography is not peculiar in focusing on unique phenomena; all sciences deal with unique events which can only be accounted for by an integration of laws from various systematic sciences, but this does not prevent – although it undoubtedly makes more difficult – the development of laws. 'It is, therefore, absurd to maintain that the geographers are distinguished among the scientists through the integration of heterogeneous phenomena which they achieve. There is nothing extraordinary about geography in that respect' (p. 231).

In the second part of his paper, Schaefer traced the exceptionalist view in geography back to an analogy drawn by Kant (1724–1804) between geography and history, an analogy repeated by Hettner and by Hartshorne (see above, p. 46). He quoted from Kant's *Physische Geographie* (vol. I, p. 8) that 'Geography and history together fill up the entire area of our perception: geography that of space and history that of time' (p. 233). But when Kant was working, Schaefer claims, history and geography were cosmologies, not sciences, and a cosmology is 'not rational science but at best thoughtful contemplation of the universe' (p. 332). Hettner followed Kant's views and developed geography as a cosmology, arguing that both history and geography deal with the unique, and thus do not apply the methods of science. Schaefer argued that this is a false position, for in explaining what happened at a certain time period historians must employ the laws of the social sciences. Time periods, like places, are undoubtedly unique assemblages of phenomena, but this does not preclude the use of laws in unravelling and explaining them. History and geography can both be sciences for: 'What scientists do is . . . *They apply to each concrete situation jointly all the laws that involve the variables they have reason to believe are relevant*' (p. 239). Schaefer argued that Hartshorne disregarded one aspect of Hettner's writing which was nomothetic in its orientation, however, and in doing this misled American geographers. (Muller-Wille, 1978, p. 55, claims that Hettner predated Christaller in the development of ideas regarding central place theory; Hartshorne made no reference to the paper by Hettner cited by Muller-Wille. On the same point, see Butzer, 1990, and Smith, 1990.)

The final part of Schaefer's paper reviewed some problems of applying a nomothetic (law-producing) philosophy to geography as a spatial, social science. He recognized difficulties of experimentation and quantification, for example, and suggested a methodology based on cartographic correlations. He argued that laws produced in geography are morphological in their format whereas those from other, 'maturer' social sciences are concerned with processes: in order fully to comprehend the assemblages of the phenomena described in geographers' morphological laws, therefore, it is necessary to derive process laws from other social sciences, a procedure which requires team work (the last point was also made by Ackerman). Geography according to Schaefer, then, is the source of the laws on location, which may be used to differentiate the regions of the earth's surface.

Hartshorne's response

Schaefer's paper did not produce much direct reaction in print, despite later claims that it was a major stimulus to work in the genre which he proposed (Bunge, 1962). But it did draw considerable response from Hartshorne, first in a letter to the editor of the *Annals* (Hartshorne, 1954b) and later in three substantive pieces (Hartshorne, 1955, 1958, 1959): the last was another major book which, although probably not as influential as the 1939 volume, showed Hartshorne's continued importance to American geographers as an interpreter of their subject's methodology and philosophy.

The purpose of Hartshorne's first (1955) paper (which subsumed the earlier letter) was to indicate the many flaws which he identified in Schaefer's scholarship (see also Gregory, 1978a, p. 31). He begins with a further discussion of the *mores* of methodological debate (Hartshorne, 1948): most of the paper was organized to illustrate that Schaefer was limited in his references, drew unsupportable conclusions, and misrepresented the views of others, so that 'In every paragraph, in nearly every sentence of this third section, there is serious falsification, either by commission or by omission, of the views of the writer discussed' (p. 236). (This statement refers to the third part of Schaefer's paper, which focused on Hettner's work.) More generally, Hartshorne claimed that Schaefer's paper 'ignores the normal standards of critical scholarship and in effect offers nothing more than personal opinion, thinly disguised as literary and historical analysis' (p. 244). Since Hartshorne himself (1959, p. 8) strongly believed that 'geography is what geographers have made it', to him all methodological and philosophical statements should be based on a close and careful analysis of others' published works.

Although most of the 1955 paper examined the nature of Schaefer's 'evidence', in the final section Hartshorne turned to the anti-exceptionalist argument. He pointed out that in coming to the conclusion that geography should take process laws from the systematic sciences and use them to produce morphological laws, Schaefer came very close to preaching the sort of exceptionalist claim that he sought to destroy, so that his critique 'is a total fraud' (p. 237). Schaefer's position is summarized as 'geography must be a science, science is the search for laws, and all phenomena of nature and human life are subject to such laws and completely determinable by them' (p. 242) . Such scientific determinism is opposed to the summary of what geographers do, set out in *The Nature of Geography*, which in any case was treated in a most cavalier way by Schaefer.

Hartshorne's second paper (1958) addressed Schaefer's claim that Kant was the source of the exceptionalist view. Literary analysis suggests that both Humboldt and Hettner reached the same position independently, being unaware of Kant's views when they were writing. May (1970, p. 9) suggests that both Hartshorne and Schaefer could have misunderstood Kant's conception of a science, however, and of the role of geography as a science, although he confirms Hartshorne's dismissal of Schaefer's interpretation of the source of Kant's ideas (see the later exchange between Hartshorne, 1972 and May, 1972).

The third and most substantial piece in Hartshorne's rebuttal was a monograph (*Perspective on the Nature of Geography:* Hartshorne, 1959) whose production was stimulated by Schaefer and by requests from colleagues that he respond in detail to Schaefer's argument, but which was also used as a vehicle for discussing a wide range of other issues raised during the two decades since the publication of the original statement (Hartshorne, 1939). The discussion used a framework of ten separate questions/topics as its organization: the aim was to provide a methodology by which geography could meet its need for 'new conceptual approaches and more effective ways of measuring the interrelationships of phenomena' (p. 9), which could only develop out of an understanding and acceptance of the subject's 'essential character'.

The first set of questions concerned the meaning of areal differentiation, the definition of the earth's surface, a discussion of the peculiar geographical interest in the integration of phenomena in 'the total reality [that] is there for study, and geography is the name of the section of empirical knowledge which has always been called upon to study that reality' (p. 33), and the determination of what is significant for geographical study; it led Hartshorne to the definition that 'geography is that discipline that seeks *to describe and interpret the variable character from place to place of the earth as the world of man*' (p. 47). He considered that human and natural factors do not have to be identified separately – any prior insistence on this was a function of the arguments of environmental determinists – and that a division into human and physical geography is unfortunate, because it limits the range of possible integrations in the study of reality.

Turning to temporal processes, Hartshorne argued that geographers need only study proximate genesis, since classification by form of appearance rather than by provenance is important for the geographical investigation of areal differentiation: as most landforms are stable, or virtually so, from the point of view of human occupance, for example, then study of their change is irrelevant to the aims of geography. According to this argument (see also p. 51 above), geomorphology, insofar as it is the study of landform genesis, is not part of geography; the study of landforms is. With regard to cultural features in the landscape, Hartshorne made an important distinction between expository description and explanatory description: 'geography is primarily concerned to describe . . . the variable character of areas as formed by existing features in interrelationships . . . explanatory description of features in the past must be kept subordinate to the primary purpose' (p. 99). Thus historical geography should be the expository description of the historical present 'but the purpose of such dips into the past is not to trace developments or seek origins but to facilitate comprehension of the present' (p. 106); studies of causal development and genesis are the prerogative of the systematic sciences.

In answering the question 'Is geography divided between systematic and regional geography?' Hartshorne developed a position different from that in *The Nature of Geography*. Thus in 1959 he accepted that studies of inter-relationships could be arranged along a continuum 'from those which analyse the most elementary complexes in areal variation over the world to those which analyse the most complex integrations in areal variation within small areas' (p. 121). The former are topical studies and the latter regional studies, but whereas 'every truly geographical study involves the use of both the topical and the regional approach' (p. 122) there is no argument that one is superior and that to which all geographers should aspire. Hartshorne thereby somewhat downgraded the regional synthesis from his earlier view regarding its centrality in the geographical enterprise.

Regarding Schaefer's important question – 'Does geography seek to formulate scientific laws or to describe individual cases?' – Hartshorne argued for the latter, largely by pointing out the difficulties of establishing such laws through geographical investigations; however, he did not argue that geographers should not seek and use general laws for the understanding of individual cases – it is an 'erroneous presumption that to focus on studies of individual places and to focus on generic concepts are opposing alternatives mutually exclusive' (Hartshorne, 1984, p. 429). Scientific laws must be based on large numbers of cases, but geographers study complex integrations in unique places; scientific laws can best be established in laboratory experiments which allow only a few independent variables to vary, but such work is rarely possible in geography; interpretation requires skills in the systematic sciences which are beyond the capability of geographers; scientific laws suggest some kind of determinism, but this is inappropriate to the human motivations which are in part the causes of landscape variations: for all these reasons, the search for laws is irrelevant to geography. But laws are not the only means to the scientific end of comprehending reality: instead

> Geography seeks (1) on the basis of empirical observation as independent as possible of the person of the observer, to describe phenomena with the maximum degree of accuracy and certainty; (2) on this basis, to classify the phenomena, as far as reality permits, in terms of generic concepts or universals; (3) through rational consideration of the facts thus secured and by logical processes of analysis and synthesis, including the construction and use wherever possible of general principles or laws of generic relationships, to attain the maximum comprehension of the scientific interrelationships of phenomena; and (4) to arrange these findings in orderly

systems so that what is known leads directly to the margin of the unknown (pp. 169–70).

This, he says, is a perfectly respectable scientific goal. (It is very similar to the overall goal of positivist work – p. 34 – and is the reason why several commentators see very little difference in ends, if not means, between Hartshorne's work and that of spatial scientists.)

Finally, in discussing geography's position within the classification of sciences, Hartshorne returned to the Hettnerian analogy of geography as a chorological science with history as a chronological science. This is valid, he argues, because it describes the way in which geographers have worked, on both topical and regional subjects, with reference to inter-relationships and integrations within areas. (This view was revived by Harris, 1971: see Earle, 1996.)

Reconciliations?

The major basis of their methodological and philosophical differences was that Hartshorne had a positive view of geography – geography is what geographers have made it – whereas Schaefer's was normative, of what geography should be, irrespective of what it had been. Over the decade after Hartshorne published his *Perspective* Schaefer's view prevailed with most geographers, on both sides of the Atlantic, although the extent of his influence via the 1953 paper was probably very slight and the real iconoclasts of the 'revolution' were those discussed in the next section. (According to Smith, 1990, Hartshorne's view that geography should be what (he said) it had always been in effect 'turned the discourse of geography into a museum and [Hartshorne] appointed himself curator': Gregory, 1994, p. 285.) Indeed in Britain, although Hartshorne's two books were clearly widely read and referenced, Schaefer's paper was apparently not. It receives no mention in Freeman's (1961; 1980a) books, none in Chorley and Haggett's (1965b) trail-blazing *Frontiers in Geographical Teaching,* and only one in their major (Chorley and Haggett, 1967) *Models in Geography* – in the chapter by Stoddart (see, however, Stoddart, 1990). Thus it is not surprising that relatively little attention has been paid elsewhere in the geographical literature to the Schaefer/Hartshorne debate (Gregory, 1978a, p. 32. See also Martin, 1951, and Jones, 1956, for a separate, British, debate). Interestingly, Schaefer is not in the index of authors referred to in the encyclopaedic *Geography in America* either (Gaile and Willmott, 1989).

Guelke (1977a, 1978) has argued that Hartshorne's and Schaefer's views were not so antagonistic as they themselves suggested (see also Gregory, 1978a, p. 31 and Entrikin, 1981, 1990). Hartshorne was in general terms very much a supporter of the scientific method as defined by the positivists, but created his own problems regarding the application of this method in geography because of his view on uniqueness. Schaefer, on the other hand, accepted the full positivist position, and showed that uniqueness was a general problem of science, and not a peculiar characteristic of geography. Thus

> In extending the idea of uniqueness to everything, Schaefer effectively removed a major logical objection to the possibility of a law-seeking geography and demonstrated that Hartshorne's view of uniqueness as a special problem was untenable for anyone who accepted the scientific model of explanation (Guelke, 1977a, p. 380).

Also Hartshorne's distinction between idiographic and nomothetic approaches was misleading and both Hartshorne and Schaefer ignored the possibility of geographers being major 'law-consumers', however; to Hartshorne, the alternatives were either law-making or the description of unique places, whilst to Schaefer geographers had to develop morphological laws, and ignore the interest in process laws which characterizes the systematic sciences.

Guelke (1977a, p. 348) claimed that Schaefer's insistence on the need for geographers to develop laws 'created a major crisis within the discipline'. Within a decade of Schaefer's paper being published, however, many human geographers, especially among the youngest recruits, had adopted at least part of his manifesto with their growing concern for quantification and law-making. They were presented with a choice between such activity and the sort of contemplation of the unique advocated by Hartshorne. As Guelke (1977a, p. 385) points out, 'Not surprisingly, most geographers opted for geography as a law-seeking science' because by then (Guelke, 1978, p. 45):

> Universities were expected to produce problem-solvers or social-technologists to run increasingly complex economies, and geographers were not slow in adopting new positions appropriate to the new conditions. Statistics and models were ideal tools for monitoring and planning in complex industrial societies. The work of the new geographers, however, often lacked a truly intellectual dimension. Many geographers were asking: 'Are our methods rigorous?', 'What are the planning implications of this model?', and not 'How much insight does this study give us?', 'Is my understanding of this phenomenon enhanced?', 'Does this study contribute to geography?'. The last-mentioned question was considered of little consequence. Yet it should have been asked, because one of the weaknesses of the new geography was a lack of coherence.

Developments in systematic geography in the United States

Whether because of or independent from the statements by Ackerman (1945), Schaefer (1953) and Ullman (1953), systematic studies became much more important in the research and teaching of American geographers during the 1950s. (On Hartshorne's influence on systematic studies, see Butzer, 1990.) This did not mean a departure from Hartshorne's contemporary views, since by 1959 he no longer gave primacy to regional studies, but the trend towards the scientific method proposed by Schaefer marked a clear break with the Hartshornian tradition.

The growing popularity of topical specialisms is illustrated in the review chapters in the collection edited by James and Jones (1954) and by the 1950s' journal literature. Very few of the investigations reported aimed at generating laws, however: indeed some could almost be categorized under the exploration paradigm, since their major purpose was the provision of new factual material: such work is best described as empiricist – it lets 'the facts' speak for themselves.

Fundamental to scientific progress in the positivist mould espoused by Schaefer is the development of theory. Several of the reviews in the James and Jones (1954) volume refer to what is in one place termed location theory (Harris, 1954a, p. 299), but very few examples are cited of empirical investigations related to that body of theory. The chapter on urban geography, for example, cites seminal pieces on central place theory, such as Ullman's (1941) original paper, and devotes two pages to the three 'models' of intra-urban spatial patterns which had been reviewed a decade previously by Harris and Ullman (1945), but there is not a single reference to any work done by geographers since in the context of those models (Mayer, 1954, pp. 152–63). Thus despite some precedents, very little work was done by geographers prior to the mid-1950s which followed the dictates of the 'scientific approach'.

Once a new idea gains circulation through the professional journals it is available to be taken up by all. Initial development is usually concentrated in a few places only, however, where pioneer teachers encourage students to conduct research within the new framework. Thus most of the methodological changes in systematic studies in geography during the 1950s can be traced back to a few centres in the United States, and that work is discussed here. The changes were largely concerned with method, and their scientific underpinning was stressed very little, although law-seeking was the clear goal. Certainly methods dominated the writings of both the pros and cons; many early contributions by the former group were in relatively fugitive, departmental publications, presumably because of difficulties in getting such 'new' material accepted by the journals.

The Iowa school

Although Schaefer was at Iowa until his death in 1953, he had little influence on the developments promoted by geographers there, who for a number of years were members of the economics department and thus open to the views and approaches of their peers in that more 'mature' social science. The group's leader was Harold McCarty, author of a major text on American economic geography (McCarty, 1940), and associated with him were J. C. Hook, D. S. Knos, H. A. Stafford and later, J. B. Lindberg, E. N. Thomas and L. J. King (McCarty, 1979; King, 1979a).

McCarty and his co-workers wanted to establish the degree of correspondence between two or more geographical patterns, akin to the morphological laws of accordance discussed by Schaefer. (Interestingly, none of their publications refers to Schaefer's paper, although they do refer to the works of, and assistance given by, Gustav Bergmann, a positivist of the Vienna school who also strongly influenced Schaefer and read the proofs of his 1953 paper: Davies, 1972a, p. 134; King, 1979a; Golledge, 1983. See also Martin, 1990.) These laws were to be embedded in a theory, thus (McCarty, 1954, p. 96):

> If we are to accept the idea that economic geography is becoming the branch of human knowledge whose function is to account for the location of economic activities on the various portions of the earth's surface, it seems reasonable to expect the discipline to develop a body of theory to facilitate the performance of this task.

Such theory could be either topically or areally focused, and in its early stages of development would probably be restricted both in its areal coverage and in the topics whose spatial inter-relationships it considered.

The purpose of theory is to provide explanations, and McCarty recognized two sorts of explanation. The first involved a search for the cause of observed locational patterns, but: 'the search for causes can never produce an adequate body of theory for use in economic geography. . . . Variables became so numerous that they were not manageable, and, in consequence, solutions to locational problems were not obtainable' (p. 96). The second, and preferred, type focuses on associations:

> Its proponents take the pragmatic view that if one knew that two phenomena always appear together in space and never appear independently, the needs of geographic science would be satisfied, and there would be scant additional virtue in knowing that the location of one phenomenon caused the location of another (p. 97).

Such laws of association are built up in a series of stages, which begins with a statement of the problem and of the necessary operational definitions, and proceeds through the measurement of the phenomena (with attendant problems of sampling in time and space) to a statement of the findings, in tabular or graphical form. These three descriptive stages precede analysis which seeks out correlations between the distributions of phenomena:

> the nub of the problem of research procedure seems to lie in finding the best techniques for discovering *a, b* and c in the 'where *a, b, c,* there *x*' hypothesis in order to give direction to the analysis. But where shall we search for its components? . . . One source of . . . clues lies in the findings of the systematic sciences. The other source lies in the observations of trained workers in the field or in the library (p. 100).

Thus geography, in its search for morphological laws, is to a considerable extent a consumer of the laws of other disciplines, which may be theoretically rather than empirically derived. According to the causal or process approach to explanation:

> Models may be created showing optimal locations for any type of economic activity for which adequate cost data may be obtained. These models may then be used (as hypotheses) for the comparison of hypothetical locations with actual locations. Divergences of pattern may then be noted and the hypothesis altered to allow for them (often by inclusion of factors not ordinarily associated with monetary costs). Ultimately the hypothesis becomes generally applicable and thus takes on the status of a principle (McCarty, 1953, p. 184).

This statement, although not referenced as such, very faithfully reflects positivists' views, and also those of Popper and Lakatos, as to how science progresses by the continual modification of its hypotheses, so as better to represent reality.

In their major demonstration of this procedure in operation, McCarty *et al.* (1956) discussed several statistical procedures for measuring spatial association and adopted the now well-known technique of multiple regression and correlation – which had been used previously, among geographers, by Rose (1936) and Weaver (1943), both apparently after contacts with agricultural economists. Their empirical context was similarity in the location patterns of certain manufacturing industries in the United States and in Japan: other studies by the group included Hook's (1955) on rural population densities, Knos's (1968) on intra-urban land-value patterns, and King's (1961) on the spacing of urban settlements. Thomas (1960) had used similar procedures in his study of population growth in suburban Chicago, presented as a Ph.D. thesis to Northwestern University (where he was a contemporary of Garrison), and he extended the methodology with a paper on the use of residuals from regression for identifying where the putative laws of association do not apply fully, thereby suggesting further hypotheses for areal associations. (This last paper developed on an earlier one by McCarty (1952) which was not widely circulated.) McCarty (1958) later expressed some doubts about the statistical validity of the procedure, but the method that he and his associates pioneered, with its focus on the testing of simple hypotheses derived either from observation or from theoretical deductions, became the exemplar for much research in the ensuing decades.

Wisconsin

The department of geography at the University of Wisconsin, Madison, has a long tradition of research with a quantitative bent. Notable among its early products was John Weaver's Ph.D. thesis on the geography of American barley production, which included a major section (published in his 1943 paper with no supporting methodological argument) using multiple correlation and regression to identify the influences of climatic variables on barley yields; Weaver later taught at the University of Minnesota, where he developed a widely adopted statistical procedure (Weaver, 1954) for the definition of agricultural regions. Other work at Madison focused on the quantitative description of population patterns (e.g. Alexander and Zahorchak, 1943). A combination of these two interests was furthered by a group led by A. H. Robinson, whose main interests were in cartography; cartographic correlations were introduced to him by his research supervisor at Ohio State University, Guy-Harold Smith (Brown, 1978). Robinson also worked with R. A. Bryson, of the department of meteorology, who was a source of statistical ideas and expertise. (Cartographic work was for a long time called 'mathematical geography' by some. It also developed statistical interests: see Blumenstock, 1953.)

Robinson wanted to develop statistical methods for map comparison, as indicated by the title of an early paper – 'A method for describing quantitatively the correspondence of geographical distributions' (Robinson and Bryson, 1957); as with the Iowa work, Rose and Weaver's lead was followed with the adoption of correlation and regression procedures. Particular attention was paid to the problems of representing areal data by points (Robinson *et al.*, 1961) and of using correlation methods in the comparison of isarithmic maps (Robinson, 1962). Like McCarty (1958), Robinson was aware of difficulties in applying classical statistical procedures to areal data, and he proposed a procedure to circumvent one of these (Robinson, 1956): Thomas and Anderson (1965) later found this proposal wanting, as it dealt with a special case only and not with the more general problems. Interestingly, though, the main early work on this topic was published by a group of sociologists, under the title *Statistical Geography* (Duncan, Cuzzort and Duncan, 1961: perhaps even more interestingly, this work was virtually ignored by geographers, even though Brian Berry – 1993, p. 439 – indicates that he developed a close working relationship with Dudley Duncan at Chicago in the late 1950s and persuaded him to give the book that title).

W. L. Garrison and the Washington school

By far the largest volume of work in the spirit of Schaefer's and McCarty's proposals published during the 1950s came from the University of Washington, Seattle. The group of workers there was led by W. L. Garrison, whose Ph.D. was from Northwestern University, where he was associated with Thomas (the two returned to Northwestern in the early 1960s; Taaffe, 1979); according to Bunge (1966, p. ix), Garrison was influenced by Schaefer's paper, although the dates of his earliest publications indicate that he was involved in applying the positivist method to systematic studies in human geography before 1953. Also involved was E. L. Ullman, who moved to Seattle in 1951 (Harris, 1977), and who had already done pioneering research on urban location patterns and transport geography (see Morrill, 1984). A large group of graduate students worked with Garrison and several became leaders in the new methodology during the subsequent decade, including B. J. L. Berry, W. Bunge, M. F. Dacey, A. Getis, D. F. Marble, R. L. Morrill, J. D. Nystuen and W. R. Tobler (Garrison, 1979). The group also benefited from a visit to Seattle by the Swedish geographer, Torsten Hagerstrand, who was developing methods for generalizing spatial patterns and processes (see below, p. 154), and from Garrison's contacts with the business school and the engineering department at Seattle (Halvorson and Stave, 1978: Gould, 1969, claims that the Scandinavian countries were a major source of new ideas). Berry (1993) and Getis (1993) both provide accounts of the intellectual activity at Seattle during the mid-1950s: Getis stresses the role of the department chairman, Donald Hudson, and his 'inferiority complex that comes from being remote' (p. 530), which led him to place Washington graduates as faculty members in several of the main mid-western graduate schools (such as Chicago, Northwestern and Michigan Universities) in order to establish his department's intellectual reputation.

Garrison and his co-workers had catholic interests in urban and economic geography. Much of their research was grounded in theory gleaned from other disciplines – notably economics – and they directed their efforts towards both testing those theories and applying them to planning problems. In developing testable theoretical statements they drew on a much stronger mathematical base than was the case at Iowa and Madison. They also searched widely for relevant statistical tests for their investigations of point and line patterns – the biological sciences provided several, such as nearest-neighbour analysis of point patterns (Dacey, 1962), whereas others, for grouping, classifying and regionalizing, were derived from psychology (Berry, 1968). Garrison (1956a, p. 429) argued that 'there is ample evidence that present tools are adequate to our present state of development. No type of problem has been proposed that could not be treated with available tools', which contradicted a claim by Reynolds (1956), although he criticized some uses of standard techniques (Garrison, 1956b). The dominant thrust of the group's work therefore involved the derivation from other systematic sciences of relevant normative theories, mathematical methods and statistical procedures with which to develop morphological laws.

The wealth of the work done at Seattle is illustrated by their major publications. Garrison (1959a, 1959b, 1960a) himself contributed a seminal three-part review article on the state of location theory. The first part reviewed six recent books – none of them by geographers – addressing the question 'What determines the spatial arrangement (structure, pattern, or location) of economic activity?' (Garrison, 1959a, p. 232). Each incorporated locational considerations into traditional economic analysis, which Garrison concluded offered valuable economic insights to traditional geographical problems.

Central place theory was the dominant location theory on which the group worked. This had several independent origins (Ullman, 1941; see also Harris, 1977, and Freeman, 1961, p. 201, who notes that findings akin to those in central place theory were reported by the 1851 Census Commissioners of Great Britain). Christaller's (1966) thesis attracted most attention, however (see Muller-Wille, 1978). Working in Germany in the 1930s, Christaller developed ideas regarding the ideal distribution of settlements of different sizes acting as the marketing centres of functional regions, within constraining assumptions relating to the physical environment and the goals of both entrepreneurs and customers: a translation became available in the late 1950s, and was published in 1966. Studies of functional regions were not novel, of course (see above, p. 46), and other geographers had tested Christaller's notions regarding a hierarchical organization of settlements distributed on a hexagonal lattice (e.g. Brush, 1953); related, more inductive, work on 'principles of areal organization' was reported by Philbrick (1957). Dacey wanted to make the analysis of these spatial hierarchies more rigorous (Dacey, 1962) whereas Berry focused on empirical investigations of the settlement pattern north of Seattle and the retail centres in the city of Spokane (Berry and Garrison, 1958a, 1958b; Berry, 1959a).

The second part of Garrison's (1959b) review article dealt with possible geographical applications of the mathematical procedures of linear programming, which identify optimal solutions to problems of resource allocation in constrained situations. He illustrated how neoclassical economic analytic procedures could be adapted for investigations of ideal solutions to the problems of where to locate economic activities and how to organize flows of goods. Six problems which could be treated by linear programming were identified:

1 the *transportation problem*, which takes a set of points, some with a given supply of a good and some with a given demand for it, plus costs of movement, and determines the most efficient flow pattern of the good from supply to demand points which minimizes the expenditure on transport;

2 the *spatial price-equilibrium problem*, which takes the same information as the transportation problem, but determines prices as well as flows;

3 the *warehouse-location problem*, which determines the best location for a set of supply points, given a geography of demand;

4 the *industrial-location problem*, which determines the optimum location for factories from knowledge of the sources of their raw materials and the destinations for their products;

5 the *interdependencies problem*, which locates linked plants so as to maximize their joint profits; and

6 the *boundary-drawing problem*, which determines the most efficient set of boundaries (i.e. that which minimizes total expenditure on transport) for, for example, school catchment areas.

If these are being used to investigate actual patterns and not as the bases for future plans, the purpose, as Losch (1954) put it, is to see whether reality is 'rational', whether decision-makers have acted in ways that would produce the most efficient solutions: efficiency is defined as cost-minimization, particularly transport-cost minimization. Investigations by the Washington group in this context included studies of inter-regional trade (Morrill and Garrison, 1960) and the optimal location, by regions, of agricultural activities in the United States (Garrison and Marble, 1957).

In the final part of his review article, Garrison (1960a) dealt with four further books on locational analysis, which were empirical in orientation and shared a common interest in the agglomeration economies reaped by industrial clusters. Several topics and techniques were discussed, such as the use of input–output matrices to represent industrial systems, and Garrison concluded by stressing the need for geographers to investigate location patterns as systems of inter-related activities.

The empirical work with a planning orientation undertaken by the Seattle group is illustrated by Garrison's large investigation of the impact of highway developments on land use and other spatial patterns (Garrison *et al.*, 1959). This book includes four studies: Berry's on the spatial pattern of central places within urban areas; Marble's on the residential pattern of the city (as indexed by property values) and relationships between household characteristics – including location – and their movement patterns; Nystuen's on movements by customers to central places; and Morrill's on the locations of physicians' offices, both actual and the most efficient. In addition, Garrison himself worked on the impacts of highway improvements, and devised indices of accessibility based on graph theory

(Garrison, 1960b: the work was continued by Kansky, 1963, after Garrison and Berry moved to work in the Chicago area). He also used the simulation procedures developed by Hagerstrand (1968) to investigate urban growth processes (Garrison, 1962), a topic taken much further by Morrill (1965).

Somewhat separate from the work of the others in the group, although aligned with their general purpose, was Bunge's thesis, *Theoretical Geography* (1962, reprinted in enlarged form, 1968). This displays a catholic view of geography, together with an acknowledged debt to Schaefer (Bunge worked at Iowa for a short period, and also at Madison, where he disagreed with Hartshorne). It is an extremely difficult book to summarize, but the basic theme is very clear: geography is the science of spatial relations and inter-relations; geometry is the mathematics of space; hence geometry is the language of geography. The early chapters establish geography's scientific credentials, in a debate with Hartshorne's published statements, especially those concerning uniqueness and predictability. As Lewis (1965) and others argued, Bunge claimed that Hartshorne confused uniqueness and singularity: he opposed Hartshorne's claim that geography cannot formulate laws because of its paucity of cases by arguing for even more general laws, and countered the argument that geographical phenomena are not predictable with the claim that science 'does not strive for complete accuracy but compromises its accuracy for generality' (p. 12).

Having established geography's scientific credibility to his satisfaction, Bunge then investigated its language. An intriguing discussion of cartography led him to conclude that descriptive mathematics is preferable to cartography as a more precise language. The remainder of the book looked at aspects of the substantive content of the science of geography, beginning with 'a general theory of movement' and then a chapter on central place theory: 'If it were not for the existence of central place theory, it would not be possible to be so emphatic about the existence of a theoretical geography ... central place theory is geography's finest intellectual product' (p. 133). Problems of testing the theory showed the need for map transformations (see also Getis, 1963), and the final chapter of the first edition clarified the links between geography and geometry: 'Now that the science of space is maturing so rapidly, the mathematics of space – geometry – should be utilized with an efficiency never achieved by other sciences' (p. 201).

The richness of the work done by this group during the mid- and late 1950s continued in various locations after it broke up (only Ullman remained, although Morrill later joined the staff at Seattle). Berry was one of the most prolific and seminal, not only in his original field of central place theory (Berry, 1967) but also over a wide range of other topics in economic and social geography. Berry's work has always had a very strong empirical and utilitarian base, whereas Dacey continued to work on the mathematical representation of spatial, especially point, patterns (e.g. Dacey, 1973) and Tobler (1995) moved into computer cartography. In total, the work of this significant group of scholars influenced the research and teaching of a whole generation of human geographers, throughout the world.

The social physics school

This group's work was initiated and developed independently from the other three, and its early publications preceded Schaefer's paper by more than a decade. The leader was J. Q. Stewart, an astronomer at Princeton University, who traced the origins of social physics in the work of a number of natural scientists who applied their methods to social data (Stewart, 1950). His own investigations began when he noted certain regularities in various aspects of population distributions which were akin to the laws of physics, such as a tendency for the number of students attending a particular university to decline with increasing distance of their home addresses from its campus. From these observations he developed his ideas on social physics, which he defined (Stewart, 1956, p. 245) as:

> that the dimensions of society are analogous to the physical dimensions and include numbers of people, distance, and time. Social physics deals with observations, processes and relations in these terms. The distinction between it and mathematical statistics is no more difficult to draw than for certain other phases of physics. The distinction between social physics and sociology is the avoidance of subjective descriptions in the former.

He established a laboratory at Princeton to investigate a wide range of such regularities. (Note Warntz's, 1984, remark that he was introduced to Stewart's ideas via a book on *Coasts, Waves and Weather* – Stewart, 1945 – which was 'prepared primarily to explain to marine and air navigators the physical environment . . . [but] Stewart could not resist the temptation to include an exotic chapter describing potential of population and its sociological importance'.)

Stewart introduced his ideas to geographers through a paper in *The Geographical Review* (Stewart, 1947). Four empirical rules were adduced: the first, the rank-size rule for cities, showed that in the United States the population of a city multiplied by its rank (from 1 for the largest to n for the smallest) and standardized by a constant, equalled the population of the largest city, New York (Carroll, 1982); the second indicated that at various dates the number of cities in the country with populations exceeding 2500 was very closely related to the proportion of the population living in such places; the third showed that the distribution of a population could be described by the population potential at a series of points, in the same manner as the potential in a magnetic field is described in Newtonian physics; and the fourth illustrated a close relationship between this population potential and the density of rural population in the United States. From these regularities, Stewart claimed that (p. 485): 'There is no longer any excuse for anyone to ignore the fact that human beings, on the average and at least in certain circumstances, obey mathematical rules resembling in a general way some of the primitive "laws" of physics.' No reasons were given why this should be so (Curry, 1967, p. 285, called this a 'deliberate shunning of plausible argument'): the rules were presented as empirical regularities which had some similarity to the basic laws of physics. Causal hypotheses were not even postulated, let alone tested.

Stewart's main collaborator was William Warntz, a graduate of the University of Pennsylvania who was later employed by the American Geographical Society as a research associate, working on what he termed 'the investigation of distance as one of the basic dimensions of society' (Warntz, 1959b, p. 449; see also Warntz, 1984). The wide range of empirical regularities which they observed (see Stewart and Warntz, 1958, 1959) was used to develop their concept of macrogeography (Warntz, 1959b, 1959c). Warntz (1959b) claimed that geographical work was dominated by micro-studies: 'The tendency of American geographers to be preoccupied with the unique, the exceptional, the immediate, the microscopic, the demonstrably utilitarian, and often the obvious is at once a strength and a weakness' (p. 447) because 'the assembly of more and more area studies involving an increase in the quantity of detail does not mean *per se* a shift from the microscopic to the macroscopic' (p. 449). Geographers were in danger of being unable to perceive general patterns within their welter of local detail, and to counter this, Stewart and Warntz suggested the search for 'regularities in the aggregate'. Stewart's concept of population potential was used to describe general distributions, and was shown to be related to a large number of other patterns in the economic and social geography of the United States. Although these findings were recognized as only empirical regularities, they could be used as the basis for theory development (Stewart and Warntz, 1958, p. 172), for geography needed theory which (Warntz, 1959b, p. 58): 'has as its aim the establishment and coordination of areal relations among observed phenomena. General laws are sought that will serve to unify the individual, apparently unique, isolated facts so laboriously collected.' The approach to theory was inductive (see Figure 3.1) rather than deductive, although there was a clear underlying belief in the importance of distance and accessibility as influences on individual behaviour. Christaller's ideas were seminal for Warntz as well as Garrison, Ullman and others (see Bunge, 1968).

These macroscopic measures, particularly population potential, were used in a variety of contexts, as in Warntz's (1959a) *Toward a Geography of Price,* which established strong relationships between the prices of agricultural commodities in the USA and measures of supply and demand potential; Harris (1954b) and Pred (1965a – Pred studied with Harris at Chicago) used the potential measure in later studies of industrial-location patterns. The 'macrogeographers' also did much work on various distance-decay functions (see Chapter 4) and extended early Russian work on cen-trographic measures (Sviatlovsky and Eels, 1937; Neft, 1966); this was done both at the American Geographical Society (where its pioneering, mathematical nature ran contrary to the general conservatism displayed by the Society) and later, under Warntz, at the Graduate School of Design's Laboratory for Computer Graphics and Spatial Analysis at Harvard University, where it was the forerunner of later developments in computer-ized cartography and GIS (see p. 117).

These lines of work contrasted markedly with that of the other three groups reviewed here, in a variety of ways. First was the topic of scale; Stewart and Warntz perhaps conformed more than any others to Bunge's call for a scientific approach which aimed at a high level of generality. Second there was the nature of the approach to theory, for macrogeography was inductive in its search for regularity rather than testing deductive hypotheses. Finally, the analogies sought for human geography were in a natural science – physics – and not in the other social sciences.

Summary

The developments outlined in this section marked the beginning of major changes in the field of human geography, changes which were rapidly taken up by others, within and beyond the United States. Although the focus was on theory and measurement, and the development of 'geographical laws', in line with the general ethos of academia in the immediate post-war decades, the work did not deviate too far from Hartshorne's expanded (1959) definition of the nature of geography (see above, p. 58). The main difference between the new work, with its focus on systematic studies, and its regional predecessor was the greater faith of geographers in their ability to produce laws, to work within the canons of accepted scientific method, to master and apply relevant mathematical and statistical procedures, and to move out of their self-imposed academic isolation (Ackerman, 1945, 1963).

Diffusion of the new approach was rapid, involving a number of gradu-ates from the four centres identified above, especially Seattle, plus other willing change-agents. Ned Taaffe, for example, trained in journalism and meteorology and, after experience of teaching economics and statistics, was involved in the development of spatial analysis at Northwestern University (on which also see Hanson, 1993) and then moved to Ohio State University 'with a mandate to build a research department': he appointed 'a group of geographers, each of whom influenced the development of spatial analysis in geography in a somewhat different way' (Taaffe, 1993, p. 423: he names Howard Gauthier, Les King, George Demko, Kevin Cox, John Rayner, Emilio Casetti, Larry Brown, Reg Golledge, John Arnfield and

Harold Moellering. Golledge, King (Taaffe's 'most influential appointment' according to Getis, 1993, p. 522), and Rayner – and also Bill Clark – were at the University of Canterbury, New Zealand when Harold McCarty was a visiting professor in the late 1950s). Taaffe was also involved in several summer workshops conducted at Northwestern University with funding under the National Defense Education Act, which introduced many other geographers to quantitative methods.

The role of certain individuals and institutions in the promotion of change was thus crucial (on which see the exchange between Morrill and Johnston: Morrill, 1993, 1994; Johnston, 1994). The model was repeated two decades later with the establishment of a strong centre at the University of California, Santa Barbara, out of which developed the National Center for Geographic Information Systems and Analysis (Getis, 1993: Abler, 1993).

Scientific method in human geography

Whether or not the changes just described initiated a 'Kuhnian revolution' in human geography, it is clear that they heralded a major reorientation in the nature of much geographical research. This reorientation was not focused on any blueprint or grand design, however. No paper or book was published to provide either a philosophy for the new approach or a detailed outline of how research should be conducted in this framework. Schaefer's paper said nothing about how geographical laws were to be stated and derived, and although McCarty and his co-workers discussed methods in both general (McCarty, 1954) and specific (McCarty *et al.*, 1956) contexts, they provided no programmatic statement for how the discipline should be practised. As Gregory (1978a, p. 47) puts it, 'geography has (with some notable exceptions) paid scant attention to its epistemological foundations', whereas Livingstone (1992, p. 328) suggests that 'Geography's confrontation with the vocabulary of logical positivism . . . was a *post hoc* means of rationalizing its attempt to reconstitute itself as spatial science'; it was a technical revolution which sought philosophical legitimation.

One piece was not central to the initial efforts, because it post-dated many of them, but it was widely quoted in the 1960s as the new ideas spread. Ackerman's (1958) essay on *Geography as a Fundamental Research Discipline* was an analysis of research organization. He indicated that the ultimate goal was integration to provide a full comprehension of reality, and that with regard to current developments, 'If any one theme may be used to characterize this period, that theme would be one of illuminating covariant relations among earth features' (p. 7). As a science, even one which is eventually an idiographic science since it deals with unique places, geography, according to Ackerman, needed to strive for 'an increasingly nomothetic component'. Its fundamental research:

need not necessarily be law-giving. . . . Much fundamental research in geography has not been law-giving in the strict sense but it has been concerned with a high level of generalization, and it has given meaning to other research efforts which succeeded it. In this sense it has a block-building characteristic (p. 17).

Such fundamental research 'is likely to rest on quantification . . . accurate study depends on quantification' (p. 30) and should 'furnish a theoretical framework with capacity to illuminate actually observed distributional patterns and space relations' (p. 28).

Ackerman's essay was a clarion call for theory development, the application of quantitative methods, and a focus on laws and generalizations to form the building-blocks for further nomothetic research. But it contained no detailed discussion of how such research should be undertaken: it lacked detail. Seven years later, the report of a National Academy of Sciences–National Research Council (1965) committee on *The Science of Geography* set out 'geography's problem and method' in the statement that:

> Geographers believe that correlations of spatial distributions, considered both statistically and dynamically, may be the most ready keys to understanding existing or developing lifesystems, social systems, or environmental changes. In the past . . . progress was gradual, however, because geographers were few, rigorous methods for analysing multivariate problems and systems concepts were developed only recently (p. 9).

Once again, however, a general statement on research orientation had no accompanying detail on research conduct and methods. And yet a paper published in 1963 had claimed that an intellectual revolution – the quantitative and theoretical revolution – had occurred in geography: 'The revolution is over, in that once-revolutionary ideas are now conventional' (Burton, 1963, p. 156). Something had become conventional, but nobody had written a full formulation for the discipline of what that something was!

Scientific method

There is no suggestion here that the groups of researchers who proposed changes in the nature of human geography lacked a clear rationale for their work; indeed, those involved were undoubtedly very clear as to both means and ends (though this may not be the case with some of their disciples), but they did not discuss these in detail in print. Nor, if their citations in the published works are any lead, did many research deeply into the philosophy that they were adopting (though a few, such as Peter Gould, regularly attended courses in philosophy throughout their subsequent careers): the exceptions are the references to Bergmann in the Iowa group's papers (see Golledge, 1983) and in Bunge's (1962) thesis. Texts on statistical and mathematical procedures were widely quoted and several were produced by and for geographers (e.g. Gregory, 1963; King, 1969b), but the first major work on the philosophy of the 'new geography' was not published until 1969, in a book which received wide acclaim (Harvey, 1969a: a parallel, but briefer statement is Moss, 1970: note that Harvey was trained in Britain and his book was published there).

Harvey identifies two routes to explanation (Figure 3.1). The first, the 'Baconian' or inductive route, derives generalizations from observations: a pattern is observed and an explanation developed from and for it. This involves a dangerous form of generalizing from the particular case, however, because acceptance of the interpretations depends too much on the charisma of the scholar involved and on the unproven representativeness of the case(s) discussed (Moss, 1970). So the preferred method is the second route in Figure 3.1 (though see Bennett, 1985b, and Barnes, 1996). This also begins with observers perceiving patterns in the world; they then formulate experiments, or some other kind of test, to prove the veracity of the explanations offered for those patterns. Only when ideas have been tested successfully against data other than those from which they were derived can a generalization be produced.

Scientific knowledge obtained via the second route is 'a kind of controlled speculation' (Harvey, 1969a, p. 85), and an increasing number of human geographers sought to apply such a procedure during the 1950s. Its philosophy, known as positivism, was developed by a group of philosophers working in Vienna during the 1920s and 1930s (Guelke, 1978). It is based on a conception of an objective world in which there is order waiting to be discovered. Because that order – the spatial patterns of variation and covariation in the case of geography – exists, it cannot be contaminated by the observer. A neutral observer, acting on either observations or reading others' research reports, will derive a hypothesis (a speculative law) about some aspect of reality and then test it: verification of the hypothesis translates the speculative law into an accepted one.

A key tenet of this philosophy is that laws must be proven through objective and replicable procedures, and not accepted simply because they seem plausible: 'the plausibility or intuitive reality of a theory is *not* a valid basis for judging a theory' (Bunge, 1962, p. 3). A valid law must predict successfully, so that having developed an idea about certain patterns the researcher must formulate it into a testable *hypothesis* – 'a proposition whose truth or falsity is capable of being asserted' (Harvey, 1969a, p. 100). An experiment is then designed to test the hypothesis, data are collected, and the validity of the predictions evaluated.

If the test results do not match the predictions, then either the observations on which the hypothesis was based or the deductions from the works

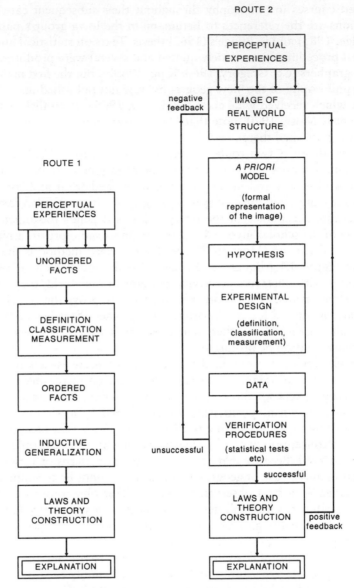

Fig 3.1 Two routes to scientific explanation (source: Harvey, 1969a, p. 34).

of others are thrown into doubt. There is thus negative feedback (Figure 3.1) and the image of the world has to be revised, creating a new hypothesis. This is the Popperian view that any hypothesis can be found wanting by a single falsification, to which Harvey (1969a, p. 39) gives only eight lines, however, preferring the more general one that only 'severe failure' – which he does not define – discredits a hypothesis totally (see, however, Moss, 1977; Bird, 1975;

Petch and Haines-Young, 1980; Haines-Young and Petch, 1985). Hay (1985a) and Marshall (1985) present cases for adopting Popper's critical rationalism in human geography. The goal is the same as that of positivism – the development of comprehensive theories which allow predictions with high degrees of certainty. The two differ on means, not ends, because critical rationalists believe that hypotheses can never be comprehensively verified, only falsified. If the researcher is a good observer and a logical thinker, falsification of hypotheses should be rare. If the test is successful, then the speculation of the hypothesis becomes an acceptable generalization. One successful test will not turn it into a law, however. Replication on other data sets will be needed since a law is supposed to be universal; and there is always the possibility of a falsification.

Bird (1989) extended the case for critical rationalism, with a procedure entitled PAME – an acronym for Pragmatic Analytical Methodology-Epistemology. 'Pragmatic' is placed first in the sequence to stress the importance of external validation of conjectures against the 'real world' (p. 236), and is followed by 'analytical' because Bird adopts a hypothetico-deductive approach rather than 'a methodology which uses consideration of individual cases *inductively*, in the Micawberish hope that something will turn up to confirm the over-arching structures already in place' (p. 237). By methodology he implies the use of paradigms as exemplars (see p. 17), and this is hyphenated with epistemology because development of a theory of knowledge requires a workable, pragmatic set of procedures. With the whole (pp. 238–9):

> The one stable element is the hypothetico-deductive nature of the methods of inquiry. And because the method is pragmatically warranted by successful correspondence to the thing in the real world, all else in the methodological-epistemological structure can be changed as experience dictates.

This open-ended procedure should be employed not in the search for 'ultimate truths' but rather for the stimulating comparison of 'ideas that we hold, always on probation' (p. 246).

According to Harvey (1969a, p. 105):

> A scientific law may be interpreted as a generalization which is empirically universally true, and one which is also an integral part of a theoretical system in which we have supreme confidence. Such a rigid interpretation would probably mean that scientific laws would be non-existent in all of the sciences. Scientists therefore relax their criteria to some degree in their practical application of the term.

After sufficient (undefined) successful tests, therefore, a hypothesis may be accorded law-like status, and fed into a body of *theory*, which comprises a series of related laws. There are two types of statement within a full theory: the *axioms*, or givens, which are statements taken to be true, such as laws;

and the deductions, or *theorems,* from those initial conditions, which are derived consequences from agreed facts – the next round of hypotheses. There is a positive feedback from the theory stage to the world view, therefore (Figure 3.1), so that the whole scientific enterprise is a cyclical procedure whereby the successes of one set of experiments become the building blocks for thinking about the next.

One stage in Figure 3.1 so far ignored is the *model,* a widely used term which has been given a variety of meanings (Chorley, 1964). Models have two basic functions: as *representations of the real world,* such as a scale model, a map, a series of equations and some other analogue (Morgan, 1967); and as *ideal types,* representations of the world under certain constrained conditions. Both are used in the positivist method to operationalize a theory, as a guide to the derivation of testable hypotheses.

Quantification is central to this scientific method. Mathematics are particularly useful in developing models, as in the linear programming procedures adopted by Garrison. Relatively few geographers have strong backgrounds in mathematics, however (this was especially true in the 1950s), and so little work involved representing the real world as sets of equations. Instead the central role was given to statistics, used in hypothesis testing. Two types of statistics are available: *descriptive statistics* can be used to represent a pattern or relationship; *inductive statistics* are used for making generalizations about a carefully defined population from a properly selected random sample; they use the same procedures. Many geographical researchers confused the two. Inductive statistics employ significance tests to show whether what has been observed in the sample probably also occurs in the parent population, so that if the data analysed are not a sample, such tests are irrelevant (some disagreement was expressed over 'what is a sample?': see Meyer, 1972 and Court, 1972). Many geographers have used inductive methods in a descriptive manner, however, using the significance tests as measures of the validity of their findings (as argued for in Hay, 1985b).

The main attraction of statistics to many early adherents of the 'new geography' was their precision and lack of ambiguity – compared to the English language – in description. This was expressed by Cole (1969), who annotated a quotation from a well-known text (Stamp and Beaver, 1947, pp. 164–5): in this the text is Stamp and Beaver's and the annotations in parentheses, to shown the ambiguities, are Cole's:

> The present distribution of wheat cultivation in the British Isles (space) raises the conception of two different types of limit. Broadly speaking (vague), it may be said that the possible (vague) limits (limit) of cultivation of any crop are determined by geographical (vague), primarily by climatic conditions. The limits so determined (how?) may be described (definition) as the ultimate (vague) or the geographical (vague) limits (Cole, 1969, p. 160).

Cole argued that the full quotation (only part is reproduced here) is so full of ambiguities that it could refer to about one million million possible combinations of some forty counties and it is impossible to reconstruct a map from that description:

> the correlations suggested are so tentative and imprecise that they leave the reader still wondering why wheat is grown where it is. The application of a standard correlation procedure ... in itself would give a more precise appreciation of the relationship (p. 162).

Similar views became widely held during the 1950s and 1960s, and quantification became the *sine qua non* of training in the new methods (LaValle, McConnell and Brown, 1967).

The scientific method increasingly adopted by geographers was a procedure for testing ideas, therefore, but a highly formalized one, about which there has been a great deal of debate among philosophers of science and others (Harvey, 1969a). Although many aspects of the method were used by geographers, their citations indicate relatively little training or exploration in depth into the full philosophy of positivism. (Positivism, as used here, refers to what is often known as 'scientific method'. It is embraced by the philosophy of logical positivism, which claims that only scientifically obtained knowledge is valid knowledge: Johnston, 1986a, 1986b.)

Reactions to scientific method

Despite (or perhaps because of) the lack of a clear programmatic statement of the 'new Theology' (Stamp, 1966, p. 18), at least until the appearance of Harvey's (1969a) book, reactions to the developments were many and varied. (James, 1965, p. 35, called the debate 'continued, bitter and uncompromising warfare'.) Two related issues were the main foci of contention: whether quantification was sensible in geographical research, and whether law-making was possible. As Taylor (1976) points out, to some extent the debate was inter-generational, of the type discussed in Chapter 10 (p. 86): some of the 'old guard' thought the proposals were just not geography and should be banished to another corner of academia.

The quantification issue was the less important, and few spoke out against it entirely, although its extent was criticized. Spate (1960a, p. 387) recognized quantification as 'an essential element':

> This is, like it or not, the Quantified Age. The stance of King Canute is not very helpful or realistic; better to ride the waves, if one has sufficient finesse, than to stake attitudes of humanistic defiance and end, in Toynbee's phrase, in the dustbin of history (p. 391).

However, he identified three dangers. The first was a confusion of ends and means; some protagonists wanted to quantify everything (after Lord Kelvin

– 'when you cannot express it in numbers, your knowledge is of a meagre and unsatisfactory kind': Spate, 1960b), but some things, like the positions of Madrid and Barcelona in Spanish thought, cannot be treated in that way. Secondly, there was the dogged analysis of trivia, producing platitudinous findings, a fault which Spate recognized as part of all academia, and especially its revolutions: 'Quantified or not, the trivial we will always have with us' (Spate, 1960a, p. 389), and the problem is usually the extreme positions taken up – Robinson's (1961) perks (the hyperquantifiers) and pokes (the hypoquantifiers). Finally, there was the quantifiers' vaunting ambition and belief that solution of the world's problems lay just around the corner.

Spate was more generous than many critics. Burton (1963) identified five types of criticism:

1 Geography was being led in the wrong direction.
2 Geographers should stick with their proven tool – the map.
3 Quantification was suitable for certain tasks only.
4 Means were being elevated over ends, with too much research on methods for methods' sake.
5 Objections were not to quantification *per se* but to the quantifiers' attitudes.

He believed that quantification had been proven to be more than a fad or fashion, however, and that geography would soon proceed beyond a stage of testing relatively trivial hypotheses with its new tools so that 'The development of theoretical, model-building geography is likely to be the major consequence of the quantitative revolution' (p. 156).

More critical to many geographers than quantification was the issue of theory, and in particular the role of laws in geography. For some, this continued the debate over environmental determinism, which was still active in Britain (Clark, 1950; Martin, 1951; Montefiore and Williams, 1955; Jones, 1956). Jones, for example, extended the debate to cover scientific determinism and its implications for human free will. Martin (1951, p. 6) had argued that possibilism is 'not merely wrong but is mischievous' because all human actions are determined in some way, so that in human geography:

> Unless we can assume the existence of laws or necessary conditions similar in stringency to those of physical science, there can be no human geography nor social sciences worth the name, but only a series of unexplainable statements of bare events . . . such laws cannot differ, except in respect of . . . far greater complication, from those of physical science (pp. 9–10).

Jones (1956) indicated the impossibility of discovering universal laws about human behaviour and pointed to the use of two types of law in physics: the determinate laws of classical physics, which apply macroscopically; and the probabilistic quantum laws, which refer to the behaviour of individual particles. The latter allow for the exercise of free will within prescribed constraints, and their application in human geography would at least allow answers to be offered to the question 'how?' if not to 'why?'. But the question of causality clearly worried many, as indicated by Lewis's (1965) counter-argument that 'it is erroneously assumed that causes compel their effects in some way in which effects do not compel their causes' (p. 26).

Golledge and Amedeo (1968) addressed this same problem, pointing out that critics of law-seeking in human geography defined a law as a universal postulate which brooked of no exception. Scientists use several types of law, however, and the veracity of a law-like statement can never be finally proven, since it cannot be tested against all instances, at all times and in all places. Four types were identified with relevance for human geographers:

1　*Cross-sectional laws* describe functional relationships (as between two maps) but show no causal connection, although they may suggest one.
2　*Equilibrium laws* state what will be observed if certain criteria are met.
3　*Dynamic laws* incorporate notions of change, with the alteration in one variable being followed by (and perhaps causing) an alteration in another. They may be either *historical*, for example showing that B would have been preceded by A and followed by C, or *developmental*, in which B would be followed by C, D, E, etc.
4　*Statistical laws* are probabilistic statements of the likelihood of B happening, given that A exists: all laws of the other three categories may be either deterministic or statistical, with the latter almost certainly the case with phenomena studied by geographers.

None of the papers just discussed were part of an ongoing debate on quantification and theory-building; they were reactions to generally held and discussed attitudes rather than to published critiques (in Britain there were none for several years: Taylor, 1976). There was one debate in the American literature, however, initiated by Lukermann (1958) as a reaction to Warntz's views on macrogeography (see above, p. 71) and to a paper by Ballabon (1957). The latter claimed that economic geography lacked general principles; it was 'short on theory and long on facts' (p. 218). McCarty had shown how to conduct research, but Ballabon stressed the use of location theory being developed by economists as a source for hypotheses. Lukermann responded that the main problem in Ballabon's and Warntz's arguments lay in the assumptions behind their hypotheses (Warntz's analogies with physics and Ballabon's with economics) which did not conform to his view of geography as an empirical science. Statistical regularities and isomorphisms with other subject matter do not provide explanations, so that hypotheses derived from such models test only the models themselves and (Lukermann, 1958, p.9: see also Moss, 1970): 'the hypotheses to be tested are neither statistically nor rationally derived; that is, they are derived neither from empirical observation nor from deductions of previous knowledge in the social, economic or geographic fields'. Berry (1959b) countered with the contention that models, for all their simplifica- tions and unreal assumptions, offer insights towards understanding the real world: 'A theory or model, when tested and validated, provides a miniature of reality and therefore a key to many descriptions. There is a single master- key instead of the loaded key ring' (p. 12). But Lukermann (1960a) was not convinced that models based on assumptions of perfect knowledge and competition, for example, could help towards understanding if they were not empirically derived: 'the crucial problem is the construction of hypothe- ses from the empirical realities of economic geography ... more light is shed and less truth is sophisticated through inventory than through hunches' (p. 2). King (1960) claimed that all laws are really only hypothe- ses, and that deviations of observed from expected values in their testing indicate where the assumptions are invalid. Lukermann responded three times. In the first paper, he showed the lack of consensus in 'explanations' of the geography of cement production in the United States (Lukermann, 1960b) because economic analyses ignored 'Historical inertia, geographical momentum, and the human condition' (p. 5). Secondly, in response to King, he presaged some of the arguments developed later by Sack, who worked with him at Minnesota (see below, p. 122), and pointed out that much of the theory being introduced to economic geography (such as Losch's) was not based on providing understanding of, and explanation for, reality (Lukermann, 1961). Finally, a discussion of several aspects of the debate concluded with the statement that (Lukermann, 1965, p. 194):

> Thus, we see scientific explanation as far removed from the context within which the macroscopic geographers would have us put it – the end product of geographic research. Science does not explain reality, it explains the consequences of its hypotheses.

He made a further call for explanations in geography to be based on observations of reality and not the import of analogies which cannot offer explanations, but only unreal assumptions. Lukermann's basic point, never fully tackled by his critics, was that tests showing conformity between empirical reality and a model were tests of the model only, and could not indicate how empirical reality was created if the assumptions on which the model was constructed were not themselves grounded in reality. (See Barnes, 1996, Chapter 9, for a full evaluation of Lukermann's work.)

This clear difference of opinion over the way in which geographers should seek explanations (which was not about the positivist scientific method itself, but about the inputs to the images of the real-world structure – Figure 3.1) suggests the sort of generation gap discussed in Chapter 1. It is doubtful whether papers such as those of Jones, of Lewis, and of Golledge and Amedeo quieted the fears of those unconvinced by the argu- ments of the 'quantifiers', any more than Berry and King convinced

Lukermann. But the differences soon became a non-issue, at least in the published papers resulting from the research activities of geographers in many topical specialisms. As Burton claimed, by the mid-1960s the changes seem to have been widely accepted, and the regional approach had certainly been ousted from its prime position in the publications of human geographers. Increasingly, quantitative and theoretical material came to dominate not only the more obvious journals, such as *Economic Geography* and *Geographical Analysis* (a 'journal of theoretical geography' founded in 1969), but also the prestigious general journals, notably the *Annals of the Association of American Geographers*. (*The Geographical Review* was an early partial 'convert' through the American Geographical Society's sponsorship of the macrogeographers, although Berry states that it rejected his early papers with Garrison as 'not geography': Halvorson and Stave, 1978; Berry, 1993.) Most of the work contributed little to theory, however. It was quantitative testing of theory- or model-derived hypotheses in some cases, but with little indication of how good the results were. In others, it was quantitative description that could inform theory and model development, but in itself was merely a series of 'factual reports'. By the 1970s, textbooks were being published which began with discussions of scientific method and quantification before proceeding to the substantive content of the 'empirical science' (Abler, Adams and Gould, 1971; Amedeo and Golledge, 1975); the 'revolution' had become the orthodoxy.

Spread of scientific method

The initial development of systematic studies using the positivist scientific method in the USA was very largely focused on economic geography and associated economic aspects of urban geography. This reflects the relative sophistication of economics within the social sciences and the existence of several approaches to 'location theory' (see p. 68), providing a model for geographers to copy, not only to advance their discipline but also to promote its cause in the search for utility to the worlds of business and government. The long tradition of empirical work in human geography meant, however, that with few exceptions research in the systematic areas mainly comprised the statistical testing of relatively simple hypotheses, with little mathematical modelling or writing of formal theory.

Contemporaneous with, and an important stimulus to, these developments in human geography was the emergence of a new discipline in the United States – *regional science*. (On regional science as stimulus, see Berry, 1995.) This was very much the product of one iconoclast scholar – Walter Isard – an economist who built spatial components into his models, to provide a stronger theoretical basis for urban and regional planning than had existed previously. In general terms, regional science is economics with a spatial emphasis, as illustrated by Isard's (1956a, 1960) two early texts, but the Regional Science Association attracted relatively more practising geographers than economists. To some, regional science and economic geography are hard to distinguish: the former can be separately characterized by its greater focus on mathematical modelling and economic theorizing, however, whereas geographical work has remained more empirical and less dependent on formal languages. (Initially Isard, 1956b, saw geographers as doing the empirical tests of the regional scientists' models: see Berry, 1995, on his reaction to this.) Over time, the interests of regional scientists broadened (Isard, 1975), but the strong theoretical base has remained. (The history of regional science, 40 years after the foundation of the Regional Science Association, is discussed in two issues – volume 17, numbers 2 and 3, 1995 – of the *International Regional Science Review*: see, in particular, Isserman, 1995.) The new discipline did not create a substantial niche within American academia, however (Garrison, 1995, likens it to Moses' 40 years wandering in the wilderness). Nor did it have a lasting, substantial impact on geography and geographers, although the latter remain a considerable proportion of the Association's membership. Berry (1995) argues that, for him, relative disillusion with regional science was engendered by Isard's categorization of geographers as the empirical workers who tested the regional scientists' theories, which implied a subservient position for geographers in a two-class academic society and would hinder the reformation of economic geography that Berry sought. (This has an interesting parallel to the debates within geography on the division of labour between 'theorizers' and 'empirical testers': see p. 87.)

The Regional Science Association has flourished internationally, however, and Isserman (1995, p. 261) claims that 'Regional science has become a mainstream group within geography and geography departments . . . [and] . . . so successful within geography that it became worthy of caricature.' He then notes major tensions between those geographers who remain committed to the goals of regional science (almost certainly a relatively small number within the profession) and those whose interests lie elsewhere, however. Warf (1995, p. 192) points to these – 'Class and gender, historical sensitivity, the politics of the state, the recovery of the living subject and everyday life, the unintended reproduction of social worlds' – and invites a *rapprochement* between regional science and a geography inspired by social theory. As illustrated later in this book, however, the distance between positivist theorizers and other geographers has widened in recent decades, and with it a substantial rejection of the value of quantitative methods.

The emphasis on statistical methods in so much of the new work in American human geography led to its partial *rapprochement* with physical geography. (One of the leading 'quantitative geographers' of the 1960s, Leslie Curry, trained as a climatologist but switched his interests to economic geography in the late 1960s.) More physical geography papers were published in the leading journals, more physical geographers were appointed to university departments, geologists such as Krumbein, Leopold, Schumm and Wolman were major sources of quantitative ideas, and there was a common interest in the training of graduate students (LaValle *et al.*, 1967). This shared concern with procedures was illustrated at a 1960 conference on quantitative geography, from which emerged two volumes (Garrison and Marble, 1967a, 1967b) on methodological developments, one for human geography and the other for physical geography. In the former, Berry introduced the family of factor analysis methods as a way of collapsing and ordering large data matrices; Dacey investigated line patterns and Beckmann the optimal location of routes; Robinson continued his work on the statistical comparison of maps; Mayfield and Thomas extended the analysis of central place patterns; Marble, Morrill and Nystuen looked at patterns of movement; and Warntz continued the work on macrogeography. Other conferences and summer schools to train geographers in quantitative techniques were held at this time (on their impact, see Gould, 1969; Taaffe, 1979), and American geographers were to the forefront in launching an International Geographical Union Commission on Quantitative Methods.

Expansion within American geography

The launch of the 'quantitative and theoretical revolutions' identified by Burton (1963) was concentrated on a few topical specialisms within American human geography only, so an early task for the 'revolutionaries' was to spread their 'new Theology' wider through the discipline, convincing others of the benefits which quantification and the associated scientific method could bring to their special interests. A major piece of advocacy was an NAS/NRC report (1965) on *The Science of Geography* which was prepared in order to chart research priorities within the discipline. The case was presented for more 'theoretical-deductive' work to balance the earlier emphasis on 'empirical inductive analysis', the detailed argument being based on four premises:

(a) Scientific progress and social progress are closely correlated, if not equated. (b) Full understanding of the world-wide system comprising man and his natural environment is one of the four or five great overriding problems in all science. (c) The social need for knowledge of space relations of man and natural environment rises, not declines, as the world becomes more settled and more complex, and may reach a crisis stage in the near future. Last, (d) progress in any branch of science concerns all branches, because science as a whole is epigenetic.

The social need for knowledge of space relations means an imminent practical need. As the population density rises and the land-use intensity increases, the need for efficient management of space will become even more urgent (p. 10).

Because the committee members (E. A. Ackerman, B. J. L. Berry, R. A. Bryson, S. B. Cohen, E. J. Taaffe, W. L. Thomas Jr, and M. G. Wolman) defined geography as 'the study of spatial distributions on the earth's surface' (p. 8) then it followed that 'Geographic studies will be irreplaceable components of the scientific support for efficient space management' (p. 10). The positivist scientific method was being sold to geographers and at the same time geography was being sold to the scientific establishment, from whom financial research support was sought.

The committee chose four problem areas to illustrate geography's potentials as a 'useful science'. The first was physical geography. The second was cultural geography, which studies 'differences from place to place in the ways of life of human communities and their creation of man-made or modified features' (p. 23), with a major focus on landscape development and the diffusion of specific cultural features over space and time: 'applying modern techniques to studying the nature and rate of diffusion of key cultural elements and establishing the evolving spatial patterns of culture complexes' (p. 24) was identified as a profitable avenue for development. The third problem area was political geography, with proposals for work on boundaries and resource management. Finally, the committee recognized location theory studies, an amalgam of work in economic, urban and transport geography in which the 'dialogue' between the empirical and the theoretical had gone furthest, 'revealing the potential power of a balanced approach when applied to other geographical problem areas' (p. 44). Location theory involved work on spatial patterns, the links and flows between places in such patterns, the dynamics of the patterns, and the preparation of alternative patterns through model-building exercises which identify efficient solutions.

In the development of the science of geography,

> A major opportunity seen by workers in the location theory problem area is that of integrating their work more closely with other geographers as they begin to deal with spatial systems of political, cultural, and physical phenomena. ... This could be achieved ... by the accelerated diffusion of techniques and concepts to other geographers, and communication on the definition of research problems. The result would be to hasten the confrontation of empirical-inductive studies by theoretical-deductive approaches throughout geography. ... Testing the theory in a variety of empirical contexts should aid in the overall development and refinement of viable theories. It should also serve to connect geographic progress to local problems more rapidly and more effectively (pp. 50–1).

The deductive-theoretical scientific methodology was central to the committee's blueprint for the advancement of geographical research, therefore. All geographers would have a role to play in this movement forward, for:

> Geographers have one other asset that should be capitalized on. Those who have been interested in the study of a specific part of the earth (regional geography) develop competences for interpreting the physical-cultural complexes of the regions that they study. Students of the way a particular part of the earth has evolved (historical geography) have other competences for interpreting the historical development and modification of a region. These two groups have students that are particularly qualified to undertake the field observation and field study of problems recognized in a more systematic way and to conduct field tests of generalizations arrived at through systematic study. ... The regional or historical geography specialist who has mastered the technique of field observation and historical study thoroughly ... can make himself indispensable if he understands the direction in which the generalizing clusters are headed and relates his work closely to their growing edges (p. 61).

A clear division of labour was being suggested, comprising theoretical-deductive 'thinkers' and empirical-inductive 'workers', a division which was apparently unequal in status and was resented by some as such (James, 1965; Thoman, 1965: see Berry, 1995 – and p. 84 above – for a similar reaction to a proposed division of labour between regional scientists and geographers).

A somewhat similar report, prepared for the Committee on Science and Public Policy of the National Academy of Sciences and the Problems and Policy Committee of the Social Science Research Council, was published five years later (Taaffe, 1970). Also the product of a committee (E. J. Taaffe, I. Burton, N. Ginsburg, P. R. Gould, F. Lukermann, P. L. Wagner), this stressed human geography as 'the study of spatial organization expressed as patterns and processes' (pp. 5–6), incorporating people–environment relationships and cultural landscapes and stressing relevance to planning and other policy issues. Much of the report is constructed to illustrate the nature of human geography, promoting a case that:

> there are many opportunities for the expansion and improvement of geographic research. If geography is to have a strong and beneficial impact on the constantly changing patterns of spatial organization of American society, it will be necessary to continue this development (p. 131).

This led to six conclusions regarding the discipline's needs: (1) greater collaboration among the social sciences; (2) alleviation of geography's manpower shortage; (3) establishment of centres for cartographic training and research; (4) development of remote sensing and related data bases; (5) greater support for foreign area study; and (6) programmes established 'to strengthen the mathematical training of geographers'. Human geography was presented as an integral component of the social sciences, increasingly sophisticated in its analytical tools, focusing on spatial organization, and offering particular skills in mapping and data acquisition.

One systematic area of geography colonized early by the new methods was that part of urban geography which dealt with the internal spatial structure of cities. Until the 1960s, little work had been done on this topic except with regard to commercial land uses (particularly the Central Business District and the relationship of the pattern of suburban shopping centres to the postulates of central place theory): almost no attention was paid to the human content of residential areas, perhaps because geography was seen as the science of places, not of people. Recognition that 'people live in cities' (Johnston, 1969) generated interest in residential areas, which gained much stimulus from the work of the urban ecology school of sociologists at Chicago (some of their works had been introduced to geographers earlier – Harris and Ullman, 1945; Dickinson, 1947 – but with little impact). Social area analysis gained in popularity as a methodology (Berry, 1964a) and a new approach to urban geography was initiated which adopted its norms from the functionalist schools of sociology. Society consists of socio-economic classes, whose nature and composition are widely accepted, and these classes come to consensual agreements about the allocation of land among competing groups (Johnston, 1971). Such urban geography became a separate systematic branch of the discipline; few people did research in it as well as in other aspects of urban geography.

The argument advanced by proponents of the new methodology focused on a common set of procedures to tackle geographical problems. Berry (1964b) argued that the geographer's viewpoint emphasizes space, with regard to distributions, integration, interactions, organization and processes. Its data can be categorized in a single matrix (Figure 3.2) in which places form the rows and characteristics the columns: each cell defines a 'geographic fact'. Berry recognized five different types of geographical study by focusing on different elements of this matrix: study of a single row (a place) or column (a characteristic); comparison of two or more rows (places) or columns (characteristics); or study of rows and columns together. Adding further matrices, one for each time-period (Figure 3.3), allows five further types of study, based on the earlier five but concentrating on changes over time. Thus, he concluded, systematic and regional geography are part of the same enterprise – a repetition of Hartshorne's (1959) arguments – with neither sufficient in itself.

Berry's matrices referred only to the characteristics of places; further matrices (Figure 3.4) show flows between places, with one matrix for

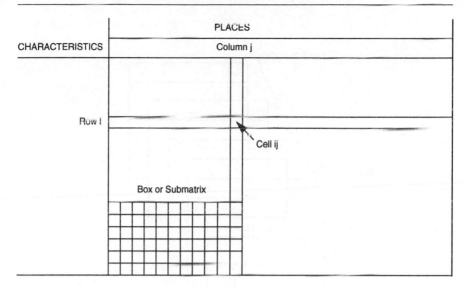

Fig 3.2 The geographic matrix: each cell – ij – contains a 'geographic fact', the value of characteristic i at place j (source: Berry, 1964b, p. 6).

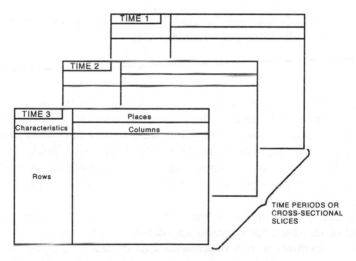

Fig 3.3 A third dimension to the geographic matrix: each cell contains a 'geographic fact', the value of characteristic i at place j at time t (source: Berry, 1964b, p. 7).

each flow category in each time-period (Clark, Davies and Johnston, 1974). Berry used this extension, though he did not formalize it, in his attempted fusion of the procedures for formal and functional regionalization (see above, p. 46) to produce a general theory of spatial behaviour – Berry's (1968) field theory, which he applied in a large study of

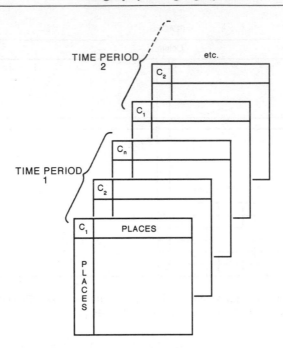

Fig 3.4 Geographic flow matrices – one per commodity (C_1-C_n) per time period (source: after Clark, Davies and Johnston, 1974).

the spatial organization of India (Berry, 1966). The techniques employed there became widely used in the 1960s, as access to high-speed computers became easier, though not universal, for university academics. They were given the umbrella term of factorial ecology (Berry, 1971), and were widely applied in many aspects of geography, resulting in a methodological unity which was previously unknown across the various systematic specialisms.

The changes introduced in the 1950s in the United States spread into several of the discipline's topical specialisms and by the mid-1960s use of statistical methods to test hypotheses was common. The methods united the specialisms, which in substance remained very separate identifiable branches within the geographical enterprise. The decline of interest in work which aimed at integrating their findings into regional syntheses, despite the efforts of Berry and a few others (see also Taaffe, 1974), meant that human geography experienced a centrifugal trend with regard to substance, contemporaneous with a centripetal one with regard to procedures. Since positivist scientific method, and the statistical techniques, were used much more widely than in human geography alone, the former trend was probably the most important.

Trans-Atlantic translation

By the early 1960s the quantitative and theoretical revolutions were having considerable impact beyond the United States, as the result of two agencies. The first was the publication of work by the American iconoclasts in the major journals. Secondly, and probably more importantly, during the 1950s and 1960s a number of British geographers went to the United States, either as postgraduate students or as visiting staff members (see Haggett, 1990); during the 1960s and 1970s many British academics were recruited to teach on the intensive summer-school courses provided by American universities for part-time, usually mature, students. Some encountered the new ideas and disseminated them 'back home', to both their students and, via the newly-established (1962) Study Group in Quantitative Methods of the Institute of British Geographers (Gregory, S. 1976), their fellow academics. (Others, including Brian Berry, stayed in North America: on why, see Berry, 1993.) They had a local base on which to build, mainly in physical geography and arising out of the early use of statistics by climatologists (e.g. Crowe, 1936): as a result, and perhaps somewhat surprisingly, the first undergraduate text in statistics for geographers was written by an English academic (Gregory, 1963). The positivist methodology is entirely implicit in Gregory's book, however: his Preface speaks only of the geographer's raw material 'becoming progressively more of a quantitative nature' and 'the need to present both data and conclusions in sound quantitative terms' (pp. xiii–xiv); statistical techniques are needed, but their use in hypothesis testing is not set out. In addition, there was some interest in location theory – at University College, London, in the early 1950s, for example (Halvorson and Stave, 1978; Berry, 1993), and in work on both settlement patterns (Dickinson, 1933; Smailes, 1946: note, however, the unwelcoming anonymous review of Christaller's thesis on central place theory in the *Scottish Geographical Magazine,* 1934) and industrial location (Smith, 1949; Rawstron, 1958).

Although statistics courses were introduced to several British university departments of geography by the mid-1960s, and aspects of the scientific methodology were taught in at least a few (Whitehand, 1970), the main focus for the introduction of the 'new geography' to Britain during the early years of the decade was the University of Cambridge. The leaders were R. J. Chorley (a geomorphologist, who had spent some time studying in the United States) and P. Haggett (a Cambridge-trained human geographer, although his early published work was in biogeography, who had also visited the United States and experienced the development there: Haggett, 1965c, p. vi; 1990: see also Haggett and Chorley, 1989). Their impact on British geography was considerable, through innovative research and teaching (Gregory, S., 1976). They worked on the adaptation of certain statistical techniques to geographical (both physical and human) problems

(Chorley and Haggett, 1965a; Haggett, 1964: Haggett and Chorley, 1969), but their most lasting contribution was probably in editing two collections of papers which resulted from courses that they directed, aimed at introducing the 'new geography' to teachers. (Note that Barnes, 1996, focuses almost exclusively on Haggett's role as a change-agent within British geography, ignoring the substantial impacts of others such as Stan Gregory, who was very influential not only in establishing the Study Group on Quantitative Methods within the IBG but also within the Geographical Association and the Joint Matriculation Board, whose examinations were used for entrance to most British universities.)

The first of the books edited by Chorley and Haggett – *Frontiers in Geographical Teaching* (Chorley and Haggett, 1965b) – was based on a 1963 course designed 'to bring teachers and like persons into the University, there to encounter and discuss recent developments and advances in their subjects' (p. xi). In it Wrigley (1965) discussed the changing philosophy of geography, identifying the increasing use of statistical techniques as the contemporary development 'of singular importance' (p. 15). He pointed out that techniques of themselves do not form a methodology and that 'Geography writing and research work have in recent years lacked any general accepted, overall view of the subject even though techniques have proliferated' (p. 17). He offered no outline of such a view, however, arguing that eclecticism in mode of analysis was likely to be most productive and that 'the best sign of health is the production of good research work rather than the manufacture of general methodologies' (p. 17). Many of the other chapters interpret geography as if the 'revolutions' had not occurred in the United States, however: Smith's (1965) chapter on historical geography, for example, is an excellent British companion to the American statement published a decade earlier (Clark, 1954).

Elsewhere in the book, Pahl (1965) introduced the models of the Chicago school of urban sociologists and suggested a social geography in which the prime factor is distance (p. 95), but only the chapters by Haggett and by Timms introduced much of the trans-Atlantic turmoil. Haggett (1965a) wrote on the use of models in economic geography, both those based on simple views of the world, such as developments on von Thunen's (Chisholm, 1962), and those derived from observations of particular cases (e.g. Taaffe, Morrill and Gould, 1963). He noted that:

Perhaps the biggest barrier that model builders in economic geography will have to face in the immediate future is an emotional one. It is difficult to accept without some justifiable scepticism that the complexities of a mobile, infinitely variable landscape system will ever be reduced to the most sophisticated model, but still more difficult to accept that as individuals we suffer the indignity of following mathematical patterns in our behaviour (p. 109).

He introduced the notion of indeterminacy at the individual level (see also Jones, 1956 and p. 81 above) and showed how random variables must be introduced to make models operational; his chapter on scale problems (Haggett, 1965b) illustrated methods of sampling and of map generalization from samples. Timms (1965) demonstrated the use of certain statistical techniques for the analysis of social patterns within cities (based on Shevky and Bell's social area analysis, and developed independently of Berry's work on this topic – see above, p. 88), pointing out that:

> The sciences concerned with the study of social variation have as yet produced few models which can stand comparison with the observed patterns or which can be used to predict those patterns. . . . Prediction rests on accurate knowledge of the degree and direction of the interrelationships between phenomena. This can only be attained by the use of techniques of description and analysis which are amenable to statistical comparison and manipulation. If the goal of geographical studies be accepted as the formulation of laws of areal arrangement and of prediction based on those laws, then it is inevitable that their techniques must become considerably more objective and more quantitative than heretofore (p. 262).

If the majority of the contributors to *Frontiers in Geographical Teaching*, almost all of whom were associated with the department of geography at the University of Cambridge, were not as committed to the 'new geography' as was Timms (later, like Pahl, to become a professor of sociology; Wrigley became an economic historian/demographer), this cannot be said of the editors, whose epilogue (Haggett and Chorley, 1965, pp. 360–1) presented a strong case for the 'theoretical revolution':

> We cannot but recognize the importance of the construction of theoretical models, wherein aspects of 'geographic reality' are presented together in some organic structural relationship, the juxtaposition of which leads one to comprehend, at least, more than might appear from the information presented piecemeal and, at most, to apprehend general principles which may have much wider application than merely to the information from which they were derived. Geographical teaching has been remarkably barren of such models. . . . This reticence stems largely, one suspects, from a misconception of the nature of model thinking. . . . Models are subjective frameworks . . . like discardable cartons, very important and productive receptacles for advantageously presenting selected aspects of reality.

This view dominated their next, and substantially more influential, volume (Chorley and Haggett, 1967).

Models in Geography, most of whose contributors were linked to the Cambridge department, presented a synthesis of most of the work completed before the mid-1960s by adherents to the 'quantitative and theoretical revolutions'. Individual authors had been asked 'to discuss the role of model-building within their own special fields of research' (Haggett and Chorley, 1967, p. 19), which resulted in a series of substantive review essays, some dealing with particular topical specialisms (urban geography and settlement location; industrial location; agricultural activity – there were similar reviews for physical geography), some with particular themes ranging across several specialisms (economic development; regions; maps; organisms and ecosystems; the evolution of spatial patterns), and some with methods and approaches (demographic models; sociological models; network models). A catholic use of the term 'model' was allowed, allowing it as a synonym for a theory, a law, a hypothesis, or any other form of structured idea (see Moss, 1970). The approach was strongly nomothetic, however: as Harvey (1967a, p. 551) expressed it,

> the student of history and geography is faced with two alternatives. He can either bury his head, ostrich-like, in the sand grains of an idiographic human history, conducted over unique geographic space, scowl upon broad generalization, and produce a masterly descriptive thesis on what happened when, where. Or he can become a scientist and attempt, by the normal procedures of scientific investigation, to verify, reject, or modify, the stimulating and exciting ideas which his predecessors presented him with.

All the contributors had clearly chosen the latter course: their focus was on models – on generalization of reality – and methods were very much secondary.

The orientation of this significant volume is given by the editors' introduction. (Its significance lay in its two uses: first, as a synthesis and argument, the volume was widely read and used by researchers and teachers as a guide; second, as a series of major reviews, when republished as a series of paperback volumes, the book was extensively employed as an undergraduate text.) Haggett and Chorley (1967, p. 24) presented the model as: 'a bridge between the observational and theoretical levels ... concerned with simplification, reduction, concretization, experimentation, action, extension, globalization, theory formation and explanation'. It can be descriptive or normative, static or dynamic, experimental or theoretical (see also Chorley, 1964). It forms the basis for a proposed paradigm, which made no attempt 'to alter the basic Hartshorne definition of Geography's prime task' (p. 38) but offered hope for much greater progress:

> the new paradigm ... is based on faith in the new rather than its proven ability. ... There is good reason to think that those subjects which have modelled their forms on mathematics and physics ... have climbed considerably more rapidly than those which have attempted to build internal or idiographic structures (p. 38).

Models in Geography stands as a statement of that faith, and as a major illustration of the expanding use of scientific methods in the systematic fields of human geography.

Although the editors and contributors to *Models in Geography* comprised many of the early active participants in the move to change British geography towards a 'more scientific' approach, others involved are not directly represented. Notable amongst them was a group who graduated at the University of Cambridge in the 1950s, having been tutored by A. A. L. Caesar. (Chisholm and Manners, 1973, p. xi, credit his role, one which continued until the end of the 1970s, with a steady stream of productive research workers from St Catherine's College.) It included Michael Chisholm, Peter Hall and Gerald Manners as well as Haggett, the only one who worked within the 'quantitative revolution'. Chisholm, for example, focused on theoretical developments in economic geography (e.g. Chisholm, 1962, 1966, 1971a) but with relatively little quantitative analysis (though see Chisholm and O'Sullivan, 1973), and both Hall and Manners were more concerned, as was Caesar, with geographical analysis of contemporary issues, although in some cases the analyses led to attempts at theory-derivation (Hall, 1981a).

Chorley and Haggett's editing, and their joint work on technical developments – such as trend surface analysis (Chorley and Haggett, 1965a) and network analysis (Haggett and Chorley, 1969) – reflected a belief in the unity of physical and human geography. This was based on an assumption that a shared interest in methods and techniques could unite the two – Haggett (1967, p. 664) writes on: 'the basic proposition that a wide range of different geographical networks may be usefully analysed in terms of their common geometrical characteristics'. As in North America, while the focus of geographical analysis remained geometry then physical and human geographers could find common cause (Woldenberg and Berry, 1967). Explanation of the geometry required the study of very different processes, however, and the two soon separated again.

The relatively untouched

The NAS/NRC (1965) report (see above, p. 85) identified two main systematic specialisms within human geography relatively untouched by the developments outlined here: cultural and historical geography (see also Darby, 1983a). In addition, despite Berry's (1964b) attempt to reframe it, regional geography remained largely apart from the changes in methodological emphasis. (Furthermore, political geography was described by Berry, 1969, as 'that moribund backwater'.) Not all cultural, historical and regional geographers ignored the changes occurring elsewhere, and some were in the 'revolutionary' vanguard: two of the chapters in *Models in Geography*, for example, were written by individuals (David Grigg and

David Harvey) who had done empirical research (e.g. Harvey, 1985c) on historical topics. But in general terms the NAS/NRC report was correct; there is little evidence of success in winning cultural, historical and regional geographers over to the new methodology.

Of the three groups, *historical geographers* were probably most concerned about their apparent isolation within the discipline. This concern was summarized by Baker (1972, p. 13) in terms of the approaches which historical geographers need to consider in greater detail:

> An assumption is necessary here: that methodologically the main advances can be expected from an increased awareness of developments in other disciplines, from a greater use of statistical methods, from the development, application and testing of theory, and from exploitation of behavioural approaches and sources. ... Rethinking becomes necessary because orthodox doctrines have ceased to carry conviction. As far as historical geography is concerned, this involves a questioning of the adequacy of its traditional methods and techniques.

All of these would have to be followed with care, and the potentials of the methodological developments assessed cautiously, but Baker clearly believed there was considerable scope for change, as had already been shown in economic and in social history, and perhaps even more so in archaeology (see Renfrew, 1981). Particular areas of historical geography, including those relating to urban settlements (e.g. Ward, 1971; see also Johnston and Herbert, 1978, p. 20), are perhaps more open to such changes than are others, if for no other reason than the better quality, as well as quantity, of available data, and increasing amounts of such work were reported (e.g. Whitehand and Patten, 1977; Johnson and Pooley, 1982; Dennis, 1984). There are possible implications in such work, however: as Baker notes:

> Studies in, for example, 'historical agricultural geography', 'historical urban geography' and 'historical economic geography' seem to offer possibilities of fundamental development, particularly in terms of a better understanding of the processes by which geographical change through time may take place. Such an organization of the subject would view historical geography as a means towards an end rather than as an end in itself (p. 28).

In the early discussions of the relationships between historical geography and the 'theoretical and quantitative revolutions', most attention focused on quantification rather than theory. Vance (1978) pointed out that the development of theory does not have to involve 'quantitative abstraction', as his work showed (Vance, 1970; see also Pred, 1977b, and Conzen, 1981). Available data can be manipulated to test theories regarding past spatial patterns (e.g. Goheen, 1970), but Radford (1981, p. 257) argued that theory is the more important: 'In the cities of the nineteenth-century

United States, a set of principles . . . was taken to something approaching a logical conclusion'. This assumes that theory is possible in historical geography. As illustrated in Chapter 6, some dispute this – for geography as a whole and not just for historical geography. The positivist method implies objectivity, but the geographer in describing a landscape is subjective:

> In describing a landscape, is he not committed by his past training and his past experiences – by his prejudices, if you will? Just as the portrait an artist paints will tell you much about the artist as well as his sitter, so the description of a countryside will tell you a great deal about the writer (Darby, 1962, p. 4).

Darby portrayed geography as both a science and an art:

> [It] is a science in the sense that what facts we perceive must be examined, and perhaps measured, with care and accuracy. It is an art in that any presentation (let alone any perception) of those facts must be selective, and so involve choice, and taste, and judgement (p. 6).

Such a position clearly separated those wedded to an implicitly idealist view of historical geography (see p. 181) from those advancing the cause of a more scientific approach, whether or not it was quantitative, though some would argue that the first quote from Darby could equally well apply to positivist work (positivistic training introduces subjectivity in choice of subject matter and approach too). In any case, S. Gregory (1976) illustrated the use of statistical procedures to test hypotheses derived from Darby's (1977) classic work on the Domesday Book.

Cultural geographers were less concerned about their apparent drift away from the mainstream of geographical activity than were historical geographers, perhaps because of the lack of any parallel developments to those affecting geography in anthropology, the discipline with which many cultural geographers had most contact (see Mikesell, 1967, for a general comparison: some anthropological work experienced major paradigmatic threats, if not changes, notably in the structural work of Claude Lévi-Strauss – see E. R. Leach, 1974 – but these were not discussed by geographers in the 1960s). Increased interest among geographers in diffusion, stimulated by Hagerstrand's (1968) work, led to contact between the spatial analysis and cultural analysis schools of thought, however (Clarkson, 1970). Nevertheless, as Mikesell (1978, p. 1) expressed it, 'Stubborn individualism and a seeming indifference to academic fashion are well-known characteristics' of cultural geographers, whose preferences are for: a historical orientation; a focus on the role of human agency in environmental change, on material culture, and on rural areas; links with anthropology; an individualistic perspective; and field work. (These are all illustrated in a book of essays on *Geography as Human Ecology*: Eyre and Jones, 1966; see also Turner, 1989.) Similarly, Porter (1978, pp. 30–1) concludes his review of 'geography as human ecology' with:

in the past 25 years those interested in the mutual relations of people and environment have taken an interesting journey in search of a satisfactory replacement for environmental determinism. Along the way they have been offered, but generally have declined to use, the shiny wares of gravity and ... [other] models. ... They have been impressed by, and at times been perhaps a bit envious of, the accomplishments of their colleagues in the analysis of spatial organization. ... [But as the] impulse [in human ecology] is cosmographical, holistic and synthetic, it tends to reject the analytical methods of normal science.

Cultural geography was very largely a preoccupation within North American human geography at that time. The Berkeley 'school', founded and led for several decades by Carl Sauer, was the main focus and training centre, stressing the morphology of landscape (Sauer, 1925) and human intervention in landscape evolution through plant and animal domestication, the use of fire, the diffusion of ideas and artefacts, and the creation of settlements, for example (all themes reviewed in Thomas, 1956, and maintained by Sauer's disciples, such as Kniffen: see Matthewson, 1993). Sauer's approach was ecological and clearly opposed the perceived determinism of other workers; Duncan (1980) argued that members of the school tended to reify culture, however, promoting a form of 'cultural determinism' rather than seeing culture as a human creation that enables and constrains human agency and is itself constantly being reworked through that agency.

Another criticism of this dominant view of cultural geography is that it very largely ignores those elements of culture which are directly related to the production of goods and services, so that cultural geographers had little to offer either the growing number of spatial scientists interested in economic and urban geography or those promoting social geography, with its major interest in the concept of social class. Spencer and Thomas's introductory textbook (1973, p. 6) provided the following definitions:

> *Culture* is the sum total of human learned behavior and ways of doing things. Culture is invented, carried on, and slowly modified by people living and working in groups as each group occupies a particular region of the earth and develops its own special and distinctive system of culture. Related single *culture traits* that go together in practice form a *culture complex:* a large assemblage of culture complexes fit together into a *culture system.* A culture system followed by a population inhabiting a specific area of the earth forms a *culture region.* A group of related culture regions is termed a *culture world.*

But their book concentrated on artefacts and certain aspects of social organization (religion and language) only, saying nothing about many of the salient aspects of 'the sum total of human learned behavior'. Zelinsky's

(1973b) *The Cultural Geography of the United States* is similarly partial, though it introduces more material on the political content of societal organization; its focus is sustained in a later compilation presented as *An Atlas of United States and Canadian Society and Cultures* (Rooney, Zelinsky and Louder, 1982).

This subdiscipline has been criticized on a number of grounds. Cosgrove (1983, p. 3) claimed that because it ignored the dialectic between nature and culture 'it dissolved into either the idealist reification of culture as an agent of change or a semi-determinism dignified by the name "possibilism" . . . This has left cultural geography theoretically impoverished, many of its studies existing in a theoretical vacuum' (p. 3) and he identified it as 'so diffuse that one is tempted to characterize cultural geography more by its refusal to adopt economic or social theory as its guiding principle than by any unity of aim or method'. The economic and social theories that he promoted (discussed in Chapters 7 and 8 of this book) are not those of spatial science. But spatial scientists, too, found cultural geography atheoretical and empiricist in explicit orientation, yet often, as Duncan (1980) indicates, implicitly determinist. The links between the two developed hardly at all and, in addition, there was what Duncan (1994, p. 401) identified as 'a civil war' within cultural geography since the early 1980s, during which:

> the older generation has been content to remain in their well-entrenched positions within the academy as the younger generation fired the occasional salvo at them to little apparent effect. The younger generation launched its attacks using an arsenal of theory-seeking weapons provided by suppliers in the humanities and social sciences.

The older generation was almost entirely North American; the younger had substantial membership from the UK.

Finally, the situation in *regional geography* was surveyed by Paterson (1974), whose essay had two main sections – 'On the problems of writing regional geography' and 'Is progress possible in regional geography?'. The first investigated six problems, including the growing shortage of subordinate materials (micro-regional studies), and the increasing submergence of regional distinctiveness, though

> only a certain amount of innovation is possible if the regional geographer is to perform his appointed task, which is to convey to his reader the essentials of his region; to illuminate the landscape with analytical light. Landforms and climate are common to all terrestrial landscapes . . . and human activities to most of them: how shall repetition be avoided? (p. 8).

'So long as contrasts between region and region [remain], and no matter to what they are attributable, there is work for the geographer to do' (p. 16).

He did not conclude that there is no possibility of progress, despite the constraints: regional geography can advance on two criteria – content and
insight. Reference to Zelinsky's (1973b) book illustrated the increased
range of content currently being introduced; discussion of Meinig's work
(e.g. Meinig, 1972) showed the ability of regional geographers to produce
fresh spatial insights, although Paterson concludes that 'Adventurousness is
not a quality that most of us associate with regional geography' (p. 9).
Thus:

> The way is open for regional studies which are less bound by old for-
> mulae; less obliged to tell all about the region; more experimental
> and, in a proper sense of the word, more imaginative than in the past,
> and covering a broader range of perceptions, either popular or spe-
> cialist (p. 23).
>
> Regional Geography['s] . . . goals are general rather than specific; it is
> not primarily problem-orientated but concerned to provide balanced
> coverage, and its aims are popular and educational rather than practical
> or narrowly professional. Such relevance as it possesses it gains by its
> appeal . . . to the two universal human responses of wonder and con-
> cern. . . . One may recall Medawar's assertion that in science we are
> being progressively relieved of the burden of singular instances, the
> tyranny of the particular, and in turn assert that there is a frame of mind
> on which the particular exercises no tyranny, but a strong fascination
> (p. 21).

Mead (1980, p. 297) strongly defended geographers who 'adopt other
lands . . . share other cultures . . . [and] make a contribution to the store of
knowledge about them' and Hart (1982, p. 29) argued that an important
element in the 'highest form of the geographers' art' requires them to
'adopt a region, to immerse themselves in its culture, to acquire a specialist
understanding of it'.

The implication is that there was a strongly perceived difference between,
on the one hand, many historical geographers and, on the other, most cultural and regional geographers with regard to the degree to which they felt
'left behind' or 'relatively untouched' by the changes that occurred during
the 1950s and 1960s in other branches of human geography. The former, at
least, seem to have been impelled to consider the possibility of making
methodological changes whereas the latter continued to work within their
established tradition (see also Mikesell, 1973). Not all historical geographers would agree with Baker's analogy from systems theory that simply
'Historical geography has a long relaxation time' (p. 11), however, and
Chapter 6 indicates the degree to which they mounted attacks against the
positivist approach.

Conclusions

Reaction to the regional approach began to take shape in the United States during the mid-1950s. The aims of research in human geography were not debated until then, and relatively traditional definitions of the field were observed: the main issues concerned means and methods. The innovations of the period involved the strengthening of the systematic and topical geographies, and their release from a largely subservient relationship to regional geography, by attempts to develop laws and theories of spatial patterns, using models of various kinds for illumination, and applying mathematical and, especially, statistical procedures to facilitate the search for generalizations. Whereas the regionalists saw geography as, at most, law-consuming, those of the new persuasion aimed at producing their own laws, which could be used to explain particular regional outcomes.

These changed means to the geographic end were rapidly accepted in many branches of human geography, particularly in those topical specialisms dealing with economic aspects of contemporary life. They were soon accepted in the growing field of contemporary social geography, but were relatively ignored in historical geography and almost completely shunned in cultural and regional investigations, which, it was claimed, focus on unique characteristics of unique places. They also spread into the corresponding fields of study across the Atlantic (and across the Pacific, too), and within little more than a decade British geographers produced a major review volume containing 816 pages of testimony to the innovators' enthusiasm and their links (largely one-way) with other social sciences. But methods are insufficient to sustain an academic revolution unless they can be applied to a coherent substantive core, and the search for such a core is the subject of the next chapter.

4

The search for a focus: spatial analysis and spatial science

The changes in human geography which emanated from several centres in the United States during the 1950s were very much concerned with methods of investigation. The ultimate aim of geographical study – as set out in Hartshorne's revised definition (p. 58) – remained the same: indeed, the definitions offered by Ackerman (1963) and the NAS/NRC committee (1965: see above, p. 85) suggest that human geographers had become even more ambitious in their hopes of explaining 'the world-wide system comprising man and his natural environment'. But within this general ethos, the proximate aims of geographical investigation were not always clear. Systematic studies were in the ascendant, and the implicit intent was to develop valid laws and theories within the unstated (perhaps unrecognized) positivist framework, but what the exact content of those laws and theories should be was not immediately apparent.

In furthering their analyses, human geographers increasingly sought a clear identity of their own within the social sciences, alongside economics, sociology and political science. Adoption of the positivist philosophy required that disciplines be identified by their content rather than their methods, however. Geographers argued that their discipline could provide a particular viewpoint and contribution to the overall goal of the group of disciplines with which they sought common cause, which involved adopting a new focus as well as a new methodology. This new focus was the spatial variable and the study of spatial systems.

Spatial variables and spatial systems

Geography is a discipline in distance, according to a Scottish professor's inaugural lecture (Watson, 1955; Johnston, 1993a), with the relative location of people and places as its central theme. Cox (1976) later argued that the contemporary importance of relative location within society stemmed from alterations in societal structure consequent upon technical change. The main interactions in primitive societies are between relatively isolated groups of individuals and their physical environments, making relationships between societies and 'a spatially differentiated nature' (p. 192) an obvious focus for geographical scholarship. With technological advancement, however, the main links are among individuals. Interdependence within and between societies increases as a consequence of the more complex differentiation between places which reflects the division of labour, so that the most important facts in modern human existence relate to spatially differentiated societies, not to a spatially differentiated nature. This interdependence between groups living in different places creates the patterns of human occupance on the surface of the earth, and provides the basic subject matter for human geographers.

The focus on spatial arrangements or spatial structures (the areal differentiation in human activities and the spatial interactions which this produces), and on the role of distance as a variable influencing the nature of those arrangements, was highlighted in the new textbooks of the 1960s and 1970s which summarized the contemporary activities of human geographers for the next generation of students. A pioneer among such texts was Haggett's (1965c), whose depiction of pattern and order in spatial structures was phrased within a decomposition of nodal regions into five geometrical elements; a sixth was added in the second edition (Haggett, Cliff and Frey, 1977).

The geometrical elements in Haggett's schema (Figure 4.1) assume a spatially differentiated society within which there is a desire for interaction;

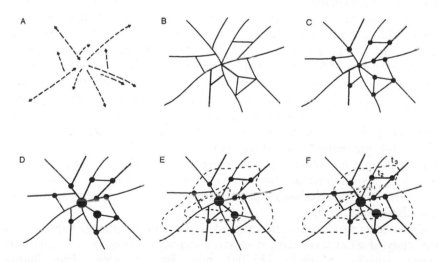

Fig 4.1 The elements in Haggett's schema for studying spatial systems: A movement; B channels; C nodes; D hierarchies; E surfaces; and F diffusion (source: Haggett, Cliff and Frey, 1977, p. 7).

people in place X want to trade with those in place Y, for example, whereas those in place Z want goods and services which they themselves cannot provide. This results in patterns of *movement* – of goods, of people, of money, of ideas and so on – between places, and so the first step in the analysis of nodal regions involves representing the patterns of movement. Some movement is unimpeded – aircraft can move in all directions – but most is channelled along particular route corridors. Thus the second element in the analysis involves characterization of the movement *channels* or networks. Networks comprise edges and vertices; in a transport system, many of the latter are the *nodes,* the organizational nexus. Their spatial arrangement forms the third element, and the fourth investigates their organization into *hierarchies,* which define the importance of places within the settlement framework. Finally, in the original scheme, there are the *surfaces,* the areas of land within the skeleton of nodes (settlements) and networks (routes) which are occupied by land uses of various types and intensities.

Patterns in the human occupation of the earth's surface change frequently, and the spatial order to such changes forms the sixth element in Haggett's revised schema. Change does not occur uniformly over space in most circumstances, however; it usually originates at one or a few localities from whence it spreads to others (as with the 'new' geography in the United States discussed in the previous chapter), along the movement channels, through the nodes, across the surfaces and down the hierarchies. The processes of change over space and time thus involve spatial *diffusion.*

Haggett stressed his conception of geography as a science of distributions, emphasizing the regularities in various elements of these distributions. The first edition was very reliant on work in other disciplines; the second illustrated the amount of work done by geographers in the following decade. In his collection of semi-autobiographical essays, Haggett (1990) continually reiterated his view of geography as an 'emphasis on space and on geometry' (p. 5), which he expressed as personal, 'though I hope not wholly ... quirky'. Maps, and their demonstrations of spatial structure, form the foundation of his geography, as abundantly illustrated by his four essays on 'Levels of resolution', 'The art of the mappable', 'Regional synthesis' and 'The arrows of space'. Thus although he recognized that 'there are other delights in geography' (p. 184), he retained a 'simple delight in the beauty of geographical structures and the challenges posed in finding them and mapping them' (see also Gould, 1993).

Other texts produced in that period took a similar general approach to the study of what were termed spatial systems, although emphasizing different aspects. Morrill's (1970a) title, for example, *The Spatial Organization of Society,* clearly emphasized his view of the role of geographical analysis in the larger task of the social sciences – 'understanding society'. According to him, the core elements of human geography are:

'Space, space relations, and change in space – how physical space is structured, how men relate through space, how man has organized his society in space, and how our conception and use of space change' (p. 3).

Space has five qualities relevant to the understanding of human behaviour in this context: (1) distance, the spatial dimension of separation; (2) accessibility; (3) agglomeration; (4) size; and (5) relative location. Put together, these can be used to build theories, such as those on which Garrison based his work:

> Virtually all theory of spatial organization assumes that the structure of space is based on the principles of minimizing distance and maximizing the utility of points and areas within the structure, without taking the environment, or variable content of space, into account. Although the differential quality of area is interesting and its effect on location and interaction is great, most of the observable regularity of structure in space results from the principles of efficiently using territory of uniform character. The theoretical structures for agricultural location, location of urban centres, and the internal patterns of the city are all derived from the principle of minimizing distance on a uniform plane (Morrill, 1970a, p. 15).

Thus all decisions about the use of land and about locations are taken in order to minimize movement costs. The spatial approach to understanding society assumes a world in which one variable – distance – is a dominant influence on human behaviour and it seeks to account for observed spatial patterns within this framework. And so in Morrill's book:

> the explanation of spatial structure proceeds from the deductive – what would occur under the simplest conditions – to the inductive – how local factors distort this 'pure' structure. To begin with, all the local variation may introduce is a risk of missing the underlying structure. Modern theory of location therefore stresses the spatial factors – above all, distance – which interact to bring about the regular and repetitive patterns (p. 20).

Morrill's organizing framework for geographical scholarship focused on distance. His summarizing 'theory' of spatial structures proceeds as follows:

1 Societies operate to achieve two spatial efficiency goals:
   (a) to use every piece of land to the greatest profit and utility; and
   (b) to achieve the greatest possible volume of interaction at the least possible cost.
2 Pursuit of these goals involves four types of location decisions:
   (a) the substitution of land for transport costs when seeking accessibility;
   (b) substituting production costs at sites for transport costs when seeking markets;

   (c)   substituting agglomeration benefits for transport costs; and
   (d)   substituting self-sufficiency (higher production costs) and trade
        (higher transport costs).
3 The spatial structures resulting from these decisions include:
   (a)   spatial land-use gradients; and
   (b)   a spatial hierarchy of regions.
These are somewhat distorted by environmental variations to produce:
   (c)   more irregular but predictable patterns of location.
Whereas over time distortion may result from:
   (d)   non-optimal location decisions; and
   (e)   change through processes of spatial diffusion.

Like Haggett, therefore, Morrill stressed the geometry of human organization of activities on the earth's surface, but whereas Haggett emphasized pattern and geometry Morrill paid more attention to the decision-making processes which would produce the most 'efficient' pattern, as an underlying basis for the imperfect examples of that pattern which are observed in the 'real world'. Other texts (such as Abler, Adams and Gould, 1971, which has a stronger focus on positivist methodology: see also Gould, 1977, 1978) followed Morrill's lead. The importance of the spatial variable is less dominant in the third edition of the book, however (Morrill and Dormitzer, 1979), and location theory is presented as providing:

> fairly simple models that permit us to highlight some essential principles and factors of human location. The real world, of course, does not correspond very closely to the patterns projected by location theory because the human landscape is the complex product of many different forces – historical, physical, cultural, political, and behavioral, as well as economic (spatial).

Later, however, in a paper on 'Some important geographic questions' Morrill (1985) returned to a strong focus on spatial variables.

These textbooks illustrate the centrality of space and distance as the major focus of geographical interest in the 1960s. The emphasis on pattern was noted in King's (1969b, p. 574) major review of 'The geographer's approach to the analysis of spatial form ... the mathematics which are used and the geometrical frameworks which are favored'. He focused on descriptive mathematics, which represent 'what is' rather than 'what should be', realizing that 'when they are pursued to their extremes in very formal terms these studies run the risk of appearing as seemingly sterile exercises in pure geometry' (p. 593; see also King, 1976, 1979b); nevertheless, geographers had by then not proceeded very far in providing process theories which would account for observed spatial patterns. The school of thought which he was reviewing involved working backwards, finding what order there was to explain rather than deducing what the world should be like from knowledge of human behaviour.

That school was also partial in its treatment of space, treating it as a continuous variable and virtually ignoring its discontinuous nature in patterns of human organization. Bounded spaces on a great variety of scales, from the largest nation-state to the 'bubble of personal space' around each individual, are probably more characteristic of human manipulation of their environments than are settlement hierarchies, but they were largely ignored by geographers promoting their discipline as spatial science, perhaps because generalizations were more difficult to identify, let alone explain. Their study had been advanced by a French political geographer (Gottmann, 1951, 1952), whose work had no impact on his American contemporaries (Johnston, 1996b): despite his periods of working there, he was 'outside the project'.

Spatial theory

Just as the methodological developments reviewed in the previous chapter lacked clear guidelines from programmatic statements, so the growth of the spatial viewpoint similarly lacked any manifesto. (Watson's 1955 paper was not widely referenced: Johnston, 1993a.) The only attempt to provide a lead – apart from general statements about geography and geometry, such as Bunge's (1962) – was a paper by Nystuen (1963) which was not widely read until its reprinting in 1968. His objective was 'to consider how many independent concepts constitute a basis for the spatial point of view, that is, the geographical point of view' (p. 35 – all page references are to the 1968 reprint) so that rather than look at the 'real world', with its many distorting tendencies, he sought clarity in considering abstract geographies.

To illustrate his deduced basic concepts, Nystuen used the analogy of a mosque completely lacking furniture (i.e. an isotropic plain) in which a teacher chooses to occupy a location at random. The students then distribute themselves so that they can see and hear him; their likely arrangement is in semi-circular, staggered rows facing and as close as possible to the teacher. This arrangement has three characteristic features:

1 *directional orientation* – they all face the teacher, to perceive expressions and to hear better;
2 *distance* – they cluster, because audibility diminishes with distance; and
3 *connectiveness* – they arrange themselves in rows, so organized that each has a direct line of sight to the teacher.

The third of these, connectiveness, is in part a function of distance and direction, but not entirely so. As Nystuen expresses it (p. 39):

> A map of the United States may be stretched and twisted, but so long as each state remains connected with its neighbors, relative position

does not change. Connectiveness is independent of distance and direction – all these properties are needed to establish a complete geographical point of view.

In addition,

> Connections need not be adjacent boundaries or physical links. They may be defined as functional associations. Functional associations of spatially separate elements are best revealed by the exchanges which take place between the elements. The exchanges may often be measured by the flows of people, goods, or communications.

Within the mosque, therefore, the connectivity between teacher and student involves not only a direct line of sight between them but also a direct flow from one to the other, in this case, of ideas.

These three concepts – direction, distance and connectiveness – were presented by Nystuen as both necessary and sufficient for the construction of an abstract geography, grounded in the study of sites (abstract places) rather than of locations (real places).

> The terms which seem to me to contain the concepts of a geographical point of view are *direction or orientation, distance*, and *connection or relative position*. Operational definitions of these words are the axioms of the spatial point of view. Other words, such as pattern, accessibility, neighborhood, circulation, etc., are compounds of the basic terms. For abstract models, the existence of these elements and their properties must be specified (p. 41).

Nystuen was unsure whether these three comprised the full set of necessary and sufficient concepts for a geographical argument (see also Papageorgiou, 1969); boundary, he felt, might be a primitive concept and not a derivative of the basic three, and bounded space was of much less interest to spatial theorists than continuous space (albeit often transformed: Haggett, 1990). Nystuen's general case that arguments in human geography could be based on a small number of such concepts was implicitly accepted by much of the work in 'the spatial tradition', although rarely explicitly referred to.

Haynes (1975) suggested an alternative approach to writing theory in human geography, based on the mathematical procedures of dimensional analysis. Five basic dimensions were defined – mass, length, time, population size, and value – and were manipulated to indicate the validity of functional relationships, such as distance-decay equations (see p. 111), by checking their internal consistency. This approach is defended by the statement that:

> Although most quantitative geographers would probably claim to be engaged in the discovery of relationships, it appears that geography has not passed the first stage [in the development of a science] with

any degree of rigor. . . . With no clear idea of which variables are relevant and which particular characteristics in a system should be isolated, it is pragmatic to define our measurement scales with regard to a particular set of observations rather than the other way round. The method of physical science . . . is a superior system, as measurements can be interpreted exactly, different results compared, and experiments replicated (p. 66).

This is clearly a deductive approach to research. It is not as closely tied to a 'spatial view' as Nystuen's but has the same goal – the derivation of a set of fundamental concepts which can form the basis for writing geographical theory to be tested in the 'real world' (see also Haynes, 1978, 1982).

These attempts at isolating human geography's primary concepts differed from most contemporary efforts at producing geographic theory which, according to Harvey (1967b, p. 212), were 'either very poorly formulated or else derivative'. Central place theory, for example, was based entirely on postulates from economics about how people behave as 'rational economic actors', to which the basic geographic concept of distance was added, producing a theory about the size and spacing of settlements. The attraction of central place theory to many was indeed that geographers could apparently contribute their own basic concept to theory development, and need not be totally dependent on other disciplines for concepts. Harvey (1970) identified a group of geographical concepts which could form indigenous elements in building integrated social science theories which drew concepts of equal status from all component disciplines – location, nearness, distance, pattern, morphology; most are compounds of Nystuen's basic terms.

An example of the development of theory involving both derivative and indigenous concepts is provided by research on spatial diffusion, which received considerable attention in the late 1960s (Brown, 1968). The basic behavioural postulate, taken from sociological research, was that word of mouth is the most effective form of communication about innovations. Geographers expanded this by introducing the effect of distance: most interpersonal communication is between neighbours, so information about innovations should spread outwards in an orderly fashion from the locations of the initial adopters. Pioneer work on this hypothesis was conducted in Sweden by Hagerstrand, who introduced it to the Washington school in the 1950s, where it was taken up by Morrill (1965, 1968), whose doctoral dissertation research was undertaken in Sweden: Hagerstrand's own major work was made available in an English translation in 1968. Much subsequent work has been done on the process of diffusion and, even more, on patterns of spatial spread which, it is assumed, result from diffusion processes (see Abler, Adams and Gould, 1971, Chapter 11); Brown (1981) provided an extended review of this literature (see also p. 115).

Work on spatial diffusion increasingly became concentrated on topics in medical geography, in which the concept of spatial spread across space and down settlement hierarchies was valid (see Thomas, 1992). The lead in much of this work was provided by Haggett and Cliff and included classic spatial epidemiological investigations of influenza, measles and AIDS as well as more general studies (e.g. Cliff and Haggett, 1988). Changes in communications technology, among others, have had substantial impacts on these flows, introducing new technical issues (Cliff and Haggett, 1995).

Social physics and spatial science

Two of Nystuen's three basic concepts – distance and connectivity – received much more attention from those advocating geography as a spatial science than did the other. Direction was relatively ignored, except for some work on migration patterns (Wolpert, 1967) which included a seminal statement by Adams (1969). (Direction is used here in cardinal terms. To some extent, any discussion of movement patterns which identified destinations more precisely than in terms of distance from origins involved directional analysis.)

By far the greatest volume of work on spatial science followed the research directions established by the social physics school (p. 70). The relationship between distance and various types of interaction – migrations, information flows, movements of goods, etc. – had been identified by several workers in the nineteenth century, such as Carey (1858), Ravenstein (1885; Grigg, 1977; Tobler, 1995), and Spencer (1892). Their impact is unclear. McKinney (1968), for example, suggested that Stewart and others were unaware of the seeds of the 'gravity model' and 'population potential' ideas in Carey and Spencer's writings, and claimed that 'current geographers could learn much' (p. 105) from their publications. Warntz (1968) retorted that Stewart was well aware of such writings, although McKinney's rebuttal pointed out that Stewart referred to them in his 1950s' papers, but not in those of the 1940s. Ravenstein's papers, on the other hand, were very influential on later research into migration patterns. (Interestingly, the pioneering work on distance-decay and the gravity model, as with much of the rest of spatial science, was done outside geography: as Tocalis, 1978, p. 124 expresses it, 'geographers' contributions to the theoretical evolution of the gravity concept were minor', although one major contributor – Alan Wilson – 'became a geographer'.)

Apart from Stewart, whose pioneering efforts were referred to in the previous chapter, the most influential figure on social physics after World War II was probably Zipf (1949), who devised a 'principle of least effort'. Individuals organize their lives so as to minimize the amount of work which they must undertake. Movement involves work, and so the minimization of movement is part of the general principle of least effort. (These principles underpinned the central place models developed by Christaller and Losch – see p. 67 – but were not identified as such by the social physicists.) To explain this, Zipf expanded on Stewart's finding that with increasing distance from Princeton, fewer students from each state attended that university. Two aspects of work are involved in going to university: (1) the work involved in acquiring information about the university; and (2) the work involved in travelling there. Thus the greater the distance between any potential student's home and Princeton, the less that is likely to be known about the university and the less prepared they will be to travel there. The validity of this distance-decay argument was tested against many other data sets: material on the contents of newspapers and on their circulation illustrated the expected distance bias in flows of information, for example, and data on movements between places showed that the greater the distance separating them, the smaller the volume of inter-place contact.

Zipf called the regularity which he had identified the P_1P_2/D relationship, and Stewart noted the analogy between this and the Newtonian gravity formula. The 'gravity model' was fitted to flow data in many research projects, as shown in Carrothers' (1956) and Olsson's (1965) reviews. To achieve reasonable statistical fits, the various elements of the equation had to be weighted, and because different weights were determined empirically in almost all studies, Olsson (1965, p. 48) claimed that the gravity model of interaction was 'an empirical regularity to which it has not yet been possible to furnish any theoretical explanation' (Wilson, 1967, offered a statistical explanation). In effect, the influence of distance apparently varied from place to place, from population to population, and from context to context. The influence of distance on flow volumes seemed virtually universal; what no theory could account for was the variability in the strength of its impact.

It was not only in social physics, and the work of Stewart, Zipf and others, that distance was receiving attention at this time. Economists and sociologists were becoming increasingly aware of its influence on behaviour (Pooler, 1977), the former in the location theories of Weber, Hoover, Losch and others, which stimulated the work of Garrison and his associates (p. 68), as well as Isard's development of regional science, and the latter in studies by the Chicago school, which stimulated a large literature on urban residential patterns (Johnston, 1971, 1980b). Thus (Pooler, 1977, p. 69):

> a number of geographers became aware of the spatial enquiries that were being undertaken in a social science context outside their own discipline and, upon realizing their relevance to geography, proceeded to emulate them. The appearance of the spatial tradition was prompted, not by discoveries from within geography, but by an awareness and acceptance of investigations external to the discipline.

The space-centred scientific enquiries of other social sciences became paradigms for geographers, simply because those enquiries, being spatial, were seen to be of relevance by some practitioners.

This fitted both the philosophical framework outlined by Schaefer (1953) and the 'quantitative revolution'. In adopting spatial viewpoints pioneered in other disciplines, however, human geographers were frequently selective in what they imported. Work on urban residential patterns concentrated on certain aspects of the urban sociology of the Chicago school, for example. Robert Park's dictum relating social distance to spatial distance stimulated much work on the measurement of residential separation (Peach and Smith, 1981), but the social Darwinism and ecological theory underlying this dictum was largely ignored (though see Robson, 1969; Entrikin, 1980) and the humanistic concerns in Park's work were only identified later when spatial science was losing popularity (Jackson and Smith, 1981, 1984).

In their distance-based analyses, various social scientists looked not only at the influence of distance but also at its meaning and measurement. Stouffer (1940), for example, established a relationship in which migration between X and Y was accounted for not so much by the distance between them but rather by the number of intervening opportunities. He measured distance in terms of opportunities; the greater the number of opportunities available locally, the less work that has to be expended in moving to one. (See Berry's, 1993, reference to contacts with Stouffer when he was a student at the University of Washington.) Others took up this flexible approach to measurement of the basic variable. Ullman (1956), for example, developed a schema for analysing commodity flows in which the amount of movement between two places was related to three factors:

1 *complementarity* – the degree to which there is a supply of a commodity at one place and a demand for it at the other;
2 *intervening opportunity* – the degree to which either the potential destination can obtain similar commodities from a nearer, and presumably cheaper, source or the potential source can sell its commodities to a nearer market; and
3 *transferability* – for complementarity to be capitalized on, movement between the two places must be feasible, given channel, time and cost constraints.

This schema was not as easily fitted statistically as the 'gravity model', to which its relationship is clear (Hay, 1979b). Fitting such models also requires accepting that the influence of distance, measured as either time and cost, varies from place to place and from time to time (Abler, 1971; Forer, 1974: Janelle, 1968, 1969).

These analyses of movement patterns were stimulated not only by their obvious relevance to geography's growing spatial science focus, and its development of location theories, but also by their clear applicability in forecasting contexts. The planning of land-use patterns and transport systems (especially road systems) became increasingly sophisticated technically during the 1950s and 1960s, first in the United States and then in Britain. Initially, data were collected to show both the traffic-generating power of various land uses and the patterns of interaction between different parts of an urban area, with the gravity model being used to describe the latter. Future land-use configurations were then designed, their traffic-generating potential derived, and the gravity model used to predict flow patterns and identify needed road systems. Later models, most based on Ira Lowry's, assessed different land-use configurations in terms of traffic flows, thereby suggesting the 'best' directions for future urban growth (see Batty, 1978; on the extent of their application, see Batty, 1989).

The demands for sophisticated planning devices stimulated much research using the gravity and Lowry models. American economists initiated this but it was later taken up by British workers, including Alan Wilson, who was appointed to a professorship of geography at the University of Leeds in 1970, thereby becoming a professional geographer although he had no training in the discipline. (He was a physicist who became interested in traffic forecasting when a city councillor in Oxford: Wilson, 1984b.) Wilson (1967) derived the gravity model mathematically, thereby giving it a stronger theoretical base, and extended Lowry's model into a more general suite concerned with location, allocation and movement in space (Birkin *et al.*, 1990: see below, p. 122; these later became the foundation of a major commercial operation at the University of Leeds – GMap Ltd.)

Collaboration between academic geographers and practising planners on these modelling exercises led to developments which paralleled those in regional science in the United States. Two new British journals catered for these research areas – *Regional Studies* and *Environment and Planning*: both attracted contributions and readers from other social sciences (Johnston and Thrift, 1993). The new geographical methodology was proving of considerable applied value, therefore. Careers as planners became extremely popular among geography graduates for a few years, notably in the late 1960s and early 1970s, and the growing desire of academic geographers for their discipline to equal that of other social sciences in its policy-making relevance seemed well on the way to fulfilment.

Spatial science and spatial statistics

The developments just outlined were characterized by a large volume of technical research, concerned with a range of problems associated with the description of spatial patterns and connectivities (for example, Haggett and Chorley, 1969). Although Bunge (1962), Harvey (1969a) and others argued that geometry provides the language for analysing spatial form, much of this work concentrated on applying descriptive and inferential statistics to spatial problems, in line with the strong empirical tradition in geographical work.

Most of the researchers involved, including Garrison (1956a), accepted that statistical procedures developed for other fields of enquiry could be adopted for geographical investigations without difficulty. Some suggested necessary modifications – as in spatial sampling (Berry and Baker, 1968) – but, despite Robinson's (1956; see p. 65) early work, it was generally assumed that there were no technical problems involved in applying standard procedures to spatial data sets. A number of student texts appeared, especially in the late 1960s, which made little reference to any peculiarities of geographical data; they differed from the texts produced by other social scientists only in the nature of their examples. (Even in the 1980s, few contributors to a book on *Recent Developments in Spatial Data Analysis* – Bahrenberg *et al.*, 1984 – discussed spatial data *per se*; most of their work referred to the use of standard statistical procedures in geographical applications.)

During the late 1960s, however, researchers at the University of Bristol began to question the widespread assumption about the relevance of standard statistical procedures in geographical investigations. Statisticians had for long recognized difficulties in applying the general linear model – notably in its regression and correlation form – to the time-series data used in economic analysis and forecasting. The major issue was autocorrelation. One of the model's assumptions is that all observations are independent of each other; the magnitude of one reading on a variable should in no way influence that of any other. Such autocorrelation clearly exists in most time series: the value of the retail price index at one date, for example, strongly influences its value at nearby later dates, and is itself influenced by earlier values. (This issue was first brought to geographers' attention in the discussion in Chisholm, Frey and Haggett, 1971, p. 465.) Because of such interdependence between adjacent observations, conventional regression methods could not be used: autocorrelation led to biased regression coefficients and cast doubt on the validity of any forecasts.

The Bristol group argued not only that this autocorrelation problem applied to spatial data sets, but also that it was much less tractable than with temporal series. Time proceeds in one direction only, but space is two-dimensional, and the independence requirement can be violated in all directions around a single point. Spatial autocorrelation can therefore involve all neighbours influencing all others with regard to the values of a particular variable, as recognized by some statisticians (e.g. Geary, 1954) and hinted at by geographers such as Dacey (1968), whose work stimulated the Bristol investigations. (Andrew Cliff worked as a graduate student with Dacey at Northwestern University in the early 1960s.) Considerable effort was expended on this problem over a number of years (Cliff and Ord, 1973,

1981), covering a wide range of spatial applications, including the well-known gravity model (Curry, 1972; Johnston, 1976a; Sheppard, 1979; Fotheringham, 1981).

The spatial autocorrelation problem severely constrains the application of conventional statistical procedures in geographical analyses, and several people argued that application of regression methods, and others based on the general linear model such as principal components and factor analyses, was invalid with spatial data sets (Haining, 1980). As Harvey (1969a, p. 347) put it:

> The choice of the product-moment correlation coefficient for regionalization problems appears singularly inappropriate, since one of the technical requirements of this statistic is independence in the observations. Since the aim of such regionalization is to produce contiguous regions which are internally relatively homogeneous, it seems almost certain that this condition of independence in the observations will be violated.

Gould (1970a) noted that spatial autocorrelation reflects the order that geographers were seeking to establish with their laws and theories, and it was somewhat paradoxical that its presence prevented its identification in the accepted technical modes. On realizing the force of the case, however, some accepted that conventional statistical procedures could not be applied in their work (e.g. Berry, 1973b) and the Bristol group omitted them from their rewrite of Haggett's major text (Haggett, Cliff and Frey, 1977). Most continued to apply the methods, however, either in ignorance of the autocorrelation case or on the grounds that the biased-coefficients problem refers only to the use of the general linear model in forecasting and prediction, and does not affect use of the procedures for description (see Johnston, 1978a) and for certain types of ecological analysis (Johnston, 1982c, 1984f).

The work undertaken by the Bristol group concentrated on spatial forecasting, however, involving the development of procedures for estimating how trends – in the spread of a disease, for example, and in prices and unemployment – would proceed through time and over space (Haggett, 1973; Cliff et al., 1975). They organized a major symposium on this (Chisholm, Frey and Haggett, 1971) and stimulated a considerable volume of work, much of it technical, on the problems of identifying and forecasting spatio-temporal trends (e.g. Bennett, 1978b). Hagerstrand's work on diffusion patterns provided the foundation for much of this research, which focused on the patterns of spread rather than on their generating processes and generated a large literature on the technical aspects of forecasting (e.g. Bennett, 1979) and on deducing 'spatial processes' from mapped patterns (Haining, 1981, 1990).

Hay's (1978) review of this spatial forecasting work challenged what he saw as its fundamental assumption, that phenomena behave coherently in both time and space. He questioned the assumed stability of inter-place inter-relationships over time in analyses such as Martin and Oeppen's (1975) on market-price variations, and argued instead for consideration of the value of catastrophe theory wherein small changes in the control variable(s) can stimulate major changes in the dependent variable being studied (such major changes are the catastrophes). If catastrophes occur, then the linear extrapolations typical of the Bristol group's work have clear limitations as forecasting procedures; there have been relatively few geographical applications of this relatively new area of mathematics, however (Wilson, 1976a, 1981a: it has received more attention among physical geographers – Thornes, 1989a, 1989b).

Technical advances in modelling and analysing geographical data continued. The spatial peculiarities of geographical data were stressed by some, arguing that particular geographical procedures are needed (see Gaile and Willmott, 1984). One of those peculiarities links the issues of aggregation and scale. Openshaw (1984a, 1984b) showed in a number of important papers how different spatial aggregations of the same data set (counties into regions, perhaps, or census enumeration districts into social areas) can produce very different correlations between two variables; indeed, his simulations suggest that with many data sets any correlation between +1.0 and −1.0 might be obtained, so that the particular aggregation employed needs careful justification (see Wrigley, 1995). Others sought to ensure proper use of the standard statistical procedures (e.g. Jones, 1984; Wrigley, 1984), with the use of statistical significance tests forming the basis of further debate (Summerfield, 1983; Hay, 1985b) and the merits of exploratory data analysis being extolled (Cox and Jones, 1981). Better ways of testing hypotheses were also expounded, notably with regard to the use of categorical (or nominal) data (e.g. Fingleton, 1984; Wrigley, 1984) but also in terms of pattern analysis (Upton and Fingleton, 1985), representing relationships through structural models (e.g. Cadwallader, 1986), and identifying place-by-place differences in general relationships using multi-level modelling (Jones, 1991). Much of this work has been criticized for its reliance on methods that impose structures on data rather than 'letting the data speak for themselves' (Gould, 1981a); the mathematics of q-analysis was proposed to meet the latter criterion (Gould, 1980; Beaumont and Gatrell, 1982; Gatrell, 1983: see also the essays edited by Macgill, 1983) but was not taken up.

Various aspects of spatial interaction continued to be a focus of formal modelling work, as in major studies of the spatial diffusion of diseases (Cliff *et al.*, 1981a; Cliff, Haggett and Ord, 1987; Cliff, Haggett and Smallman-Raynor, 1993; Smallman-Raynor, Cliff and Haggett, 1992). Bennett and Haining (1985; Bennett, Haining and Wilson, 1985) placed such work at the forefront of methodological developments in the discipline. Much of it was summarized by Wilson and Bennett's (1985) volume in a *Handbook of Applicable Mathematics* series; the inclusion of a book on *Mathematical Methods in Human Geography and Planning* is testimony to the substantial achievements in the field of spatial analysis – though by a relatively small group of workers.

These technical developments have been aided, indeed stimulated, by developments in data-handling capacity with high-speed computers (Rhind, 1989). Traditional data sources, such as censuses and surveys, have been added to by a wide range of others collected by traditional means (such as market research) and those made available by new collection methods such as remote sensing from satellites and other airborne devices. Openshaw (1994, 1995) has argued that the computational power now available can transform the practice of human geography, providing that (1994, p. 504): 'you are neither blinded by past prejudices about science, not scared of the words computational human geography, nor too hyped up by an overenthusiasm for AI or infected by the neural net virus' (on which see Fischer and Gopal, 1993). His approach is empiricist, calling for (p. 503): 'greater intelligence by first becoming less sophisticated in our analysing and modelling technology and then more than compensating by computational intensity'. He rejects (1995, p. 161) 'a qualitative, linguistically defined understanding of how a whole system works in vague and general terms' and promotes 'a detailed and precise computer model of one small part of it', using fuzzy set logic as 'a new scientific paradigm for doing geography' (Openshaw, 1996). Thus he calls for a HPC (high performance computing) culture within geography focusing on HSM (human system modelling), and the University of Leeds has established a Centre for Geographical Computation to advance this work (as illustrated in Openshaw and Rao, 1995).

Spatial data capture, storage, integration, display and analysis have been advanced since the early 1980s by the development of *Geographical Information Systems* (GIS: Curran, 1984; Green *et al.*, 1985; Chrisman *et al*, 1989), which bring together rapid advances in computer cartography and data collection in dedicated machines (comprising both hardware and customized software) for the analysis of the three main types of spatial data: those referring to points (such as mapped locations), to areas (such as towns and fields), and to lines (transport routes, for example). Increasingly data are geocoded (i.e. given unique spatial references which facilitate spatial analysis), and technological advances are leading to the introduction of computers with which it is possible to gather, store, display and analyse spatial data.

Development of this technology in the United Kingdom was heralded and promoted by a major project initiated to mark the nine-hundredth anniversary of the 1086 Norman Domesday Book census. A GIS to provide a modern equivalent was developed and marketed as a teaching aid for schools and other education institutions (Goddard and Armstrong, 1986); its massive data base can be used to display a wide range of maps (including the Ordnance Survey 1:50 000 series for the whole of Great Britain) and pictures, and can also be analysed interactively (Openshaw *et al.*, 1986).

GIS are a major research tool which allows many problems to be addressed that previously could not be afforded in terms of the time and resources necessary for data collection, analysis and display: they are the basis for much applied research work (see p. 143). The British Economic and Social Research Council established a network of Regional Research Laboratories to conduct research into both their potential and their advancement as research tools in the late 1980s, most of them located in university departments of geography (see the October 1988 issue of the *ESRC Newsletter,* which is devoted to the topic 'Working with geographical information systems') and the United States National Science Foundation invested 5.5 million dollars into a National Center for Geographic Information and Analysis, based at three sites (see the announcements in the Association of American Geographers' *Newsletter* for August 1987 and October 1988 and Fotheringham and MacKinnon, 1989). Their use was promoted in teaching (Maguire, 1989; Fisher, 1989a); an Association for Geographic Information was established in the United Kingdom, following a major report by a House of Lords Select Committee for which a geographer (Rhind) acted as scientific advisor (Rhind, 1986; Rhind and Mounsey, 1989); several specialist journals have been launched (such as the *International Journal of Geographical Information Systems*); a bibliography of over 1000 items was published in early 1990 (Bracken *et al.*, 1990); and a major encyclopaedic survey was published in the following year (Maguire *et al.*, 1991: a second, much revised and rewritten edition is due in 1997: for a review of work in urban studies alone, see Sui, 1994).

In his 1989 presidential address to the Canadian Association of Geographers, Roger Tomlinson (1989, p. 298), a pioneer in the development of GIS from the 1960s on, referred to their growing use in the following terms:

> Geographers have a crucial role to play in integrating a wide variety of technologies into new forms of 'earth description' which will act as a foundation for geographical methodology and open the way to richer forms of spatial analysis and geographical understanding.

Spatial problems will multiply and become more complex in the future, and geographers' ability to handle and analyse large bodies of data in the search for solutions, using GIS, should demonstrate the strength of their 'integrating science', as part of a 'data are good ethic' (p. 292: see also p. 349 below). Whereas those involved in the development of GIS have largely focused on the resolution of the many technical issues involved with integrating data bases and their statistical manipulation and cautioned against over-optimism regarding their potential (see Goodchild, 1995), some of its promoters have made grand claims, however, as illustrated by

Openshaw's (1991) reaction to Taylor's (1990a) separation of geographical information systems from 'geographical knowledge systems' (see the discussion on p. 312).

Others have expressed concerns regarding the potential power being sought for applications using these information systems (see Curry, 1994, and several of the essays in Pickles, 1995a). Pickles (1995b, p. 2), for example, argues that GIS are much more than 'merely more efficient counting machines' providing more accurate descriptions and visual images; their 'virtual representations will produce illusions that will be so powerful it will not be possible to tell what is real and what is not real' (p. 10). For others, their use, whether in sophisticated marketing procedures (i.e. geodemographics: see Batey and Brown, 1995; Birkin, 1995; for a critique see Goss, 1995) or military applications (N. Smith, 1992) raise moral and ethical issues (none of which are specific to GIS but apply to any application of 'knowledge'): to Curry (1995) GIS involve ethical inconsistencies which are necessarily built into systems which can be used to advance surveillance (whether by the state or by other bodies) very substantially.

In responding to some of these critiques, Sui (1994) has categorized them into those which stress one or more of the following:

1 *ontological inadequacy* – GIS present a limited, Cartesian representation of reality, ignoring social and cultural representations;
2 *epistemological insufficiency* – GIS are deeply rooted in positivist modes of thinking;
3 *methodological insufficiency* – GIS applications lack coherent theory and are biased by their use of secondary data; and
4 *ethical inconsistency* – the values embedded in GIS are ignored, along with any recognition of subjective differences in representations of reality.

He claimed that the critiques have (Sui, 1994, p. 268): 'at best, been refuted by the GIS community or, at worst, have been totally dismissed as anti-progress and anti-science nuisances'. Examples are presented of GIS researchers who have addressed each of the four issues, showing that (p. 271):

> the GIS community has realized that the implementation of GIS should go beyond mere technical decisions justified by matters of efficiency and give the ethical use of this information a serious consideration. . . . GIS, if not a fertile ground for common search, at least has initiated a search for common ground.

(The last sentence draws on the title of Golledge, Couclelis and Gould's, 1988, collection.) Nevertheless, he concluded by noting that 'so far, GIS enthusiasts and GIS opponents have been mutually hostile or at least dismissive of each other's views' (p. 272) and suggested that closer cooperation is needed in which 'GIS enthusiasts must avoid the imperialistic claim that science is the only guarantor of objective truth. Post-modernists must relinquish the playful cynicism in their critiques on the scientific chauvinism of GIS.'

The advances in computer technology, including GIS, have had very substantial impacts on the practice of cartography and other forms of graphic information display. Only a small number of geographers are involved in research on cartographic methods (see MacEachren, 1995), but their work on visualization has allowed major advances in the computerized presentation of large and complex data sets (see Dorling, 1995).

Dobson (1983a) has argued that dealing with these massive data sets will require the development of *automated geography*, an 'integrated systems approach to geographic problem solving in which the problem is defined, the appropriate methods are chosen, and the tools are selected from a broad repertoire of automated and manual techniques' (p. 136). This would be highly dependent on computer hardware and software developments (sufficient of which are already available; though see Cowen, 1983) and would facilitate participation in public policy research on a large scale. Most commentators on Dobson's paper (in the August 1983 issue of *The Professional Geographer*) argued that automated geography was a misnomer, since the main benefits of the hardware and software that he described related to the scale and speed of data manipulation only. They preferred less emotive terms such as 'computer-assisted geography' or 'computer-assisted geographic systems', and some, like Poiker (1983), argued that the case was no more than one for another set of tools, just as the case for quantification turned out to be decades earlier: 'our tool seems to attract an inappropriate number of prophets' (p. 349). Dobson (1983b) disagreed, however, arguing that 'what I am talking about is not a system or collection of systems. It is a *discipline* which uses human and electronic cybernetic systems to further understanding of physical and social systems' (p. 351) with some decisions (e.g. with regard to how data are displayed) taken by the computers, not by people. Thus computers should be programmed so that they incorporate 'the character of our discipline', presenting the results of geographical analysis without the detailed intervention of geographers (see Couclelis, 1986b; Fisher, 1989b; Haines-Young, 1989).

Dobson's arguments were revisited a decade later. He claimed that (Dobson, 1993, p. 431):

> GIS has become a *sine qua non* for geographic analysis and research in government, business and academia. This advance has been hailed widely as a technological revolution, but I proposed a more exciting prospect beyond technology. The strength of GIS innovation and diffusion suggests that science and society are in the beginning stage of a technological, scientific and intellectual revolution as

profound as earlier revolutions brought on by the printing press or the computer.

Nevertheless, he expressed concerns over the amount of geographical research done using GIS, other than in mapping and survey (p. 432), suggested that 'the GIS community is still searching for a greater sense of purpose' (p. 433), and claimed that 'the current state of GIS is woefully deficient when compared to our predecessors' ultimate goal of representing landscapes as three-dimensional geographic units with interactions of multiple phenomena in space and time, with order' (p. 434). He ended with a call for geographers to become more involved in what was clearly, to him, an unfulfilled revolution (p. 438):

> The technological revolution, understandably, may be led by private companies and government laboratories, and technical leadership may come from many fields. But academic geographers are essential to the scientific and intellectual revolutions. Geographers are needed as consummate leaders in conceptual design and in geographic analysis employing GIS.

Others agreed, while noting that hardware and software developments, although substantial, were still far from those needed for completing Dobson's revolution (Armstrong, 1993; Cromley, 1993). But whereas some were optimistic about the growing use of computers in all aspects of geography (and not just those involving numerical applications: Monmonier, 1993), others were pessimistic. Marble and Peuquet (1993, p. 116), for example, argued that since 1983 'the impact of the computer upon geographic research appears to have been even less than our most pessimistic estimates', in part because of 'a substantial turning away from those activities that GIS can do best' (p. 447). Goodchild (1993, p. 445) appeared to agree: 'Whether or not GIS and automated geography have anything to offer to geographic research, we cannot escape the fact that GIS has had a significant impact on many kinds of human activity.' This was reflected in the subtitle of Sheppard's (1993) contribution – 'What kind of geography for what kind of society?' – and in Pickles' (1993) rehearsal of arguments regarding the impact of technology on geographical and other practices (see p. 306).

The technical advances have not been associated with parallel developments in substantive geographic theories it is argued, therefore, in large part because (as discussed in later chapters) spatial analysis has been somewhat marginalized within the discipline. In some of the continuing work, the 'traditional' models associated with the developments of the 1950s/1960s have provided the basic underpinning, focusing on optimum distributions of points and lines and on flows between those points along the lines. Wilson (1984a) claimed that mathematical development enables us to build new models by combining old problems (for example von Thunen's and Weber's) with new methods, claiming that comprehensive models can now be built combining macro-population backcloth, macro-economic backcloth, spatial interaction, the location of population activities, the location of economic activities, and the development of economic infrastructure into representations of spatial pattern or settlement structure. Two papers 'demonstrate the ability of models of spatial interaction and structure to reproduce the results of classical industrial location theory' (Birkin and Wilson, 1986, p. 305) and they are the foundation for the applied work undertaken at Leeds (Birkin, Clarke and George, 1995), but relatively few workers now concentrate on such modelling, preferring the behavioural approaches detailed in the next chapter. Even more prefer non-positivistic approaches, leaving the substantial GIS community somewhat isolated within geography as a whole, with its focus largely external rather than internal: indeed, Obermeyer (1994) claimed that GIS now meets all of the criteria for identification as a separate profession.

Countering the spatial separatist theme

Opposition to arguments for human geography as a spatial science developed largely as counters to claims for its separate status. Crowe (1970), for example, portrayed using the spatial variable in nomothetic studies as naive spatial determinism, like the earlier environmental determinism, whereas others based their criticisms on the implicit divisibility of the social sciences in the spatial claims. The latter criticisms are the core of the present section.

The most sustained argument against geography as a spatial science – what he calls the 'spatial separatist' theme – came in a series of papers by Sack, a former associate of Lukermann at the University of Minnesota (see p. 82). Reality has three dimensions – space, time and matter – and geography, according to the spatial separatist view, is the science of the first. Sack contended that space, time and matter cannot be separated analytically in an empirical science concerned to provide explanations, however. His first paper argued that geometry is not an acceptable language for such a science (Sack, 1972). Geometry is a branch of pure mathematics which is not concerned with empirical facts; its laws are static laws, with no reference to time, and they are not derivable from any dynamic or process laws. Geographic facts have geometric properties (locations) but if, as Schaefer proposed, geographical laws are concerned only with the geometries of facts, then they will provide only incomplete explanations of these facts. (To illustrate this contention, Sack used an analogy of chopping wood. If the answer to the question 'why are you chopping wood?' is 'because the force of the axe on impact splits the wood' then it is a static, geometric law, but if the answer is 'to provide fuel to produce heat' then it is an instance of a process law, which incorporates the geometric law. In this analogy, a process law is equated with the intention behind an action.) The laws of geometry are sufficient to explain and to predict geometries, according to Sack, so that if geography aimed only to analyse points and lines on maps it could be an independent science using geometry as its language. But 'We do not accept the description of the changes of its shape as an explanation of the growth of a city' (p. 72) so that 'Geometry alone, then, cannot answer geographic questions' (p. 72) leading to the conclusion that:

> To explain requires laws and laws (if they are valid) explain events. Since the definition of an event implies the delimitation of some geometric properties (all events occur in space), the explanation of any event is in principle an explanation of some geometric properties of events (p. 77).

Thus geography is closely allied with geometry in its emphasis on the spatial aspects of events (the instances of laws), but geometry alone is insufficient as a basis for explanation and prediction since no processes are involved in the derivation of geometries.

Bunge (1973a) responded, claiming that spatial prediction was quite possible with reference to the geometry alone, as instanced by central place theory and Thunian analysis. Such geometries provide 'classic beauty', and 'purging geometry from geography reduces our trade to no apparent gain' (p. 568). Sack (1973a) replied that the static laws espoused by Bunge are only special cases of dynamic laws having antecedent and consequent conditions, and that:

> Although the laws of geometry are unequivocally static, purely spatial, non-deducible from dynamic laws, and explain and predict physical geometric properties of events, they do not answer the questions about the geometric properties of events that geographers raise and they do not make statements about process (p. 569).

Sack did not argue, as Bunge supposed, that geometry should be purged from geography, but only that space should not be considered independently from time and matter. He further developed this theme with the contention that (Sack, 1973b, p. 17):

> for a concept of physical space to conform to the rules of concept formation and be useful in a science of geography every instance of the geometric or spatial terms must be connected or related to one or more instances of non-geometric terms (to be called substance terms).

Thus physical distance is not a concept in itself: it is necessary to know the terrain which a road crosses, for example, in order to assess the significance of its length in a gravity model – geometry alone is not enough (hence the considerations of the meaning of distance referred to above – p. 112; see also Mackay, 1958, on the influence of boundaries on movement patterns and Taylor, 1971a, 1971b). Since there is no such thing as empty physical space so there are no frictions of distance *per se*. There are frictions which demand work in crossing a substance, but the substance itself and the context in which it is being crossed create the frictions, not simply the distance: 'There are frictions and there are distances, but there is no friction of distance' (p. 22). Geography, according to Sack, is concerned to explain events and so it requires substantive laws: such laws may contain geometric terms, such as the frictions of crossing a certain substance, but these terms alone are insufficient to provide explanations.

The spatial separatist approach proposes an independent position for geography within the social sciences based on its use of geometry, but Sack (1974b, p. 446) contended that 'The spatial position's aim of prying apart a subject matter from the systematic sciences by arguing for spatial questions and spatial laws does not seem viable'. Instead, two types of law relevant to geographical work must be identified (Sack, 1974a). *Congruent substance laws* are independent of location: statements of 'if A then B' are universals which require no spatial referent. *Overlapping substance laws,* on the other hand, involve spatial terms: 'if A then B' in such cases contains a specific reference to location. Both are relevant and necessary in answering geographical questions, so no case can be made for a necessary 'spatialness' to the substance laws of human geography. Further, 'Space is an essential framework of all modes of thought' but 'geographic space is seen and evaluated in different ways at different times and in different cultures' (Sack, 1981, pp. 3–4). His book illustrated this with examinations of different approaches to the study of space: that of the social scientist concerned with objective meanings of space; that of the social scientist concerned with subjective meanings; the practical view of people who live in and learn about space (e.g. children); the mythical and magical views of space; and the societal conceptions within which organizations and institutions structure and use space. Only the first of these is the concern of 'spatial separatists', whose approach takes space out of its relational context, a consequence of which is that 'Ignoring spatial relations or conceiving them non-relationally will hinder the discovery and confirmation of social science generalizations' (p. 85).

May (1970, p. 188) also argued against Schaefer's claim that geography is the study of spatial relations:

> If we extend Schaefer's argument to include time, and assign the study of temporal sequences or relations to the historian, then the only conclusion respecting this matter that can be drawn is that economic, social, political, and other relations must be non-spatial and non-temporal. Hence economics, sociology, political science, etc. are

non-spatial, non-temporal sciences. But this is absurd ... insofar as economics qualifies as a science possessing empirical warranty, then its generalizations must apply to given spatio-temporal situations.

If all sciences have a spatial content, what is left with which to define a separate discipline of geography? May lists five possibilities.

1 Geography is a 'super-science' of spatial relations, 'a generalizing science of spatial relations, interactions and distributions' (p. 194) drawing on the findings of other sciences. This would leave the latter truncated and their studies unfinished. Such an approach has already proved unsuccessful; 'the issue of the conception of geography as a generalizing or law-finding science that somehow stands above the social sciences and history is not even appropriately debatable' (p. 195).

2 Geography is a lower-level science of spatial relations, applying in empirical contexts the laws of higher-level, generally more abstract sciences. (This may be a valid description of much of the geography produced in the 1960s.) This again truncates the latter sciences and raises the question 'what differentiates economic geography from economics, and vice versa?'.

3 Geography is the study of geographical spatial relations. This implies that there are spatial phenomena not studied in the other social science disciplines and which can therefore be claimed as geography's: May could conceive of no objects which are purely geographical (or, in the parallel argument, purely historical either).

4 Geography is the study of 'things in reality' spatially. Yet again, this abstracts from other sciences; May admits that certain 'bits and pieces' are not studied elsewhere, but argues that they do not offer a satisfactory empirical foundation for a separate discipline.

5 'Geography is not a generalizing or law-finding science of spatial relations' (p. 203).

The first four of these indicate that, because of the analytical indissolubility of time, space and matter, all social sciences are concerned with spatial relations, so that May, like Sack, argued that geography cannot claim an independent status on the basis of the spatial variable and the geometrical aspects of space alone. Moss (1970, p. 27) reached a similar conclusion at about the same time:

> geometrical relationships must be assigned economic, social, physical, or biological meaning before they can in any sense become explanatory ... though geometries may be important tools in geographical study and research, they cannot be a source of theory since their analogy with geographical phenomena is simply through particular logical structures, and not through explanatory deduction ... such an application implies that space, area, distance, etc., are important in and of themselves, quite independently of any implications they may have in

terms of diffusion, of cost, of time, or of process. This is manifestly false (p. 27).

Gregory (1978b, 1980) also criticized the extremely narrow, even superficial, view of spatial processes which he identified in many spatial scientists' work. He presents their claims as instrumentalist, involving theories which cannot be validated conclusively but which can only be evaluated pragmatically against the real world. Bennett (1974), for example, accepts that his models do not mirror actual processes, but assumes that they do in order to allow policy formulation (and therefore produce self-fulfilling prophecies): because they can postdict the world as it presently is, it is assumed that they explain it, and so can be used to predict the future.

Sack's arguments against the 'spatial separatist' theme did not lead him to reject a 'spatial viewpoint' at the core of human geography, and later works (Sack, 1983, 1986) promoted a theory of territoriality as a basis for understanding certain aspects of human behaviour. Territoriality was defined as 'a human strategy to affect, influence and control' (1986, p. 2) and he showed how:

> under certain conditions territoriality is a more effective means of establishing differential access to people, or resources, than is non-territoriality. It can be bounded thus readily communicated; and it can be used to displace personal relationships, between controlled and controller, by relationships between people and 'the law of the place'. (Sack, 1983, p. 57)

He exemplified this with a number of case studies, concluding that (1986, p. 215):

> it is clear that territoriality alone cannot alter social relations to the point of changing the complexion of an entire society, but it can, through its own internal dynamics, set in motion heretofore unforeseen, and often undesirable, social consequences. This was true with ancient civilizations. It was true with the Catholic Church. It was the case with the American territorial system and the work place. By all accounts it is true in socialist countries and one can expect it will be true in attempts at establishing more utopian communal organizations. Territoriality's effects are multiple, important, and must be reckoned with.

Sack's replacement for the 'spatial separatist' theme involved a theory in which space is used by people, individually and collectively, to promote social goals, an argument taken up by others. (See Johnston, 1991, for an extension of Sack's work. Wolch and Dear's, 1989, book *The Power of Geography* is subtitled '*How Territory Shapes Social Life*', but territoriality as defined here is not widely discussed therein, and the focus is largely on what is discussed below – p. 241 – as locality research. Cox's, 1989, chapter was relevant to the territoriality theme, however, emphasizing what he terms 'the politics of turf'.) As noted above (p. 107), territoriality had been promoted earlier by Gottmann (Johnston, 1996b); Taylor (1994) characterized its use as arguing for a 'world of containers' (see also Adams, 1995, and Mann, 1996).

The arguments reviewed in the present section are critical of much of the work undertaken by geographers in the fashion which dominated the 1960s, impelled by the 'Victorian myth of the supremacy of the natural sciences' (Gregory, 1978a, p. 21). The case for spatial analysis continued to be argued, however. Gatrell (1983), for example, countered Sack's critique of spatial science – while accepting his case that the separation of space from substance is untenable. He did not confine spatial analysis to a positivist philosophy, however, stating that:

> My response to both structuralists and humanists is that, since they too deal with *relations* (among individuals or social groups, or between man and his environment), they cannot avoid the notion of space, since any relation defines a space. Moreover, because every relation has a geometry associated with it . . . they cannot avoid the fact that geometry underlies much of what they deal with. Structures . . . *are* intrinsically spatial, but not in any simple geographical sense (p. 5).

His definition of space is much broader than simple distance, therefore, and he promoted spatial analysis not as a separate paradigm of geography but as an arsenal of tools to be used in all empirical research; much of that research, he argued elsewhere (Gatrell, 1985, p. 191), involves portraying objects arranged in space, investigating the role of distance as a constraint on human spatial organization, and achieving efficiency in locational arrangements.

Systems

The study of systems, as the term is currently understood, was introduced to the geographical literature by Chorley in 1962 although Foote and Greer-Wootten (1968) claimed that systems analysis was promoted in Sauer's (1925) programmatic statement *The Morphology of Landscape* with the words 'objects which exist together in the landscape exist in interrelation': Garrison's (1960a) review paper made a similar point. To some extent, the adoption of a systems approach involved putting 'old wine into new bottles', although the holistic approach currently espoused differs very much in format from Sauer's. More generally, the notion of a system has a long history, as Bennett and Chorley (1978, pp. 11–14) point out: teleological traditions, for example, postulate the world as 'a vast system of signs through which God teaches man how to behave' (p. 12), whereas functionalism links observed phenomena together as 'instances of repeatable and predictable regularities' of form.

The keystone of the study of systems is connectivity. As Harvey (1969a, p. 448) points out, reality is infinitely complex in its links between variables, but systems analysis provides a convenient abstraction of that complexity in a form which maintains the major connections. A system comprises three components (p. 451):

1   a set of elements;
2   a set of links (relationships) between those elements; and
3   a set of links between the system and its environment.

The last component may be non-existent, creating a closed system. Closed systems are extremely rare in reality, but are frequently created, either experimentally or, more usually in human geography, by imposing artificial boundaries, in order to isolate a system's salient features. Thus, just as an internal-combustion engine comprises a set of linked elements which receives energy from its environment and returns spent fuel to that environment, so a set of settlements linked by communications networks forms a spatial system, with links to settlements outside the defined area of the system providing the contacts with the environment. A system's elements have volumetric qualities and material flows along the links; as the system operates, so the various quantities may change.

Systems terminology was widely adopted by human geographers in the 1960s, and formed the basis, for example, of Haggett's (1965c) pioneering text. Not all applications involved the study of explicitly spatial systems. In many, the elements were phenomena whose spatial locations were not studied, and the links between the phenomena were functional relationships. In such contexts, geographical study of causal systems was identical to the study of similarly structured problems in other disciplines, but because the phenomena occupied locations and because the links involved crossing space, this suggested the need for an extra, geographical dimension which led to investigations of causal, spatial systems.

The early geographical literature on systems analysis was programmatic rather than applied; it suggested how the terminology might be applied in research and teaching contexts, often reinterpreting old material (McDaniel and Eliot Hurst, 1968). Relatively few applications were reported, and more than a decade later much of the literature assessed in a major review was written by other scientists (Bennett and Chorley, 1978). Nevertheless, Harvey (1969a, p. 479) wrote that:

> If we abandon the concept of the system we abandon one of the most powerful devices yet invented for deriving satisfactory answers to questions that we pose regarding the complex world that surrounds us. The question is not, therefore, whether or not we should use systems analysis or systems concepts in geography, but rather one of

examining how we can use such concepts and such modes of analysis to our maximum advantage.

In seeking answers to this question, two variants on the systems theme have been employed. The first is *systems analysis;* the second is *general systems theory,* which is an attempt to provide a more unified science than current disciplinary boundaries allow.

Systems analysis

Several typologies of systems and systems analyses have been suggested for studying geographical systems, whose definition is often difficult (Harvey, 1969a, pp. 445–9). Chorley and Kennedy (1971) identified four types of system (Figure 4.2):

1 *Morphological systems* are statements of static relationships – of links between elements: they may be maps showing places joined by roads, or equations describing the functional relationships between variables. Much of the spatial analysis outlined earlier in this chapter described such morphological systems.

2 *Cascading systems* contain links along which energy passes between elements: factories are cascading systems, for example, with the output from one element forming the input for another. Each element may itself be a system (linked departments within a factory, for example), producing a nesting hierarchy of cascading systems, as with Haggett's (1965c) nodal regions and the input–output matrix representation of an

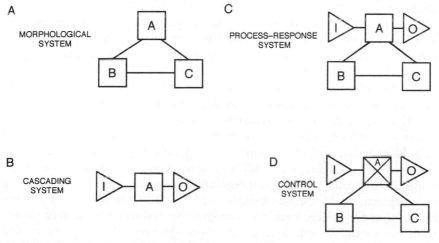

Fig 4.2 Types of system. A, B and C indicate system elements, I represents input, O represents output, and in the control system A is a valve (source: Chorley and Kennedy, 1971, p. 4).

economy (Isard, 1960). Berry (1966) linked these two examples of cascading systems in his inter-regional input–output study of the Indian economy. Within each element in a cascading system, the material flowing through is manipulated in some way (the industrial process in a factory, for example). The nature of the manipulative process may be ignored entirely in the investigation, with focus on the inputs and outputs only: such a representation of the element is termed a black box. White-box studies investigate the transformation process, whereas grey-box analyses make a partial attempt at their description.

3 *Process-response systems* are characterized by studies of the effects of linked elements on each other. Instead of focusing on form, as in the first two types, these are studies of processes, of causal inter-relationships. In systems terms these may involve, for example, the effects of one variable, X, on another, Y; in the analysis of spatial systems they could involve the effect of variable X in place a on variable Y in place b, as with the transmission of a disease from one area of a country to another (Cliff *et al.*, 1975).

4 *Control systems* are special cases of process-response systems, having the additional characteristic of one or more key elements (valves) which regulate the system's operation and may be used to control it.

Attention has focused on process-response and control systems. Langton (1972), for example, suggested that the former provide an excellent framework for studying change in human geography. He identified two sub-types. *Simple-action systems* are unidirectional in their nature: a stimulus in X produces a response in Y, which in turn may act as a stimulus to a further variable, Z. Such a causal chain merely reformulates 'the characteristic cause-and-effect relation with which traditional science has dealt' (Harvey, 1969a, p. 455); in another language it is a process law. More important, and relatively novel to human geography, is the second sub-type, *feedback systems*. According to Chorley and Kennedy (1971, pp. 13–14): 'Feedback is the property of a system or sub-system such that, when change is introduced via one of the system variables, its transmission through the structure leads the effect of the change back to the initial variable, to give a circularity of action.' Feedback may be either direct – A influences B which in turn influences A (Figure 4.3A) – or it may be indirect, with the impulse from A returning to it via a chain of other variables (Figure 4.3B). With *negative feedback* the system is maintained in a steady state by self-regulation processes termed homeostatic or morphostatic: 'A classic example is provided by the process of competition in space which leads to a progressive reduction in excess profits until the spatial system is in equilibrium' (Harvey, 1969a, p. 460). But with *positive feedback* the system is characterized as morphogenetic, changing its characteristics as the effect of B on C leads to further changes in B, via D (Figure 4.3D).

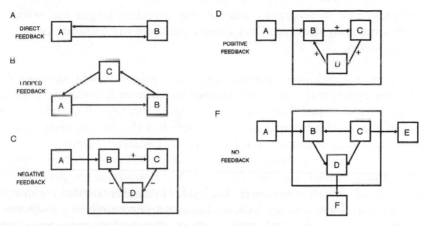

Fig 4.3 Various types of feedback relationships in systems (source: Chorley and Kennedy, 1971, p. 14).

The concept of feedback, with the associated notions of homeostasis and morphogenesis, provides 'the nuclei of the systems theory of change' (Langton, 1972, p. 145): as a consequence, Langton argued that the nature of feedback should be the focus of geographical study. In many spatial systems, feedback may be uncontrolled, but others may include a regulator, such as a planning policy (Bennett and Chorley, 1978). There are few geographical studies of such feedback processes, however. For homeostatic systems, Langton cites investigations of central place dynamics in which the pattern of service centres is adjusted as the population distribution changes, to reproduce the previous balance between supply and demand factors (Badcock, 1970). Morphogenetic systems are illustrated by Pred's (1965b) model of the process of urban growth, in which expansion in a sector generates, via a series of links, further expansion there, as in Myrdal's (1957) more general theory of cumulative causation: such systems modelling has been used to predict urban futures (Forrester, 1969). In most of these studies the input from systems theory is slight, however, leading Langton (1972, pp. 159–60) to the paradoxical conclusions that:

> First, there is little correlation between the extent of the penetration of the terminology of systems theory and the rigorous application of its concepts. The empty use of terminology, which is typified by the use of the term feedback as an explanatory device rather than as a description of a fundamental research problem, must be counterproductive in a situation in which the terms themselves may be given many subtle different shades of meaning. ... Second, somewhat paradoxically, many of the concepts of systems theory are

already used in geography without the attendant jargon and without apparently drawing direct inspiration from the literature of systems theory.

This second conclusion suggests that geographers should develop the necessary conceptual framework themselves without introducing relevant concepts from related disciplines, but Langton disputes this, arguing that systems theory clarifies questions in established theories, focuses directly on the processes of change, and forces careful analytical study.

One of the most substantial attempts to apply systems theory to a problem in human geography was Bennett's (1975) study of the dynamics of location and growth in northwest England. Having represented the system – its elements, links, and feedback relationships – he estimated the influence of various external (i.e. national) events on the system's parameters, isolated the effects of government policy (Industrial Development Certificates) on the system's structure, and forecast the region's future spatio-temporal morphology. He developed the forecasting aspects of this methodology in later papers (Bennett, 1978a, 1979), suggested how an optimal distribution of government grants could be achieved (Bennett, 1981b), and outlined the likely spatial variation in the impact of a new tax (the 'poll tax': Bennett, 1989b).

An area of investigation which has firmly adopted the systems approach covers the intersection of human and physical geography. The *ecosystem* is a process-response system involving energy flows through biological environments, most of which include, or are affected by, people. It is also a control system whose living components regulate the energy flows: 'they further represent a major point at which human control systems must intersect with the natural world' (Chorley and Kennedy, 1971, p. 330). Most naturally occurring ecosystems are homeostatic for much of the time (see Chapman, 1977, Chapter 7), but human 'interference' often transforms them into morphogenetic systems, with potentially catastrophic effects (Johnston, 1989b).

Stoddart (1965, 1967b) argued that the ecosystem should be employed as a basic geographical principle, but despite other programmatic statements (e.g. Clarkson, 1970), including two based on the associated concept of community (Morgan and Moss, 1965; Moss and Morgan, 1967), Langton's conclusion about relatively little substantive research in human geography based on systems' thinking remains valid (see Grossman, 1977). Similar attempts, but involving less consideration of the biotic environment, have been made in allied disciplines, and have occasionally been imported into the geographical literature. Sociologists' human ecosystem models (e.g. Duncan, 1959; Duncan and Schnore, 1959) have been used as frameworks for investigating migration (Urlich, 1972) and urbanization (Urlich Cloher, 1975), for example, and the operational research techniques of economists, with their important feedback mechanisms, have stimulated work in transport geography (e.g. Sinclair and Kissling, 1971).

The most comprehensive attempt to forge a systems approach to geographical study was Bennett and Chorley's book (1978), written to provide 'a unified multi-disciplinary approach to the interfacing of "man" with "nature"' (p. 21), with three major aims:

> First, it is desired to explore the capacity of the systems approach to provide an inter-disciplinary focus on environmental structures and techniques. Secondly, we wish to examine the manner in which a systems approach aids in developing the interfacing of social and economic theory, on the one hand, with physical and biological theory, on the other. A third aim is to explore the implications of this interfacing in relation to the response of man to his current environmental dilemmas. . . . It is hoped to show that the systems approach provides a powerful vehicle for the statement of environmental situations of ever-growing temporal and spatial magnitude, and for reducing the areas of uncertainty in our increasingly complex decision-making arenas.

This mammoth task (in a mammoth book of 624 pages!) involved elucidating not only the 'hard systems' of physical and biological sciences but also the 'soft systems' characteristic of the social sciences. The latter cover a very large and fertile literature concerned, first, with the cognitive systems describing people as thinking beings and the decision-making systems used by humans, as individuals and in groups, and, second, with the socio-economic systems made up of very many of these interacting individuals and groups. They then attempt to interface the two types, because:

> in large-scale man-environment systems the symbiosis of man as part of the environment of the system he wishes to control introduces all the indeterminacies of socio-economic control objectives. . . . In particular, we need to ask what are the political and social implications of control, and for whom and by whom is control intended? (p. 539).

Not surprisingly, their essay ends with a discussion of the many substantial problems involved in such interfacing, although many of these are firmly tackled in the text.

Use of systems analysis is based on the assumption (usually implicit) underlying much positivist work in human geography, that valid analogies can be drawn between human societies on the one hand and both natural phenomena-complexes and machines on the other. Individual elements in a system have predetermined roles, and can act and change in certain ways only – depending on the system's structure and its inter-relationships with the environment. As a descriptive device, this analogy allows the structure and operation of society and its components to be portrayed and analysed, providing a source of ideas from which hypotheses can be generated (see Coffey, 1981). And once a system has been defined and modelled, systems analysis can be used as a predictive tool, to indicate the nature of the elements and links following certain environmental changes (such as the introduction of new elements and/or links, as in the classic Lowry model used to predict the impact of new land-use configurations on traffic flows: Batty, 1978).

The potential fertility of this analogy has been examined by several authors. Wilson (1981b), for example, discussed methods of analysing environmental systems in which he defines environment as 'natural, man-nature, and manufactured "systems of interest"' (p. xi), in all of which 'the main concern is with *complicated* systems whose components exhibit high degrees of interdependence. The behaviour of the "whole" system is then usually something very much more than the sum of the parts' (p. 3). He argued that moorland ecosystems, water resource systems, and cities (his three initial examples) can all be studied in the same way and his book set out the available methods; a further book (Wilson, 1981a) presented the mathematics for studying systems in which the rates and directions of change suddenly alter. Huggett (1980) similarly argued that systems analysis has wide applications in both human and physical geography, plus the interface where people and environment interact (see also Huggett and Thomas, 1980). Once a system has been successfully modelled, it can be manipulated using control theory which 'is a *dynamic* optimization technique ... [that] permits optimal allocation over long time horizons ... [and] shifts emphasis from mere model construction to model use' (Chorley and Bennett, 1981, p. 219). Such a combination of models describing systems with a theory of system control has a wide range of potential applications, according to Chorley and Bennett, in such fields as pollution control, catchment management, inter-area resource allocation and urban planning. It suggests a commonality of interest, focused on methods, between applied physical and applied human geography.

Systems theory, information and entropy

A system has been defined so far here as a series of linked elements interacting to form an operational whole. This was challenged by Chapman (1977, p. 6) who opened his book with the statement that:

> I do not think that the concept of a system will have any great operational consequences in geography for a long time yet. It represents an ideal that the real world does not fully approach. On the other hand, in conceptual terms I think the concept is extremely important and

useful, and that it has a great and immediate role to play for those who are about to plan the strategy of their research. As a framework for analysis, it has no current peers (p. 6).

For him, a system comprises a series of elements which can take alternative states, and his definition – following Rothstein (1958) – is:

A system is a set of objects where each object is associated with a set of feasible alternative states: and where the actual state of any object selected from this set is dependent in part or completely upon its membership of the system. An object that has no alternative states is not a functioning part but a static cog (p. 80).

An example is a number of farms, each comprising a series of fields: every farmer has to decide how to use those fields. Each decision will in part reflect the farm's general operations and the uses to which all other fields are put; in part, it will be a function of the external market and the decisions made by other farmers regarding their own fields. Thus there is a large number of possible states for the system of fields – different configurations of land uses. According to Chapman, systems analysis should involve investigating those configurations and placing the observed pattern within the context of the alternatives:

to theorize merely about what does exist is not very useful. If we restrict ourselves to that alone, all explanation will be merely historical accidental. At all stages it is most important to include consideration of what else could have been. The definition of organization in a system even explicitly requires the assessment of what else could have been (pp. 120–1).

Taylor and Gudgin (1976) argued in the same way in a different research context: instead of simply asking 'is there a bias in the electoral districting of a borough?' they ask 'what is the likelihood of a bias occurring, given the constraints of the system?', and they set the study of electoral districting, the aggregation of spatial units into constituencies and its consequences, on a strong statistical footing (Gudgin and Taylor, 1979).

Such analyses focus on one configuration of the system as a sample from a set of possible configurations, and among their key concepts is *entropy*. A system's entropy is an index of uncertainty. In the second law of thermodynamics an increase in system entropy involves an increase in uncertainty. A good example is the introduction of a layer of hot water to a body of cold water. Initially, the two are separate, and one can be completely certain about the location of the hot molecules, for example. But with no external influence the two slowly mix, until all are at the same temperature. As the mixing proceeds, so the entropy increases.

Social scientists have drawn their usage of entropy from two separate, though linked, definitions. Thermodynamic entropy relates to the most

probable configuration of the elements, within the constraints of the system's operations. In information theory, entropy refers to the distribution
of the elements across a set of possible states, and is an index of element
dispersion. One can be completely certain about a distribution, in terms of
predicting where one element will be, if all elements are in the same state;
conversely, one will be most uncertain when elements are equally distributed through all possible states, so that prediction of the location of any
one element is most difficult. (An example of this use is in Johnston's,
1976a, work on the pattern of international trading partners; see also
Webber, 1977, and Thomas, 1982 on the relationship between entropy and
uncertainty.)

In its simplest form, the information-theory measure of entropy is
another descriptive index, but it can be developed in a variety of ways.
Chapman (1977) illustrates three uses:

1   as a series of indices of variations in population distributions;
2   as an index of redundancy in a landscape, where redundancy is defined
    as relating to a regular sequence so that it is possible to predict the land
    use at place *a*, for example, from knowledge of the land uses at neigh-
    bouring places; and
3   as a series of measures of reactions to situations in states of uncertainty.

In general terms, although not in detailed methodology, these continue the
tradition of Stewart and Warntz's macro-geography (p. 71): the aim is to
describe a pattern rather than to explain it, although the nature of the constraints used to derive the entropy measures provides an input to explanatory modes of analysis (Webber, 1977).

The use of entropy as developed in statistical mechanics rather than in
information theory was introduced to the geographical literature by Wilson
(1970). His initial example was a flow matrix. The number of trips origi-
nating in each of a series of residential areas is known, as is the number
ending in each of a series of workplace areas, but the entries in the cells of
the matrix – which people move from which residential area to which
workplace area (the flow pattern) – are unknown. What is the most likely
flow pattern? Even with only a few areas and relatively small numbers of
commuters, the number of possibilities is very large. Wilson defines three
system states. The *macro-state* comprises the number of commuters at each
origin and the number of jobs at each destination. The *meso-state* consists
of a particular flow pattern: five people may go from zone A to zone X, for
example, and three from the same origin to zone Y, but it is not known
which five are in the first category and which three in the second. Finally, a
micro-state is a particular example of a flow pattern – one of the many possible configurations of eight people moving from zone A, five to X and
three to Y. Entropy-maximizing procedures find that meso-state associated
with the largest number of micro-states:

the most probable distribution is that with the greatest number of micro-states giving rise to it. Thus the distribution corresponds to the position where we are most uncertain about the micro-state of the system, as there are the largest possible number of such states and we have no grounds for choosing between them (p. 6).

This, too, follows the macro-geography tradition. It is not an attempt at explanation and Wilson sees his work illustrating:

the application of the concept of entropy in urban and regional modelling; that is, in hypothesis development, or theory building. ['Model' and 'hypothesis' are used synonymously, and a theory is a well tested hypothesis.] ... the entropy-maximizing procedure enables us to handle extremely complex situations in a consistent way (pp. 10–11).

The hypothesis that the entries in the flow matrix conform to the most likely distribution can be tested against 'real' data. If it is falsified, either entirely or in part, it can be refined by building in more constraints. Wilson does this with his intra-urban transport models, for example, by introducing travel-cost constraints, different types of commuters (class, age, etc.), different types of jobs, and so on. The aim is to describe the most likely system structure from a given amount of information, which is incomplete.

Wilson subsequently developed both the theory of his modelling and the substantive applications (see Wilson, 1981b). His general text (Wilson, 1974) introduced a whole family of models that can be used to represent, and then forecast, the various components of a complex spatial system such as an urban region. Some of these models were expanded (e.g. Rees and Wilson, 1977, on demographic accounts) and applied, with varying success, to the West Yorkshire region (Wilson, Rees and Leigh, 1977).

Wilson's work has been applied and developed by others (see Batty, 1976, 1978, for example): it has been extended to other areas of study such as Johnston's (1985b) estimation of spatial variations in voting behaviour in England. In reviewing Wilson's (1970) presentation, Gould (1972) termed it 'the most difficult I have ever read in geography' but continued 'he has planted a number of those rare and deep concepts whose understanding provides a fresh and sharply different view of the world' (p. 689). Webber (1977) extended Wilson's argument that the purpose of entropy-maximizing models is to draw conclusions from a data set which are 'natural' in that they are functions of that data set alone and contain no interpreter bias. For him, they provide a convenient way of organizing processes of thinking about a complex world, and he identifies an 'entropy-maximizing paradigm' (p. 262) focusing on location models (the probability of an individual being in a particular place at a particular time), on interaction models (the probability of a particular trip occurring at a particular time), and on joint location/interaction models. At Leeds, Wilson's work on the entropy-maximizing approach was extended into the field of micro-simulation (Birkin and Clarke, 1988, 1989; Clarke and Holm, 1988) which produces statistically reliable estimates of micro-level characteristics of a population (the number of people aged 25–34 with cancer, for example) from macro-data (the number of people in each age group and, separately, the number with cancer).

In emphasizing aggregate patterns, this work is macro-geographic; according to Webber (1977): 'The entropy-maximizing paradigm asserts . . . that, though the study of individual behavior may be of interest, it is not necessary for the study of aggregate social relations' (p. 265). The patterns predicted by the models are functions of the constraints (which are the information provided at the meso-state), so that knowledge of them means that 'the entropy-maximizing paradigm is capable of yielding meaningful answers to short-run operational problems' (p. 266) and thus of immense value for immediate planning purposes. But:

> in the longer run, much of the economic system is variable: the constraints and the spatial form of the urban region may change . . . the research task facing entropists is (1) to identify the constraints which operate upon urban systems, which is partly an economic problem; (2) to deduce some facets of the economic relations among the individuals within the system from the use of the formalism; and (3) to construct a theory which explains the origins of the constraints. Only when the third task has been attempted may the paradigm be adequately judged (p. 266).

The entropy-maximizing model acts not only as a 'black-box' forecasting device (p. 130), therefore, but also as a hypothesis: if a system's operation is to be understood, the axioms – the constraints – must themselves be explained. Given the nature of the constraints (in Wilson's initial example, why people live where they do, why people work where they do, and why they spend a certain amount of time, money and energy on transport), the task is a major one: entropy-maximizing models aim to clarify it and indicate the most fruitful avenues for investigation.

The study of systems allows dynamic processes to be incorporated within geographical analyses, instead of focusing on static patterns which are the outcomes of such processes. Thus much of the mathematical development reviewed above (p. 113) has been concerned with such processes, to advance understanding of change and the ability to forecast it. (On the use of the term 'process' in this context, see the critique by Hay and Johnston, 1983.) Dynamical systems theory and analysis were the focus of a major research programme conducted at the University of Leeds by Alan Wilson and his associates from the mid-1970s, for example, during which a number of major advances were made; in particular, attention was directed away from relatively straightforward linear modelling to complex representations in which change is presented as discontinuous and not necessarily unidirectional (see Wilson, 1981a).

Many geographical analyses of dynamical systems have focused on static spatial patterns or structures, representing these as equilibrium or steady-state situations within the ongoing dynamic processes. Change is then handled, as Clarke and Wilson (1985b, p. 429) describe it 'by forecasting (in some other theory or model) the independent variables associated with a system and then calculating the new equilibrium or steady state'. Their applications of dynamical systems analysis suggest that this approach is not tenable, however, because in complex systems 'There are too many possibilities of transition to different kinds of equilibrium or non-equilibrium states' (p. 431) which casts doubts on the validity of the traditional approach to forecasting and hence the contribution of geographical modelling to planning. The nature of that contribution must be rethought, therefore, because conditional forecasting is of little value – 'there are simply too many possible futures for this to be useful' (p. 446). The new contributions, they argue, could involve the following:

> First, it may often be possible to recognize the 'nearness' of some instability or structural shift to an undesirable state. Policy can then be focused on conservation. Secondly, it may be possible to see how to bring about a shift to a new *desired* state by changing policy in order to move a parameter through some critical value. Thirdly, both the ideas of dynamical analysis and the capabilities of modern computer technology lend themselves to the construction of planning systems focused on information retrieval . . . and monitoring; so that at least planners, and policy makers are in a position to respond more rapidly when difficulties are identified (p. 446).

Geographers thereby come more to terms with the inherent complexity, and hence unpredictability, of the world.

The theory of systems and general systems theory

Like regional science and the impact of Walter Isard, the development of General Systems Theory (GST) was very much tied up with the academic career of one man – Ludwig von Bertalanffy (see von Bertalanffy, 1950). He attempted to unify science via perspectivism, removing the usual division of science through reductionism. He focused on isomorphisms, the common features among the systems studied in different disciplines, and whose 'subject matter is the formulation and derivation of those principles which are common for systems in general' (Walmsley, 1972, p. 23). The goal was a meta-theory with rules that apply in a variety of contexts; application is usually by analogy from one discipline in order to advance understanding in another (as in Chappell and Webber's, 1970, use of an electrical analogue of spatial diffusion processes and in work on artificial intelligence using computer modelling: Couclelis, 1986a). GST offers geographers an organizing framework; GST itself is an empirical exercise using inductive procedures to fashion general theories out of the findings of particular disciplines (see Coffey, 1981).

Although it was claimed that there have been no advances in either the theoretical base or the empirical application of GST (Greer-Wootten, 1972), Woldenberg and Berry (1967) drew analogies between the hierarchical organization of rivers and of central-place systems, for example; Berry (1964a) argued that cities are open systems in a steady state, as exemplified by the stability of their behaviour-describing equations; and several authors (e.g. Ray, Villeneuve and Roberge, 1974) applied the concept of allometry – that the growth rate of a component of an organism is proportional to the growth of the whole.

According to its proponents, the advantages GST offers human geography lie in its inter-disciplinary approach, its high level of generalization, and its concept of the steady state of an open system (Greer-Wootten, 1972; Walmsley, 1972), but they also contend that geography's strong empirical tradition means that it has more to contribute than to take from GST. One critic claimed that 'General systems theory seems to be an irrelevant distraction' (Chisholm, 1967, p. 51), however, an argument based largely on Chorley's (1962) paper on the Davisian system of landscape development. Chisholm summarizes the case for GST as follows:

1 there is a need to study systems rather than isolated phenomena;
2 there is a need to identify the basic principles governing systems;
3 there is value in arguing from analogies with other subject matter; and
4 there is a need for general principles to cover various systems.

He argues that something as grand as a meta-theory is unnecessary in order to convince people of the need to understand what they study, of the value of inter-disciplinary contact, and of the potential fertility of arguing by analogy.

Moving forward

Much of the remainder of this book is concerned with alternative approaches to human geography to those outlined here; most are based on critiques of 'positivist spatial science', and some have attracted large numbers of adherents. If the paradigm model discussed in Chapter 1 is relevant to geography, then as a consequence interest in the approaches set out in the present chapter should have waned in the 1970s and 1980s, if not disappeared totally. In relative terms there has undoubtedly been some decline, as the alternative world views on offer (and discussed later in this book) have attracted substantial attention. But a considerable amount of work is still being done which is firmly based within the world view described here. Some of the adherents have sought to accommodate the critiques (e.g. Wilson, 1989a) while others have counter-attacked and argued for the validity and vitality of their point of view. There is a general consensus among the counter-attackers that modelling is a viable approach to human geography, though there is considerable variability in emphasis as to the way forwards.

Recent reviews suggest that modelling within human geography has led to 'dramatic advances' within the discipline (Wilson, 1989b, p. 29), and that it has 'a substantial contribution to make in the long term' (Clarke and Wilson, 1989, p. 30). Wilson (1989b, p. 30) accepts the limitations of the approach – models 'have relatively little to offer in relation to individual behaviour directly', but are important tools which can 'help to handle complexity in a variety of situations'. His review focuses on work (much of it his own) which has sought to extend the 'classics' of location theory (von Thunen, Weber, Burgess and Hoyt, and Reilly: see also Clarke and Wilson, 1985b), and which he considers has provided, through its mathematical sophistication, substantial advances in understanding. (Wilson, 1989b, claims that the use of mathematics to model complexity distances his work from positivism: statistics are needed to calibrate and test models, but 'a purely statistical approach to theory and model-building is limited in scope and is also more directly connected to positivism', p. 41.)

The modellers' achievements are similarly lauded by Macmillan (1989a, 1989b), who equates the activity with 'quantitative theory construction'; to him theories are empirically testable statements of 'universal empirical propositions, law-like generalizations' (1989a, p. 93), so that modelling involves the development of theories which contain 'quantifiable propositions' that can be tested. It has already 'helped geography to achieve widespread and significant advances' (1989b, p. 292), and the feelings within the modelling community are of 'frustrated optimism rather than pessimism' (p. 291). He resists a return to focusing on the uniqueness of places (see Chapter 7), arguing that (p. 305): 'if we want to understand famines, we must look at famine processes wherever they occur and test our theories about them accordingly'. Several authors have argued for a shift in emphasis within the approach, however, as a consequence of recent developments in information technology and data availability. Macmillan (1989b) identifies three separate areas of modelling activity: statistical; mathematical; and data-based, which is largely inductive in orientation. The case for the last is forcefully put by Openshaw (1989, p. 71) who terms 'the application of a mathematical tool kit to urban and regional systems modelling within a hypothetico-deductive framework ... [as in Wilson's work] ... noble objectives'. He claims that the available models have not been much used, however (see also Batty, 1989), leading to the situation whereby 'so much of the intellectual capital of quantitative human geography [is] encapsulated in theoretical models that seemingly serve no useful purpose outside of geography itself' (p. 72). The deductive route to understanding, and eventually to applied work, 'has been taken too far and has not been particularly successful' (p. 73):

> There is no doubt about either the quality of the work or of the utility of the mathematical methods; rather, criticism is focused on the lack of applied relevancy, the lack of attention to empirical study, and the absence of explicit geography.

The recent explosion in data collection, and of data provision in machine-readable form, offers many new research opportunities (though see Openshaw and Goddard, 1987, on the problems of much of that data being held and treated as commodities in the private sector). Openshaw proposes to capitalize on that wealth of data with inductive work which he defines as 'Data-driven computer modelling in an information economy'.

According to Openshaw (1989), the prime concern of data owners (including those in the public sector who are increasingly required to justify the costs of data collection by their sale to users) is to 'add value' to their information, by analyses which make it more readily usable, and hence saleable. Theory is much less relevant than the ability to conduct spatial analysis of data, which relies much more on inductive approaches than deductive, a situation which he accepts is pragmatic but which he proposes as a means of ensuring the short-term survival of geography as an academic discipline (p. 81):

> never before in human history has so much information about so many people and their spatial behaviour patterns been stored in computers and therefore, theoretically speaking, accessible for analysis. Yet it appears that if geographers want access to such rich data sources as now exist they will in future either have to pay or join the data-keepers in providing information services (pp. 81–2).

He has chosen the latter way forward, using the analytical skills of geographers, including GIS (p. 117), not to advance theoretical understanding but rather to improve business and predictive efficiency by inductively seeking methods that work – as with answers to questions such as 'Given the population profile of an area, what pattern of spending can be expected there?' and 'Can house prices in that area be predicted with accuracy?'. Data bases should be explored to identify patterns that provide useful, and hence saleable, information. In this context, he developed a Geographical Analysis Machine (Openshaw, Charlton, Wymer and Craft, 1988), which has been used to identify clusters of disease outbreaks and to suggest hypotheses regarding their causes (Openshaw, Charlton, Craft and Birch, 1988), and he used proto-GIS technology to explore issues regarding the potential impact of nuclear attacks within Great Britain (Openshaw, Steadman and Greene, 1983), the siting of nuclear power stations

(Openshaw, 1986), and the disposal of nuclear waste (Openshaw, Carver and Fernie, 1989).

Openshaw's call for a return to inductive methods as a way of adding value to data, and thereby to prove the utility of geography within modern society, is backed by others, who identify technical ability with quantitative methods as one of the geographers' strongest selling points (e.g. Beaumont, 1987). Thus Rhind (1989, p. 189), who later became director-general of the British Ordnance Survey when it was being 'commercialized', argued that geographers should develop as 'gatekeepers' to usable information, as a way of staying 'central to the action'. These are clearly calls for more applied geography (which are discussed in more detail below: p. 318), which sit uneasily alongside the technical arguments of Haining (1989) for more sophisticated analytical work, though applications such as those of Cliff and Haggett (1989; see also Smallman-Raynor and Cliff, 1990) on the spread of diseases indicate how sophisticated spatial analysis based on deductive modelling can be put to use.

The division between the inductive 'number-crunchers' and the 'deductive modellers' suggests a growing split within the body of human geographers concerned with quantitative analysis: on the one hand, there are those who promote widespread application of well-known, usually relatively straightforward and easily applied methods; and on the other, there are those who advance the development of sophisticated procedures, whose use demands considerable technical expertise. Cox (1989) argues that the latter have had a very limited impact on geography as a whole, because: first, they have shown that spatial data are 'messy', requiring special analytical procedures which are not generally available, and so set 'intellectual and practical difficulties [which] have limited widespread understanding and application of the more correct procedures that have been produced' (p. 206); and second, with few exceptions the procedures have not been applied to problems that are in the mainstream of attempts at geographical understanding, have little basis in known 'human processes or behaviour', and so are peripheral to most geographers' interests. Thus much quantitative analysis has been marginalized within geography, while supporters of Openshaw's approach and the use of GIS have promoted their saleable expertise (Cox, 1989, p. 207):

> It is clear that geographers specializing in geographical information systems will survive only by combining considerable technical expertise with a taste and talent for the *Realpolitik* of grants, contracts, committees, administrators, and entrepreneurs. A canny awareness of commercial and political realities will be as necessary as any particular intellectual qualities.

For Cox, this carries the danger that 'preoccupation with data will omit imagination and creativity' (p. 208) but, as the later section on applied geography shows, many see it as a major way to protect the discipline in a context where utility and saleability are the crucial indicators of the worth of its work.

While a small number of geographers continue to promote the development of analytical methods within geography, and some at least link this to better explanatory procedures and predictive methods that will lead to a wider applicability and application of geographers' skills, many more are relatively unconcerned with either the sophistication of method or the search for grand generalizations. As Flowerdew (1986) forcefully pointed out, their work is very largely empiricist in orientation, having as its goal the portrayal of the 'objective' world as they perceive it, and of the salient elements which they identify as needing (or offering interesting) research. A great deal of it describes the changing geography of the world, and the amount and rapidity of change have called forth a great volume of published work. On the global scale, for example, major changes in the geography of productive activity have been recorded (as in Dicken, 1986; Knox and Agnew, 1989; Wallace, 1989); more locally, changes in the geography of individual countries have been the subject of substantial attention (as with the United Kingdom – Johnston and Doornkamp, 1982; Johnston and Gardiner, 1990; Lewis and Townsend, 1989 – and the United States – Knox *et al.*, 1988). Similarly, changes in the structure of society and their geographical ramifications have been a major focus, as with the growth of service industries (Daniels, 1982, 1985; Price and Blair, 1989).

This work is not entirely atheoretical – as critics of the empiricist and positivist positions on value-freedom and neutrality point out, the choice of what to study (i.e. what is significant) and how to study it must be theoretically informed, however implicitly – nor is it undertaken in isolation from the theoretical and methodological debates and developments taking place elsewhere in the discipline. A large number of geographers prefer not to get embroiled in those, however, but rather to draw on them, as they see fit, to inform their own work which in its orientation is close to Hartshorne's well-known call 'to describe and interpret the variable character from place to place of the earth as the world of man'. The debates and developments charted in this book have engaged directly only a minority of practising geographers, certainly to more than a passing degree, but they have all been directly or indirectly influenced by them as they follow their personal bents within the world of geography.

Conclusions

As stressed in the previous chapter, much of the force of the developments in human geography during the 1950s and 1960s concerned methodology and the approach to traditional geographical questions. In addition, however, attempts were made to inaugurate and press a particular geographical point of view, with two main themes. The first – the spatial- science viewpoint – was fairly widely and rapidly accepted, and many geographers placed the spatial variable at the centre of their research efforts: as will be indicated later, their work came under increasing attack in the mid-1960s. The second – the systems approach – has received much less detailed attention, despite frequent approving gestures, except in studies of environmental systems. Compared with the spatial-science view, which could be rapidly assimilated within the developing statistical methodology (although note the critique discussed on p. 114), the systems approach was technically much more demanding, and perhaps for that reason attracted fewer active researchers (though see Coffey's, 1981, defence). They have remained the source of fertile ideas, however, and continued to publish major works throughout the 1970s, paralleling, and in some cases (e.g. Bennett and Chorley, 1978) responding to, the academic movements outlined in the following chapters.

Human geography as spatial science was inaugurated in North America in the 1950s. By the end of the 1960s it was dominating many of the journals published throughout the English-speaking world. Most research was positivist in its tone, if not its detail, seeking to describe patterns of spatial organization and to account for these as consequences of the influence of distance on human behaviour. Much of that work used quantitative methods, and it contributed to bodies of theory, either about spatial organization in general or certain aspects of it in particular (as with industrial location: Smith, 1981). It has influenced other academic disciplines and the planning profession.

Since the 1970s, positivist spatial science has been under considerable attack among Anglo-American human geographers. This attack had little initial effect on the volume of work done in this paradigm, as illustrated by a major review of *Quantitative Geography* by British workers (Wrigley and Bennett, 1981); this work has been stimulated and aided by advances in data-handling technology and display (Rhind, 1981). The approach was also fostered outside the Anglo-American realm, especially in Western Europe where the reaction to quantification has been stronger than in Britain and North America (see Bennett, 1981c).

The partial disappearance of work in the spatial-science mould from many of the discipline's mainstream journals has suggested to some that it is no longer a viable and vibrant area of work. The field of GIS, for example, has spawned its own suite of outlets and professional societies, much like remote sensing before it, and other specialist journals, such as *Location Studies* and *Geographical Modelling*, have been launched for those who continue to apply mathematical and statistical procedures to large spatial data sets. This is part of the specialization and fragmentation that now characterize human geography (Johnston, 1996a) and, given the large volume of work done in other parts of the discipline, means that much of what is being reported can readily go unnoticed.

This great volume of work suggests that there has been 'progress' within the positivist approach. There have been substantial advances in the development and adaptation of sophisticated modelling and analytical techniques (on the former, see Papageorgiou, 1976), although some thought these advances, although 'scholarship of the highest order', may reflect 'a misdirection of effort' (King, 1979b, p. 157) because they have not illuminated spatial organization as much as they might. New techniques continue to be advanced, assisted by massive technological advances in data handling. Empirical work in the spatial-science context has very much taken the form of case studies set in a general theoretical context, however, and – despite the efforts of some general surveys (e.g. Haggett, Cliff and Frey, 1977) – it is not clear whether such case studies have provided much general understanding. What is without doubt, however, is the volumetric impact of spatial science on human geography.

5

Behavioural geography

The changes outlined in the previous chapter had achieved a substantial impact on the practice of human geography by the middle of the 1960s (as Burton's, 1963, coverage of 'the quantitative revolution and theoretical geography' implied), not only in the United States, Canada and the United Kingdom but also in other countries whose academic traditions and life are closely tied to one or more of those North Atlantic countries. The changes were resisted by some geographers, most of whom defended the (usually implicit) philosophies and methodologies within which they had been academically socialized: as suggested in the previous chapter, however, the 'new' very largely prevailed over the 'old'.

By the end of the 1960s, other critiques of 'quantitative and theoretical geography' had been articulated, based not on the split between the 'old' and the 'new' but rather on the latter's perceived failures to provide viable paths to understanding. Several proposed yet other philosophies and methodologies, some of them linked (weakly in a few cases) to previous approaches within human geography and others promoting more radical views: they are discussed in detail in later chapters. There was also a substantial rebuttal from within the spatial-science approach, from those who had tried the 'new' and were attracted to many of its tenets but were disappointed by its 'accomplishments'. Modifications were suggested, as ways of improving work within the paradigm.

These proposed modifications to the spatial-science approach stimulated work that became known as 'behavioural geography', whose birth was announced in a seminal book of essays (Cox and Golledge, 1969) and whose maturity was reviewed twelve years later in a companion volume (Cox and Golledge, 1981). Its essential ingredients, as set out by Golledge and Timmermans (1988), are:

1   a search for models of humanity which were alternatives to the economically and spatially rational beings of normative location theory;

2 a search to define environments other than objective physical reality as the milieux in which human decision-making and action took place;
3 an emphasis on processural rather than structural explanations of human activity and the physical environment;
4 an interest in unpacking the spatial dimensions of psychological, social and other theories of human decision-making and behaviour;
5 a change in emphasis from aggregate populations to the disaggregate scale of individuals and small groups;
6 a need to develop new data sources other than the generalized mass-produced aggregate statistics of government agencies which obscured and over-generalized decision-making processes and consequent behaviour;
7 a search for methods other than those of traditional mathematics and inferential statistics that could aid in uncovering latent structure in data, and which could handle data sets that were less powerful than the traditionally used interval and ratio data; and
8 a desire to merge geographic research into the ever broadening stream of cross-disciplinary investigation into theory-building and problem-solving.

This chapter reviews the field as they define it. (For a commentary from practitioners in the related field of psychology, see Spencer and Blades, 1986.)

Towards a behaviouristic spatial science

The main ground for disillusion from within the positivist camp was a growing realization that the models being propounded and tested provided poor descriptions of reality; progress towards the development of geographical theory was painfully slow and its predictive powers consequently weak. The large body of work based on central place theory, for example, was built on axioms regarding human behaviour with regard to choice between spatial alternatives, from which a settlement pattern was deduced. But the deductions were often only vaguely reflected in settlement morphologies, which suggested that the axioms provided a weak foundation for understanding this aspect of the spatial organization of society. The theory suggested how the world would look under certain circumstances of economic rationality in decision-making; that those circumstances did not prevail suggested that the world should be looked at in other ways in order to understand how people behave to structure their spatial organization (a position made clear by Losch, 1954.). As Brookfield (1964, p. 285) put it with regard to the models then popular:

> We may thus feel that we have proceeded far enough in answering our questions when, by examination of a sufficient number of cases,

we can make assertions such as the following: population density diminishes regularly away from metropolitan centres in all directions; crop yields diminish beyond a certain walking distance from the centres of habitation; air-traffic centres lying in the shadow of major centres do not command the traffic that their populations would lead us to expect. ... Such answers, which represent the mean result of large numbers of observations, whether statistically controlled or otherwise, are valuable in themselves, and sufficient for many purposes. But each is also an observation demanding explanations which may seem self-evident, or which may in fact be very elusive. Furthermore, there will be exceptions to each generalization, and in many cases, there are also limits to the range of territory over which they hold true. Both the exceptions and the limits demand explanation.

The issues being raised, therefore, were concerned not with the basic goal of positivist work – the establishment of generalizations and theories – but rather with the particular route being taken to that goal. The criticisms focused on the models being applied, particularly their basic axioms regarding rational economic behaviour on the basis of perfect information. The hypotheses then derived were, at best, only weakly verified in empirical studies. Better models were needed, and the search for them took a more inductive route than that previously followed. Rather than base hypotheses on assumptions about behaviour, behaviour was to be investigated inductively, and models then constructed which replicated it. (This involved creating a new 'image of real world structure': Figure 3.1.)

Rationality in land-use decisions

Some of the earliest attempts by geographers to explore behaviour inductively, as a prelude to later modelling, was the series of investigations into human responses to floods and other environmental hazards organized at the University of Chicago during the late 1950s and early 1960s. Its director was Gilbert White, whose own thesis on human adjustment to floods was published in 1945, and who has been described as 'the outstanding geographer in the man-land tradition, in the study of natural resources and hazards, and the study of the human environment' (Kates and Burton, 1986b, p. xi). Water resources were the main focus of his own work, and in their study 'he found himself a leader in the newly developing geography of perception, the world inside people's minds'. (A collection of his writings and of valedictory essays is provided by Kates and Burton, 1986a.) His associates developed a behaviourist approach for studying reactions to the hazards, based on Herbert Simon's (1957) theories of decision-making. Roder (1961), for example, categorized

Topeka residents according to their attitudes to the probability of future floods there, concluding that:

> Flood danger is only one of the variables affecting the choices of the flood-plain dweller, and many considerations operate to discourage a resident from leaving the flood plain, even when he is aware of the exact hazard of remaining (p. 83).

Such behaviour did not fit easily into the notions of profit-maximizing decision-making on which geographical theories were currently being built.

A major exponent of the behaviouristic approach was Kates (1962), whose study of flood-plain management began with the statement that 'The way men view the risks and opportunities of their uncertain environments plays a significant role in their decisions as to resource management' (p. 1). Kates developed a schema for studying such decision-making, which he claimed was relevant to a wide range of behaviours. It was based on four assumptions:

1 People are rational when making decisions. Such an assumption may be either prescriptive – describing how people should behave – or descriptive of actual behaviour. The latter seemed most fruitful, both for understanding past decisions and for predicting those yet to be made. Kates suggested adoption of Simon's concept of bounded rationality, according to which decisions are made on a rational basis, but in relation to the environment as it is perceived by the decision-maker, which may be quite different from either 'objective reality' or the world as seen by the researcher. Rational decision-making is constrained, therefore, and is not necessarily the same as the maximum rationality assumed in the neo-classical normative models discussed in earlier chapters of this book; people make decisions in the context of the world as they observe and interpret it, which may differ from other's perceptions (including those of geographers studying them).

2 People make choices. Many decisions are either trivial or habitual so that they are accorded little or no thought immediately before they are made. Some major decisions regarding the environment and its use may also be habitual, but such behaviour usually only develops after a series of conscious choices has been made, which leads to a stereotyped response to similar situations in the future.

3 Choices are made on the basis of knowledge. Only very rarely can decision-makers bring together all of the information relevant to their task, however, and they are frequently unable to assimilate and use all that is available.

4 Information is evaluated according to predetermined criteria. In habitual choice the criterion is what was done before, but in conscious choice the information must be weighed according to certain rules. Some normative theory prescribes maximizing criteria (of profits, for example);

descriptive theory may use Simon's notion of satisficing behaviour, involving decision-makers who seek a satisfactory outcome (a given level of profit, perhaps) only.

Models based on such behavioural axioms are therefore likely to differ very substantially from those which assume not only rationality but also complete information, perfect decision-making ability, and common goals. As Kates (1962, p. 16) describes his model: 'men bounded by inherent computational disabilities, products of their time and place, seek to wrest from their environments those elements that might make a more satisfactory life for themselves and their fellows'. As a consequence (p.19):

> Thus, a descriptive theory of choice must deal with the well informed and the poorly informed and the choices that men make under certainty, risk or uncertainty ... such a theory must deal with the eventuality that not only do the conditions of knowledge vary, but the personal perception of the same information differs.

People behave rationally, but within constraints, therefore, many of which are imposed by their situations – the cultures in which they have been socialized to make decisions.

The results of decision-making which do not match the predictions from the theories employed by normatively inclined spatial scientists do not imply irrational behaviour, therefore. Most decisions are made rationally on the basis of a, probably non-random, selection of information, are intended to satisfy a goal which does not imply making a perfect decision, and are based on criteria which vary somewhat from individual to individual. Having learned a satisfactory solution to a given class of problems, decision-makers will apply it every time such a problem occurs, unless changed circumstances require a re-evaluation.

Kates wanted to understand why people chose to live in areas which are prone to flooding. Their information was based on their knowledge and experience, and they varied according to the certainty of their perceptions regarding further floods. In justifying their decisions, most were boundedly rational; they had made conscious choices in order to satisfy certain objectives. Similar work was reported by others who worked with White and, as he points out, their findings had some impact on public policy formulation in the United States (White, 1973). It covered a wide range of environmental hazards, providing inputs for both national policies and programmes of international cooperation (Burton, Kates and White, 1978). Their initial impact on the wider geographical enterprise was not great, however, especially in the early years of their work, perhaps because they were operating close to the boundaries between human and physical geography, which few American researchers approached. Later work in other fields brought Simon's ideas more forcefully before the geographical audience.

This area of research, more than any other in recent years, has seen the sub-disciplines of human and physical geography come closest together. This is desirable, according to Cooke (1985a, 1992), not only because of the importance of topics such as human response to hazards (as in Hewitt, 1983) and society-induced modifications of the physical environment (Goudie, 1986a; Cooke, 1985b), but also because of the integrating role that geographers could play between the physical and the social sciences. Little of the work on these important topics does integrate the two types of science, however (Johnston, 1983a, 1986a), and though Cooke (1985a, p. 146) argues that:

> some extremists . . . in writing the post-war history of geography . . . will be tempted to ignore or underplay, for example, those geographers who have contributed most influentially to studies of the relations between communities, cultures and the physical environment. The names of Carl Sauer and Gilbert White, and their numerous students, will not figure prominently in their reviews.

It is difficult to find anything in the writings of Sauer and White that could be categorized as physical geography and reviewed as such (e.g. in K. Gregory, 1985).

Research by Sauer, White (see Kates and Burton, 1986a) and Cooke (1985b) himself throws considerable light on how societies interpret and use the natural environment (Blaikie, 1985, sets it in a realist framework – see p. 216 below), while others have explored their impact on the environment (e.g. Goudie, 1986a; Huggett, 1994). Such interactions have long been of interest to geographers, and a strong case has been made for retaining them as a major focus of geographical teaching (Pepper, 1987); Douglas (1983) has illustrated their importance in the urban context. But study of the interactions does not necessarily integrate the scientific studies of physical and human processes (for an exploration of the semantics of this argument, see Johnston, 1986b; see also Graham, 1986): nor is such integration needed, for according to Sayer (1983, pp. 55–6):

> in human geography . . . we may be interested in the causes of flooding and this will often include 'social' as well as 'natural' events. But although floods may be the effects of social actions this does not make the floods social . . . Understanding the social character of the actions which caused the floods . . . would not be essential for understanding the latter.

The work by White and many others demonstrates that people act according to 'manufactured' views of environmental resources in their conflict 'over who should benefit from the exploitation of the natural resource base' (Rees, 1985, p. xv) and geographers may well, as Pepper (1987) argues, seek to alter those views. This may eventually produce an interesting outcome in terms of future behaviour, but at present there is little research that perceives the need to provide an understanding of physical processes in order to advance an appreciation of human behaviour. (Flowerdew, 1986, contends that 'basically . . . most human geography is essentially irrelevant to physical geography and vice versa', p. 263.)

Wolpert and the decision process in a spatial context

Many human geographers were probably introduced to the behaviourist alternative by Julian Wolpert's (1964) paper in the *Annals of the Association of American Geographers*. (It was based on his Ph.D. thesis submitted to the University of Wisconsin, and is further testament to the innovative qualities of the geographers there, and of their contacts with other social scientists, including agricultural economists. His research was conducted in Sweden.) The normative theory espoused by many geographers at that time assumed a rational economic decision-maker (so-called 'economic man') who:

> is free from the multiplicity of goals and imperfect knowledge which introduce complexity into our own decision behavior. Economic Man has a single profit-maximizing goal, omniscient powers of perception, reasoning, and computation, and is blessed with perfect predictive abilities . . . the outcome of his actions can be known with perfect surety (p. 537).

In the study of spatial patterns, however, 'Allowance must be made for man's finite abilities to perceive and store information, to compute optimal solutions, and to predict the outcome and future events, even if profit were his only goal' (p. 537). Farmers face an uncertain environment – both physical and economic – when making land-use decisions, which in aggregate produce a land-use map. Wolpert suggested that differences between these decisions and those that would be made by 'economic man' should reflect aspects of the farmers' economic and social environments.

Comparing the observed labour productivity of farms in an area of Sweden with what could have been achieved under optimizing decision-making, Wolpert decided that the farmers were probably satisficers, although such a conclusion is difficult to verify without detailed knowledge of aspiration levels. How they acted was undoubtedly contingent upon their available information, and he argued that clear spatial variations in the levels of potential productivity achieved were the consequences of parallel spatial variations of knowledge. Only conspicuous alternatives are considered, it was suggested, and the result is rational behaviour, adapted to an uncertain environment.

Wolpert (1965) continued this theme with studies of migration, aiming to model the decision-making which lies behind the patterns of migration reported in census volumes. He contended that the gravity model is inadequate to represent such flow patterns; indeed, 'Plots of migration distances defy the persistence of the most tenacious of curve fitters' (p. 159). Boundedly rational individuals make sequential decisions: first, whether to move, and second, where to, and do so on the basis of place utilities – their evaluations of the degree to which each location, including that currently occupied, meets defined needs. Not only is the information on which these utilities are computed far from complete; for many places people have none. Thus each individual has an action space – 'the set of place utilities which the individual perceives and to which he responds' (p. 163) – whose contents may deviate considerably from that portion of the 'real world' which it purports to represent. Once the first decision – to migrate – has been made, then the action space may be changed as the would-be mover searches through it for potential satisfactory destinations and, if necessary, extends the space if no suitable solution to the search can be found (see Brown and Moore, 1970).

Wolpert's papers heralded – certainly in timing and to some extent in influence too – the development of what was termed behavioural geography (Cox and Golledge, 1969, pp. 1–3), an approach: 'united by a concern for the building of geographic theory on the basis of postulates regarding human behaviour ... upon social and psychological mechanisms which have explicit spatial correlates and/or spatial structural implications'. Early work in this framework focused on topics related to decision-making in spatial contexts, much of it involving researchers associated with the Ohio State University. Golledge (1969, 1970), for example, looked at models of learning about space and of habitual behaviour therein, and with Brown investigated methods of spatial search (Golledge and Brown, 1967). Others researched into the information flows on which decisions are based, indicating the influence of local context on behaviour (Cox, 1969), and Brown and Moore (1970) extended Wolpert's place utility and action space concepts for the study of intra-urban migration.

The fundamental aim of behavioural geography, according to a review by Golledge, Brown and Williamson (1972), is to derive alternative theories to those based on 'economic man', 'more concerned with understanding why certain activities take place rather than what patterns they produce in space' (p. 59). This involves 'the researcher using the real world from a perspective of those individuals whose decisions affect locational or distributional patterns, and ... trying to derive sets of empirically and theoretically sound statements about individual, small group, or mass behaviors'. Individuals are thus active decision-makers, not reactors to institutionally created stimuli (Cox and Golledge, 1981; Thrift, 1981). In evaluating such behaviouristic endeavours, Golledge, Brown and Williamson indicated the seminal influence of Hagerstrand's (1968) use of the concept of a mean information field (analogous to the action space of a place's residents) to model migration flows and the adoption of innovations. The initial interest in resource management decisions by White and his co-workers was extended from environmental perception and decision-making into aspects of attitudes and motivation, and was applied to studies of migration, the diffusion of innovations, political behaviour (especially voting), perception, choice behaviour, and spatial search and learning. By studying behavioural processes in these contexts, the aspiration was to increase geographers' understanding of how spatial patterns evolve, thereby complementing their existing ability to describe them. Morphological laws are insufficient in themselves to provide understanding; the amalgamation of concepts about decision-making taken from other social sciences with geography's spatial variable would allow development of process theories that could account for the observed morphologies.

Further developments included another series of pioneering papers by Wolpert and his associates on political decision-making. Regarding the distribution of certain artefacts in the landscape, Wolpert (1970, p. 220) pointed out that the location of, for example, a public facility in an urban area is frequently the product of policy compromise:

> Sometimes the location finally chosen for a new development, or the site chosen for a relocation of an existing facility, comes out to be the site around which the least protest can be generated by those displaced. Rather than being an optimal, a rational, or even a satisfactory locational decision produced by the resolution of conflicting judgements, the decision is perhaps merely the expression of rejection by elements powerful enough to enforce their decision that another location must not be used . . . These artefacts are rarely 'the most efficient solutions', and frequently not even satisfactory neither for those responsible for their creation nor for their users.

(This argument avoids considering any definition of optimal, in either economic or political terms.) Such decisions involve what Wolpert terms maladaptive behaviour. Kates and others had suggested that decisions are adaptively rational, within the constraints of uncertainty, utility, and problem-solving ability. Coping strategies under the mutual exchange of threats between interested parties can involve decision-making which does not involve the careful and methodical investigation of alternatives until a satisfactory solution is found, however. Instead decisions are the consequences of conflict between groups with different attitudes and motivations, and are not the result of joint application of criteria on whose relevance there is a consensus.

> This formulation then lays the framework for the interpretation of locational decisions which appear to be more the product of press responses than the end result of a dispassionate and considered selection of alternatives posited by the classical normative approaches or even the Simon scheme of bounded rationality (p. 224).

The approach was applied in a variety of contexts, such as the routes for intra-urban freeways and the siting of community mental-health facilities (Wolpert, Dear and Crawford, 1975).

Mental maps

One aspect of behavioural analysis enthusiastically adopted by a number of workers was the concept of a mental map of the environment which guides a decision-maker's deliberations. The term 'mental map' was not new to the geographical literature, having been used in Wooldridge's (1956) descriptions of the perceived environments within which farmers make their land-use decisions. Gould (1966) revived it in a seminal paper, which included his guiding belief that:

> If we grant that spatial behavior is our concern, then the mental images that men hold of the space around them may provide a key to some of the structures, patterns and processes of Man's work on the face of the earth (p. 182; reference to 1973 reprint).

Increasingly, he argued, location decisions are taken with regard to perceived environmental quality, so it is necessary to know how people evaluate their environments, and whether their views are shared by their contemporaries. To investigate such questions, Gould asked respondents in various countries to rank-order places according to their preferences for them as places in which to live, and these rankings were analysed to identify their common elements – the group mental maps (Gould and White, 1974, 1986). Such maps, he argued, are useful not only in the analysis of spatial behaviour but also in the planning of social investment – such as offering differential salaries to attract people to less desirable areas.

Those who followed Gould's lead investigated a range of methods for identifying and analysing spatial preferences (Pocock and Hudson, 1978). Their results made little impact on theoretical development, however, and Downs (1970, p. 67) wrote that:

> Even the most fervent proponent of the current view (that human spatial behaviour patterns can partially be explained by a study of perception) would admit that the resultant investigations have not yet made a significant contribution to the development of geographic theory.

Apart from Gould's rank-ordering procedures, Downs identified two other major approaches to the study of environmental images: the structural approach, which inquires into the nature of the spatial information stored in people's minds and which they use in their everyday lives – Lynch's (1960) book was a model for such work – and the evaluative approach, in which 'The question is, what factors do people consider important about their environment, and how, having estimated the relative importance of these factors, do they employ them in their decision-making activities' (Downs, 1970, p. 80).

With this latter approach geographers moved into the wider field of cognitive mapping – 'a construct which encompasses those cognitive processes which enable people to acquire, code, store, recall and manipulate information about the nature of their spatial environment' (Downs and Stea, 1973, p. xiv). They worked alongside both psychologists, who were becoming increasingly interested in the individual's relationship to a wider area than the proximate environment and the development of relevant non-experimental research techniques, and designers concerned with the creation of more 'liveable' environments. The journal *Environment and Behavior* was launched in 1969 to cater for this inter-disciplinary market, but despite some interest (e.g. Tuan, 1975a; Downs and Stea, 1977; Pocock and Hudson, 1978; Porteous, 1977) the general field has not made a major impact within human geography.

The concept of 'mental map' and the associated process of 'cognitive mapping' – which 'seems to imply the evocation of visual images which possess the kinds of structural properties that we are familiar with in "real" cartographic maps' (Boyle and Robinson, 1979, p. 60) – became the centre of considerable debate both among behavioural geographers and between them and outside critics. Gould's initial work, and that which it stimulated, was criticized as the study of space preferences only (Golledge, 1981a; see also Golledge, 1980, 1981b; Guelke, 1981a; Robinson, 1982). But, as Downs and Meyer (1978, p. 60) make clear, 'perceptual geography' – 'the belief that human behaviour is, in large part, a function of the perceived world' – extends much further than the elicitation and mapping of space preferences. The fundamental arguments of behavioural geography are that:

1 people have environmental images;
2 those images can be identified accurately by researchers (as in landscape evaluation: Penning-Rowsell, 1981); and
3 'there is a strong relationship between environmental images and actual behaviour' (Saarinen, 1979, p. 465).

The nature of those images – whether they are maps in the generally understood sense of that word and whether they can be apprehended by researchers – remained a problem: the concept of a 'mental map' may be a red herring but, to behavioural geographers, the argument on which it is based is not. (For a review of research into cognitive mapping, see Golledge and Rushton, 1984.)

Time-geography

One area of work that is sometimes identified as behavioural, but which extends into the humanistic and realist philosophies discussed in later chapters, was developed by Hagerstrand from the late 1960s on and introduced to a wider audience by Pred (1973, 1977a). As interpreted by Carlstein *et al.* (1978), time and space are resources that constrain activity. Any behaviour requiring movement involves individuals tracing a path simultaneously through space and time, as depicted in Figure 5.1 where movements along the horizontal axis indicate spatial traverses and those along the vertical signify the passage of time. All journeys, or lifelines, involve movement along both and are displayed by lines that are neither vertical nor horizontal: vertical lines indicate remaining in one place; horizontal lines are not possible for people, though they are (or virtually so) for the transmission of messages (see Adams, 1995).

Movement in space and time is constrained in three ways, according to Hagerstrand:

1 *capability constraints*: over time, these include the biological need for about eight hours sleep in every 24, whereas movement across space is constrained by the available means of transport;
2 *coupling constraints* require certain individuals and groups to be in particular places at stated times (teachers and pupils in schools, for example) and thus limit the range of mobility during 'free time'; and
3 *authority constraints* may preclude individuals from being in stated places at defined times.

Together, these three define the time–space prism (Figure 5.1) which contains all of the possible lifelines available to an individual who starts at a particular location and has to return there by a given time.

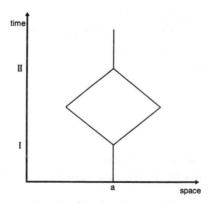

Fig 5.1 The time–space prism. In this simple example a person cannot leave place a until time I and must return there by time II; the prism between those two times indicates the maximum available spatial range (in miles or some other metric).

Pred (1977, p. 207) claimed that time-geography 'has the potential for shedding new light on some of the very different kinds of questions customarily posed by "old-fashioned" regional and historical geographers, as well as "modern" human geographers' because

It is ... a great challenge ... to cease taking distance itself so seriously ... to accept that space and time are universally and inseparably wed to one another; to realize that questions pertaining to human organisation of the earth's surface, human ecology, and landscape evolution cannot divorce the finitudes of space and time ... it is a challenge to turn to the 'choreography' of individual and collective existence – to reject the excesses of inter- and intradisciplinary specialization for a concern with collateral processes (p. 218).

Thus time-geography was in part a critique of spatial analysis as currently practised by human geographers. For Hagerstrand, it was much more, because the problems that he wished to address were concerned with the quality-of-life implications of packing people together in space and time. This is made clear in some of his later writings: for example (Hagerstrand, 1984, p. 375), we are enabled to:

see existing together what we otherwise have chosen to see apart. Neighbour meets neighbour in some sort of association. My concern now will be with how their presence together has come about, how they go along together and what is going to happen ahead in time.

(For autobiographical details on Hagerstrand, whose lifeline strongly influenced his academic work, see the chapter on him in Buttimer, 1983.) To explore what Pred (1977a, p. 213) calls the 'principle of togetherness', Hagerstrand employed three concepts:

1 *Paths* (or lifelines in the earlier terminology) are 'successions of situations' (Hagerstrand, 1982, p. 323) traced by individuals. To study paths alone is just to map outcomes (in the same way that other behavioural geographers map migrations or shopping trips); investigation of the path alone may reveal very little about the purpose and meaning behind the events that it incorporates, because it tells nothing about the: 'living body subject, endowed with memories, feelings, knowledge, imagination and goals – in other words capabilities too rich for any conceivable kind of symbolic representation but decisive for the direction of paths' (p. 324).
2 *Projects*, goals and strategies for project fulfilment are the activities which generate paths, whose intersections produce situations, particular moments in the flow of history in particular places.
3 Geographical study of such situations has traditionally involved the

concept of landscape, devised to represent 'the momentary thereness and relative location of all continuants' (p. 325). Hagerstrand claimed that this concept insufficiently incorporates 'the human body subjects, the keepers of memories, feelings, thoughts and intentions and initiators of projects' (p. 320) and preferred the concept of *diorama*, normally used to denote static museum displays which depict people and animals in their usual environments. The concept implied to Hagerstrand that: 'All sorts of entities are in touch with each other in a mixture produced by history, whether visible or not ... [we] appreciate how situations evolve as an aggregate outcome quite apart from the specific intentions actors might have had when they conceived and launched projects out of their different positions' (p. 320).

Hagerstrand illustrated his concepts with a description of his childhood home, demonstrating the importance of coupling and authority, as well as capability, constraints on the flow of daily life and the dioramas (particular situations) that resulted. (See also Pred's, 1979, application of the concepts to his academic career.) The study of paths alone is insufficient if time-geography is adequately to portray the 'real life of real people' (Hagerstrand, 1982, p. 338); thus autobiographical material (as in Buttimer, 1983; Billinge, Gregory and Martin, 1984) provides an important source on projects and dioramas because (Hagerstrand, 1982, p. 338): 'Only one's own experience is able to provide the kind of intimate detail which can bring the study of project and situation into any real depth. He is after all an expert on his own networks of meanings.'

Most interpretations and applications of Hagerstrand's initial work focused on the study of paths, and were set in the positivist mould of other behavioural geography enterprises. Indeed, this aspect of time-geography was the only one presented in an introductory review by Thrift (1977), who emphasized that a '"physicalist" approach is the backbone' (p. 4). With it, for example, Parkes and Thrift (1980) sought 'to place time firmly in the minds of human geographers' (p. xi) in an approach that they termed chronogeographical, arguing that 'Together, territories, societies and times are the principal components of urban and social geography' (p. 34).

The concept of diorama, and the 'togetherness' of a situation that it denotes, were absent from these early presentations, apart from passing vague references such as Thrift's (1977, p. 7), quoting Hagerstrand (1975), that 'every situation is inevitably rooted in past situations'. Projects were also mentioned, as in Thrift's (1977) statement that 'All human beings have goals. To attain these they must have projects, series of tasks which act as a vehicle for goal attainment and which, when added up, form a project' (p. 7), but it was implicit that those projects can be treated in the same sort of aggregate data analysis and policy prescription that characterize most behavioural geography (see Palm and Pred, 1978). Van Paassen

(1981) argued that Hagerstrand's work is essentially humanistic, however, aiming 'to provide insight into what is specifically human in man's nature and . . . [to] elucidate the specific human situation' (p. 18), a case accepted by Hagerstrand (1982) in his reference to human intentions. More recently, the concept of diorama has led authors to associate Hagerstrand's work not so much with the humanistic approaches outlined later in this chapter but rather with the realist proposals discussed in Chapter 7. The social theorist Giddens (1984), for example, draws heavily on Hagerstrand in his arguments for a structurationist approach (see below p. 238), though he too stresses the physicalist and behavioural elements when noting that 'Hagerstrand's approach is based mainly upon identifying sources of constraint over human activity given by the nature of the body and the physical contexts in which activity occurs' (p. 111). (Note, however, Gregson's, 1986, critique of Giddens' attempts to incorporate time-geography into structuration theory, but Gregory's, 1985a, p. 329, contention that time-geography 'must draw on substantive theories of structuration for its explications of contingency and necessity'.) In all of these applications, however, authority constraints and their associations with bounded spaces have been largely ignored, with continuous time-space the dominant concern.

Most criticisms of time-geography focused on the physicalist description of paths. Thus Baker (1979, p. 563) argued that:

> While space and time may usefully be considered as resources whose competitive allocation gives rise to patterns of use which may be observed empirically and modelled theoretically, the nature of that human struggle to control and structure time and space – the process underlying the form – should be of paramount concern rather than descriptions of temporal or spatial organisation.

As with other aspects of spatial science, description of the outcomes, however sophisticated, cannot elucidate the processes involved in their production. (Processes here are defined as mechanisms, the products of human agency, and not just sequences: see Boots and Getis, 1978; Haining, 1981; Hay and Johnston, 1983.) Thrift and Pred (1981) responded to Baker by denying his physicalist interpretation; to them

> time-geography is much more than that. It is a discipline-transcending and still evolving perspective on the everyday workings of society and the biographies of individuals. It is a highly flexible and growing language, a way of thinking about the world at large as well as the events and experiences, or content, of one's own life (p. 277).

They argued that time-geographers are necessarily concerned with underlying processes, with the ideological uses of time and space as devices to channel individual paths, and with the crucial role of human agency in the production of particular situations. They tie time-geography into Giddens' ideas on structuration and also with developing marxist humanism, concluding that:

> Some see the graphs used in time-geography as just neat pieces of art but others, in turn, are able to internalize the perspective represented by the graphs and use the path and project language as a way of thinking about themselves and the world. This will we believe be the lasting legacy of time-geography (p. 284).

Baker's (1981) response was largely positive, accepting that time-geography could be of value in a reorientation of geographical work: 'we should be examining the social organisation of space and time, not the spatial and temporal organisation of society for this is to put the cart before the horse' (p. 440). Gregory (1985a) was more cautious, however, arguing that Hagerstrand too readily focuses on paths rather than on the people whose projects fashion those paths, thereby failing to explore the meanings that are hidden beneath the tasks that define the biographies.

Hagerstrand's seminal papers were written before the 'information technology revolution' which enabled the almost instantaneous transmission of information to multiple sites around the globe (and beyond!) and promised virtual reality, whereby people in one place could operate as if in another. This led Adams (1995) to promote a reconsideration of what he termed 'personal boundaries in space-time', using Janelle's (1973) concept of 'personal extensibility', which he defines as (Adams, 1995, p. 267): 'the ability of a person (or group) to overcome the friction of distance through transportation or communication'. As that ability increases and 'distant connections become easier to maintain' (p. 268) so interaction patterns alter, to the extent that: 'Some believe that time-space and cost-space convergence have reached the point that one's location is of little or no significance for an increasing variety of interactions.' Extensibility promotes the transcendence of place, breaking down the crucial role of spatially bounded locales (see below, p. 238) as interaction containers. Thus people are both point-entities and extensible persons. They comprise (p. 269):

> A) a body rooted in a particular place at any given time, bounded in knowledge gathering by the range of unaided sensory perception and, in action, by the range of the unaided voice and grasp; and B) any number of fluctuating, dendritic, extensions which actively engage with social and natural phenomena at varying distances. This dynamic entity depends on media.

Acceptance of this argument requires a reworking of the importance of the presence/absence distinction in much analysis of human behaviour, and a new form of social geography (heralded, Adams argues, by Webber in 1964). It implies the need to distinguish between the geography of the body from that of the person: the former is grounded in space-time, but technology removes that constraint for the latter, with many implications for the exercise of power and responsibility.

Methods in behavioural geography

Whereas the theory- and model-builders of the spatial-science school of human geographers derived much of their stimulus from neo-classical economics, in some cases via regional science, the alliance for behavioural geography was largely with other social sciences, notably psychology and sociology. The behaviouristic approach is largely inductive, aiming to build general statements out of observations of ongoing processes. The areas studied were very much determined by the work in the spatial-science school, however. As Brookfield's statement quoted at the beginning of this chapter suggests, the spatial scientists' models and theories, such as central place theory, raised many of the queries which the behaviourists sought to follow up, stimulated by their observations of the failings of such theories when matched against the 'real world'. In terms of the accepted route to scientific explanation (Figure 3.1), therefore, behavioural geography involved moving outside the accepted cyclic procedure to input new sets of observations on which superior theories might be based. In doing this, the behaviourists did not really move far from the spatial-science ethos. Indeed, many of their methods were those of their predecessors (many of them started their research careers as spatial scientists); Gould's mental map studies, for example, used the technical apparatus of factorial ecology (p. 90).

Somewhat away from this general orientation of behavioural work, Pred (1967, 1969) presented an ambitious alternative to theory-building based on 'economic men' in a two-volume work *Behavior and Location*. His critique of location theory was based on three groups of objections concerned with: logical inconsistency – it is impossible for competing decision-makers to arrive at optimal location decisions simultaneously; motives – maximizing versus satisficing behaviour; and human ability to collect, assimilate, manipulate and use all possible information. To him (Pred, 1967, p. 17):

> Bunge's theoretical geography is easily distinguished from geographical location theory because its optimal final goals are disassociated from the interpretation of real-world economic phenomena ... and because these same goals can only yield a body of theory that for all intents and purposes is totally abehavioral and static rather than dynamic.

Location and land-use decisions are made with imperfect knowledge by fallible individuals and as a consequence there is bound to be disorder in the ensuing spatial patterns. (For another critique, see Barnes, 1988.)

As an alternative to the forms of theory-building criticized, Pred proposed using a behavioural matrix (Figure 5.2), whose axes are quantity and quality of information available and the ability to use that information: completely informed rational decision-makers are located in the bottom right-hand corner. Because of the nature and importance of information flows, decision-makers' positions on the first axis partly depend on their spatial locations; positions on the second reflect aspiration levels, experience and the norms of any groups to which the individuals belong. Different people in the same matrix position could vary in their decisions, therefore, if they acted on different information and in the pursuit of different goals, even if their quantitative stores of knowledge and ability to use them were commensurate. Similarly, people occupying different positions could make the same decision, based on different information and different ways of using it.

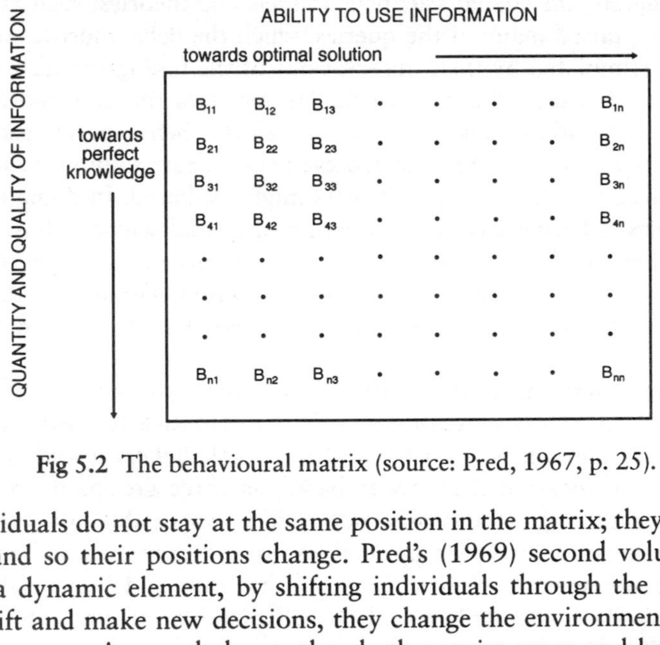

ABILITY TO USE INFORMATION

QUANTITY AND QUALITY OF INFORMATION

towards optimal solution ⟶

B_{11}	B_{12}	B_{13}	•	•	•	•	B_{1n}
B_{21}	B_{22}	B_{23}	•	•	•	•	B_{2n}
B_{31}	B_{32}	B_{33}	•	•	•	•	B_{3n}
B_{41}	B_{42}	B_{43}	•	•	•	•	B_{4n}
•	•	•	•	•	•	•	•
•	•	•	•	•	•	•	•
•	•	•	•	•	•	•	•
B_{n1}	B_{n2}	B_{n3}	•	•	•	•	B_{nn}

towards perfect knowledge

Fig 5.2 The behavioural matrix (source: Pred, 1967, p. 25).

Individuals do not stay at the same position in the matrix; they learn and react, and so their positions change. Pred's (1969) second volume introduced a dynamic element, by shifting individuals through the matrix; as they shift and make new decisions, they change the environment in which others operate. As people learn, they both acquire more and better information and become more skilled in its use; they shift towards the bottom right-hand corner of the matrix, some of them in advance of others, who thereby benefit from the experience of the 'decision-leaders'. The unsuccessful are gradually eliminated, so that a concentration of 'good' decision-makers close to the optimum position evolves, although new entrants to the matrix will probably not be located there. Changes in the external environment (such as market prices for farmers or the closure of some outlets for shoppers) generate parametric shocks, however, which result in decision-makers becoming less informed and less certain; as a consequence they are shifted back towards the upper left-hand corner and another learning cycle begins. As long as parametric shocks occur relatively frequently then optimal location patterns will never emerge, except perhaps by chance: the environment will have changed before all people reach the bottom right-hand corner of the matrix.

Pred (1969) presented the behavioural matrix as a 'gross first approximation' (p. 91), arguing that any theory is better than none (p. 139), even if the model itself is literally untestable (p. 141: see also Pred and Kibel, 1970). Harvey (1969b) treated it to a scathing review, however, calling the two dimensions vaguely defined, ambiguous, unoperational, and an over-simplification of the complex nature of the factors influencing behaviour. Indeed, Harvey (1969c) was generally sceptical about the potential of any behavioural location theory, a view shared by Olsson (1969) who pointed out the difficulties of studying processes and demonstrated that much behavioural geography involved only inferring processes from aggregated data on individual behaviour. Others argued that such inference could be very strong, however. Rushton (1969), for example, accepted that any one pattern of behaviour – what he termed behaviour in space – was largely a function of the spatial structure within which it occurred (the choice of shopping centre to patronize, for example), but general rules of spatial behaviour could be deduced from examining the types of preferences displayed within a particular pattern: 'To say that these preferences do not exist independently of the environment where the decision is made is to argue that environments could exist about which the person would be unable to reach a decision' (p. 393). Thus a distance-decay pattern reflects the details of the environment in which it is observed, and its production involves decision-making based on certain rules, which Rushton claimed his analytical procedure could isolate. Its validity has been queried, however (Pirie, 1976), and, like Pred's behavioural matrix, it has not attracted many disciples. (Rushton's arguments suggest activity in a 'taken-for-granted world': see p. 195.) Few others took up the explicit notion of the behavioural matrix, but many of the ideas that it was founded on were explicit in other studies, such as those industrial geographers who studied the spread of 'best practice' methods.

Harvey (1969c) suggested two alternatives to behavioural location theory – further development of normative theory and the construction of a stochastic location theory – both of which offered him more immediate pay-offs for understanding spatial patterns than did behavioural theory, because of the latter's conceptual and measurement problems. A stochastic approach was also favoured by Curry (1967), who argued that large-scale patterns are the outcomes of small-scale indeterminacy; individual choices may be random within certain constraints, but when very many of them are aggregated they may display considerable order (see above, p. 136). Similarly, Webber (1972) attempted to model locational decision-making processes in states of uncertainty using normative approaches, concluding that 'uncertainty increases agglomeration economies' (p. 279), thereby leading to greater concentration of economic activities and people into cities than predicted by models based on 'economic men' (see also Richardson, 1973; Johnston, 1976c; Scott, 1988). (Game theory is a mathematical procedure developed to handle decision-making in uncertainty which has received little attention from human geographers (Gould, 1963); the most ambitious attempt to use it, which had very little impact, is in Isard *et al.* (1969).)

For some critics, therefore, micro-analysis at the level of the individual is one or more of impossible, unnecessary or misleading; macro-analysis provides sufficient insight to the behaviour that produces aggregate patterns. Both are needed according to Watson (1978), however: macro-analysis is the first step towards providing an overview that poses questions which can only be answered by behavioural study. (See also Sayer, 1984, 1992a, on extensive and intensive research: p. 219 below.)

The behavioural approach did not stimulate a major revolution away from the spatial-science focus within human geography, therefore, and in effect generated an extension to it. Whereas many normative approaches start with simplifying assumptions about human behaviour and deduce what spatial patterns follow from those axioms, behaviourists were employing inductive procedures to identify the rules of behaviour, and then using these to predict (and therefore explain) spatial patterns (Gale and Golledge, 1982). Their approach involves a sequence of inter-related investigations. An individual is faced with a decision, with either a direct spatial input or spatial consequences. To make it, the individual sets criteria and collects information which is evaluated against the criteria. That may lead to a behavioural decision; alternatively, no satisfactory outcome could be identified so the individual may than alter the criteria, collect more information, or both. (This pragmatic approach to problem-solving is similar to the procedures of the scientific method outlined in Figure 3.1.) Many investigations have focused on certain aspects of this sequence, such as the flow of information, from which characteristics of other elements may be inferred.

Whereas much of the work done in the spatial-science mould could be conducted using either published data sources (such as censuses) or relatively small field-collection exercises, the behavioural approach generally requires specific data collection from the individual decision-makers. The need for social surveys of various kinds furthered the growing links between geographers and sociologists, psychologists and, to a lesser extent, political scientists, and led to an expansion in the data-handling procedures necessary for the training of geographers. One of the problems in their use is that many of the topics studied in human geography involve very large numbers of individual decisions – as in migrations, journeys to work and to shop, voting decisions, and so on. Very large sample surveys may be necessary to produce valid generalizations about such behaviour, especially across a range of places, but resource limitations have meant concentration on both small selections and only limited segments of the full behavioural sequence. Thus Brown and Moore's (1970) schema for the study of intra-urban migration decisions has mainly been tested in part only (e.g. Clark, 1975, 1981). More success with use of the behavioural sequence has probably been achieved in those branches of human geography that deal with topics involving relatively few decision-makers. In the study of diffusion, for example, Brown (1975) attempted to divert attention away from overall patterns of spatial spread and the reasons for adoption (or not) to the decision-making which brings certain innovations to places; most of these innovations involve selling products, which he terms 'consumer innovations' (see also Brown, 1981). Similarly, a number of industrial geographers have moved from investigations of aggregate patterns, which could be compared to those predicted by application of neo-classical economic analysis (Smith, 1971), to the study of decision-making behaviour within firms (e.g. Hamilton, 1974; Carr, 1983; Hayter and Watts, 1983: Schoenberger, 1993).

The analytical procedures employed within behavioural geography became much more sophisticated during the 1980s, paralleling developments in statistical procedures for analysing categorical data. Most of the research involved analysing patterns of behaviour within the framework set by the spatial-science approach and its general positivist orientation. The goal was to explain, through mathematical modelling and statistical analysis, variations in an aspect of behaviour (choice of travel mode for the journey to work, for example) in terms of variations in a number of independent variables (such as the characteristics of the decision-makers and the milieux in which their decisions were made: see Kitchin, 1996). Data for such studies are usually categorical in form, involving classifications (which travel mode was used, for example, and the gender of the person concerned) rather than variables measured on interval or ratio scales. Wrigley (1985) provided a major overview of the relevant statistical procedures for analysing such data; Davies and Pickles (1985) explored the important issue of inferring secular trends from cross-sectional data, while others promoted the analysis of longitudinal data comprising repeated interviews with the same individuals (Dale, 1993). Methods for quantifying attitudes and other aspects of human characteristics and behaviour have been explored, involving both increasingly sophisticated survey instruments for investigating how people cognize and learn about their spatial environments (Golledge and Timmermans, 1990) and technical procedures for representing those cognitions quantitatively (see Aitken *et al.*, 1989).

Many of the choice models underpinning work in behavioural geography, which have become increasingly explicit in the more sophisticated work, are based on theories of utility maximization. This approach was heralded by Cadwallader (1975), exemplified in fuller detail by

Wrigley and Longley (1984), and became the basis for a burgeoning literature (as displayed in great detail in Golledge and Timmermans, 1988,
and reviewed in Timmermans and Golledge, 1990). Such theories propose that decision-makers select from within the choice sets presented to
(and perceived by) them, according to the utility which they allocate to
each of the alternatives evaluated: as Timmermans and Golledge (1990)
express it, all 'are based on (variants of) a conceptual model that explicitly relates choice behaviour to the environment through consideration of
perceptions, preference formation and decision-making' (utility being a
measure of preference). A full analysis of discrete choices thus involves
knowing each of:

1 the available choice set;
2 the elements of the choice set considered by each individual;
3 the criteria on which each member of the choice set was evaluated by
   individual decision-makers; and
4 the relative importance they attached to each criterion.

Bird (1989), following Desbarats (1983), illustrates this (Figure 5.3). The
process starts with a listing of all possible opportunities, some of which are
then discarded to form an 'objective choice set'; evaluation results in some

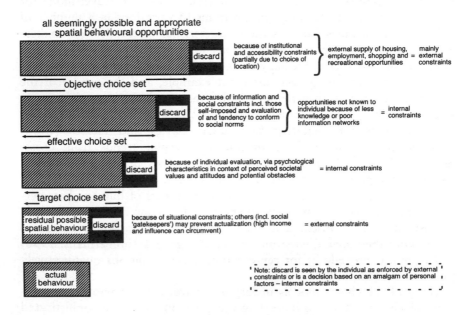

Fig 5.3 Choice sets in the spatial decision-making process (source: Bird, 1989,
p. 145, after Desbarats).

of the elements being discarded as unacceptable, to produce an 'effective choice set'; and further evaluation leads to its reduction to a 'target choice set' from which the final selection is made. A potential house-buyer in a particular town 'discards' many of the available homes through ignorance, for example, and then whittles down those perceived as viable purchases to a final list from which the selection is made (with, in some cases, a recursive loop to identify further viable buys if none in the final choice set either proves acceptable or can be bought).

Other work within this general framework combines categorical data with other forms, such as indicators of the local context within which decision-makers operate. Studies of voting behaviour, for example, have investigated the decision of which party to support according to both the characteristics of the individuals concerned and various perceived and objective indicators of their local milieux. (See, for example, Pattie and Johnston, 1995; Fieldhouse, Pattie and Johnston, 1996.) Methods developed by educational sociologists for analysing variations in children's achievements according to their personal characteristics, plus those of the classes and schools in which they are taught, have been adapted for such work (e.g. Jones, Johnston and Pattie, 1992), as well as investigations of variations in morbidity and mortality. These enable elucidation of the relative importance of 'nature' and 'nurture' – or compositional and contextual influences (see below, p. 238).

Behavioural geography is now widely accepted within the positivist orientation. It seeks to account for spatial patterns by establishing generalizations about people–environment inter-relationships, which may then be used to stimulate change through environmental planning activities that 'modify the stimuli which affect the spatial behavior of ourselves and others' (Porteous, 1977, p. 12). Gold (1980) relates this approach to four main features:

1 the environment in which individuals act is that which they perceive, which 'may well differ markedly from the true nature of the real world' (p. 4);
2 individuals interact with their environments, responding to them and reshaping them;
3 the focus of study is the individual, not the group; and
4 behavioural geography is multidisciplinary.

The research methods vary substantially but the general orientation – inductive generalization leading to planning for environmental change – remains. Eventually, it is hoped, a 'powerful new theory' (Cox and Golledge, 1969) will emerge, and Golledge (1981a, p. 1338) argued that substantial advances in understanding spatial behaviour have already been made by studying 'individual preferences, opinions, attitudes, cognitions, cognitive maps, perceptions, and so on' (p. 1339) – what he terms process variables.

Approaches to behavioural geography

Apart from the technical issues, the positivist foundation of most behavioural geography has been subject to considerable debate, as part of a wider consideration of the relevance of that philosophy of social science to human geography. Golledge, a leading practitioner of behavioural geography, has made major contributions to this debate, arguing that positivism is a constraining philosophy, and that (Couclelis and Golledge, 1983, pp. 333–4) 'behavioural geography has evolved by gradually shedding the tenets of the philosophy out of which it was born, as these have come to be seen as barriers to further progress and understanding'. He prefers the term 'analytical research' to describe his preferred approach, which is: 'not so much a philosophy as a distinct mode of discourse, a space of possibilities for theoretical languages that meet the criteria of clarity, coherence, inter-subjective validity, and a concern never to lose sight of experience' (p. 334). Nevertheless, he retains one of the central tenets of the positivist programme, as illustrated by references to 'a search for generalizations' (Golledge and Couclelis, 1984, p. 181), 'significant generalizations ... about particular sub-groups' (Golledge, 1980, p. 16) and the need to shift research away from 'the more narrow perspective of behaviour in space' towards 'a general understanding of spatial behaviour' (Golledge and Rushton, 1984, p. 30; see also p. 165 above).

Golledge and Stimson claimed that their approach, which they called *analytical behavioural geography*, is based on 'what is truly positive in positivist thought' (Golledge and Stimson, 1987, p. 9). They rejected the 'classic positivist separation of value and fact' and argue for a positivist position which is able to 'interpret values and beliefs in a scientific manner'. This they term a transactional or interactionist position, characterized by:

(a) the importance of logical and mathematical thinking; (b) the need for public verifiability of results; (c) the search for generalization; (d) the emphasis on analytic languages for researching and expressing knowledge structures; and (e) the importance of hypotheses testing and the importance of selecting the most appropriate bases for generalization or theorizing.

The goal is clearly to develop quantitatively verified theory (Macmillan, 1989b).

This continued emphasis of key elements of the positivist philosophy links to the second, much less important in volumetric terms, type of work within the behavioural approach. Pipkin (1981) notes that much effort has been expended on identifying behaviour patterns but little has been done to explore the mental constructs that underpin them. Understanding why people behave as they do has not been advanced very far, therefore; as Greenberg (1984, p. 193) expressed it: 'For the most part, the intention of behavioural-perceptual geographers has not been to explain the spatial organisation of society, but to illuminate the spatial behavior of individuals' (p. 193). The field of Artificial Intelligence (AI) was presented as offering much to the study of cognitive processes (T. Smith, 1984), using computer-modelling procedures to represent the decision-making processes and thereby gaining insights (by analogy) to the nature of the human brain; Couclelis (1986a, p. 2) indicates that this involves the 'human computer' metaphor according to which: 'cognitive functions such as problem-solving, pattern-recognition, decision-making, learning, and natural language understanding are investigated by means of computer programs that purport to replicate the corresponding mental processes'. The ability to replicate the processes by predicting their outcomes (as in Smith, Clark and Cotton, 1984) cannot be equated with understanding those processes (see p. 126 on the instrumentalist approach): Couclelis (1986a, p.111) concluded a detailed discussion of AI's utility in the study of human behaviour noting that: 'reliable predictions can be made about intentional systems even by theories which ... are totally vacuous psychologically'. Whether one wants to know how or why people behave is crucial: AI can reproduce the former, it is claimed, but can it assist in the latter? Nystuen (1984, p. 358) doubts the former claim, however, arguing that:

> I see little potential in AI methods available today in addressing problems considered important in the spatial decision-making literature, such as the decision to migrate ... These processes would require elaborate models of spatial cognition and tradeoff behavior whereas even the simplest model of a child's wayfinding is complex and contains major unresolved methodological problems.

Nevertheless, he would welcome the claim being sustained; it would provide only a constructed explanation (see the discussion of Lukermann's arguments above, p. 81) but:

> I am struck by the fact that careful empirical analysis by biologists describing the anatomy and behavior of bats did not lead to the discovery of how bats navigated in the dark. The explanation was beyond imagination until a purely human system (radar, followed by sonar) was invented and by analogy applied to the behavior and anatomy of the bat. Then all the things fell into place (p. 359).

In other words, if you can reproduce a process, then you may well gain some appreciation of it or, as Nystuen put it:

> If a constructed computer program can repeatedly resolve an issue under varying spatial conditions in a way that is considered useful to geographers, then one might say that we understand the issue. This is a sufficient claim: the problem has been solved by whatever logic or

capacities the program has at hand. There is no need to claim that this is necessarily the way human spatial decision-making works (p. 359).

Others reject such an instrumentalist argument, however (e.g. Gregory, 1980), giving rise to the cases against positivist work outlined in other parts of this book. (See the discussion of automated geography: p. 120.)

Moving on

Thrift (1981, p. 359) argued at the end of the 1970s that:

> The halcyon days of behavioural geography are long gone. With them have passed the days when behavioural geographers made inflated claims for the explanatory power of their subject area. But the subject area still has its place in human geography.

He recognized two criticisms of behavioural work: those which perceived it presenting the individual decision-maker as little more than an automaton responding to stimuli in a programmed way; and those which claimed that it ignored the characteristic of society as a whole greater than the sum of its individual parts. Thus he argued that behavioural geography might be presented as 'half-blind': 'But to say that behavioural geography is therefore half-blind is not to say that it can see nothing at all. Its explanations may be limited. That does not mean that they are therefore non-existent.' Behavioural geography has provided a useful methodology that can be validly employed when relevant.

If Thrift's conclusion were valid, one would anticipate a slowing-down in the pace of work within behavioural geography. It could be argued that the opposite has occurred; a substantial volume of work was conducted and reported during the 1980s (and summarized in reviews such as Golledge and Rushton, 1984; Golledge and Timmermans, 1990; and Timmermans and Golledge, 1990) which, in the context of the normal science model outlined above (p. 19), has achieved substantial progress. But it could also be claimed that behavioural geography has not only become a minority interest but has also become increasingly isolated within human geography, because of both the theoretical position and analytical sophistication achieved by its leading practitioners and what Cloke, Philo and Sadler (1991, p. 67) call its 'partial treatment of people' (which they contrast with their 'complete neglect in spatial science'). Many human geographers have accepted the need to collect and analyse individual data through questionnaire and similar methods, in order to portray how people learn about, represent and behave in space, but relatively few have kept pace with the methodological developments reviewed above. Others have rejected that approach entirely, because it cannot provide telling insights into individuals – thus making it 'somewhat limiting', 'dehumanising', 'pallid' and 'horribly reductionist' according to Cloke, Philo and Sadler. Behavioural geography is just one of many components of an increasingly diverse discipline: it made a considerable general impact upon geographical practice in its early years and has since been advanced by a small group of specialists somewhat apart from the mainstream.

The continued vitality of that group of specialists is illustrated by Golledge and Stimson's (1987) text on *Analytical Behavioural Geography*. They introduce the approach, and its origins in the 1960s, as a recognition that (p. 1):

> in order to exist in and to comprehend any given environment, people had to learn to organise critical subsets of information from the mass of experiences open to them. They sense, store, record, organise and use bits of information for the ultimate purpose of coping with the everyday task of living. In doing this, they create knowledge structures based on information selected from the mass of 'to whom it may concern' messages emanating from the world in which we live. Different elements from these various environments are given different meanings and have different values attached to them. It was the explicit recognition of the relationship between cognition, environment and behaviour that initially helped to develop behavioural research in geography.

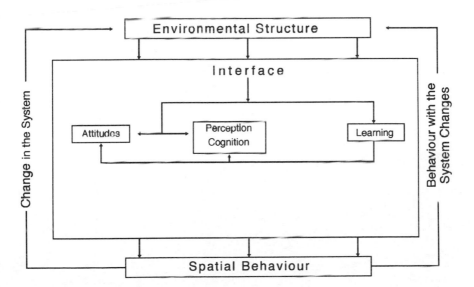

Fig 5.4 The society–environment interface (source: Golledge and Stimson, 1987, p. 11).

They then illustrate how the work developed, and structure the sub-discipline's approach through a diagram which places individual decision-makers at the interface between environment and behaviour, learning and acting within the environment and changing it as they act (Figure 5.4). The organization of their book, covering both the methods used by behavioural geographers and their substantive achievements in various topical areas, provides a clear view of a vital sub-discipline; as does the size of their bibliography, along with the large number of other pieces listed in the review papers on which parts of this chapter have drawn.

|6|

Humanistic geography

Behavioural geography involves a reorientation of the work undertaken within the spatial-science approach, but maintains its generally positivist framework. Alongside it, from the early 1970s on, though with roots extending much further back, a fundamental critique of the positivist approach emerged, including behavioural geography as promoted by Golledge and others. It argued instead for a focus on the individual as decision-maker, but denied the goal of explanation and prediction that was inherent to the behavioural approach. There was no coherent body of opinion in the alternatives that it proposed to positivism, but the critics were united in their opposition to what was being promoted as the 'new geography' (as illustrated in Cosgrove's 1989a, 1989b, cogent presentations of that case). That opposition was clarified by Entrikin's (1976, p. 616) definition of what became known as humanistic geography as:

a reaction against what they believe to be an overly objective, narrow, mechanistic and deterministic view of [the human being] presented in much of the contemporary research in the human sciences. Humanist geographers argue that their approach deserves the appellation 'humanistic' in that they study the aspects of [people] which are most distinctively 'human': meaning, values, goals and purposes.

Cloke, Philo and Sadler (1991, p. 69) express the critique more fully:

It almost goes without saying that the geographers referenced here were unhappy about the tendency of spatial science to treat people as little more than dots on a map, statistics on a graph or numbers in an equation, since the impression being conveyed was of human beings 'whizzing' around in space – travelling from place X to place Y; shopping in centre X rather than in centre Y; selling produce in market X rather than at market Y – in a fashion little different from the 'behaviour' of stones on a slope, particles in a river or atoms in a gas.

Indeed, it was complained that such exercises ... effectively con-
verted human beings into 'dehumanised' entities drained of the very
'stuff' (the meanings, values and so on) that made humans into
humans as opposed to other things living or non-living.

The nature of the criticisms, the alternatives and the developing practice are
the subject of the present chapter.

Cultural and historical behavioural geography

As indicated in earlier chapters, cultural and historical geographers were
neither closely involved in nor attracted by the so-called quantitative and
theoretical revolutions. There was some application of statistical procedures to make work in these fields appear more 'modern'; most of these
attempts were by non-historical (e.g. Pitts, 1965) and non-cultural
geographers and few changes of substance were adopted by adherents
working within the traditional approaches (despite the efforts of the *ad hoc*
committees established to chart ways forward for the 'new geography': see
above, p. 85). But by the 1970s, historical and cultural geographers had
taken the initiative, and were proposing alternative philosophies to that of
positivism, philosophies which were humanistic in their orientation:
according to Hugill and Foote (1994, p. 12), prior to World War II 'cultural
geography was more advanced in its theories than was the rest of human
geography' but the challenge of the 'revolutions' outlined in the previous
two chapters meant that 'Cultural geography ceased to be the most theoretically informed of the subdisciplines and embarked on a long search for
theory that is still underway' (p. 14).

The beginnings of these changes can be traced to two papers, one of
which had more impact than the other on the geographical discipline at
large. In the first, John K. Wright (1947, p. 12) introduced the term *geosophy*, defined as the study of geographical knowledge:

> it covers the geographical ideas, both true and false, of all manners of
> people – not only geographers, but farmers and fishermen, business
> executives and poets, novelists and painters, Bedouins and Hottentots
> – and for this reason it necessarily has to do with subjective concep-
> tions (p. 12).

Wright conceded that study of such subjective ideas could not employ the
strict scientific principles of physical geography, but claimed that it provided indispensable background and perspective to geographical work:

> geographical knowledge of one kind or another is universal among
> men, and is in no sense a monopoly of geographers . . . such knowl-
> edge is acquired in the first instance through observations of many
> kinds. . . . Its acquisition, in turn, is conditioned by the complex

interplay of cultural and psychological factors . . . nearly every important activity in which man engages, from hoeing in a field or writing a
book or conducting a business to spreading a gospel or waging a war,
is to some extent affected by the geographical knowledge at his disposal (pp. 13–14).

These words could well have heralded an earlier start to behavioural geography than that chronicled here, but the lack of reference to them in later
published works suggests that they had little impact until taken up by
Wright's colleague at the American Geographical Society, David Lowenthal
(1961, p. 259), in a widely cited paper 'concerned with *all* geographic
thought, scientific and other: how it is acquired, transmitted, altered and
integrated into conceptual systems'. Lowenthal argued in a wide-ranging
survey that the world of each individual's experience is intensely parochial
and covers but a small fraction of the total available. There are consensus
views about many aspects of the world, but individuals will often mistak-
enly assume that their view is the consensus. We all live in personal worlds,
which are 'both more and less inclusive than the common realm' (p. 248).
Our perceptions of these worlds are personal too; they are not fantasies,
being firmly rooted in reality, but because 'we elect to see certain aspects of
the world and to avoid others' (p. 251) behaviour based on such perceptions must have its unique elements. Different cultures have their own
shared stereotypes, however, which are often reflected in language, and
attempts are made to create environments fitting into these stereotypes:

> The surface of the earth is shaped for each person by refraction
> through cultural and personal lenses of custom and fancy. We are all
> artists and landscape architects, creating order and organizing space,
> time, and causality in accordance with our apperceptions and
> predilections (p. 260).

His ideas were implemented in papers concerned with the interpretation of
landscapes as reflections of societal norms and tastes (e.g. Lowenthal,
1968; Lowenthal and Prince, 1965), thereby belatedly bringing Wright's
ideas before a wider, and perhaps more readily appreciative, audience.

The second of the original papers was by a British geographer, although
it was published in India (Kirk, 1951; it is reprinted in Boal and
Livingstone, 1989: see also Johnston, 1993c): the main arguments were
reiterated in a later article (Kirk, 1963). Kirk stressed that the environment
is not simply a 'thing' but rather a whole with 'shape, cohesiveness and
meaning added to it by the act of human perception' (Kirk, 1963, p. 365):
once this meaning has been ascribed, it tends to be passed to later generations. Thus he recognized two, separate but not independent, environments; a phenomenal environment, which is the totality of the earth's
surface, and a behavioural environment, which is the perceived and interpreted portion of the phenomenal environment: 'Facts which exist in the

Phenomenal Environment but do not enter the Behavioural Environment of a society have no relevance to rational, spatial behaviour and consequently do not enter into problems of the Geographical Environment' (p. 367). Since much geography is concerned with decision-making and its consequences, appreciation of the behavioural environment should be central to its study; indeed, according to Ley (1977a) one cannot proceed without such awareness of what is in the behavioural environment. Even an apparently neutral statement such as 'Pittsburgh is a steel town' is, he argued, a value-laden view of a geographer-outsider, which may not accord with the perceptions of the resident-insiders, so that:

> Too often there is the danger that our geography reflects our own concerns, and not the meanings of the people and places we write of. . . . The geographical fact is as thoroughly a social product as the landscape to which it is attached (p. 12).

Scientific disciplines, as indicated in Chapter 1, usually comprise a number of 'invisible colleges', groups of scholars working on the same topic who refer to each other's publications. Wright, Lowenthal and Kirk were not members of any major college during the early 1960s, and so had very little impact on the first phase of behavioural work identified in the previous chapter. None of the three is in Pred's (1967, 1969) bibliographies, for example, nor referred to by Wolpert. Golledge. Brown and Williamson (1972, p. 75) make only a passing reference to Lowenthal's work:

> Pursued by insightful researchers, the analysis of literary and other artistic data of past and present can have strong explanatory power. The subjective element in these attempts to assess the impact of spatial perception is acknowledged, but its presence in many other studies is more subtle and potentially damaging.

Kirk (1978, p. 388) referred to his influence on non-geographers who adopted the concept of the behavioural environment, however, and Spate (1989, p. xix) commented on Kirk's initial paper, and its discussion at an IBG Conference, as 'the Catalytic Crystal in the Saturated Solution' (for a detailed exegesis of Kirk's work, see Campbell, 1989). The behavioural work of the mid-1960s was in the positivist mould; that of the three authors just discussed was not, and they were ignored by the former, who were in the ascendancy.

Several cultural and historical geographers took up the concept of the behavioural or perceived environments. One of the leaders was Brookfield, a British geographer with field experience, by the mid-1960s, of South Africa, Mauritius, New Guinea and several Pacific Islands. In reviewing work by cultural geographers on alien societies, he noted (Brookfield, 1964, p. 283):

> A difference of approach is apparent between those who have an overtly chorographic purpose, who scarcely ever seek explanations in

matters such as human behaviour, attitudes and beliefs, social organization, and the characteristics and interrelationships of human
groups, and those whose inquiries are not primarily chorographic,
and who are more inclined to undertake a search for processes as a
means of reaching explanation.

Social organization provides the key to many explanations, he argued, so
that:

> when an individual human geographer is sitting down in one small
> corner of a foreign land, and seeks to interpret the geography of that
> small corner, then it is difficult for him to do so without trying to
> comprehend the perception of environment among the inhabitants
> (p. 287).

Geographers had largely failed to delve into such details of social organization, however, because of the broad areal scale at which they had tended to
work, their concern with distributions rather than with processes, and their
avoidance of what he terms 'micro-geography'. Inquiry in human geography should involve three stages:

1 general statements about areal patterns and inter-relationships;
2 detailed local inquiries which follow up the questions about processes
   raised by these general statements; and
3 organization of the general and local material to produce explanatory
   generalizations.

Brookfield argued for more micro-geographical studies at the second of
these stages, to provide the foundation for the development of comparative
methods with which generalizations could be forged (Brookfield, 1962: on
comparative study in social science, see Taylor, 1996).

Brookfield's (1969, p.53) later literature survey showed that 'decision-
makers operating on an environment base their decisions on the environment as they perceive it, not as it is. The action resulting from decision, on
the other hand, is played out in a real environment.' Referring to the 'modern' behavioural work, as well as studies by cultural geographers, he
pointed out the great problems involved in isolating the perceived environment – something which is 'complex, monistic, distorted and discontinuous, unstable and full of unwoven irrelevancies' (p. 74) – and of building it
into an analytical methodology. Further data are needed, too, on such topics as work organization, time allocation and budget allocation, on the
meaning of consumption and of distance – all necessary tasks for the full
understanding of people–environment systems.

The concept of the perceived environment has a considerable pedigree in
historical–geographical scholarship, though without the current terminology, as illustrated by Glacken's (1956, 1967) seminal survey of societal attitudes to environments (see also N. Smith, 1984, and Pepper, 1984).

Historical geographers, Prince (1971a) suggested, must study a trilogy of worlds:

1 the *real world*, as recorded in documents and in the landscape (though see Harley's, 1989, 1990, seminal papers on maps as documents);
2 the *abstract world*, as depicted by general models of spatial order in the past; and
3 the *perceived world*: 'Past worlds, seen through the eyes of contemporaries, perceived according to their culturally acquired preferences and prejudices, shaped in the images of their assumed worlds' (p. 4).

Using all three, it may be possible to explain landscape changes, which cannot be done by assuming processes from investigations of continua of data over time (see also Moodie and Lehr, 1976): thus 'it [is] the province of the intellect to observe the facts, to reduce them to order and to discover relationships among them, but it [is] the imagination which [gives] them meaning through the exercise of judgement and insight' (Prince, 1961–2, p. 21). Reconstruction of past environments is extremely difficult, however, for it involves seeing the written record through the cultural lens of the writer:

> A study of past behavioural environments provides a key to understanding past actions, explaining why changes were made in the landscape. We must understand man and his cultures before we can understand landscapes; we must understand what limits of physical and mental strain his body will bear; we must learn what choices his culture makes available to him and what sanctions his fellows impose upon him to deter him from transgressing and to encourage him to conform (Prince, 1971a, p. 44).

Perhaps the enormity of such a task explains why most successful reconstructions have concerned the relatively recent past, such as the perceptions which guided the settlement of the American West (e.g. Lewis, 1966), although Wright (1925) essayed a similar task for Europe at the time of the Crusades. More recently, representations of nineteenth-century Egypt have been explored (Godlewska, 1995; Gregory, 1995a) to illustrate how representations of that country were implicated in the development of imperial ideologies (see p. 291).

Not all perceived worlds refer to either past or present; some landscapes have been fashioned out of Utopian views of the future (Porter and Lukermann, 1975; Powell, 1971). In general, however, and whether of past, present or future, geosophy has not become a popular field of study, despite some intriguing essays (see Lowenthal and Bowden, 1975). But directly or indirectly, it has led to arguments for alternative approaches in human geography to those of the positivist, and those arguments form the material for the rest of this chapter.

The attack on positivism and the humanistic approaches

From the early 1970s, on some cultural and historical geographers attacked the positivism of spatial science. To replace it, a variety of humanistic approaches has been proposed, focusing on decision-makers and their perceived worlds and denying the existence of an objective world which can be studied by positivist methods. The intent is to reorient human geography towards a more humanistic stance, to resurrect its synthetic character, and to re-emphasize the importance of studying unique events rather than the spuriously general.

Anti-positivism, idealism and historical geography

The various humanistic approaches have much in common, but they can be separated into different proposals in the present context. The first discussed here is associated with two workers who were together at the University of Toronto at the end of the 1960s; both are historical geographers.

The initial paper's basic theme (Harris, 1971) was that geography is a synthetic discipline, concerned with particular assemblages of phenomena and not with the science of spatial relations (see Cosgrove, 1996). Thus:

When the history of North American geography in the 1950s and 1960s is written, a paradox with which it will have to deal is how, with little argued, logical justification, so many geographers came to see their subject as a science of spatial relations (p. 157).

With May and Sack (p. 122ff.), Harris saw the spatial perspective producing a dismemberment of geography, as specialists communicate more with their contemporaries in other disciplines than with other geographers, and develop theories which are descriptions of how the world might operate under certain conditions, rather than of how it actually works.

The difficulty in conceiving of geographical theory comes down to this. The development of theory is necessarily an exercise in abstraction and simplification in which the complexities of particular situations are eliminated to the point which common characteristics become apparent. But if geography is thought to have a particular subject matter, it is certainly not individual phenomena or categories of phenomena which other fields do not study. Rather it is a whole complex of phenomena, many or all of which may be studied individually by other fields but which are not studied elsewhere in their complex interactions (p. 162).

The clear parallel is with history, for: 'Few historians would attempt to develop a general theory of revolutions. In so doing they would lose grasp of the type of insight that characterizes good historical synthesis' (p. 163). The goal of both history and geography is synthesis, therefore. In developing syntheses, positivist methods may be applicable: historians may be law-consumers, applying the generalizations of other social scientists to particular events, and geographers could operate likewise; alternatively, both historians and geographers could apply the idealist method, arguing that all activity is based on personal theories. Thus Harris (1978, p. 126) wrote of a 'historical mind':

> Such a mind is contextual, not law-finding. Sometimes it is thought of as law-applying but, characteristically, the historical mind is dubious that there are overarching laws to explain the general patterns of human life.

It is open and eclectic, he argued, uses no formal research procedures, sees things in context, is sensitive to motives and values, excludes little, and is wary of sweeping generalizations. Its goal is understanding, not planning, and this should be the case with the 'geographical mind' too. To understand an event is to appreciate why it took place, which is the humanistic goal: to explain an event is to predict it, as an instance of a general law or suite of such laws, and that is the positivist goal.

Developing his theme of the parallel between history and geography, Harris (1971, p. 167) identified four points of agreement concerning the nature of history:

1 its primary concern is with the particular;
2 explanation may take into account the thoughts of relevant individuals;
3 explanation may make use of general laws; and
4 'explanation in history relies heavily on the reflective judgement of individual historians.'

From these, he concluded that: 'If geography aims to describe and explain not so much particular events or peoples, as particular parts of the surface of the earth, then these points of agreement about history also apply to geography' (p. 167). (The term 'particular parts' can be widely interpreted, it would seem, and Harris (1977) himself sought to understand the nature of northwestern European colonizing societies 'by a model': the implication is that colonists had a common reaction to the 'new world'.) The landscape results from actions; behind those actions lie thoughts; study of thoughts allows understanding of landscape. Thus synthesis is crucial, since:

> the idea of synthesis itself becomes more important as it becomes obvious that our larger problems transcend narrow subject-matter fields . . . integration . . . in a larger understanding is still achieved, however aided by statistical methods and computers, by the

judgement of wise men who have cultivated the habit of seeing things together (p. 170).

Geographers are presumably to be those wise individuals, which was not an original claim for, according to Buttimer (1978a, p. 73), the basis of Paul Vidal de la Blache's work was that:

The task which no other discipline with the possible exception of history claims is to examine how diverse phenomena and forces interweave and connect with the finite horizons of particular settings. Temporality and spatiality are universal features of life so historical and geographical study belong together.

Positivist work seeks the same end – interweaving parts of a whole – but its parts are instances of general laws, not unique events.

Many of Harris's arguments were extended by his colleague, Leonard Guelke, whose first paper (Guelke, 1971) was a strong criticism of the 'narrowly conceived scientific approach' (p. 38) to geography using the positivist method. He argued against geography as a law-seeking activity by asking the proponents of the positivist approach to indicate how their laws would meet the basic standards of scientific acceptability, particularly with regard to prediction. Whereas they might be able to produce generalizations concerning the phenomena which they actually studied he felt it very unlikely that they could generate laws applicable to all examples of the relevant phenomena. Statistical regularities are not laws and

Until the new geographers have shown that the laws that might conceivably be discovered in geography will be more than generalizations, which describe common but non-essential connections between phenomena, their claims must be treated cautiously . . . there is little cause for optimism, especially as the statistical methods widely employed by geographers cannot be considered appropriate law-finding procedures (p. 42).

Regarding geography as a law-applying science, Guelke argued that laws of human behaviour are virtually impossible to conceive in anything but the most generalized form, because so much behaviour is culturally specific; an *a priori* statement of the determining conditions for their operation is thus not feasible so that: 'Human geographers cannot consider themselves to be law-applying scientists . . . because they have no laws to apply' (p. 45).

Turning to the use of theories and models in geography, Guelke pointed out that for them to serve a valid purpose in the pursuit of understanding, criteria must be specified which indicate how such devices are testable against reality (see also Newman, 1973, on the vague use of the term 'hypothesis'). Such criteria have not been, and cannot be, stated, Guelke claimed; studies purporting to test central place theory seemed to operate on the rule that 'one counts one's hits but not one's misses' (p. 48; see also

Guelke, 1978, p. 50), for example. Too often, failure to reproduce reality is explained by claims that the test environment was not entirely suitable, and *ad hoc* hypotheses are frequently adduced to account for observed disparities. Models and theories may have heuristic value for human geographers, clarifying certain aspects, therefore, but they can have no explanatory power.

Guelke (1971, pp. 50–1) concluded that:

> The new geography . . . has not yet produced any scientific laws and . . . appears unlikely to produce them in the future. . . . The theories and models . . . are not amenable to empirical testing. . . . The new geographers insisted on . . . logical and internally consistent theories and models. Yet, none of their theoretical constructs were ever complex enough to describe the real world accurately. They had achieved internal consistency while losing their grip on reality.

His offered alternative to that so discredited procedure was the idealist approach mentioned by Harris, which is 'a method by which one can rethink the thoughts of those whose actions he seeks to explain' (Guelke, 1974, p. 193). All actions, according to the idealist, are the result of rational thought, the parameters of which are constrained by a theory, which is 'any system of ideas that man has invented, imposed, or elicited from the raw data of sensation that make connections between the phenomena of the external world' (p. 194). Many such theories are part of the individual's society and culture; they include its religions, myths and traditions. Using them, 'the explanation of an action is complete when the agent's goal and theoretical understanding of his situation have been discovered. . . . One must discover what he believed, not why he believed it' (p. 197; but see Sayer, 1981). Thus human geographers do not need to develop theories, since the relevant theories, which led to the action being studied, already exist (or existed) in the minds of the actors involved. The analyst's task is to isolate those theories (a task of considerable difficulty, according to Curry, 1982a, because people cannot always identify the reasons for their actions). Some of them may be unique to particular individuals, but most are shared in large part by (sometimes large) numbers of actors; they represent the order which people themselves have stamped on the world, and do not require further theories in order to be understood.

Guelke's argument was challenged by Chappell (1975), who pointed out that by focusing on the individual actor alone the idealist omitted any reference to the environmental constraints and influences on that person's actions (see also Gregory, 1978a). Guelke (1976) accepted the existence of such constraints and influences, but claimed that their investigation lay outside the geographer's domain. Study of environmental causes would, he felt, lead into physiology and psychology and deviate attention from 'the most critical dimension of human behaviour, namely the thought behind it' (p. 169). Chappell (1976) responded that 'to go so far as to say that there is no possible respectable theory to explain man's rational theories and the actions which flow from them' (p. 170) is to be myopic: 'paradigms not only explain facts but they guide the research of whole disciplines' (p. 171). To him, Guelke's contention that the ultimate causes of actions lie outside the scope of human geography places geographers in an inferior position in the academic division of labour.

In a further essay, Guelke (1975) promoted his ideas on idealist approaches to historical geographers, as a counter to the arguments that they should adopt the approaches and techniques of positivism (see p. 96). He argued that:

> It is obvious that quantitative techniques will often be useful. . . . Statistical methods put in harness with positivist philosophy are a dangerous combination. . . . Historical geographers need to rethink not their techniques but their philosophy. . . . This can best be achieved by moving from problem-solving contemporary applied geography towards the idealist approach widely adopted by historians (p. 138).

Gregory (1976) agreed with the first part of this statement, but not with the proposed solution. Like Chappell, he saw the need to investigate individual action within its constraining structures (see below, p. 238 and Curry, 1982a).

Positivist approaches in human geography were defended against the idealist attack by Hay (1979a), who both responded to the criticisms and raised points of contention with the proposed alternative. He argued that Guelke's case is ill founded and rests on misconceptions of the nature of positivism, such as:

1 that all theory must be both normative and based on conceptions of optimal decision-making;
2 that to be scientific is to be nomological; and
3 that prediction is the same as prophecy (rather than simply testing, from the known to the unknown).

He also claimed that Guelke presented an anti-positivist argument by using a positivist test, and failed to realize the value of *ad hoc* hypotheses in the improvement of theory (the basis of Lakatos's concept of a research programme and its positive heuristic – p. 18): Guelke should not, according to Hay, ask 'does this theory explain Y?' but rather, 'does this theory contribute to an understanding of Y?'.

Regarding the idealist alternative, Hay raised the problem of studying groups rather than individuals. To Guelke (1978, p. 55):

> The assumption that thought lies behind human action is not related to the numbers involved. . . . If thousands of people drive motor cars to their places of work the idealist assumes that each of these journeys

is a considered action involving thought. In such situations the inves-
tigator will not be able to look at each case individually, but he will
seek to isolate the general factors involved in typical circumstances
. . . [for which he] might well make use of statistical procedures . . .
the value of statistical analysis will largely depend on its successful
integration in the general interpretation or explanatory thesis being
developed.

This procedure is akin to that employed by the behaviourist geographers.
Such a procedure, Hay contends, does not give ontological status to groups
as collections of individuals in which, as with 'traditional' regions, the
whole is greater than the sum of the parts. Second, Hay points out that
objective facts must influence behavioural outcomes, in addition to the
thoughts of the actors: Columbus found America, because it was there
(although he didn't recognize it as such). Third, he claims that the idealist
position ignores the possibility of either unconscious or subconscious
behaviour. In sum, idealism is reductionist, but the world is more than a
large number of independent decision-makers.

Idealism was also criticized by Mabogunje (1977) who claimed that
'Such a retreat from objective theory formulations as a means of seeking
explanation to certain events would exclude from our consideration the
exploration of the consequences of societal actions' (p. 368) and that
instead of retreating to a focus on particular cases – 'seek[ing] special
explanation for each situation in which a different value system can be
shown to be operative' (p. 370) – geographers should attempt to build
better theories encompassing these differences in value systems. Others
asked how an idealist interpretation can be verified. (Mercer and Powell,
1972, raised the same question regarding phenomenology – p. 190; they
wondered whether two phenomenologists can ever have the same 'intuitions' of a phenomenon, or indeed know whether they have.) Guelke
was prepared for this argument, presenting the analogy of the court of
law, rather than the positivist's laboratory, for his Popperian procedure.
A well-verified idealist explanation is one in which a 'pattern of behaviour can be shown to be consistent with certain underlying ideas. Where
data are presented which are not in accordance with a proposed explanation a new hypothesis will be needed' (Guelke, 1978, p. 55), which
contradicts somewhat his earlier statements that theory should not be
imposed from without by the observer. Even so, 'one cannot guarantee
mistake-free interpretations. The complex nature of human societies and
lack of pertinent data make it inevitable that many idealist interpretations will be of a tentative character' (p. 55), a position which is very
similar to Moss's (1977) outline of deductive procedures in historical
explanation.

The idealist philosophy, according to Guelke (1981b, p. 133), combines
two positions: a metaphysical argument that 'mental activity has a life of its

own which is not controlled by material things and processes'; and an epistemological argument 'that the world can only be known indirectly through
ideas ... all knowledge is ultimately based on an individual's subjective
experience of the world, and comprises mental constructs and ideas. There
is no "real" world that can be known independently of mind.' Positivist
spatial science is criticized because it believes in the existence of a 'real'
world, whose nature it seeks to explain via general laws of behaviour.
Behavioural geography is similarly criticized, not for its acceptable (to
Guelke) premise that behaviour is a response to perceived images and subjective evaluations, but rather because of two assumptions within this field:
'that identifiable environmental images exist that can be measured accurately ... [and] that there are strong relationships between revealed images
and preferences and actual (real-world) behaviour' (Bunting and Guelke,
1979, p. 453). Such an approach traps human geographers into single-
cause models, much like environmental determinism, and even if one
accepted this model, research in behavioural geography has failed to validate it. Guelke argued for an idealist perspective, which focuses on overt
behaviour and its interpretation: 'In searching for the truth a scholar con-
ducts a critical dialogue with his evidence and in due course he puts the
results before his colleagues for their appraisal' (p. 458). In response,
Downs (1979), Rushton (1979), and Saarinen (1979) claimed that Bunting
and Guelke misrepresented much behavioural geography and argued that
although the study of overt behaviour may give clues as to the answers to
'why?', only the study of decision-making can provide understandings of
value in planning, as in the work on environmental hazards. Golledge
(1981a, p. 1328), also responding to Bunting and Guelke, claimed that to
qualify as behavioural geography research may focus on the overt act, but
more importantly the set of explanatory variables must include 'one or
more process variables'.

Curry (1982a) criticized Guelke's idealist programme on a number of
grounds. Central to Guelke's argument is the claim that positivist science
methods are irrelevant to the study of human behaviour; the latter results
from rational thought, and thus can only be explained by reconstructing
that thought. There is then a single answer to the question 'why did you do
that?', which forms the explanation. But, according to Curry,

> First, in our daily lives we do many things for which, at the time, we
> consciously entertain no reason; at the same time, we hardly consider
> these actions to be non-rational. Second, we often, both at the time
> and after the fact, give reasons that are not the 'real' reasons for our
> actions. And, third, we can often attribute reasons to behaviour in
> which they were patently not involved (p. 43).

Those reasons, which may not be readily explored, may be examples of
individuals following rules, whether consciously or not, and such rules
need not be determinate – as in chess, where the rules identify what

moves are possible in a given situation but do not define which will be followed in any particular game. Thus we must study behaviour in its context:

> Without appeal to rules human action would be random and, hence, incomprehensible. The idealist geographer ... must consider the human world as a whole as imbued with the same sort of normative significance as that found in the more limited area of rules or maxims of actions (p. 46).

In response, Guelke (1982) restated his position that geography is an ideographic discipline seeking to understand the 'complexity of human activity on the land' (p. 52), drawing the boundaries such that: 'The geographer is not concerned with explaining fluctuations in wheat prices or the level of interest rates, but he is concerned with the impact that these factors might have on, say, farming in western Canada' (p. 53). In seeking to understand the reasons for any action, he accepted that stated reasons may not always help and that a 'historical reconstruction of thought' is necessary to explicate the learned response which is 'part of an individual's cultural heritage' (p. 54); this does not require exploring the subconscious forces that may have influenced the composition of that heritage, however. Nevertheless, Guelke accepted much of Curry's case, claiming that 'for both of us human geography is concerned with understanding the meaning of human activity on the earth's surface in its unique cultural contexts' (p. 57). Curry (1982b) disagreed, however, arguing that 'we remain very much farther apart than he believes' (p. 59), because he wishes to understand the processes that produce the thoughts behind actions that are Guelke's sole interest. Pickles (1986, p. 34) followed Curry in arguing that idealism, as promoted by Guelke, is not concerned with understanding that is 'empathetic in an emotional sense' but rather requires conformity 'to rules of inference and evidence'. Guelke remained committed to his cause, however, and two decades later wrote in a review of Gregory's (1994) discussion of European colonization in *Geographical Imaginations* that (Guelke, 1995, p. 185):

> The geostrategic thinking of both the colonists and the colonized needs to be reconstructed in a way that provides a convincing explanation of how one group subdued others and how colonization itself was sustained within ideologies of white supremacy. This task involves explorations of how people saw themselves in societies and in relation to the earth, and of the contrasts that existed between different systems of knowledge and territorial control and exploitation as they actually articulated themselves in specific geographical circumstances.

Phenomenology and related approaches

Phenomenology attracted more attention from human geographers than idealism. The first direct advocacy for a phenomenological approach was by Relph (1970), who was also associated with the department of geography at the University of Toronto. Despite a variety of specific interpretations, he noted that phenomenology's basic aim is to present an alternative methodology to the hypothesis-testing and theory-building of positivism, an alternative grounded in people's lived world of experience. Phenomenologists argue that there is no objective world independent of human existence – 'all knowledge proceeds from the world of experience and cannot be independent of that world' (p. 193). Thus, according to Entrikin (1976), 'phenomenologists describe, rather than explain, in that explanation is viewed as [an observer's] construction and hence antithetical to the phenomenologist's attempt to "get back" to the meaning of the data of consciousness' (p. 617). (Seamon, 1984, p. 4, defines phenomenology as a 'descriptive science'.)

Phenomenology in human geography is concerned with what Kirk termed the phenomenal environment (p. 177). That environment's contents are unique to every individual, for each of its elements is the result of an act of intentionality – it is given meaning by the individual, without which it does not exist but through which it influences behaviour. Phenomenology is the study of how such meanings are defined. It involves researchers identifying how individuals structure the environment in an entirely subjective way; the researchers are presuppositionless, using no personal ideas in seeking to understand their subjects' ideas. (Thus the subjectivity is that of the focus of the study, not that of the researcher who, in the positivist method, imposes a personal subjective view of the world: Ley, 1980.) Phenomenologists may be satisfied with such empathetic understanding, but some seek to go further, however, and identify essences – elements in individual consciousness which control the allocation of meanings (Johnston, 1983b; Pickles, 1988, p. 252, distinguishes between 'transitional essences' and 'invariant and universal structures (understood carefully)'). Phenomenology, then, studies human appraisals. It works at the individual level, but may search for the common (imprinted not agreed) elements among those appraisals.

Relph's paper was followed by one from another geographer associated with the University of Toronto, Yi-Fu Tuan (1971), for whom geography is a mirror, revealing the essence of human existence and human striving: to know the world is to know oneself, just as careful analysis of a house reveals much about both the designer and the occupant. The study of landscapes is the study of the essences in the societies which mould them, in just the same way that the study of literature and art reveals much of human life. Such geographical study has its foundations in the humanities, rather than the social or physical sciences. Tuan (1974, 1975b) illustrated it in a number of essays, giving, for example, insights into such topics as the sense of place (Tuan, 1976, pp. 266–7):

> Humanistic geography achieves an understanding of the human world by studying people's relations with nature, their geographical behaviour as well as their feelings and ideas in regard to space and place ... Scientific approaches to the study of man tend to minimize the role of human awareness and knowledge. Humanistic geography, by contrast, specifically tries to understand how geographical activities and phenomena reveal the quality of human awareness.

Tuan exemplified this with five themes: the nature of geographical knowledge and its role in human survival; the role of territory in human behaviour and the creation of place identities; the inter-relationships between crowding and privacy, as mediated by culture; the role of knowledge as an influence on livelihood; and the influence of religion on human activity. Such concerns are best developed in historical and in regional geography; their value to human welfare is that they clarify the nature of the experience (see also Appleton, 1975, 1994). Indeed, Tuan (1978, p. 204) claimed that 'The model for the regional geographers of humanist leaning is ... the Victorian novelist who strives to achieve a synthesis of the subjective and the objective', and quoted the first two pages of E. M. Forster's *A Passage to India* as a paradigm example.

Tuan's corpus of work has not involved philosophical explorations, and he rarely claims allegiance to any particular approach. Rather, he has been involved in a variety of explorations of the inter-relationships between people and environments:

> my point of departure is a simple one, namely, that the quality of human experience in an environment (physical and human) is given by people's capacity – mediated through culture – to feel, think and act ... I have explored the nature of human attachment to place, the component of fear in attitudes to nature and landscape, and the development of subjective world views and self-consciousness in progressively segmented spaces (Tuan, 1984, p. ix).

Thus, for example, he shows that fear is both a representation of the environment and an influence on the creation of environments (Tuan, 1979), that the creation of residential segregation reflects the retreat of individuals from wholes to segmented parts (Tuan, 1982), and that the construction of gardens reflects a desire to dominate the environment (Tuan, 1984). Such works are implicitly phenomenological in that they suggest the existence of general essences, or behavioural stimuli, but the term is not in the index to any of his books.

Among other advocates for phenomenology, Mercer and Powell (1972, p. 28) argued that the use of positivism in geography 'left the subject with too many technicians and a dearth of scholars'; they claimed that land-use patterns can never be understood 'by the elementary dictates of geometry and cash register' (p. 42) and the world can only be comprehended through people's intentions and attitudes towards it. In a lengthy discussion of phenomenology and its development in other disciplines, notably sociology, they pointed to 'a very real danger of the research worker assuming that concepts which are cognitively organized in his own mind "exist" and are equally clearly organized in the minds of his respondents' (p. 26); they argued instead for research methods involving empathy between researcher and researched, which for geographers involves 'that we make every effort to view problems and situations not from our own perspective, but from the actor's frame of reference' (p.48) – a position that they term 'disciplined naiveté'.

Buttimer (1974) made a similar case for geographers studying the values which permeate all aspects of living and thinking. The order, precision and theory produced by positivists are dearly bought – 'we often lose in adequacy to deal with the values and meanings of the everyday world' (p. 3) – and behavioural geography similarly fails to break away from the mechanistic, natural scientific view of humans as preconditioned responders to stimuli. Ley (1981) also identified the positivist elements in behavioural geography and the suggestions of operand conditioning: people are assumed to act in predetermined ways to particular stimuli, and so can be manipulated accordingly. On the other hand: 'An existentially aware geographer is ... less interested in establishing intellectual control over man through preconceived analytical methods than he is in encountering people and situations in an open, inter-subjective manner' (Buttimer, 1974, p. 24). Such activity results in 'a meditation on life', with geographers providing more comprehensive mirrors than their colleagues from more specialized disciplines can achieve, thereby clarifying the structural dynamics of life. Prediction would be impossible, apart from 'the most routinized aspects of experience' (p. 29), but the deeper appreciation achieved would allow much more vital social action and planning than is possible using spatial science and behavioural approaches.

Buttimer (1976) also directed geographers' attention to the concept of 'lifeworld', an amalgam of the worlds of facts and affairs with those of values in personal experience – 'the pre-reflective, taken-for-granted dimensions of experience, the unquestioned meanings, and routinized determinants of behavior' (p. 281). Positivism should be rejected for analysing the lifeworld because it separates the observer from the object of study and thus constrains appreciation of the human experience. Idealism should also be rejected, because it accepts the existence of a 'real world' outside the individual's consciousness. Phenomenology, on the other hand, provides a path to understanding, on which informed planning can be built:

It helps elucidate how . . . meanings in past experience can influence
and shape the present . . . extremely important as preamble not only
to scientific procedure, but also as a door to existential awareness. It
could elicit a clearer grasp of value issues surrounding one's normal
way of life, and an appreciation of the kinds of education and socialization which might be appropriate for persons whose lives may
weave through several milieux (p. 289).

The result is an understanding of actions as those involved understand
them, rather than in the terms of abstract, outsider-imposed models and
theories. Having achieved that understanding, human geographers can
transmit it, thereby helping the subjects of their investigation to understand
themselves better and realize their potential. In this way, the applied geographer acts as a provocateur, stimulating human development but not forc-
ing it (Buttimer, 1979).

Berry (1973b, p. 9) backed this phenomenological orientation, calling
for: 'a view of the world from the vantage of *process metageography*. By
metageography is meant that part of geographic speculation dealing with
the principles lying behind perceptions of reality, and transcending them,
including such concepts as essence, cause and identity' (p. 9). But not all
accepted that phenomenology can entirely replace the positivist approach.
Walmsley (1974), for example, accepted the merits of Buttimer's case
because so many human decisions are based on 'experiential' rather than
'factual' concepts, but argued that the scale of geographical enquiry, and its
long tradition of certain types of empirical work, required maintaining the
positivist orientation. That the perceived world is not necessarily the same
as the real world must be realized, but 'logical consistency and empirical
truth will remain central to geographical enquiry provided the importance
of values is recognized' (p. 106).

Gregory (1978a) criticized both positivism and phenomenology. Those
favouring the former were condemned for making 'social science an activity
performed *on* rather than *in* society, one which portrays society but which
is at the same time estranged from it' (p. 51) and for supporting a procedure which, because it so often assumes *ceteris paribus* in testing its models,
can never be sure why these fail to replicate reality (p. 66). He recognized
the need for humanistic approaches, but argued that these would not be
sufficient to provide a satisfactory foundation, because they ignore the
'constraints on social action which are so much part of the taken-for-
granted life-world of the actors' (Gregory, 1978b, p. 166). Thus: 'A geography of the life-world must therefore determine the connections between
social typifications of meaning and space-time rhythms of action and
uncover the structures of intentionality which lie beneath them' (Gregory,
1978a, p. 139). But 'A major deficiency . . . is [the] restricted conception of
social structure: in particular, it ignores the material imperatives and consequences of social actions and the external constraints which are imposed on

and flow from them' (*ibid.*). Phenomenology and idealism must be incorporated with investigations of those imperatives and constraints; such incorporation produces a critical science, whose nature is discussed in Chapter 7.

Whereas a key feature of the positivist/spatial-science approach has been a great numerical superiority of practitioners over preachers, the phenomenological movement (like the idealist) was initially characterized by the converse – much preaching but relatively little practice (Relph, 1981a, pointed to the absence of substantive applications of the phenomenological approach in geography, though see Jackson and Smith, 1984, p. 44):

> There is an essential difference between the contemplative intentions of this transcendental philosophy and the practical concerns of a social science, so that it is scarcely surprising that . . . geographers' . . . efforts have been directed towards the destruction of positivism as a *philosophy* rather than the construction of a phenomenologically sound *geography* (Gregory, 1978a, pp. 125–6).

Tuan's interpretative essays exemplify its use, and Relph, who 'would much prefer to see substantive applications rather than discussions of the possible uses of phenomenology' (1977, p. 178), published his thesis on *Place and Placelessness* in which – implicitly, he says – phenomenological methods are used 'to elucidate the diversity and intensity of our experiences of place' (Relph, 1976, p. i): his essential themes are the sense of place and identity in the human make-up and the destruction of this through the growing placelessness of modern design (see also Porteous, 1988). Other work generally quoted as phenomenological includes pieces on European settlement of the New World. Powell (1972), for example, has written on images of Australia; his major work on the settlement of Victoria's western plains (Powell, 1970) examined the conflict between official and popular environmental appraisals, the dialogue between these, and the learning process which resulted in the final settlement pattern (see also Powell, 1977, where he presents the study of eiconics, or image-making, in the context of colonization processes). Billinge (1977, p. 64) queried whether such works are really phenomenological, however:

> the idea has spread that since certain branches of our discipline are less susceptible to quantitative reduction (and, so the argument continues, by false extension, to scientific analysis), we can justify our partially formulated hypotheses, exploit the atypicality of our data, cease worrying about the validity of our reconstruction and within some weakly articulated framework label the whole exercise phenomenological.

The study of perceived environments represents an 'important and vigorous movement' (p. 65) but Billinge argued that phenomenology is not just the study of such environments: its method is presuppositionless, concerned with the human consciousness and not only its output, thus 'phenomenological we have by no means become' (p. 67) since geographers have paid little attention to consciousness. Much work claimed as phenomenological is probably closer to the idealist position, since it does not seek to isolate essences.

Geographers' attempts to adopt a phenomenological approach were trenchantly criticized by Pickles (1985), not because he opposed them in principle but rather because he believed that those undertaken were misconceived; he accepted that, if for no other reason than its attack on positivism, 'It cannot be denied that the founding and guiding intuitions of a phenomenological approach in geography as it exists at the moment are in the main sound and well intentioned' (p. 68) but argued that work such as Buttimer's and Relph's, 'is ungrounded method, unfounded claims, and the actual imposition of unexamined propositions' (p. 71). His criticisms led him to contend that: 'we now need to move from what passes for phenomenology in the geographical literature, *towards* what is actually the case in phenomenology itself' (p. 89). Quoting substantially from Husserl's and Heidegger's original works, he argued that phenomenology is not concerned to explicate subjective meanings as an end in itself, but rather 'to be the science of science through explicating the science of beginnings' (p. 97). Its goal is to identify the essences that underpin individual meanings:

> The essential relation between an individual object and its essence – such that to each object there corresponds its essential structure, and to each essential structure there corresponds a series of possible individuals as *its* factual instances – necessarily leads to a corresponding relationship between sciences of fact and sciences of essence (p. 111).

The subjectivity of the lived world is to be explored for the insights that it can provide on those essential structures, the underpinnings of knowledge itself.

According to this interpretation, phenomenology has much in common with certain forms of structuralism in its basic concern with neither empirical appearances nor actual decision-making but rather with the deep structures of consciousness (see p. 214; Johnston, 1986f: Pickles, 1988, p. 252, implies this too). Thus it does not fit easily into the humanistic concerns of most geographers who have espoused phenomenology, which Pickles identifies as the search for those essences that give rise to the necessity for geography, as the empirical science concerned with particular facets of behaviour. (Each empirical science should be grounded in such an essence.) Thus:

> we seek an ontological, existential understanding of the universal structures characteristic of man's spatiality as the precondition for

any understanding of places and spaces as such. That is, we seek to clarify the original experiences on the basis of which geography can articulate and develop its regional ontology if geography as a *human* science, concerned with *man*'s spatiality, is to be possible at all (p. 155).

Spatiality is that essence, he claims, best represented by the German noun *Raum* and verb *raumen* 'which means to clear away, to free from wilderness or to bring forth into an openness. *Raumen* is thus a clearing away or release of places, a making room for the settling and dwelling of man and things' (p. 167). In this context, space and place are closely related concepts (see also Gould, 1981b).

Phenomenology is closely associated with existentialism, and some geographers have experienced difficulty in separating the two (Entrikin, 1976). Whereas phenomenology assumes the primacy of essence – the allocation of meanings results from the existence of consciousness – for existentialists the basic dictum is 'being before essence – or man makes himself'. The process of defining oneself (creating an essence) involves creating an environment. Thus Samuels (1978, p. 31) argued that environments can be read as biographies – 'for every landscape or every existential geography there is someone who can be held accountable'. Generalization may be possible from analysing such landscapes. Appleton (1975), for example, suggested that landscapes reflect two primal needs – prospect (the need to search for the means of survival), and refuge (the need to hide from threatening others) – although individuals and groups may satisfy these in particular ways (as he illustrates from his autobiography: Appleton, 1994). Lowenthal (1975; see also Lowenthal, 1985) argued that individuals rewrite their biographies, and those of their ancestors, by their choices of what to preserve in the landscape.

All of this suggests that humanistic geography is concerned either with the study of individuals and their construction of, plus behaviour in, phenomenal environments (as in Rowles, 1978) or with the analysis of landscapes as repositories of human meaning. As such, it is separate from the subject matter of much human geography, notably behavioural geography and its investigation of everyday activity within environments. But the phenomenological perspective has been adapted to the latter type of work also, in the application of Schutz's writings on the 'taken-for-granted' world (Ley, 1977b). Much everyday behaviour is unconsidered, in that it involves no original encounters with new situations. The behaviour is habitual, because all of the stimuli encountered can be processed as examples of particular types. Those types are not externally defined for the individual, but are personally created. The phenomenology of the taken-for-granted world is the study of those individually defined typifications – of the unconsidered 'world of social reality' rather than 'a fictional non-existing world constructed by the scientific observer' (Ley, 1980, p. 10, quoting Schutz): see also Curry's (1982a, p. 38) discussion of 'the ordinary, everyday actions of individuals' which create for the geographer 'a complex world of complex places and actions' in which the role of the individual decision-maker 'can be determined only on an individual basis, case by case'. Interaction within communities may lead to shared typifications, which quantitative methods may be used to identify descriptively. (Quantification is not tied to positivism, except when it is used to suggest laws and other generalizations: Johnston, 1986a.)

Humanistic geography is based on a profound critique of positivist work, which it claims makes major unwarranted assumptions about the nature of decision-making and seeks inductive laws of human behaviour that can be scientifically verified (Ley and Samuels, 1978; Powell, 1980b). Its counter-argument promotes understanding of the individual as a 'living, acting, thinking' being. To some, however, it is just a criticism – Entrikin (1976, p. 616) argues that:

> humanist geography does not offer a viable alternative to, nor a pre-suppositionless basis for, scientific geography as it is claimed by some of its proponents. Rather the humanist approach is best understood as a form of criticism. As criticism the humanist approach helps to counter the overly objective and abstractive tendencies of some scientific geographers.

For others, however, the human condition can only be discerned through humanistic endeavour, for attitudes, impressions and subjective relations to places (the 'sense of place') cannot be revealed by positivist research. As Pickles (1986, p. 42) put it 'The value of humanism has been its resilience in consistently raising questions which do not fit within other frameworks . . . humanism has been the voice of man against reason, against science.'

Some critics argue that humanistic geography focuses on relatively trivial matters and not with the major foci of an applied geography concerned with improving the world. To Buttimer (1979, p. 30) the latter form of applied geography is managerial, manipulating individuals and their environments rather than seeking to advance the process of 'human becoming'. To Relph (1981b, pp. 139–41) it implies geographism: 'the view that people should behave rationally in geographical, two-dimensional space . . . that cities and industries and transportation routes should be arranged in the most efficient way' which when used as a basis for planning 'will diminish the distinctiveness and individuality of . . . communities and places. Geographism involves the imposition of generalizations onto specific landscapes; it breeds uniformity and placelessness.' Both of them argued that planning should emphasize subjectivity and individuality. Scientists, engineers and planners may seek to improve well-being, but in so doing they make people rootless and deny them individuality. Planning must be allied with environmental humility, by which (Relph, 1981b, p. 201): 'places and communities would increasingly become the responsibility of those who live and work in them instead of being objects of professional disinterest'. In this context, humanistic geography is not only a reaction against the dehumanizing treatment of people in spatial science and behavioural geography but also an argument against an applied geography which imposes that treatment on the landscape, and for a form of anarchism in which individuals are encouraged to realize what they are and how they can control both themselves and their environments (see also Pickles, 1986, p. 47). Seamon (1987) has reviewed a substantial volume of phenomenological work, illustrating that it is 'a learning tool which can help us to discover more about ourselves, others, and the world in which we live' (p. 21), and which may then have practical value in environmental design.

The practice of humanistic geography

Much of the practice of humanistic geography has been concerned with exploring and explicating the subjectivity of human action and its base in meanings (both individual and shared), with relatively little concern for the claims of idealist, phenomenological, existentialist and other philosophies. (Curry, 1996, p. 16, claims that although the earliest figures involved in the break from positivism all referred to phenomenology, for example, 'it seems fair to say that none took it very seriously'.) To the extent that such work is explicitly influenced by philosophical and methodological writings, the (albeit usually implicit) stimulus is more often the *pragmatism* (and symbolic interactionism) developed by the Chicago school of sociologists in the 1920s and 1930s (and largely overlooked by human geographers who paid much more attention to the relatively small amount of spatial analysis undertaken by that school, as typified by the Burgess model of the internal structure of cities: Johnston, 1971). Pragmatism portrays life as a continuous process of experience, experiment and evaluation through which beliefs are continually reconstructed; such reconstruction is a social process, whereby individuals learn and behave in the context of the beliefs of those with whom they interact (hence the term interactionism: see Jackson and Smith, 1984, Chapter 4: Smith, 1988, and Jackson, 1988, illustrate this in their own research).

Understanding social life within this broad framework involves participatory field work (as detailed in Evans, 1988), the methodology of which (according to S. Smith, 1984) is 'a hallmark of much geographic humanism' (p. 353). This, she continues,

> requires a commitment to fieldwork, with the aim of securing data lodged in the meanings ascribed to the world by active social subjects. The strength of this strategy derives from the unique insight it offers into 'lay' or folk' perceptions and behaviors. True to the pragmatic

maxim, the method allows the truth of a social reality to be established in terms of its consequences for those experiencing it
(pp. 356–7).

How local societies work is thereby explicated, which was the goal of the
ethnographic work of the Chicago school (Jackson, 1984, 1985); Ley's
(1974) detailed portrayal of a Philadelphia community illustrates this, as
does his general textbook on urban social geography (Ley, 1983).

Field studies, involving interaction with residents over long or short
terms, provide only one source of information about how people structure
their lives. There are other *texts*, other repositories of meanings: thus
Pocock (1983) described humanistic strategies ranging 'from library search,
to the observational, to the experiential' (p. 356). The landscape, the creation of those who live/have lived in it, is an important text, and Lowenthal
(1985) provides a perspective on its study: he portrays it not as a mirror of
the past but rather as an insight to the present:

> Traditions and revivals dominate architecture and the arts; school-
> children delve into local history and grandparental recollections;
> historical romances and tales of olden days deluge all the media.
> The past thus conjured up is, to be sure, largely an artifact of the
> present. However faithfully we preserve, however authentically we
> restore, however deeply we immerse ourselves in bygone times, life
> back then was based on ways of being and believing incommensu-
> rable with our own. The past's difference is, indeed, one of its
> claims: no one would yearn for it if it merely replicated the pre-
> sent. But we cannot help but view and celebrate it through present-
> day lenses (p. xvi).

The present landscape is a conglomerate of relics from many different
periods in most cases, and by preserving only parts of it we bias our representation of the past:

> Every act of recognition alters survivals from the past. Simply to
> appreciate or protect a relic, let alone to embellish or imitate it, affects
> its form or our impressions. Just as selective recall skews memory and
> subjectivity shapes historical insight, so manipulating antiquities
> refashions their appearance and meaning. Interaction with a heritage
> continually alters its nature and context, whether by choice or by
> chance (p. 263).

Thus when use of the landscape as a text involves reading the outcome of a
long sequence of selective retentions of earlier forms, so that we learn about
what parts of their history people wanted to build into their own presents
and futures: 'We must concede the ancients their place . . . But their place is
not simply back there, in a separate and foreign country; it is assimilated in
ourselves, and resurrected into an ever-changing present' (p. 412). The past

does not exist independently of those who seek to interpret it (as Taylor, 1988, makes clear in a very different context); the landscape may tell us more about the past which people wanted to preserve than about the past as it was experienced. (It also tells us much about the people who describe it, as Porteous, 1986, illustrates in his discussion of body imagery as a metaphor for landscape description.) Study of current processes of land-scape-creation illustrate this process, as with the creation of various types of 'living museum' and the promotion of places as 'cities of spectacle' (as in Crang, 1994; Jacobs, 1994).

An increasingly used text for the explication of meanings is literature – or 'creative literature' in White's (1985) term; Porteous (1985, p. 117) refers to 'imaginative literature', within which geographers have been highly selective:

> Plays are not considered, poetry is but occasionally used, the novel reigns supreme. The advantages of the novel lie in its length (meaty), its prose form (understandable), its involvement with the human condition (relevant), and its tendency to contain passages, purple or otherwise, which deal directly with landscapes and places in the form of description (geographical).

He argued that when using novels as texts, geographers have concentrated on nineteenth-century, rural contexts (as in Darby's, 1948, paper on Hardy's Wessex). To correct this, he proposed a two-variable categorization of situations according to whether the subject is an insider or an outsider and whether the place being described is 'home' or 'away'. The 'home-insider' provides material on sense of place, whereas the much less frequently reported 'away-outsider' refers to those experiencing alienation in a placeless world. (White's, 1985, use of novels to describe migrants' situations fits readily into this category.) 'Home-outsiders' are people who fail to develop insider relationships with their milieux, whereas 'away-insiders' are travellers reporting their experiences (as in 'road, tramp, and down-and-out novels'; Porteous, 1985, p. 119). Such distinctions are reflected in a collection of essays about literature and migration (King, Connell and White, 1995), which distinguishes between migrants' autobiographical accounts and general fiction about migration (Duffy, 1995). These are used to explore the renegotiation of identity that accompanies migration experiences (White, 1995), but as texts they are subjected to less critical scrutiny than that undertaken by deconstructionists (p. 298).

Much humanistic geography has been written to describe and appreciate the variety of the human condition, as it is experienced. (Eyles's, 1989, essay, in a volume designed to 'introduce some of the most exciting challenges of the contemporary subject to a wider audience' writes of place and landscape without reference to philosophy.) It has been relatively unconcerned with philosophical issues, such as the origins of meanings in human consciousness, and has focused almost entirely on the empirical worlds of experience, even if these have to be interpreted from secondary sources, although Watson (1983), himself a prize-winning poet, claimed that literature is not a secondary source but 'primary source material for the whole world of images' (p. 397) that illustrates the 'soul' of a place. Its relevance, according to Pocock (1983), is that:

> it attempts to unravel the nature of being-in-the-world, as it explores the existential significance of place as an integral part of human existence. In short, it is a geographical contribution to the most fundamental of questions, 'what is man?' (p. 357).

Meinig (1983, p. 325) expresses this more vigorously:

> By limiting ourselves to describing and measuring and analyzing certain aspects of the world as it seems geographers have denied themselves the possibility of probing very deeply into what it all means. Being unable to convey what it means we cannot help shape it toward what it might become.

Explicating what it means involves practising geography as art – placing the discipline firmly in the humanities as well as the sciences and the social sciences.

One problem for such work, brought home to geographers by Olsson in a series of iconoclastic essays (Olsson, 1978, 1979, 1982: see also Pred, 1988), is that the medium of the text is constraining as well as enabling. He focuses on language as the most frequently used medium, with the constraints that word definitions and usages put on their application, hence the modes of thought that they support. Thus (Olsson, 1982, p. 227):

> any social scientist is handicapped by the methodological praxis which requires him to be more stupid than he actually is. Thus, in the interests of discipline, verification and communication he relies mainly on the two senses of sight and hearing: what counts is what can be counted: what can be counted is what can be pointed to; what can be pointed to is what can be unequivocally named. Accumulation of knowledge about the nameable is consequently the point of the scientist's game.

Thus ambiguity is translated into certainty; describing a phenomenon in words allocates it to a category, and can over-simplify its complexity. Hence our ability to think is made possible by the richness of language yet also constrained by the categorizations that it imposes; the former aids our understanding but the latter can hinder it (so that, as critical theorists have made clear – p. 235 – language, as the major medium of communication, can be employed ideologically to promote certain forms of understanding: Held, 1980).

Geographers' involvement in humanistic work involves more than interpretation of texts, however, since they themselves create new texts in transmitting their appreciations to others, usually in writing but also through maps, lectures and other media. This involves what is sometimes referred to as a double hermeneutic. Initially developed for the exegesis of biblical texts, *hermeneutics* was developed by Dilthey (Rose 1980, 1981) to embrace all studies which involve scrutinizing an author's intentions when evaluating a text. This involves what Dilthey termed *Verstehen* (interpretative understanding), which not only enables students of a text to appreciate its author but also to increase their own self-awareness (Giddens, 1976, p. 17):

> Understanding a text from a historical period remote from our own
> . . . or from a culture very different from our own is . . . essentially a
> creative process in which the observer, through penetrating an alien
> mode of existence, enriches his own self-knowledge through acquiring
> knowledge of others.

(Thus an applied humanistic geography promotes both self-awareness and awareness of others.) A double hermeneutic is involved because it is not only (Rose, 1981, p. 124): 'the job of human geographers to interpret such texts as a spectator in order to make certain statements about actors operating within the texts' but also 'to communicate the meanings of such phenomena as he should deem important back to the actors involved' as well as to others who wish to understand those meanings. Thus the reader of a research paper by a humanistic geographer is only interrogating those meanings through their interpretation by an intermediary – the humanistic geographer. The latter's selection of those items 'deemed important' means that different geographers may transmit different interpretations of the same text, in the same way as two biographers might interpret a novelist's texts differently and two artists may differ in their portrayal of the same landscape.

Humanistic and cultural geography

Detailed debates about philosophy and methodology in geography have engaged only a minority of the discipline's practitioners, so far as the empiricist/positivist approaches are concerned, whereas a majority have conducted their own empirical inquiries informed by those debates but only marginally connected to them. To a considerable extent this has been the case with humanistic geography too, with Hugill and Foote (1994, p. 18) contending that whereas humanistic geographers

> turned to such philosophies and methodologies as ethnomethodology,
> existentialism, idealism, phenomenology, symbolic interactionism,
> and transactionalism . . . [although] successful applications of each of

these can be found in the geographical literature . . . none captured
the imaginations of more than a handful of geographers at a time.

A small number of writers have been concerned with philosophical underpinnings and the relative merits of different humanistic approaches, but
more have accepted the general tenor of the argument, described by Boal
and Livingstone as (1989, pp. 7–8):

> We can give up the need to find direct empirical connections
> between terms and objects in the world; we can view knowledge
> not as presenting the world in some correct way, but as just helping
> us to get along in it, or to change it. Truth, to repeat, has nothing
> to do with accurately representing, or, as Rorty has it, mirroring
> reality; it is just, according to Rose, 'what we are well advised,
> given our present beliefs, to assert'. The purpose of geography, then,
> is not to tell us about how the world 'really' is: it tells us nothing
> about regions or landscapes or economic structures or human
> agency, because these are mere linguistic fictions; it is just the search
> for the right vocabulary, the right jargon, the best discourse in
> which to pursue the kinds of account which help us, in the most
> basic sense, decide what to do.

(The quote is from Rose, 1987.) Doing that, according to Pickles (1986,
p. 29), involves 'archaeology', uncovering the layers of human behaviour to
identify the experience that underpins it, which involves the explication of
texts – the 'written records, cultural artefacts, urban landscapes or whatever' (Boal and Livingstone, 1989, p. 15) – which are the repositories of
human meanings, and whose interpretation is the goal of humanistic geography.

Some of the work undertaken within this humanistic framework, and
without the explicit philosophical underpinnings, goes under the rubric of
cultural geography, which has a long and distinguished background,
notably in the United States, based on the work of Sauer. The suggestion of
a break between an 'old' and a 'new' cultural geography was denied in the
review in *Geography in America*: Rowntree, Foote and Domosh (1989,
p. 215) argued that 'Although traditional cultural geography has preferred
topics with historical depth, there is increasing interest and emphasis on the
study of everyday life and landscapes.' They recognized that trends origi-
nating in Britain, as promoted by Cosgrove and Jackson (1987), differed
substantially from the more traditional concerns (as represented in the
chapter on cultural ecology in the same volume: Butzer, 1989; see also
Turner, 1989). 'Whether the British trajectory becomes an integral part of
North American cultural geography remains to be seen': quoting Kofman
(1988), they suggested that 'the "new" cultural geography is more talked
about than done' (p. 209). Hugill and Foote (1994, p.18), however, see the
introduction of humanistic approaches (as exemplified by Ley and

Samuels', 1978, anthology) as 'an attempt to define a "new" cultural geography' in an 'attempted coup' which failed. Furthermore, they also portray a conflict between traditional cultural geography as practised in North America and that promoted in the United Kingdom:

> a large number of [American cultural] geographers had continued to cultivate the traditional themes of the Berkeley school . . . These geographers felt little need to accept humanistic geography and resented its new agenda. Humanistic research had little to say about traditional concerns. The unfortunate mismatch between American cultural geography and British social geography compounded problems of assimilation. The contributions of British social geography to recent advances in humanistic cultural geography have been increasing . . . but the overlap is slight.

Duncan (1994, p. 401) refers to this debate as:

> a civil war [that] has been going on in cultural geography since the early 1980s. This struggle has largely, though not exclusively, had an intergenerational character, younger cultural geographers trained in the late 1970s and 1980s assailing the positions of an older generation trained in the 1950s and 1960s.

The main issue, he argued, was theory:

> The younger generation launched its attacks using an arsenal of theory-seeking weapons provided by suppliers in the humanities and social sciences. Some direct hits were scored on the lightly camouflaged theories of the older generation, but such hits proved indecisive in the war for a number of reasons. Many of the older generation were unaware that they had any theories in their camp: others insisted that these were relic theories long since abandoned. Still others claimed that the theoretical targets had in fact been surreptitiously inserted into their camp by the younger generation to discredit them. At any rate, they argued that no meaningful losses had been sustained.

Others (Price and Lewis, 1993a) argued that the older generation was far from homogeneous in its views, however, and that 'traditional cultural geography' was being misrepresented by those advancing an alternative position (see also p. 270 below).

Cosgrove and Jackson (1987) pointed not only to the continued vitality of cultural geography as the interpretation of past and present landscapes and other texts (for which they appropriated the term iconography, as in Cosgrove and Daniels, 1988: see also Powell, 1977, to whom they do not refer) but also to growing contacts with contemporary social geography. (Soon after they wrote, the Social Geography Study Group of the IBG was renamed the Social and Cultural Geography Study Group.) The field of contemporary cultural studies offered 'alternative ways of theorising culture without specific reference to the landscape concept' (p. 98), they claimed, with its emphasis on contemporary sub-cultures and their political struggles rather than 'the elitist and antiquarian predilections of traditional cultural studies'. This is illustrated in a special issue of *Society and Space* (Gregory and Ley, 1988), and in Jackson's (1989) text *Maps of Meaning* which begins with a definition of culture that characterizes the 'new' cultural geography (p. ix):

> This book employs a more expansive definition of culture than that commonly adopted in cultural geography. It looks at the cultures of socially marginal groups as well as the dominant, national culture of the elite. It is interested in popular culture as well as in vernacular or folk styles: in contemporary landscapes as well as relict features of the past.

The work of Sauer and the Berkeley school (p. 48 above) is criticized for being relatively narrow; culture is instead linked to the contested concept of ideology, and substantive chapters cover popular culture, gender and sexuality, racism and language. An agenda for future work suggests an even wider coverage, into aspects of social relations that impinge upon the economic organization of society and thus link cultural geography more firmly with aspects of radical geography (as discussed below: p. 307). Work in this mould will presumably further the challenge to the content of traditional cultural geography as 'innocent' (Rowntree *et al.* 1989, p. 214).

More recently, the concept of culture has itself been examined. The definition adopted by Duncan and others, in contrast to the 'superorganicism' of the Berkeley school and its followers, is described as (Mitchell, 1995, p. 102): 'socially constructed, actively maintained by social actors, and supple in its engagement with other "spheres" of human life and activity'. For those who adopt this position, culture exists as, according to Cosgrove and Jackson (1987, p. 95), 'the very medium through which change is experienced, contested and constituted'. To Mitchell (1995, p. 103) this means that:

> Culture, therefore, can be specified as something which both differentiates the world and provides a concept for understanding that differentiation. Culture itself is a sphere of human life every bit as important as, yet somehow different from, politics, economy and social relations. It is an important ontological category which must be theorized and understood if we hope to understand human differentiation, behaviour, experience and contest.

Against this, he argues that 'there is no such (ontological) thing as culture. Rather there is only a very powerful *idea* of culture, an idea that has developed under specific historical conditions and was later broadened as a means of explaining material differences.' As with the superorganic conception, he contended, the concept of culture promoted by the 'new cultural geography' involves reification and its adherents have reached 'something of a dead end' (p. 104): 'While important empirical work exploring the social creation of many aspects of life continues, none of this work has been able adequately to explain what culture is. Cultural geography has remained incapable of theorizing its object.'

Attempts to define culture, according to Mitchell, involve an infinite regress, using terms which themselves are neither internally coherent nor inclusive (see the later discussion on metaphors: p. 298). The regress is ended by claims that cultures exist and the term is used as part of a strategy to define and control.

> The naming and representation of cultures create partial, yet globalizing, truths. By localizing social interaction into discrete cultures . . . contentious activities are abstracted into the partial truth contained in the idea of culture: namely that there are true and deep differences between people.

This strategy is crucial in many aspects of life, such as geopolitics (see p. 234) and the promotion of consumerism (p. 307). Thus (p. 112): 'like "race", "culture" is a social imposition on an unruly world. What *does* exist, and very importantly, is the historical development of the *idea* of culture as a means of ordering and defining the world . . . Culture is an idea that integrates by dividing'; it is a strategy for control, like territoriality, and its study calls for investigations of how the concept of culture has been used to promote both positive and negative views of societies (of 'us' and 'them'). As such, many cultural geographers have been deeply involved in developments during the 1990s characterized below as 'the cultural turn' (Chapter 8).

Whereas traditional cultural geographers pay little explicit attention to methodological concerns, those linked to the more recent developments have addressed a range of issues involved in the collection of 'data', through the interrogation of both texts and people's lived experiences. 'Doing humanistic, or cultural, geography' involves a variety of methods often characterized as ethnographic and contrasted to the positivist approach (Cook and Crang, 1995, p. 4) as 'reading, doing and writing . . . thoroughly mixed up' during the course of a piece of research rather than the 'conventional *read*-then-*do*-then-*write* sequence' of other approaches. Cook and Crang note that one author has identified 43 different ways of conducting ethnographic research (Tesch, 1990), but focus on just four:

1 *participant observation*, in which the researcher(s) lives in and/or works among a selected community in order to appreciate its values and ways of life, and then interprets that culture to a wider audience – the process involves researchers and community developing mutual (or inter-subjective) understandings;

2 *interviewing*, which involves conversations between researcher(s) and researched, which can vary in the degree to which they are structured with pre-determined questions by the interviewer;
3 *focus groups*, whereby researchers initiate information-collection by bringing together groups who interact as they discuss the issues raised by the researcher (and perhaps others which lead from them), thereby treating people as members of interacting communities which influence their behaviour rather than as isolated individuals; and
4 *filmed approaches*, wherein visual material obtained from the research subjects indicate meanings and values which are not readily expressed in words.

Preparing for each of these approaches is a major task, as is the follow-up construction of the information obtained. A number of ethical issues are involved (geographers have paid much less attention to these than other social scientists, such as anthropologists, psychologists and sociologists, who depend on such approaches for their information, and whose learned societies have developed ethical codes of practice for field investigations) and the continued relationship between researcher(s) and the researched is of considerable importance. As Cloke, Philo and Sadler (1991, p. 92) put it, 'doing ethnography' lacks neither rigour nor academic merit: it is not, as some imply, a 'soft option' compared to the 'hard' spatial science:

> rather, it is to engage very honestly with both the enchantment and the problems associated with researchers trying to gain an insight into the worlds of other peoples in other places, and it is also to insist that this insight emerges not by supposing (as might a philosophically inclined phenomenologist or existentialist) that these peoples are basically the same as us but by letting them and their version of humanity simply be different.

Conclusions

A common thread links the material discussed in this and the preceding chapter; both are concerned with positive rather than normative investigations, with attempts to uncover how humans behave in the world rather than with contrasts between actual patterns of behaviour in space and those predicted from normative theories. Both are also part of a more general trend towards an anthropocentric focus within the social sciences, which reflects reorientations in the external environment. From the mid-1960s on, there was a growing disillusion with science and technology, especially among students, and the popularity of the social sciences boomed. Within the latter, there was a shift in emphasis from study of the aggregate to the individual, an increase in the relative volume of research conducted at the micro-scale, and growing unease about the role of social scientists in planning mechanisms. Both behavioural (or behaviourist) geography and the various humanistic approaches reflect these trends.

Beyond the general concern with the individual as a decision-maker, little else connects the two types of work, however. Behavioural geography has maintained strong ties with the positivist/spatial science tradition. Data are collected from individuals, but these almost all concern the conscious elements in action and their outcomes; they are usually aggregated in order to allow statistically substantive and significant generalizations to be made about spatial behaviour, almost certainly in the context of the normative models of the spatial-science school. In the humanistic strand, on the other hand, the intent has been to understand and to recognize the dignity and humanity of the individual, but there has not been a widespread embracing of the philosophies discussed in this chapter. Many (perhaps most) investigations in human geography from 1970 onwards have involved some behavioural concepts other than those of 'economic man', but relatively few human geographers have become humanistic geographers.

The two approaches are very much antithetical. Behavioural geography treats people as responders to stimuli, and investigates how different individuals respond to particular stimuli (and also how the same individual responds to the same stimulus in different situations): by isolating the correlates of those varying responses it builds models that can predict the probable impact of certain stimuli. The end product is input to processes aimed at either providing environments to which people respond in a preferred way or at changing behaviour by changing the stimuli. (Such environmental planning is considered social control by some.) The other approach – humanistic geography – treats the person as an individual constantly interacting with the environment and with a range of communities, thereby continually changing both self and milieu. It seeks to understand that interaction by studying it, as it is represented by the individual and not as an example of some scientifically defined model of behaviour. And then by transmitting that understanding, it seeks both to reveal people to themselves, enabling them to develop the interactions in self-fulfilling ways, and to promote their appreciation by others.

The two approaches are not readily combined, since the concept of behavioural laws is not in sympathy with the humanist view that people are alone responsible for their circumstances and destinies, and are free – if they so choose – to exercise that responsibility as they will. Few adherents of either approach totally denigrate the other. Humanists recognize that behavioural geography and spatial science may be valid approaches to certain aggregate phenomena – such as trade flows (Ley, 1981; see also Tuan, 1977) and the planning of transport systems – and behaviouralists recognize that their methods are irrelevant to the study of emotions and aspects of landscape (as suggested in the two main ways of studying space: Sack, 1981). But the two have little in common and are competing for central positions as the philosophy of *human* geography.

For D. M. Smith (1988a), humanistic work involves a 'new movement', and the chapters in the book that he edited with Eyles (p. 266)

> have provided persuasive accounts of ways of interpreting the geographical world which are capable of challenging if not displacing prevailing orthodoxy or at least of providing a convincing alternative. The positivist and empiricist paradigm fixated by mathematical process modelling with a promise of policy relevance . . . is already yielding to approaches more sensitive to actual human achievement.

For him, the struggle between paradigms continues, with humanistic interpretative work now being rich in empirical example as well as strong in philosophical criticism; as subsequent chapters illustrate, it has also been integrated with other approaches to provide a wider appreciation of 'individuals in society'.

|7|

'Radical' approaches

Since the late 1960s a group of related approaches have developed within human geography which lacked an all-embracing single, descriptive adjective. The term initially favoured was 'radical' (see Peet, 1977, 1978, 1985b), but this became less popular in the 1980s as the term was applied to the political movements associated with the 'new right', 'Thatcherism' and 'Reaganomics'. Structuralist, realist, critical theory and other terms were proposed and adopted by some, whereas to yet others the radicalism was clearly associated with the adoption of a marxist (or marxian; on the difference see, Harvey, 1973) approach. But although Walker (1989a, p. 135) has argued that 'While not every realist is a marxist as regards theory of society . . . every marxist must be a realist', the latter term has not been generally accepted, and the so-called 'radical' approaches lack a consensus definition and coherence.

More recently, the approaches have been categorized in two new ways. In his evaluation of their contribution to American geography, Walker (1989b) writes of 'Geography from the left', which involves 'bringing the analytic framework and progressive social agenda of marxism and allied schools of thought into most of the traditional subject areas of the discipline' (p. 619). The goal of achieving a 'more explicitly spatialized theory of capitalist societies' involved some scholars developing marxist theory more fully, whereas others concentrated on methods and welcomed 'the clarifications that realism, critical theory, and structuration theory might add to the understanding of social processes and how to grasp them' (p. 620); others focused on 'middle level' theories of such topics as local labour markets. Walker argued that:

> Left geographers can be proud of their achievements in a discipline that is not always noted for its explanatory depth or overriding concern with human oppression and liberation. The left can claim a good deal of credit for broadening the intellectual respectability of the geographic enterprise.

His review clearly illustrates the breadth of enquiry undertaken by 'the left'.

Rather than use 'the left' as the collective term within which to group a wide range of work, and no longer favouring the adjective 'radical', Peet and Thrift (1989, p. 3) selected 'political economy':

> to encompass a whole range of perspectives which sometimes differ from one another and yet share common concerns and similar viewpoints. The term does not imply geography as a type of economics. Rather economy is understood in its broad sense as social economy, or way of life, founded in production ... Clearly, this definition is influenced by Marxism ... But the political-economy approach in geography is not, and never was, confined to Marxism ... So while political economy refers to a broad spectrum of ideas, these notions have focus and order: political-economic geographers practise their discipline as part of a general, critical theory emphasizing the social production of existence.

This group of approaches began as 'radical geography' in the 1960s, as a (lukewarm, they claim) reaction to the crises of capitalism made clear in both the response to the Vietnam War and the US Civil Rights movement; in the 1970s, led by David Harvey, it explored marxism in depth; and in the 1980s it became 'more sober and less combative' (p. 7) as marxism was subjected to criticism, the recession of the 1980s led to more disciplined inquiry, greater knowledge of the problems of socialist economies made the prospects of revolutionary change less likely, geography became more narrowly professional, and some of the 'radical, anti-establishment Young Turks' joined the establishment (see Barnes, 1996, p. 49). Nevertheless, they conclude (p. 7) that: 'The political-economy approach ... has survived counterattack, critique, and economic and professional hard times, and has matured into a leading and, for many, *the* [their emphasis] leading school of contemporary geographic thought.' This chapter focuses on that work and the reactions to it.

Radical beginnings

Peet (1977) has claimed that the early 'radical' work by geographers in the late 1960s was liberal in its attitudes (p. 142):

> Radicals investigated only the surface aspects of these questions – that is, how social problems were manifested in space. For this, either we found the conventional methodology adequate enough or we proposed only that existing methods of research must be modified to some extent if they are to serve the analytic and reconstructive policies of ... radical applications (Wisner, 1970, p. 1) ... we

were fitting into an established market ... we were amenable to established ways of thinking ... we were useful in providing background ideas for the formulation of pragmatic public policy directions, and so could not, and were not, engaging in radical analysis and practice (p. 245).

This was illustrated by his own paper on poverty in the United States (Peet, 1971) – in which, like Morrill and Wohlenberg (see p. 335), he argued for a series of growth centres in the poverty areas to stimulate economic development and job creation – and also by the tenor of the articles in early issues of the radicals' journal, *Antipode* (which was launched at the 1969 AAG meetings: see p. 323). Morrill (1969, pp. 7–8), for example, argued

against the 'New Left' premise that a revolution is the only route to progress ... the dreams of revolution are naive ... the 'New Left' vastly exaggerates potential support ... a 'revolutionary program' is hopelessly dated and simplistic ... the 'New Left' underestimates the capacity of our society for change. ... All revolutions seem to have been betrayed by incompetents who preferred exercising power to executing reform.

Again, he argues (Morrill, 1970b, p. 8) that

A simple marxist-type change in the ownership of business from private to a government (or union) bureaucracy would in all probability decrease production, and would not necessarily bring any improvement in basic conditions. The key is to retain the institution of private property while instituting social control over its exchange and circumscribing its power over people.

The case for a marxist approach was first presented formally by Folke (1972) in a critique of Harvey's (1972) paper on ghetto formation and counter-revolutionary theory. Folke represented geography and the other social sciences as 'highly sophisticated, technique-orientated, but largely descriptive disciplines with little relevance for the solution of acute and seemingly chronic societal problems ... theory has reflected the values and interests of the ruling class' (p. 13). Liberal cases like Morrill's were dismissed as unlikely to succeed. Morrill's two papers in *Antipode* argued for change to be brought about by persuasion, producing a capitalist–socialist convergence. But this is the social democrat method practised in Sweden

where it has been shown over and over again that the idea of equal influence for employers and employees is an illusion. After half a century of social-democratic rule injustices and inequalities still prevail. ... No small group of experts can accomplish anything ... when it runs counter to the interests of the dominant social forces. These are not interested in equality or justice, but in profit (Folke, 1972, p. 15).

Radical change requires mass mobilization, and so, to Folke, Harvey's call for a new paradigm within geography was insufficient. What is needed is a new paradigm for a unified social science, containing geography, which deals with problems in all their complexity and provides not only theory but the basis for action: 'Revolutionary theory without revolutionary practice is not only useless, it is inconceivable . . . practice is the ultimate criterion of truth' (p. 7).

The major contribution to the case for a marxist-inspired theoretical development within geography was made by David Harvey, initially in his collection of essays *Social Justice and the City* (Harvey, 1973). The book is presented as autobiographical, illustrating the evolution of Harvey's views towards an acceptance of Marx's analysis:

> as a guide to enquiry . . . I do not turn to it out of some *a priori* sense of its inherent superiority (although I find myself naturally in tune with its general presupposition of and commitment to change), but because I can find no other way of accomplishing what I set out to do or of understanding what has to be understood (p. 17).

The first part of the book, entitled 'Liberal Formulations', comprises essays which analyse problems of inequality within societies in terms of income-allocating mechanisms; the role of accessibility and location in those mechanisms is stressed. This leads to an attempt at defining territorial social justice, which separates the processes allocating incomes from those which produce them. Only in the second part of the book – 'Socialist Formulations' – is it:

> finally recognized that the definition of income (which is what distributive justice is concerned with) is itself defined by production. . . . The collapse of the distinction between production and distribution, between efficiency and social justice, is a part of that general collapse of all dualisms of this sort accomplished through accepting Marx's approach and technique of analysis (p. 15).

The transition in Harvey's approach is marked by a paper on the ghetto (Harvey, 1972). He begins with a critique of Kuhn's model of scientific development (p. 12), asking how anomalies to the current paradigm arise and how they are translated into crises. The problem with Kuhn's analysis, he claims, is the assumption that science is independent of its enveloping material conditions, when in fact it is very much geared to its containing and constraining society. Recognition of this point is important for geographers because:

> the driving force behind paradigm formulation in the social sciences is the desire to manipulate and control human activity in the interest of man. Immediately the question arises as to who is going to control whom, in whose interest is the controlling going to be exercised, and

if control is going to be exercised in the interest of all, who is going to take it upon himself to define that public interest? (Harvey, 1973, p. 125).

To Harvey, marxist theory provides 'the key to understanding capitalist production from the position of those not in control of the means of production ... an enormous threat to the power structure of the capitalist world' (p. 127). It not only enhances an understanding of the origins of the present system, with its many-faceted inequalities, but also propounds alternative practices which would avoid such inequalities: 'we become active participants in the social process. The intellectual task is to identify real choices as they are immanent in an existing situation and to devise ways of validating or invalidating these choices through action' (p. 149). In such a context, geography can no longer be simply an academic discipline, isolated in its 'ivory towers'. Its practitioners must become politically aware and active, involved in the creation of a just society which involves not reform but replacement of the present one. The remainder of Harvey's (1973) book does not make this commitment clear, however; it contains one essay on land use and land-value theory, investigating the difficult concept of rent, and another on the nature of urbanism, presenting a marxist interpretation of the process of urbanization. His later (1984: see p. 328) historical materialist manifesto is a very clear statement of that goal, however (though see Eliot Hurst's, 1985, critique), and his (1982) *The Limits to Capital* is presented as an attempt to extend Marx's theory.

Harvey's work continued to develop these themes during the 1980s and 1990s. Two further volumes of essays extended his analyses of urbanization under capitalism (Harvey, 1985a, 1985c; see also Harvey, 1996b); other essays addressed issues of geopolitics (Harvey, 1985b) and relationships with nature (Harvey, 1994). He then returned to issues of justice (Harvey, 1993a, 1996a), whilst at the same time defending his approach against a variety of critiques (see p. 278). Throughout, he has advanced the case for marxism, as a holistic approach which necessarily puts material issues at the core of its analysis (Harvey, 1989b).

Peet also moved from a liberal to a marxist position, replacing his earlier paper on poverty (Peet, 1971) by a marxist interpretation (Peet, 1975a) based on the assumption that inequality is inherent in the capitalist mode of production. This led to a 'metatheory dealing with the great forces which shape millions of lives' (p. 567) within which 'Environmental, or geographic, theory deals with the mechanisms which perpetuate inequality from the point of view of the individual. It deals with the complex of forces, both stimuli and frictions, which immediately shape the course of a person's life' (pp. 567–8). Environmental resources act as constraints because they define the milieux within which individuals are socialized and presented with opportunities for participation within the capitalist system (see also Soja, 1980). Redistribution of income through liberal mechanisms, based on taxation policies, will not solve the problems of poverty, therefore; according to Peet, alternative environmental designs, with removal of central bureaucracies and their replacement by anarchistic models of community control, are needed and geographers should work towards their creation. (Harvey – 1973, p. 93 – disagreed with the latter, pointing out that unless resources are equalized among communities and territories, community control will only result in 'the poor controlling their own poverty while the rich grow more affluent from the fruits of their riches'.)

To Eliot Hurst (1980), however, embracing marxism would be insufficient to make geography relevant. In a subsequent essay he argued that by circumscribing their discipline geographers were severely constraining what they could contribute towards understanding the world: their discipline was presented as a technical practice only which 'is merely descriptive of self-selected phenomena and is scientifically bankrupt' (Eliot Hurst, 1985, p. 62). He examined eight separate definitions of geography, all of which were guilty of spatial fetishism (after Anderson, 1973), and he followed Slater (1977) in arguing that geography cannot be incorporated within marxism but has to be transcended and superseded by it: to become marxist, geographers as a professional group must commit suicide. He concluded that 'geography is irrelevant to contemporary society on two basic grounds' (Eliot Hurst, 1985, p. 85):

1   in terms of substantial content, it has developed 'certain types of ideography, method/technique-strong, but lacking theory in a strict sense' to provide the sort of material which underpins contemporary capitalism; and
2   in philosophical terms it occupies 'an untheorized point of entry to knowledge' because its dominant 'categories of concern, such as space, spatial differentiation, spatial interaction and uneven development, are mere fetishisms and unscientific'.

Radical revolution is needed in order to promote full understanding of capitalist society, not just of geography but of the entire body of 'bourgeois social science'.

Structuralism

Marxism is sometimes presented as a variant of the philosophy of structuralism (of which there are many different forms: cf. Johnston, 1983b). It is based on three separate levels of analysis:

1   the level of appearances, or the *superstructure*;
2   the level of processes, or the *infrastructure*; and
3   the level of imperatives, or the *deep structure*.

Only the first can be directly apprehended and studied empirically. The superstructure comprises a society's economic, social, cultural, political, and spatial organization, but it cannot be used to account for its own existence. The salient processes in its creation and maintenance are in the infrastructure, which cannot be observed: its nature can only be theorized and compared with the superstructure's contents.

Most forms of structuralism have no deterministic relationship between the infrastructure and the superstructure. A process can result in a considerable number of outcomes, depending on a variety of enabling and constraining factors, most of them within the superstructure. To illustrate this, the French social anthropologist Claude Lévi-Strauss used the analogy of the cam-shaft driving a machine for cutting the outlines of jig-saw puzzles (Figure 7.1). The machine is constrained to certain movements only but the order in which they come is random, which means that it can produce a large number of unique puzzles. Study of one of these puzzles alone could not reveal the nature of the machine, only study of the machine could do that. Structuralism is the study of the machines but, unlike the analogy, these are not available for investigation. Structuralist researchers must develop a theory of the machine, and test whether the contents of the superstructure are consistent with that theory; the contents of the superstructure cannot be predicted, however, because, to continue the cam-shaft analogy, the particular operation of the machine cannot be foreseen.

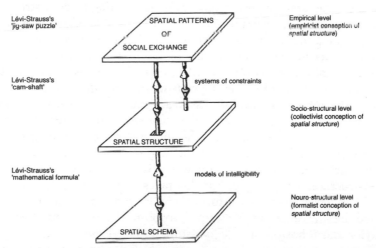

Fig 7.1 A geographical interpretation of Lévi-Strauss's cam-shaft analogy for structuralism (source: Gregory, 1978a, p. 100).

In some forms of structuralism, such as Lévi-Strauss's social anthropology (E. R. Leach, 1974), it is further argued that the processes operating in the infrastructure reflect imperatives located in the deep structure. Theoretical study of the patterns in the superstructure should reveal the nature of the deep structure, therefore (in somewhat similar fashion to the identification of essences in phenomenology: p. 189). Thus, according to Lévi-Strauss, all incest taboos are variants of a basic taboo which is imprinted in everybody's conscience. That basic form occurs in different forms in separate societies and cultures, but careful comparison of all taboos should reveal their common elements, and thus that aspect of human neural structure.

The search for deep structures has not characterized work in human geography, although it could be claimed that certain spatial concepts – like sense of place, territoriality and orientation – investigated by some human-istic geographers are realizations of deep structures. (In his important essay on territoriality, for example, Sack, 1983, decides to 'skirt the issue of whether human territoriality is a biological drive or instinct': see also Johnston, 1986c.) The relevant work of the Swiss structuralist, Piaget, has received a little attention from geographers. He claims the existence of a deep structure which allows the individual to assimilate and accommodate new material, such as language; Golledge (1981a) argued that this approach informs some of the work in behavioural geography, but it has had no major impact.

Marxism and realism

Marxist structuralism largely ignores issues relating to deep structures and focuses on the infrastructure and superstructure. Marxism has a materialist base, arguing that the infrastructure comprises a set of economic processes. The contents of the superstructure at any one place and time reflect the operations of the infrastructure, within a milieu representing the outcomes of previous operations; the superstructure provides the context within which the infrastructural processes are enacted. There is no economic determinism, therefore, because the superstructure is created, maintained and changed by human actors, who interpret the economic processes in a variety of ways; there is no one 'right' way for a process to be realized – it is widely accepted that a state apparatus is a necessary component of the infrastructure, for example, but this takes very different forms across the world and even within each form there are substantial variations, as illus-trated by the different Western European electoral and party systems (Taylor and Johnston, 1979; Johnston, 1984a, 1986d). The interpretation must be consistent with the processes, which not only enable but also con-strain human decision-making. The results of those interpretations will probably constrain the environment of the future – once an electoral system is in place it influences further developments, for example: the milieu informs the processes, but does not deflect them.

Marxist analysis searches for the processes operating in the infrastruc-ture, which it relates to the patterns in the superstructure. In human geography, this means deriving general theories of historical materialism that can account for particular patterns. As the cam-shaft analogy reveals, the patterns themselves cannot be used to identify the processes, although they may offer clues as to their contents. The processes must be identified theoretically, and compared with the superstructure to see if the outputs are consistent with the postulated economic driving forces. Within human geography, for example, Harvey (1975a) has suggested why socio-economic segregation takes place in urban residential areas, as a means of reproducing the class system and defusing inter-class conflict. The process of distancing (Johnston, 1980b) can account for the existence of segregation, but not the particular pattern that it takes: suburbanization of the rich is a sufficient but not a necessary response to this process (Walker, 1981a). Similarly for political geography, Taylor (1982, 1985b) has provided a materialist framework for the operation of the world economy, within which states act. Again, this provides a theory of what the state is, and what it does, but not of how it reacts to a specific (time-place particular) stimulus (Johnston, 1982a).

The philosophy of *realism* (sometimes known as transcendental realism to distinguish it from direct realism), within which much marxist work can be located, is based on the separation out of three domains (Bhaskar, 1978: see also above, p. 214):

1 *the domain of the empirical* is concerned solely with experiences, with the world as it is perceived;
2 *the domain of the actual* is concerned with events as well as experiences, accepting that an event (a particular item of human behaviour, for example) may be interpreted in different ways by individuals (by the actor, perhaps, and by another person experiencing it); and
3 *the domain of the real* is concerned with structures that cannot be apprehended directly, but which contain the mechanisms that lead to the events and their empirical perception.

Realists argue that all science incorporates these three domains, but the natural and the social sciences differ in the type of mechanism with which they deal. In particular, natural science deals with 'closed systems' in which the same inter-relationships can be observed on innumerable occasions, allowing the development of universal laws based on replicable results. Social science rarely if ever deals with such closed systems, however, both because of the difficulties of abstracting independent parts from the whole (most attempts to do this produce what Sayer terms 'chaotic conceptions') and because some components of the system – people – learn from their experiences, so that experiments can never be replicated.

The goal of realism, like that of positivism, is to explain events, to discover their causes by 'finding out what *produces* change, what *makes things happen*, what *allows or forces* change' (Sayer, 1985a, p. 163). It differs from positivism because it contends that: 'what causes something to happen has nothing to do with the matter of the number of times it has happened or been observed to happen and hence with whether it causes a regularity' (p. 162). Indeed 'regularities' (i.e. laws and law-like generalizations) are very unlikely to occur in the social sciences. Sayer (1984) argues that two conditions are necessary if regularities are to occur:

1 the mechanisms must be invariant; and
2 the relationships between the mechanisms and the conditions in which they occur must be constant.

If both hold, then scientists will be studying closed systems, with identical conditions leading to identical events (either naturally or in laboratory conditions). If one or both is absent, however, the object of study is an open system, and regularities will then not occur.

The social sciences study open systems:

> we can interpret the same material conditions and statements in different ways and hence learn new ways of responding, so that effectively we become different kinds of people. Human actions characteristically modify the configuration of systems, thereby violating the ... [first] condition ... while our capacity for learning and self-change violates the ... [second] condition (Sayer, 1984, p. 113).

The mechanisms underpinning the operations of human societies are interpreted in different ways, therefore, with the outcomes of those interpretations creating new sets of conditions within which future interpretations will be constrained. The mechanisms are abstract concepts which are actualized in contingently related conditions as, for example, with:

> the law of value, which concerns mechanisms which are possessed necessarily by capital by virtue of its structure (as consisting of competing and independently directed capitals, each producing for profit and being reliant on the production of surplus value, etc.), produces effects which are mediated by such things as the particular kinds of technology available, the relative power of capital and labour and state in[ter]vention. No theory of society could be expected to know the nature and form of these contingencies in advance purely on the basis of theoretical claims (pp. 128–30).

Thus whereas the mechanisms can be theorized, their particular realizations cannot, since they depend on human agency, on individuals interpreting the mechanisms in the context of their interpretations of the empirical world (see Johnston, 1989a). Furthermore, those individuals learn, with every decision and its outcomes feeding forward into future decision-making, so that even if the external milieu does not change, the nature of the decision-maker (and his or her culture) does (as Pred argued with his behavioural matrix concept: see p. 164).

The study of causation in a realist context involves what Sayer and Morgan (1985) term *intensive* research programmes, whose questions take the form 'What did the agents actually do?'; answering them involves examining causal processes in a particular case or group of cases. The answers cannot be generalized beyond those cases, in contrast to those produced in *extensive* research programmes, commonly adopted in empiricist/positivist human geography, which are 'mainly concerned with discovering some of the common properties and general patterns in a population as a whole' (p. 150). The latter almost certainly do not investigate closed systems, however, so generalizations are impossible; they are descriptive and synoptic only, whereas intensive research programmes are explanatory and 'Because intensive studies allow the identification of causal agents in the particular contexts relevant to them, it provides a better basis than extensive studies for recommending policies which have a "causal grip" on the agents of change' (p. 154). Sayer (1992a) later argued that generalization and the search for regularities (statistical or otherwise) can never be used to identify causes: using an epidemiological example, he claimed that the methods used in such work 'fail to address the problem of finding a mechanism which generates the disease, as opposed to a factor which merely covaries with it . . . neither common nor distinguishing properties need be causally relevant' (p. 115).

Massey and Meegan (1985) argued that extensive research seeks to identify common outcomes through large-scale studies which cancel out idiosyncratic behaviour in aggregate data, whereas intensive research views idiosyncrasy as the complex interactions of necessary relations with contingent conditions; intensive research teases out those necessary relations. Chouinard, Fincher and Webber (1984, p. 374) have identified similarities between positivist and realist research strategies, however:

> Because social scientists (by definition) cannot guarantee an invariant relation between specific causal mechanisms, processes and empirical events, the 'laws' posited by theory must be treated as tendencies and not as empirical regularities . . . even non-realist human geographers have adopted these interpretations of testing and scientific laws in practice.

Those non-realists assume that sufficient research will allow behaviour to be predictable, once all the possible interactions are accounted for; realist social scientists do not accept that argument, however, because the interactions are continually changing.

The time-place specificity of the contents of the superstructure provide a major element of the realist critique of positivism. Positivism, in geography as elsewhere, seeks to identify laws of general applicability; realists deny that this is possible (Sayer, 1979, 1984). Quantification is not the focus of the criticism, only its use in the search for positivist goals (although see Sayer's, 1984, criticism, and the response by Johnston, 1986a); as a descriptive tool, leading to extensive research programmes that can identify causes, it is perfectly acceptable (Walker, 1981b; Taylor, 1981a). Nevertheless, in the second edition of his important book Sayer (1992a) retained his critique of the use of both mathematics ('an acausal language': p. 178) and statistical methods (which 'offer limited forms of description which may usefully supplement qualitative description': p. 191): he does not respond to Pratt's (1989, p. 114) argument that: 'statistical inference and, more generally, quantification, are not susceptible to several of the criticisms that have been directed toward them by historical materialist and humanistic geographers'. Pratt continues by arguing that some of the valid criticisms

> serve to pare quantitative research down to size, to the status of a technique for organizing one type of information, information that needs necessarily to be complemented by a more relational, contextual understanding, as well as more abstract theoretical development.

Positivist work in human geography has been criticized by realists, including marxists, because it seeks 'laws' of the superstructure which are unrelated to the processes in the infrastructure, and which in any case cannot exist because of the change that is inherent in the infrastructure. Behavioural and humanistic work is subject to similar criticism, for apparently giving individual actors complete freedom and therefore removing them from the context in which they operate. Regarding the first point, for example, Cox (1981, p. 275) argued that 'behavioral geography . . . unwittingly accepts the assumption of the separation of individual and society'. Regarding the latter, N. Smith (1979, p. 467) claimed that 'The phenomenological approach . . . fails to take seriously the society external to the individual', and Wagner (1976, p. 84) argued that 'Existentialism and its close relatives in the Phenomenological camp seem to abdicate concern with the processes of history and the wider panoramas of geography, and so perhaps lack direct relevance'; Warf (1986, p. 279) similarly wrote that 'Phenomenology's exuberant voluntarism overstates the efficacy of intentional actions and assumes a fixed set of social relations asserting that consciousness is produced in an historical vacuum.'

In both positivist and humanistic geography, therefore, the individual is not treated as an actor constrained by society:

> In 'classical' location theory, agents were individual, autonomous, rational economic automatons. In behavioural theory, the main agents of interest were decision-makers whose prime position is exaggerated by a disregard for economic conditions, thereby producing voluntaristic accounts where 'locational preferences rule' (Sayer, 1982, pp. 80–1).

The realist alternative emphasizes that behaviour is constrained by economic processes, but it does not deny human agency. The infrastructure provides 'determinants of activity of which actors are unaware' (p. 81), but within which individuals make decisions, that then alter the constraints for future decisions: the infrastructure is thus both constraining and enabling, it restricts yet stimulates choice.

The acceptance of the concept of 'knowing actors' operating within structural constraints means that realists, like adherents of other forms of structuralism depicted by the analogy in Figure 7.1, do not argue that the contents of the superstructure are *determined* by the processes of the infrastructure. Rather they contend that the former can only be understood by the development and 'testing' (Sayer, 1982, pp. 85–7) of theories relating to those processes. Marxism, for example, is a theory of economic processes. It is not an empirical discipline – its lack of an apparent focus on the details of the superstructure led Jones (1980, p. 257) to criticize its 'withdrawal from the patterns of distribution which are usually deemed the starting-point of geographical enquiry'; it provides a framework within which empirical details can be studied and without which their understanding is impossible. Marxist geography accepts that the environment influences the superstructure's contents, creating cultural regions (Gregory, 1978a, ended his book with a call for a renewed regional geography focusing on spatial variations in the realization of economic processes). According to Peet (1979, p. 167), in a statement that Duncan and Ley (1982, p. 37) called irresponsible:

> neither environment nor space is passive as social relations pass over and through them. . . . As class relations move over space, they pick up qualities from the regions of a social formation. . . . Class relations become infused with the direct and indirect contents of regions and environments. . . . They are transformed, thereby, into socio-spatial-environmental relations. Spatial relations are at base class relations; class relations contain the effects of space and environment.

Class conflict governs capitalist economic processes. Variations in the interpretation of this conflict from place to place create subtle variations in the superstructure, however; trades unions have been much stronger in some corporate-welfare states than others, for example (Johnston, 1992a). These variations may become self sustaining, creating separate cultures within the overall capitalist framework that stimulate further spatial differences in actors' interpretations of the ongoing class conflict. (Culture never becomes a determinant, however – Duncan, 1980 – only a secondary constrainer and enabler to the demands of economic processes.) The spatial/environmental dimension is thus important to the marxist analysis of empirical variations (see Quaini, 1982, and also N. Smith, 1984, on the marxist analysis of capitalist concepts of nature).

Classical marxist analysis adopts a dialectical approach to its subject matter, since this represents how societies develop. It involves the resolution of binary opposites, usually represented by the trilogy of:

thesis – antithesis – synthesis

in which each synthesis becomes the next thesis; thus there is no permanent resolution, each synthesis containing within itself the seeds of its own destruction. Dialectic processes operate at a variety of spatial and temporal resolutions. Many empirical studies, for example, have identified 'long waves' in the economic development of capitalist economies, comprising fifty-year sequences of boom and slump (within which there are shorter-term, business cycles). These long waves (often termed Kondratieffs, after their Russian identifier) have been used to structure geopolitical as well as economic understanding (Taylor, 1989, 1990b); indeed, the two are linked, as global political and military hegemony has traditionally been linked to economic domination, as in the twentieth century with the growth to world power status of the United States (see Kennedy, 1988). Kondratieff waves are central to analyses of economic geographies because of the argument that emergence from the fifty-year slump involves major restructuring of economic activity round new products, mechanisms and state forms. Thus the late twentieth century is presented as a shift from Fordist methods, based on mass production of consumer goods in large factories, to post-Fordism, which is characterized by greater flexibility in production processes, more contracting and sub-contracting, shorter production runs for 'niche markets', and new forms of employer–employee relationships: it is termed 'disorganised capitalism' by Lash and Urry (1987).

Marxism is a form of realism, therefore, which seeks to relate the empirical world of appearances to a set of infrastructural determinants – economic processes. It is also, for many of its adherents, the basis for a political programme aimed at changing those economic processes. According to Peet and Lyons (1981, p. 205): 'If we want to understand what is happening to us, the people of the capitalist world, . . . we need a powerful, logical, sequential, political form of analysis to give an accurate, deep portrayal of the causes of events.' Marxism, including marxist geography, in doing this provides 'a powerful theoretical and political base for resistance' to the dominance of the capitalist imperatives on individual action. Its goal is based on Marx's humanism. He argued that people are alienated by the capitalist system; in particular the proletariat is exploited and has its human dignity removed through the process of selling its labour. To restore this dignity and give individuals full control over self and destiny, capitalism must be overthrown and replaced by communism. The argument is that (Relph, 1981b, p. 122): 'truly human relationships can be achieved only when everyone can take responsibility for the conditions of their own lives and when there is freedom from the ideologies and actions of a bourgeois professional class'.

Capitalism can only be overthrown by revolution, many argue, which will involve the proletariat realizing their exploitation and determining to end it. Achieving that realization requires emancipation, whereby individuals are made aware of their exploited position and given the knowledge with which to counter it (Johnston, 1988). Such achievement is the goal of *critical theory* (Gregory, 1978a) which combines a marxist realism with the development of communicative skills (hermeneutics: see above, p. 34) to enable individuals to understand the real processes operating in society – to appreciate the infrastructure. Human agency is encouraged by increasing awareness. (There is little reference to the deep structure that is assumed to exist in some forms of structuralism – as well as in phenomenology – p. 194. Marxists believe that all human behaviour patterns are socially created, hence their difficulty in dealing with the concept of territoriality when it is presented as a basic human drive.)

This political goal of marxist work leads to a major element of its critique of positivism – as indicated by Harvey's identification of three types of theory (p. 327). Positivist empiricism describes the present situation in the superstructure as a series of laws whose application in various forms of planning will maintain the present situation of exploitation while changing its superstructural form; positivist planning furthers exploitation by retaining the unjust capitalist structure rather than moving it towards the desired social and economic change. Such work, in any case, rarely understands what is happening in the superstructure, because it only describes it. Application of such incomplete understanding is termed instrumentalism (Gregory, 1978a; 1980). Positivist planning involves the imposition of laws on society, rather than recognizing portrayals of the present as descriptions of instances of a dialectically changing process that never exactly reproduces itself (Marchand, 1978).

Marxist and related realist work therefore suggests to human geographers that:

1 explanation of patterns of spatial organization and of society–environment relations within the superstructure can only be understood as realizations of economic processes operating in the infrastructure;
2 those economic processes cannot be apprehended directly, but only appreciated through the development of theories that are consistent with the outputs in the superstructure;
3 those economic processes are continually changing, and with them the outputs – and as a result universal laws of the superstructure cannot be derived;
4 class conflict (bourgeoisie v. proletariat) is central to the economic processes;
5 how the processes are realized in the superstructure reflects the actions of individuals operating within the constraints set by the processes and

their preceding realizations – individuals are constrained yet enabled, not determined;

6 any attempt to retain the present superstructure, by planning procedures based on positivist examination of the present, can only help the present unjust system to survive;

7 understanding the processes and their realizations requires much closer cooperation among, if not integration of, the social sciences than heretofore; and

8 the goal of academic work must be emancipation, leading to social change.

Adoption of this programme involves altering the model of modern western society which underpins its analysis. Eyles (1974, p. 39) identified two main models:

> one based on consensus, the other on conflict. The first would stress that social life is based on co-operation and reciprocation with norms being the basic element. Such societies would also be characterized by cohesiveness, integration and persistence and life within them would depend on the recognition of legitimate authority, mutual commitments and solidarity. The conflict model sees sectional interests as being the basis of social life which is, therefore, divisive, involving inducement and coercion as well as generating the structural conflicts . . . [that are] the central element in a social system.

Conflict between classes in a capitalist society and power – unequally distributed in favour of the bourgeoisie – are the key elements in the latter model, of which marxism is a particular form. Understanding that conflict and how power is distributed and used is basic to understanding how resources are allocated in a society. Its spatial organization is the result of the competition and conflict whose ethic (Eyles, 1974, p. 64):

> implies winners and losers and the winners in economic competition will be those with the powers of ownership and control in the productive process. In this way, growth under such a system will most probably lead to greater inequalities. It would seem obvious, therefore, that poverty and the distribution of real income in a spatial system cannot be understood without reference to power and inequality.

Other analysts accept much of the realist argument regarding the three levels of analysis (p. 217), but not the particular marxist form of economic analysis. Barnes (1996), for example, argues that both Harvey's marxist-based approach and Sayer's based on realism are untenable because each has at its foundation what he terms an essentialism. In Harvey's case, he claims, this involves a search for 'universality and absolute truth' (Barnes, 1996, p. 23) whereas in Sayer's it involves a belief in 'natural necessity ... the claim that the world is made up of nonapprehendable necessary causal relations'. Both, he argues, are contrary to an approach which (p. 49): 'abandons any notion of progress, accepts that subjects are made not given, avoids a homogenizing and totalizing portrayal of the world, and discards any essentialist notions of knowledge'. This involves him advocating a difficult transition (p. 78): 'from a nineteenth-century ironclad, rule-bound form of inquiry to a twentieth-century piecemeal approach that has no essential rules ... But not to make that move is in effect to deny the variety that as geographers we seek to explain.'

A major issue for those adopting a realist approach (including marxists) concerns the role of space – does it, as asked in the title of Sayer's (1985b) essay, 'make a difference'? In realist terms, that question can be pitched at the level of either the necessary or the contingent – is space important in theoretical work, or only in concrete realizations of the outcome of general processes? Regarding the former, he noted that 'Abstract social theory need only consider space insofar as *necessary* properties of objects are involved, and this does not amount to very much' (p. 54). All matter has a spatial location and so all events occur in space, but while recognizing this abstract theory need say very little about actual spatial forms: 'Hence, while it is important for abstract theory to be aware of the existence of space, the claims that can be made about it are inevitably rather indifferent ones' and refer to concrete realizations only. (See also Massey's, 1985, essay on the same topic, which comes – reluctantly – to a similar conclusion that 'equating geographical with the concrete seems to be a very minimalist position': p. 18.) Tackled on this issue, in a review of the second edition of *Method in Social Science* (Johnston, 1993e), Sayer (1994a) responded by arguing that social (including economic, political, etc.) processes necessarily have spatial properties but have considerable flexibility in how they operate spatially – hence the contingent element (p. 108): 'Consequently, while we can rightly assert the spatial character of social objects (e.g. that states are irreducibly territorial) we cannot say much as theorists about their actual shapes.' He concludes that to claim that space is contingent in its operation does not mean that it is unimportant: 'The difference that space makes in concrete situations can never be ignored, and therefore in that sense "geography matters"' (p. 109).

The political events of the late 1980s and early 1990s in Eastern Europe provided a major difficulty for some marxist analysts. As Sayer (1995, p. 13) notes:

'For the Right they were a vindication: communism was defeated, capitalism had won. ... Strangely, it was assumed by many that it also meant that Marxism was defeated, as if Western Marxists ought to have recanted as soon as the statues of Lenin and Marx started being toppled. This ignored the fact that the vast majority of Marx's work and most Western Marxism is concerned with *capitalism* and that most Western Marxists were overwhelmingly critical of the state socialist regimes.

This, he argued, diverted attention to capitalism's continued failings and to the failure of what he termed radical political economy to suggest viable alternatives. Hence his detailed analysis of work in that genre, which he located within (p. 12) 'Marxism's lack of a sufficiently materialist under-standing of the social division of labour and its associated division and dispersion of knowledge in advanced economies.' Instead of attacking this problem, researchers (in the context of the 'cultural turn' – see the next chapter) have switched their attention away from issues of power, domination and subordination linked to class and capital (p. 13):

> This seems an entirely reasonable response, as analysis of these mat-ters was long overdue. The uncharitable explanation is that faced with the challenge of the New Right and the weakening of Marxism, radi-cals shifted their attention to new concerns which did not require them to make any painful concessions to the Right. Socialism as an alterna-tive political system was increasingly difficult to articulate and defend, but in any case there were other important issues to turn to which pro-vided a convenient escape. While understandable, this escapist response is surely less defensible. In my view, both of these explanations of the shift from economy to culture – charitable and uncharitable – are right. Although post-Marxism has been generally a progressive development in terms of broadening radical interests it has neglected political-economic theory. There is therefore much unfinished business.

His approach to that unfinished business concentrates on the complexity of the social division of labour within contemporary capitalism (see also Sayer and Walker, 1992). Social divisions occur between enterprises involved in the creation of a vast range of products and services, whereas technical divisions (on which Marx and his followers concentrated) occur within enterprises, and are concerned with their ownership and structure. Individual enterprises can be planned in terms of how much will be pro-duced and marketed, by how many people, and at what cost; whole con-glomerations of enterprises – what Sayer follows Hayek in terming catallaxies – cannot be so planned because of their internal variety and complexity, however; their activities must be coordinated in other ways, which at present means the market economy celebrated by the New Right's identification of the 'end of history' (Fukuyama, 1992; Peet, 1993; Johnston, 1994). But the inequalities that follow pose major challenges (Sayer, 1995, p. 210):

> Political economy . . . has to address issues of division of knowledge and catallaxy, horizontal control, coordination, allocation and econo-

mizing, which generally elude democratic control . . . To ignore these issues is simply to duck the most serious problems of political economy.

So that (p. 252)

Radical political economy cannot continue to follow Marxism in standing apart from the debates of normative political theory, nor can it embrace a postmodernist celebration of fragmentation and rejection of the search for a better social framework. It is now more clear than ever that struggles which are directed against domination and oppression but which lack any normative direction in terms of alternative frameworks are unlikely to be successful. Though it has recently suffered from neglect we need a radical political economy more than ever before.

The events of 1989 in the former Soviet Union and its client regimes in eastern Europe posed a considerable problem for geographers of a marxist persuasion, as many of their opponents argued that the collapse of the communist state indicated the fallibility of their political programme. (Smith, 1991, p. 406, noted that 'it has become a common argument in left circles that the crumbling of Communist Party control throughout Eastern Europe and the Soviet Union provides definite proof that planning and state control of the economy don't work and can't work'.) Folke and Sayer (1991) argued that to claim that 'they weren't really socialist' anyhow is an unsatisfactory response, although they recognize that the events mean that 'genuine socialist visions and utopian dreams will have a hard time in Eastern Europe in the coming years' (p. 248): new blueprints of how to manage complex societies in the interests of all are needed (see Sayer, 1995), and they must recognize, as Smith (1991, p. 416) argued, that '1989 teaches us the profound depth of human resistance to all forms of economic, social and political repression': successful revolutions must be broadly based and sustained. Sayer (1992b, p. 217) responded that:

The problems that the left now faces are not just a consequence of the rise of the right, or of mistakes made in the 1920s; they are also a result of the vacuum on the left concerning socialist alternatives which are feasible as well as desirable. I think we would do better to confront that fatal weakness.

The uses of realist approaches

The research tradition generated by the introduction of realist and related approaches has several components. The first is the critique of positivist spatial science, behavioural geography and humanistic geography, exemplified by Rieser (1973) and Massey (1975). The second is the provision of general theoretical frameworks, within which empirical work can be set (as in Taylor, 1982, 1985b; Harvey, 1982; and two major collections of essays on urbanization and industrialization – Dear and Scott, 1981; Scott and Storper, 1985). Third, there is work that seeks to establish how individuals act within the structural imperatives (Johnston, 1983c). Eyles (1981, p. 1386), for example, has argued that although 'no geography or social science can be complete without Marxism', this is insufficient. It must focus on the world of lived experience, the micro-world of 'family, network, community, neighbourhood' if it is to be complete, which requires integrating structuralist marxism with humanistic philosophies. A growing number of writers have claimed that this involves appreciation of the importance of places (frequently termed 'locales' or 'localities') as the context within which people live and act, which has stimulated calls for a 'new regional geography' (Gilbert, 1988; Pudup, 1988).

Initially, the purpose of much writing in the emerging realist framework was to demonstrate the shortcomings of positivist-inspired research, especially that categorized here as behavioural geography; their work can best be described as managerialism. Gray (1975), for example, presented a critique of studies of the operations of housing markets which mapped people's residential moves and inferred that these were the outcomes of choice and the expressions of preferences. He argued that such research:

1 assumes that people are free to choose in which homes (including where) they live, when most are constrained to particular types of dwellings in certain parts of urban areas only;
2 accepts that residential patterns are the consequence of a large number of residents' decisions rather than those of a few developers and institutional managers; and
3 implies that the study of consumers provides the key to understanding the structure of urban areas.

People are not free to choose, however: 'Instead, many groups are constricted and constrained from choice and pushed into particular housing situations because of their position in the housing market, and by the individuals and institutions . . . controlling the operation of particular housing systems' (p. 230). Research in line with Gray's critique focused on the controllers of access to housing, individuals who became known as 'urban managers'. According to Pahl's (1969) seminal paper, urban residential patterns are the consequence of two basic sets of constraints: the spatial constraints on access to resources and facilities, which are usually expressed in terms of time/cost distance; and the social constraints – the bureaucratic rules and procedures operated by gatekeepers – that govern access and reflect the distribution of power in society. Study of the latter focused on the key managerial groups, such as building societies and their lending policies (Boddy, 1976; Dingemans, 1979), the managers of local government housing stock (Gray, 1976; Taylor, 1979), the local authority officials who influence the redevelopment of private housing (Duncan, 1974a, 1975), and the real-estate agents who structure access to housing sub-markets (Palm, 1979).

Pahl (1975) later revised his ideas, pointing out that planners, for example, are merely 'the bailiffs and estate managers of capitalism, with very little power' (p. 7; see, however, Pahl, 1979) so that focusing on them and other gatekeepers tends to

> view the situation through the eyes of disadvantaged local populations and to attribute more control and responsibility to the local official than, say, local employers or the national government ... such 'a criticism of local managers of the Caretaking Establishment' and 'of the vested interest and archaic methods of the middle dogs' may lead to an uncritical accommodation to the national elite and society's master institutions (pp. 267–8).

The managers are important actors in the production of empirical worlds, therefore, but are not independent agents: in the analogy of Figure 7.1 they can be equated with the cam-shaft. Without care, treatment of managers outside the constraints within which they must operate could become as reductionist as some behavioural geography studies of the consumers of the housing that they allocate (see also Williams 1978, 1982; Leonard, 1982).

Whereas the managerial approach focuses on the agents who interpret and activate the real mechanisms, producing patterns of residential segregation, Harvey paid much more attention to the mechanisms themselves (though with reference to particular outcomes: Harvey, 1974e). Residential separation is one means of reproducing class differences in a capitalist society. It is produced by those who manage finance capital, interpreting the basic mechanisms of the capitalist mode of production (i.e. the need to generate profits), and it results in a series of spatially separated housing sub-markets within which individual households may express their preferences, producing micro-scale migration patterns.

> But there is a scale of action at which the individual loses control of the social conditions of existence ... [sensing] their own helplessness in the face of forces that do not appear amenable, under given institutions, even to collective political mechanisms of control (Harvey, 1975a, p. 368).

In the American context the promotion of suburbanization was a major means of stimulating consumption (of housing, automobiles, consumer durables, etc.) at a time of dangerous latent over-production (Harvey, 1975b, 1978; see also Walker, 1981a; for the particular outcomes, see Johnston, 1984b). Thus urban geographers who accepted the realist argument saw their basic task as integrating discussions of the mechanisms that drive capitalist (and other) societies, the interpretations of those mechanisms by key agents (the managers and gatekeepers), and the empirical outcomes and experiences. Thus they focused on how and why society operates, not on generalizations regarding the empirical outcomes (as in Bassett and Short, 1980; Badcock, 1984; Johnston, 1980b).

The study of urban residential patterns is just one element of the study of urbanization, which Scott (1985, p. 481) relates directly to the nature of capitalism:

> the mechanisms of production, the interlinkages of firms, and the formation of local labor markets ... combine to create a process in which the profit-seeking (cost-reducing) proclivities of producers lead directly to the dense spatial massing of units of capital and, as a corollary, of labor.

Cities are both labour markets and places for the reproduction of labour, involving what Castells (1977) and others term 'collective consumption' of commodities produced by the state rather than purchased in the market place (see also Pinch, 1985); some of them are also 'global cities' which are the control centres of the international corporate economy (see also Johnston, 1987; Knox and Taylor, 1995). Scott (1986, 1988) outlined a theory of both the initiation of capitalist urbanization and recent urban trends within the 'advanced industrial' countries.

Urbanization is one element of the continuing process of uneven geographical development under capitalism (see Harvey, 1985a; N. Smith, 1984). Study of patterns of development has traditionally been the concern of economic rather than urban geographers, however, and for some years they were concerned with patterns and processes (i.e. changing patterns: Hay and Johnston, 1983) of modernization, statistically manipulating variables chosen to represent aspects of social and economic change (according to capitalist, technological definitions) to produce maps of what were termed modernization surfaces (e.g. Gould, 1970b; Riddell, 1970; Soja, 1968). Such work presented Drysdale and Watts (1977, p. 41) with the analogy that: 'At times, geographers have resembled spectators who, seated on a hill overlooking a battlefield, are fascinated by the direction in which clouds of smoke are blowing' (p. 41). Brookfield (1975, p. 116) was also very critical, arguing that such value-laden depictions, though useful as descriptive exercises, were redolent of:

> a superb craftsman ignorant of the material with which he is working. The folly of such an approach was never better demonstrated than on the prolonged failure of ... [Gould's] group to make its real contribution in an area where the direct participation of geographers is increasingly wanting.

That participation would involve geographers becoming involved in policy prescription, but Brookfield found attempts in that direction even more worrying. Berry (1972a), for example, had produced a paper comparing the spread of development to the diffusion of innovations and of diseases within central place hierarchies; Brookfield described this as 'a highly mathematical paper on hierarchical diffusion employing his well-known, but doubtfully relevant data on the diffusion of TV in the United States' (p. 110). Blaikie (1978) criticized such diffusion studies as a 'spatial cul-de-sac' (see also Blaut, 1987) and Slater (1973, 1975) used them to launch a critique of the positivist approach – what he termed the 'Anglo-Saxon mainstream' abstracted empiricism.

Against the empiricism of the students of modernization surfaces and diffusion patterns, scholars of realist (including marxist) persuasions have sought to develop theories of uneven development, emphasizing (Slater, 1975, p. 174) that: 'A crucial factor in the development of any spatial structure is the way in which surplus is circulated, concentrated and utilized in space.' The major contribution to this was Harvey's (1982) *The Limits to Capital* – a substantial reworking of Marx's economics to incorporate the spatial factor largely ignored in the original formulations. The key element of the operations of a capitalist system on which he focused was 'those seemingly irreconcilable contradictions that lead capitalism into the cataclysms of crises' (p. xvi). Having portrayed the economic causes of such crises (in both inflation and the falling rate of profit, for example) he examined the geographical mobility of capital and labour to show 'how the contradictions of capitalism are, in principle at least, susceptible to a "spatial fix" – geographical expansion and uneven geographical development hold out the possibility for a contradiction-prone capitalism to right itself' (p. xvii). Such 'spatial fixes' are not permanent crisis-avoidance mechanisms, however – 'Indeed spatial configurations are as likely to contribute to the problem as resolve it' (p. 429) – so although capital is becoming increasingly mobile (hypermobile according to some: Ross, 1983) in attempts to avoid it, eventually global crises will emerge (see also the essays in Johnston and Taylor, 1986a, 1989). Analyses of gentrification provide empirical examples of those spatial fixes (as in Smith and Williams, 1986), while Harvey (1985b) showed how they generate competition between places which can degenerate into geopolitical conflict.

The spatial structuring of development has attracted a range of alternative theoretical formulations to the realist-marxist (for a critique, see Corbridge, 1986; Barnes, 1985, 1989a, 1989b used a neo-Ricardian approach which he contrasted to the marxist and neo-classical). Harvey's (1982) theoretical analysis implied that uneven spatial development is a necessary precondition for the processes of capitalist accumulation; others suggested that it is an inevitable consequence. Browett (1984, p. 156) contested both of these positions, however, claiming that:

Whilst it is recognized that existing regional imbalances may well be taken advantage of and be exacerbated by capital in the process of restructuring, it is nonetheless maintained that they are not *necessary* to nor *systematic* consequences of, the logic of the capitalist mode of production.

He argued that the real exploitative relationship is not between developed and under-developed places (to him a reification of space) but between labour and capital, and that focusing on space diverts attention from that fundamental issue. N. Smith (1986) termed this 'reverse fetishism', as distorting as the spatial fetishism that Browett attacks, and presented the counter-case (developed more fully in N. Smith, 1984) 'that uneven geographical development at different (and differing) spatial scales is a necessity of the logic of capital accumulation' (N. Smith, 1986, p. 97).

Corbridge's (1986) detailed critique of much of what he terms 'radical development geography' is based on a perception that too much of it follows 'the path of dogma and determinism (wherein capitalism and its law of motion are assumed to remain essentially unchanged in time and space)' (p. 245) rather than a 'path of critical engagement (wherein the emphasis is on capitalism's temporal and spatial variation and its conditions of existence)': he was unsure which path will eventually be followed. Harvey's (1982, 1985b) work is criticized for following the former path, for example. The problems of over-accumulation are first tackled by a 'temporal fix' – the credit system – and when this fails capitalism turns to a 'spatial fix' – imperialism. But imperialism will eventually fail too, and bring global war in its wake.

> This sort of determinism is not at all what I have in mind when I talk of time, space and conditions of existence . . . If we are really to grasp the fundamentally different impact that the capitalist world system has had on (say) Brazil and Taiwan, then we must have done with models of capitalist development which oppose a fixed core and a fixed periphery (neo-Marxism) or which theorise the Third World in terms of the needs . . . of the imperialist powers alone (structural Marxism). More positively, our accounts of differential development must recognise that the dynamics of a changing capitalist world economy are always mediated by conditions of existence (population growth rates, gender relations, state policies and so on) which vary in space and time and which are not directly at the beck and call of a grand 'world system' (pp. 246–7).

(Similar arguments were made about locales and structuration; see p. 238.) Following this argument, considerable attention has been paid to the role of money in the production and reproduction of uneven development (see the essays in Corbridge, Martin, and Thrift, 1994) and the role of major 'world cities' as nodes in the world financial systems (Knox and Taylor, 1995).

Watts (1988, p. 163) characterized Corbridge's arguments as 'rash accusations of determinism' and countered that much radical development geography is concerned with the issues that Corbridge raises. The latter (1988, p. 239) accepted that there is 'a continuing dialogue *within* Marxist development studies' (he cites Watts, 1989), but maintained his contention that 'the theoretical heart of Marxism is seriously deficient and must always be rendered problematic' (p. 254). For this reason he anticipated greater benefits from studies within the French regulationist school, and their conception of different regimes of accumulation which embrace the wider systems of governance that sustain a particular realization of the mode of production – such as the Keynesian welfare-corporate state which sustained the Fordist regime of accumulation but which has been challenged as part of the shift to a new regime, usually termed post-Fordist or flexible accumulation (Scott, 1988).

In a later essay, Corbridge (1989) reviewed the development of what he termed 'a more tolerant post-Marxism' (p. 225) which is less deterministically economistic than the classical variety and its offshoots, is sympathetic to the case regarding temporal and spatial variations in the process of capitalist accumulation, and thereby links development studies to the trends categorized as the 'new regional geography' (see p. 247). This post-marxism draws on marxist analysis for its foundations, but is critical of its basic organizing concepts. The links to marxism are especially strong in its materialist base, its emphasis on inequalities in the distribution of assets and power, its acceptance of contradictions within the accumulation process, and its agreement with the tenet that people make their own history but not in conditions of their own choosing. Post-marxism differs from marxism, however, in its refusal to accept that any non-marxist concepts are incommensurate with it, and Corbridge argued for a '*careful* wedding of concepts from Marxism and non-Marxism' (p. 246). He opposed the view that economic concerns are necessarily of primary importance, was sceptical of the utility of the labour theory of value on which classical marxism is founded, drew insights for his putative general theory of power and civil society from other approaches, such as feminism, and argued that commitment to revolution is not necessarily the sole acceptable political practice. Work in this mould has already been reported within development geography, he argued, though an agenda covering seven tasks for future work was set out.

Peet (1991) sought to sustain the 'classical' marxist approach in his book *Global Capitalism*, in which he claimed to recognize a tendency in 'radical social thought which basically accepts what is and no longer dreams of what could be' (p. xiv). For him, marxism has not been defended vigorously enough:

> It is time to counter this tendency by re-stating some of the themes of Marxist development theory, pointing again to the truths they

contain, the events and characteristics they explain, and positing once
more their potential for further elaboration.

He concluded that the alternatives promoted by his critics are not viable
(p. 182):

> The challenge for those critical of a development process founded on
> class and patriarchy, which destroys or damages natural environ-
> ments, and relies on intersocietal domination, is to retain the
> dynamism of capitalism while changing the exploitative form of its
> dynamic. Yet this entire book has exemplified how developmental
> dynamics are rooted in social relations which last for centuries if not
> millennia.

For Peet, a mode of production which embraces markets is unacceptable:
he believes that there must be a 'better way' based on socialism, but we
cannot be sure, because it has yet to be tried. In a defence of his position,
which contains only a passing reference to Peet, Corbridge (1993, p. 466)
continues to express 'mistrust of grand political projects and their associated "metanarratives"'.

The role of the state in the production of geographies of uneven development was among the stimuli to renewed interest in political geography. The
nature of the state has attracted considerable theoretical and empirical
interest, with attempts to appreciate its necessity to the dynamic of capitalist accumulation, its relative autonomy as an institution within the superstructure of capitalist social formations, and its particular territorial
identity (see, for example, Clark and Dear, 1984, for a full discussion of
these issues; Taylor and Johnston, 1985, look at the British state in a similar context). The state's crucial role is recognized by Harvey (1982), for
example, who linked uneven development to global geopolitics, and by
Taylor (1985b, 1989), who proposed reorganizing much of political
geography's basic subject matter around an appreciation of the state's functions in promoting and legitimating capitalism. This allows spatial variations in state operations to be appreciated (as in Taylor, 1986) in a
non-deterministic way; in line with the realist approach, the actions of
those exercising power within the state apparatus are those of knowing
agents interpreting their roles in particular ways (as both Clark, 1985, and
Johnston, 1984b, illustrate with regard to the courts in the USA; see also
Johnston and Taylor, 1986b).

A new sub-field of critical geopolitics has also developed, focusing on the
inter-relationships among states and how images of various states are created and sustained. Foreign policy, for example, involves the use of territory to form identities (sense of belonging to sovereign territories) and
images of the residents of other states, separated by clearly defined boundaries (Dodds, 1994a, p. 204): 'Foreign policy, like security policy, has usually been understood in spatial terms as moves of exclusion. The articulation

of the "foreign" depends on geographical depictions which invoke boundaries and exclude others.' Those depictions are constructed and transmitted
in a variety of ways, as illustrated by Sharp's (1993) analysis of the role of
the American magazine *The Reader's Digest* in the creation of both positive
self-images and negative constructions of the 'others' (the Russians). Thus
critical geopolitics involves deconstructing the ways in which political elites
have depicted and represented places in their exercise of power in the international arena (Dodds and Sidaway, 1994), although O Tuathail (1994,
p. 525) refers to it as 'a paradoxical promissory declaration' – the adjective
'critical' suggests a rethinking of the traditional ideas associated with
geopolitics, thereby questioning the current exercise of power, whereas the
practice of geopolitics has long been associated (notably in Germany in
the 1930s) with the reactionary use of power. Using the techniques of discourse analysis (see p. 295), this involves (O Tuathail and Dalby, 1994,
p. 513):

> constructing theoretically informed critiques of the spatializing prac-
> tices of power; undertaking critical investigations of the power of
> orthodox geopolitical writing; investigating how geographical rea-
> soning in foreign policy in-sights (enframes in a geography of
> images), in-cites (enmeshes in a geography of texts), and, therefore,
> in-sites (stabilizes, positions, locates) places in global politics, and
> examining how this reasoning can be challenged, subverted and
> resisted.

Study of uneven development also includes work on the relationships between
society and nature. As noted below (p. 338), liberal concerns over environmental issues increased substantially during the late 1960s and early 1970s.
Harvey (1974d) introduced a marxist perspective to this issue, arguing that
contemporary statements on resource–population ratios are ideologically
based, an argument sustained through examination of the works of Malthus,
Marx and Ricardo. Resources are not defined outside the context of societal
appraisals of nature, and those appraisals reflect the current mode of production. Thus, predictions such as those of the *Limits to Growth* study
(p. 339) represent a *status quo* theoretical perspective (p. 327), whereas
employment of other perspectives may lead to different conclusions:

> let us consider a simple sentence: 'Overpopulation arises because of
> the scarcity of resources available for meeting the subsistence needs of
> the mass of the population'. If we substitute our definitions [of subsis-
> tence, resources and scarcity] into this sentence we get: There are too
> many people in the world because the particular ends we have in view
> (together with the form of social organisation we have) and the ma-
> terials available in nature, that we have the will and the way to use,
> are not sufficient to provide us with those things to which we are
> accustomed (p. 272).

With the first version of the sentence, the policy option seems clear – population reduction. The second allows Harvey to identify three possibilities, however: (i) changing both the desired end and the social organization; (ii) changing technical and cultural appraisals of nature; and (iii) changing views about the capitalist system and its scarcity basis. The Ehrlich–Commoner arguments regarding the origins of current environmental problems (p. 340) accept the first version of the sentence, however, and Harvey claimed that they have been grasped by the elite of the capitalist world to provide an underpinning for a popular ideology (the need for birth control). Similarly, Buchanan (1973) argued that the arguments for birth control in the Third World are part of the 'white north's' imperialist policy of ensuring its continued access to the resources of the international periphery (even to the extent of stimulating environmental destruction and famine: Bradley, 1986; Blaikie, 1986), and Johnston, Taylor and O'Loughlin (1987) have explored the geography of what Buchanan (1973, p. 9), along with other peace scientists, terms structural poverty: 'The poverty which is regarded as symptomatic of reckless population growth is rather a *structural poverty* caused by the irresponsible squandering of world resources by a small handful of nations.'

Under capitalism, the environment (or nature) is treated as a commodity, to be bought, exploited, and sold. N. Smith (1984) points out that nature is presented in capitalism's ideology as external to society, as the 'antithesis of human productivity . . . the realm of use-values rather than exchange values' (p. 32). But nature is produced by capitalism:

> In its uncontrolled drive for universality, capitalism creates new barriers to its own future, it creates a society of needed resources, impoverishes the quality of those resources not yet devoured, breeds new diseases, develops a nuclear technology that threatens the future of all humanity, pollutes the entire environment that we must consume in order to reproduce, and in the daily work process it threatens the very existence of those who produce the vital social wealth (p. 59).

Nature, like spatial organization and reorganization, is a product of the continuing processes of capitalist accumulation (Fitzsimmons, 1989); as an abstract concept it is of no value to society and it is of interest only as a sphere of human activity (Pepper, 1984). This representation of nature as a social construction is not new: Zimmerman's (1972) classic work on natural resources defined a resource as only existing when it was perceived as such, and Glacken's (1967) classic survey shows how cultural conceptions of nature have varied. In his adaptation, however, Pepper (1984) separates the 'historicity of nature' (how the concept of nature in a particular time and place is related to the human activities there) from the 'naturalness of history' (the relationship between natural conditions and human activity – a relationship crudely expressed in theories of environmental determinism, p. 40). Part of the creation of a society involves its historicity of nature:

> Labour is the means whereby man converts nature into forms useful to him. In the main nature does not offer ready-made subsistence to man, neither does man take direct possession of nature's resources. He has to transform them. This process of transformation is a *social* one – it is done with other people who are organized in a particular way ... Thus through shaping nature, men shape their own society and their relations with their fellows (p. 162).

Thus, just as the creation of social relations (the means of exploiting surplus value from labour by capital) is part of the making of a society, so too is the creation of society–nature inter-relations; and just as those social relations contain within them the seeds of crisis for capital accumulation (the class conflict) so too do the society–nature inter-relations (environmental problems). Johnston (1989b, 1996c) has carried this argument forward in an account of society–nature relationships under various modes of production and of the role of the state in their regulation (see also Taylor, 1994): the essays in Blaikie and Brookfield (1987) illustrate the general theme.

Locales, structuration and a 'new regional geography'

In the 1980s, marxist and other realist work on topics such as those reviewed above led to the growing integration of geographical understanding of the nature of capitalist society with that of other social scientists. This involved the realization that although it is a global phenomenon, capitalism operates at a variety of inter-dependent spatial scales, from which Taylor (1981b, 1982) extracts three. The global world economy is the *scale of reality*; capitalism has little respect for the constraints of international boundaries, and its driving mechanisms (occupying the domain of the real) are global in their reach. The scale of reality is beyond the apprehension of those subject to global capitalism, however, whose daily lives form their *scales of experience*: they do not encounter capitalist forces directly, but only as they are played out in the spatially restricted areas that form their lifeworlds. Between these two scales is the state – the *scale of ideology* – which is the major ideological force linking people's experiences of capitalism to its reality. These three scales provide the trilogy of the subtitle to Taylor's (1985b, 1989) *Political Geography – World-Economy, Nation-State and Community* (see also Short's, 1984, trilogy – capital, state and community) – which captures his basic thesis that capitalism is organized globally, justified nationally, yet experienced locally. (Mann, 1996, suggests five scales by dividing the last of Taylor's trilogy into international transnational and global.)

Structuration and locale

The scale of experience is central to much recent writing on the spatial structuring and restructuring of capital. Giddens (1984, p. 118), for example, portrays it through his concept of a *locale*.

> Locales refer to the use of space to provide the settings of interaction, the settings of interaction in turn being essential to specifying its *contextuality* . . . Locales provide for a good deal of the 'fixity' underlying institutions . . . It is usually possible to designate locales in terms of their physical properties, either as features of the material world or, more commonly, as combinations of those features and artefacts. But it is an error to suppose that locales can be described in those terms alone . . . A 'house' is grasped as such only if the observer recognizes that it is a 'dwelling' with a range of other properties specified by the modes of its utilization in human activity.

Locales can vary in size, from rooms to state territories; they provide the settings within which interactions are organized, and are thus the contexts of economic, social and political life.

Recognition of the contextuality of life, of its structuring and restructuring in locales (increasingly referred to as its 'situatedness'), has led to the development of both geographically informed social theory and cases for a restructuring of the discipline of geography. With regard to social theory, Thrift (1983a) has developed an important distinction, derived from Hagerstrand, between compositional and contextual theory. *Compositional* theories classify individuals on certain criteria and allocate beliefs and behaviour patterns to them accordingly (as with both marxist and Weberian conceptions of class). *Contextual* theories, on the other hand, focus on the role of locales as the settings within which people learn how to act as human agents; their interpretations of their compositional categories are learned in particular places. Thus *structuration* – a theoretical approach developed by Giddens (1984) to account for the ways in which people learn about and transform social structures – is a place-bound (and time-bound) process; one's context (including one's language: see p. 301) is a major influence on one's individual development. (For an exegesis and application by geographers, see Moos and Dear, 1986; Dear and Moos, 1986.)

Structuration, as an example of a contextual approach, clarifies the falseness of any distinction between social relations and spatial structures; as Gregory and Urry (1985, p. 3) express it 'spatial structure is now seen not merely as an arena in which social life unfolds but rather as a medium through which social relations are produced and reproduced'. Together they produce the process of *spatiality* (or socio-spatial dialectic), used by Soja (1980, 1985) to characterize the conjoint social production of space (*à la* Harvey, 1982) and the spatial construction of society. Giddens and others identify much common ground between his conception of structuration and Hagerstrand's of time geography (though see Gregson, 1986), and Pred has used the two in a number of essays to illustrate their relevance to understanding the unfolding of both his own career (Pred, 1979, 1984a) and the development of particular places (Pred, 1984b, 1984c). Thus

> The meanings of interpretations I have imposed upon the past, the there and then in Berkeley and Sweden, are a result of the knowledge, attitudes and values I hold now, the ever fading here and now in Berkeley. Yet, the knowledge, attitudes and values I hold here and now are rooted in my past path, my past participation in projects and the institutional and societal context which generated those projects (Pred, 1984a, p. 101).

Also 'Biographies are formed through the becoming of places, and places become through the formation of biographies' (Pred, 1984b, p. 258).

Relatively few authors have adopted structuration as a research methodology. Gregson (1987b), for example, identified 'some quite major, and unresolved, problems in front of us before Giddens' theory of structuration can be demonstrated to be of real import for empirical enquiry' (p. 89). She discussed Moos and Dear's (1986) use of two levels of analysis: analysis of the individuals involved in the production of a particular event (level 1); and the higher-order analysis of the structural properties within which those agents are operating (level 2). They argue that these are tied together by Giddens' concept of the 'duality of structure', the inter-dependent relationships among structure and agency, which they term 'abstraction'. Gregson doubts whether they have provided a way of moving 'between what are often universal theoretical categories and the specifics of the events which occur in particular times and places' (p. 81): they have provided 'abstraction without specification'. Further, she argued that (p. 83): 'it would be unreasonable to expect structuration theory to generate either empirical research questions or appropriate categories for empirical analysis as it stands, and ... to transfer structurationist concepts directly into empirical analysis is misconceived'. Structuration is about second-order questions, questions *about* social science that cannot be answered by an appeal to methods of obtaining facts, rather than about first-order questions concerning the domain of the actual.

Gregson's argument that 'structurationist concepts and categories will not, and should not, be expected to specify particular empirical research projects' (p. 84) appears to have been adopted by others, who take it as an organizing

framework within which empirical work can be set rather than a methodological protocol. Locales are presented as the social systems that provide
the contexts within which individuals become 'knowledgeable actors'. Sarre
et al. (1989, p. 43) define 'knowledgeability' as: 'individuals' ability to infer
from the complex and contradictory situations and processes of society both
the prevailing rules and the most advantageous strategies and tactics'. The
ability to apply knowledge reflects 'capability' to act in the ways that they
wish. Their locale is thus both enabling and constraining: enabling in its provision of resources – knowledge – on which they can base action, and constraining because it limits how they can act (through both the knowledge
provided and the milieux in which it is interpreted and used). As they act,
they reproduce the local knowledge, and thus ensure that the social system
continues to constrain and enable further actions (theirs and others'). But
their actions and choices will change the knowledge somewhat, creating a
(probably very slightly) different set of enabling and constraining conditions
for future action. In this way, 'The course of history is therefore neither determined nor open, but a process in which rules and resources are both reproduced and to some degree changed' (p. 45).

Regarding the empirical applicability of such a theoretical framework,
Sarre *et al.* (1989, p. 46) claim that:

> Giddens offers few prescriptions about methods, though he stresses
> that the task is essentially hermeneutic, indeed it is doubly hermeneu-
> tic: the social scientist must interpret a social reality which crucially
> involves the actor's own interpretations. The way-in to study is an
> immersion in a particular area of society which allows the observer to
> get to know how to be able to act in it. However, this must not result
> in assimilation – the observer must be more aware than the actors of
> the nature of the rules and resources involved and of the way particu-
> lar situations relate to wider structures.

Thus researchers interested in actions in a particular locale must both
appreciate its social system and understand how that is interpreted by its
residents: researchers must both set the scale of experience in its wider
context and study how the actors do the same, which involves adopting a
realist approach as developed by Sayer (1984; see also Sarre, 1987). The
practicalities of what it means to do realist research are unclear, however,
so in their own work Sarre *et al.* (1989, p. 53; see also Sarre, 1987, p. 10)
report that: 'Our response was to utilise familiar methods of data gathering and analysis but to seek to interpret the data we collected in the light
of emerging realist and structurationist views.' Lovering (1987) similarly
argued that realism allows a synthesis which avoids the empiricism, eclecticism and reductionism of other approaches (as demonstrated in his diagram: Figure 7.2). Nevertheless, some commentators have equated realist
research with empiricism (e.g. Bennett and Thornes, 1988), and there
have been substantial arguments among those who accept the basic

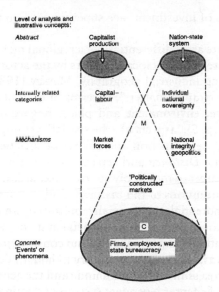

Fig 7.2 A realist conception of the defence industry. The abstract categories form the domain of the real (p. 217); the mechanisms form the domain of the actual; and the events form the domain of the empirical (source: Lovering, 1987, p. 298).

realist-structurationist case, as clearly illustrated by one major research programme.

'Localities'

The localities research programme originated from Massey's (1984a) work on the changing geography of economic activity in the UK. (A measure of the seminal importance of *Spatial Divisions of Labour* is the series of ten commentaries produced by Australian-based authors in Volume 12, Number 5 (May 1989) of *Environment and Planning A*.) Her central argument was that understanding changing locational patterns of industries requires appreciation of the links between economic and social change. Local social structures vary in how the labour process is organized, and from comparing two case study regions she concluded that (p. 194):

> although both ... are now being drawn into a similar place in an emerging wider division of labour, their roles in previous spatial divisions of labour have been very different; they have different histories. They bring with them different class structures and social characteristics, and, as a result, the changes which they undergo ... are also different.

(Without the terminology, this argument parallels that of structuration theory.) Thus changing industrial geography is linked to a changing social geography as new 'layers of investment' are superimposed on those of earlier eras.

Those processes generate areal differentiation, a regional mosaic of unique areas reflecting the interpretations of local contexts by the actors involved in creating the UK's changing industrial geography. Massey (1984b) generalized this in her essay 'Geography matters', in terms of each of its three traditional concerns – space, environment and place. Regarding space, for example, she argued that (p. 5): 'aspects of "the spatial" are important in the construction, functioning, reproduction and change of societies as a whole and of elements of society. *Distance* and separation are regularly used by companies to establish degrees of monopoly control.' Regarding the environment, she argued that conceptions of the 'natural', and thus interpretations of local environmental potentials, are socially produced, and could vary among areas as a reflection of their separate social systems, whereas with regard to place she contended that general processes can have particular outcomes in unique areas, in exactly the same way that Giddens does. Thus the study of geography involves unravelling the unique and the general: 'the fact of uneven development and of interdependent systems of dominance and subordination between regions on the one hand, and the specificity of place on the other' (p. 9) is thus a central concern for the discipline.

Massey illustrated her general contention by a number of case studies in *Spatial Divisions of Labour* (1984a), and provided others in an essay on women in the labour force (McDowell and Massey, 1984: see also Rose, 1987). She was also instrumental in the creation of a major research programme, financed by the British Economic and Social Research Council, into the Changing Urban and Regional System (CURS): this was introduced by its co-ordinator (Cooke, 1986) and summarized in a volume which reported on the programme's seven case studies (Cooke, 1989a). The substantive focus was the spatially varying processes of economic restructuring taking place in Britain during the 1980s. Thus (Cooke, 1989a, p. ix):

> The overall objectives of the programme were to explore the impact of economic restructuring at national and local levels, and to assess the role of central and local government policies in enabling or constraining localities, through their various social and political organizations, to deal with processes of restructuring.

(The terminology is direct from Giddens.) In addition: 'an important dimension of the research involved seeking to establish the conceptual status of the idea of "locality" by taking account of a wide range of social scientific theory and research'.

The concept of *locality* was chosen after both the traditional term 'community' and Giddens' own – 'locale' – were rejected:

> There is a gap in the social science literature when it comes to a concept dealing with the sphere of social activity that is focused upon

place, that is not only reactive or inward-looking with regard to place, and that is not limited in its scope by a primary stress on stability and continuity (Cooke, 1989a, p. 10).

Locale was rejected because its spatial scope is vague, it suggests a passive rather than an active context for action, and it lacks any specific social meaning. A locality, on the other hand, is (p. 12): 'the space within which the larger part of most citizens' daily working and consuming lives is lived' and in which their citizenship rights are defined. Citizenship, according to recent debates, involves individuals having obligations to their community (via the state) as well as entitlements from it – or duties as well as rights (Smith, 1989; see also Bennett, 1989a). The contest over rights and duties is fought out in places, on a variety of scales: their nature involves (Painter and Philo, 1995, p. 111): 'the construction of an identity, complete with a related package of known rights and obligations, which posits residence in a definable place or (commonly quite sizeable) territory as the basis for the nurturing and preservation of this identity'. We become what we are because of where we are and the categories used in the structuring of civil society there (and we are implicated in those local struggles over our rights, obligations and identities).

At the end of his summary volume on the localities project, Cooke argued that the seven case studies sustained his earlier contention regarding the crucial role of localities in the restructuring process and in the creation and recreation of uneven development: they illustrate (Cooke, 1989a, p. 296):

the argument that the relationship between the different scales is not simply a one-way street with localities the mere recipients of fortune or fate from above. Rather localities are actively involved in their own transformation, though not necessarily as masters of their own destiny. Localities are not simply places or even communities: they are the sum of social energy and agency resulting from the clustering of diverse individuals, groups and social interests in space. They are not passive or residual but, in varying ways and degrees, centres of collective consciousness. They are bases for intervention in the internal workings of not only individual and collective daily lives but also events on a broader canvas as affecting local interests.

This conclusion, and the path towards it, have been the subject of considerable debate, however, much of it unfavourable on conceptual grounds because what is necessarily local and what is contingently so has not been defined (Cox and Mair, 1989).

One of the initial critiques was Neil Smith's (1987a) argument that the CURS programme as formulated was likely to be submerged in a morass of statistical information; it contained within itself the potential for producing no more than those earlier empiricist studies of particular places 'which deliberately examined individual places for their own sake, and [did] not attempt to draw out theoretical or historical conclusions' (p. 62; see also Jonas, 1988). He was also concerned about the vagueness of the spatial scale in defining localities, but welcomed the attempt to blend theoretical analysis with local understanding. Cooke (1987a) refuted the charge of empiricism, arguing that the objective of the CURS initiative was 'theorised interrogation' (p. 75) of available data. His general position was supported by Urry, who initially (1986, p. 239) argued that 'there are some significant locality-specific processes', which he followed with a list of ten different ways in which social scientists have addressed them (Urry, 1987; see also Urry, 1985).

Cochrane (1987, p. 355) wondered whether CURS was 'just a cover for structural Marxism with a human face, or . . . the cover for a return to empiricism with a theoretically sophisticated face', and concluded that the programme contained the danger that as a guide to political action it might suggest that local struggle could suffice (what he terms 'micro-structuralism') rather than the realization that parts cannot readily be isolated from wholes. Gregson (1987a) was perhaps even less sanguine, arguing that the theoretical rationale for the seven case studies was far from clear, thus making the likelihood of falling into the empiricist trap high; without a properly articulated theoretical core 'CURS simply replicates the mistakes of previous local studies; with such a core it could be so much more' (p. 370). Beauregard (1988) claimed that the programme lacked any clear directions for practice, for using the radical theory to achieve social change.

Duncan's (1989) full critique accepted that the concept refers to something important – spatial variability and specificity – but he concluded that 'the locality concept is misleading and unsupported' (p. 247: he concluded elsewhere – Duncan and Savage, 1989, pp. 202–4 – that it is 'confused, unsatisfactory, and largely redundant . . . a mystification'). He accepted that 'space makes a difference', in three ways:

1 social processes are constituted in places, which may differ because of previous 'layers of investment' (to use Massey's term);
2 actions take place locally and so can vary spatially; and
3 spatially varying actions can create spatially varying contexts.

But the concept of locality implies 'social autonomy and spatial determinism' (p. 247), both of which he rejects. He is not convinced that local differences are very important in creating uneven development, relative to more general processes (p. 248): 'Locality is . . . only important if and when locality effects are part of the causal group explaining any event. And locality may well not be important' which implies a verdict of 'not proven'. Cooke's (1989b, p. 272) response (to the Duncan and Savage paper in particular) was trenchant:

Local social processes are clearly an abiding feature of contemporary social life. Duncan and Savage's injunction to ignore them and settle on the structural level, supra-local, supra-national or whatever, in order to describe spatial variation in terms which deny agency to the social groups comprising localities, is both dated and redundant.

And he concluded that: '"Locality" can be seen to be a fascinating, complex concept of considerable value to geographical theory and empirical research.'

Griffiths and Johnston (1991) outlined a clearer framework for studying how localities differ, using three components of local economic, social and political structures drawn from their study of local responses to the 1984–1985 National Union of Mineworkers' strike in the UK. Johnston (1991) later applied this framework to a range of studies of uneven development at a number of scales, including the southern United States.

The debate over 'localities' was part of a much wider range of work which addressed the role of place in the creation of social scientific understanding. Agnew and Duncan (1989, p. 1), for example, opened their introduction to *The Power of Place* with the statement that the book is: 'an attempt to make the case for the intellectual importance of geographical place in the practice of social science and history . . . [through] bringing together what can be called the geographical and sociological "imaginations"'. Agnew (1989) analysed why place has been devalued in 'orthodox social science' in recent decades, focusing on the confusion of place with community (itself a term with many definitions: Bell and Newby, 1976). The orthodox treatment identifies a decline in community with 'modernisation' and its replacement by a 'placeless' society, a transition seen as 'natural, lawful, and universal' (Agnew, 1989, p. 16), with nationalism growing as a 'place-transcending ideology'. Similarly, the alternative view derived from marxism devalued place with its emphasis on 'freeing people from places' (p. 22). (Entrikin, 1989, provides a parallel, elegant argument.) The goal of *The Power of Place* is to correct those tendencies, and, through both theoretical argument and empirical illustration, 'argue against the prevalent tendency in history and social science to overvalue the sociological imagination at the expense of the geographical' (Agnew, 1989, p. 7: see also Agnew, 1990, on the similarities and differences between his structuration-based focus on locales and Hartshorne's arguments – p. 44 – for geography as the study of areal differentiation/variation).

Agnew's (1987b) monograph on political behaviour set out his arguments at greater length in a detailed critique of political sociology. He answered his question 'Why adopt the place perspective?' in the following way:

1 First, adoption of the place perspective allows abstract categories such as class to be analysed in the context of everyday life. As a mobilizing

force, class may be present or absent in a particular place, and where it is present it may be weak or strong as a focus of social organization. Class, like other categories such as religious affiliation, means different things in different places, and understanding behaviour requires an appreciation of its nature in the milieux under consideration.

2　Second, by adopting place as the context within which structuration occurs, one avoids the search for laws of behaviour that are universal in both space and time, which is central to the positivist approach. The place perspective allows one to recognize the uniqueness of places without abandoning a commitment to causation: 'Thus, the structuration of social relations in everyday life contains many similar elements from place to place (e.g. class, central–local government relations, etc.), but produces many different outcomes in different places' (p. 42).

3　Third, one can resolve the structure-agency problems because a focus on place 'recognises human action as both motivated and intended but, at the same time, as both mediated by social structure and generative of it'. Agnew realizes that this can easily lead into either voluntarism or determinism (what Johnston, 1985d, called, respectively, the 'singularity trap' and the 'generality trap').

4　Fourth, recognition of differences between places means that the division of history into stages can be avoided, because societies evolve differently, and at different rates, in separate localities whilst under the same general operative processes.

5　Finally, recognition of local differences which reflect cultural variations provides a counter to arguments founded in economic determinism. Cultural phenomena are not merely superstructural reflections of economic determinants, but rather reflect the *modus operandi* established by people as they develop practices within which the economic imperatives can be pursued. The 'practical nature of everyday life' (p. 43) provides the context within which people act, and is the environment that they recreate as a consequence of their actions.

Agnew illustrated the benefits of this approach with a number of case studies, notably one on the geography of support for Scottish nationalism which he claims that contemporary political sociology has failed to deal with satisfactorily. He argued that his approach combines the 'geographical and sociological imaginations' in a micro-sociological procedure in which the macro-sociological outcome (the pattern of voting for the Scottish National party, for example) is the sum of the micro-sociological processes in the area considered. Such a perspective is based on the tenet (p. 233):

> that political order is produced and reproduced through microsociological routines (locale and sense of place). Whatever the specific nature of power relationships, they cannot be separated from the realm of action and everyday practices. The macro-order (location) is represented in routines and practices of people in places.

Thus Agnew was promoting not only a rejuvenation of the study of place (or locality) within geography but also a firmer integration of that work with the other social sciences.

A 'new regional geography'

The debate over locality research is part of a wider literature on the importance of studying the specific characteristics of places. The numerous works using an implicit contextual approach include Agnew's (1984, 1987b) argument that voting behaviour is strongly influenced by the electorate's local context (see also Johnston, 1986e) and a detailed analysis of variations between local states in their provision of public housing, which is also informed by the realist perspective (Dickens et al., 1985). Some translated this general concern into a plea for a revived, but restructured, regional geography. Massey (1984b, p. 10), for example, ended such a plea with the argument that it is

> necessary to reassert the existence, the explicability, and the significance, of the particular. What we [must do] ... is take up again the challenge of the old regional geography, reject the answers it gave while recognizing the importance of the problem it set (p. 10).

This echoes Gregory's (1978a, pp. 171–2) contention that:

> Ever since regional geography was declared to be dead ... geographers, to their credit, have kept trying to revive it in one form or another ... This is a vital task ... we need to know about the constitution of *regional* social formations, of *regional* articulations and *regional* structures. ... [producing] a doubly human geography: human in the sense that it recognizes that item concepts are specifically human constructions, rooted in specific social formations, and capable of – demanding of – continual examination and criticism; and human in the sense that it restores human beings to their own worlds and enables them to take part in the collective transformation of their own human geographies.

This call was not immediately taken up, although Fleming (1973), Steel (1982) and Hart (1982) all called for a revival of traditional regional geography, which to Hart meant 'producing good regional geography – evocative descriptions that facilitate an understanding and an appreciation of places, areas and regions' (p. 2). More recently, however, people have followed Gregory's lead (see also Gregory, 1985b), and argued – as do Lee (1984, 1985) and Johnston (1984c, 1985a) – for a reconstituted regional geography which recognizes (Lee, 1985) that:

1 social processes operate in historically and geographically specific circumstances, so their understanding requires a sensitivity to geographical
   variations (regional mosaics);
2 society is not a fixed phenomenon but something that is constantly
   being recreated by human actions. Since those actions occur in historically and geographically specific contexts, then societal recreation is
   similarly historically and geographically variable;
3 those local transformations occur in the context of wider social relationships; and
4 the regions that emerge are not fixed divisions of territory but changing
   social constructions.

A major goal of geography should be to uncover the nature of those
regions, as illustrated by some of the essays in Johnston, Hauer and
Hoekveld (1990).

The emphasis on change in these arguments implies the absence of a distinct niche for historical geography within the discipline's overall programme. The claim that a major route to understanding the present lies in
study of the past is not new, but adoption of approaches based on structuration and contextual theory clarifies the importance of an historical perspective. Not surprisingly, therefore, historical geographers have made
major contributions to the debates (see Baker and Gregory, 1984), drawing
on such sources as the writings of the French *Annales* school of history
(Baker, 1984; Pred, 1984d, draws on Braudel, for example). Their interpretations of the long-term evolution of a society (e.g. Dunford and Perrons,
1983) and its regional components (Langton, 1984), of basic transformations in a region, whether industrial (e.g. Gregory, 1982a) or agricultural
(Pred, 1985, 1986), and of the constitution of particular places (as in
Harvey's, 1985c, detailed analysis of nineteenth-century Paris as an example of how consciousness is created in a particular context) all draw upon
the basic realist conceptions, and illustrate how changes take place as the
result of general tendencies being played out in specific milieux by particular human agents. (Not all historical geographers agree, however: see
Meinig, 1978, especially p. 1215.) Similarly, cultural geography is no
longer clearly distinguished from other aspects of the discipline. For long,
this has been a peculiarly North American sub-discipline, focusing on
human artefacts in the landscape and, as Duncan (1980) argued, paying
relatively little attention to the concept of culture itself. If culture is interpreted as a community's entire heritage – material and non-material – then
a region's 'contents' can be equated with its culture, providing the foundations for a revived cultural geography (Cosgrove, 1983; Thrift, 1983b;
Jackson, 1989), while not precluding the continued study of landscapes as
reflections of cultures and contributors to their recreation (Cosgrove,
1984).

The differences between the 'new' and the 'traditional' approaches to regional geography are clarified by Pudup (1988, p. 374), who characterized the latter as empiricist: 'Theoretically neutral observations are the basis for areal description.' A reconstructed regional geography has foundations that 'rest in a clarified status of regions as objects of study – put simply, why geographers bother to study regions in the first place' (p. 379). The answers to that 'why?' question are provided in works such as Pred's (1984c) on southern Sweden and Gregory's (1982a) on West Yorkshire: regions are territorial entities, produced, reproduced and transformed through human agency. This argument is similar to Taylor's (see p. 237) on the scale of experience and Giddens' (p. 238) on structuration; regions are the places in which people learn a culture, and contribute to its continuation (what Thrift, 1983a, calls 'settings for interaction': p. 40). The nature of those processes is appreciated through a theorized approach, with the appreciation being provided through a narrative which draws on a defined vocabulary and permits 'theory to speak through subsequent empirical accounts' (Pudup, 1988, p. 383: see also Sayer, 1989a, 1989b). What those empirical accounts should focus on is still debated, and the need for a sub-discipline called regional geography has been contested. Warf (1988, p. 57) contended that with the replacement of traditional regional geography by positivism, 'a geography of "regions without theories" quickly became a geography of "theories without regions"', but Johnston (1990a, p. 139) argued that whereas 'we do not need regional geography . . . we do need regions in geography'.

Pudup's characterization of the 'new' regional geography was extended by Gilbert (1988), who identified three separate concerns with regional specificity in recent writings:

1 a concern with regions as *local responses to capitalist processes*, which she identified as probably the most prominent among English-speaking writers, who set the study of local variations within a political economy (usually marxist) framework;

2 a concern with the region as *a focus of identification* (or 'sense of place'), which is especially strong among French writers concerned with the analysis of culture. To them, appropriation of a place (or region) is part of the creation and recreation of cultural identity; and

3 a concern with the region as *a medium for social interaction*, playing 'a basic role in the production and reproduction of social relations' (p. 212). The work on 'localities' (p. 240) fits into this concern, though Gilbert shows that it has been developed by French- as well as English-speaking geographers.

All three represent a break from 'traditional' regional geography, she claimed, through their recognition that the persistence of regional diversity in the face of the homogenizing tendencies within capitalism (see also Peet, 1989) provides regional geography with a practical significance, in the mobilization of resistance to those tendencies. The 'new' work depends on structural theory (as illustrated in Agnew's, 1987a, book on the United States within the capitalist world economy); on the recognition that regional processes rely on dialectical rather than naturalistic theories; and on the importance of human agency in the creation, recreation and transformation of regions. Together, these suggest a mode of study committed to understanding and achieving social change, which provides the challenge of making 'geography a science useful for society' (p. 223).

Radicals in debate

The 'radical camp' is not a united body of scholars, therefore, and although marxism has remained the focus of some workers' activities it has been attacked by others, who are themselves associated with 'radical' stances. In 1986, for example, Saunders and Williams equated the recent literature in urban studies with an unchallenged 'taken for granted orthodoxy . . . [that] thrives on an unspoken and largely unexamined political and theoretical consensus' (p. 393). They claimed that this has opposition to positivism as its linchpin, and arguments within it occupy 'a very narrow spectrum embracing left Weberianism and the different varieties of marxism, but excluding almost everything else'. They concluded that, as a consequence, urban studies currently:

> makes little pretence of being 'value-free' or 'ethically neutral' and . . . is sheltered from the possibility of empirical disconfirmation. This in turn means that approaches (such as the philosophy of the so-called 'New Right') which cannot be subsumed under the orthodoxy can be dismissed almost *a priori*. Such work is rarely read, still less seriously considered on its own terms. Rather a pejorative label is attached to it (for example, 'Thatcherism', 'Reaganism', or 'authoritarian populism') which enables us to pigeonhole it within our existing conceptual apparatus . . . without ever having to engage with its intellectual content. In this way alternative ideas are dismissed but never discussed, explained away but never critically evaluated (pp. 393–4).

Elements of both marxism and realism were then criticized, with the latter being characterized as a 'justification for subordinating history to theory' (p. 394) and the source of 'causes which can only be identified theoretically and which are guaranteed immunity from falsification even where there is no manifest evidence for their existence' (p. 395): works which are explicitly anti-positivist are therefore castigated for not being subjected to positivist and critical rationalist procedures!

According to this critique, theories of structures have primacy in realist and marxist approaches and as such are outdated: the major changes in British society over the last century are ignored, it is claimed – 'we still employ an essentially outdated class theory in our analyses' (p. 397) – and the many changes are dismissed as:

> having done little or nothing to change the essential features of capitalism. As social change takes place under our noses, so we risk a situation where our methods and theories ensure that we pay it little heed. The orthodoxy is safeguarded and reproduced as the society changes.

Even the challenge of feminism and the renewed interest in ethnicity issues have largely been disregarded, according to Saunders and Williams, so that the conceptual tools of the 'radical camp' are not re-examined, and urban studies were censured as 'safe rather than innovative, conservative rather than critical'.

One of the strongest adherents to the marxist position, and a major focus of the Saunders–Williams critique, is Harvey, who characterized the goal of. 'my academic concerns these last two decades ... [as] to unravel the role of urbanisation in social change, in particular under conditions of capitalist social relations and accumulation' (Harvey, 1989b, p. 3). He responded to Saunders and Williams by noting 'a marked strategic withdrawal from Marxian theory within the field of urban analysis and a broadening reluctance to make explicit use of Marx's conceptual apparatus in articulating arguments' (Harvey, 1987, p. 367), and launched an attack against those whom he identified as abandoning the 'tough rigour of dialectical theorizing and historical materialist analysis', because:

> The case for retiring Marx's *Capital* to the shelves of some antiquarian bookstore ... is not yet there. Indeed, in many respects the time has never been more appropriate for the application of Marx's conceptual apparatus to understanding processes of capitalist development and transformation. Furthermore, I believe the claim of Marxian analysis to provide the surest guide to the construction of radical theory and radicalizing practices still stands.

His critics have caved in too readily to right-wing pressures, he argued.

Harvey (1987, p. 368) agreed with Saunders and Williams that realism and the conceptualization of agency and structure: 'are nothing more than weak disguises, soft versions of a traditional left orthodoxy ranging from left Weberianism to "different varieties of Marxism"'. But he disagreed with their identified way forward. He claimed that three 'myths' have pervaded the critique of marxist scholarship. The first is *economism*, as illustrated by Duncan and Ley's critique (1982: see below, p. 259). The second is the claim that 'The abstractions of Marxian theory cannot explain the specificities of history and the particularities of geography' (p. 370), to which he took great exception. Harvey believes that 'it is in principle possible to apply theoretical laws to understand individual instances, unique events' (p. 371), particularly since Marx's most interesting law-like statements were about capitalist processes, not events. He illustrated this with the chapter in *Capital* on 'The Working Day', arguing that such a detailed empirical description of how humans respond to their conditions on a daily basis allows: 'categories like money, profit, daily wage, labour time, the working day, and ultimately value and surplus value [to] arise through an examination of historical materials' (p. 372) thereby illustrating how theory is both derived from and developed through the unravelling of particular situations.

Harvey saw the second myth being advanced from 'within the ranks of the left itself' (p. 373); Massey and Sayer's 'very deep and serious concerns for the particularities of places, events and processes' were noted. Sayer's realist approach was presented as involving a combination of 'wide-ranging contingency with an understanding of general processes', but:

> The problem with this superficially attractive method is that there is nothing within it, apart from the judgement of individual researchers, as to what constitutes a special instance to which special processes inhere or as to what contingencies (out of a potentially infinite number) ought to be taken seriously. There is nothing, in short, to guard against the collapse of scientific understandings into a mass of contingencies exhibiting relations and processes special to each unique event (p. 373).

This, Harvey feared, is a path to 'simple empiricism' (p. 374), which he believed Saunders and Williams (1986) were also promoting; he characterized their agenda as 'nothing short of an abrogation of scientific responsibility and a caving in of political will'. Against that, Harvey insisted on 'the viability of the Marxist project', with its focus on 'universalising statements and abstractions' (p. 375) and its ability to guide political practice.

Finally, the third myth (the 'never-never land of nontotalizing discourses into which Marxists cannot enter because they insist on talking about totalities': p. 374) was presented as an attack on 'totalizing discourses', meta-theories and meta-narratives. While rejecting the notion that marxism is 'inherently totalizing', Harvey nevertheless argued that: 'nothing appears more totalizing to me than the penetration of capitalist social relations and the commodity calculus into every niche and cranny of contemporary life'. If social theory can look only at the parts, then the whole can never be apprehended and challenged (part of Harvey's argument against postmodernism: see p. 276). To Harvey (p. 375): 'every single local study I read ... points to how locality is caught up in universal processes of financial flows, international divisions of labour, and the operations of global financial markets.' Claims to the contrary are dangerous, for they involve avoiding confrontations with 'the realities of political economy and the circumstances of global power'.

Harvey's stance was supported by Smith (1987b), who defended marxism as providing both a broad analytical framework and a 'quintessentially political discourse'; he found that the 'realist project ... has become the theoretical justification for the belief that there can be no general theory *at all* concerning questions of geographical space, and that any attempt to devise such a theory is fundamentally misconceived' (p. 379). Others were less supportive, however; Ball (1987, p. 393), for example, took issue with Harvey's:

> total dismissal of anyone who does not repeatedly declare their Marxist label, who does not believe that everything Marx said is unambiguous and correct, and who fails directly to apply the most abstract propositions of Marxist theory to the empirical situations they are investigating.

Sayer (1987, p. 395) also resented the attack and its impugning of motives, and responded to both Saunders and Williams and Harvey that 'even if it means breaking off from what I had hoped and still hope for is a broad but common project, the search for a social science with an emancipatory potential'. Sayer agreed very much with Harvey that empirical research enables a clarification of theoretical understanding, it being 'partly responsible for making me revise my abstract ideas about the nature of capital, competition, class and the division of labour' (1987, p. 397: see also Sayer, 1995). He also defended realism against the charges of both reductionism and theoreticism (exemplified by Saunders and Williams), on the one hand, and empiricism (Harvey), on the other. Regarding the latter, he argued that, because of the impossibility of theoretically neutral observation, claims that realism (or any other approach) is atheoretical and empiricist cannot be sustained; instead, 'capital-labour relations, class, gender, as well as many other phenomena of interest, are historically constituted in particular localities (though not only in them) and ... the manner of this constitution must be explained' (p. 399). Thus the abstract categories of Marx's general theorizing about capitalist processes can be employed to appreciate the nature of particular realizations (which, he argues elsewhere – Sayer, 1989b – calls for a combination of humanism and historical materialism; see also Storper, 1987). Cooke (1987b) also argued for study of the interactions between universal and local processes, and for using the understanding so achieved to forge local political practice: 'I find it distinctly odd that thinking globally and thinking or acting locally should be thought somewhat "parochial"' (p. 412). Thrift (1987) believed that marxian political economy has been so successful in Britain that 'it now forms a vital subtext to most theorising' (p. 401): further, he argued that 'the realist project is the reconstitution of dialectical materialism so that it preserves the best of the Enlightenment tradition but incorporates the verities of twentieth-century developments in social theory' (p. 405). To him, social theory such as marxism is a hand torch that helps to illuminate particular instances, but not the 'searchlight flooding every nook and cranny of society with light' (p. 405), which is his interpretation of Harvey's case; Thrift argued this involves the study of agency, in places, as well as of structure.

Saunders and Williams (1987, pp. 427–8) began their response with the following categorization:

> Harvey eschews fraternization with the enemy, and he adopts a quasi-religious, almost messianic tone in delivering his epistle. He tells us . . . of his unswerving *belief* that Marxism provides the *surest guide* to radical salvation . . . Harvey's statement provides a good example of precisely that tendency in contemporary urban studies which we suggested could stifle fresh initiatives and hamper intellectual debate. If you believe, as Harvey apparently does, that the eternal verities have largely been established by Marx's *Capital*, then you have effectively closed off the possibility of open debate, and even more the possibility of learning from others who disagree.

(Gould, 1988, similarly criticized marxist writers in general and Harvey in particular for what he termed their 'claim to exclusiveness'.) Saunders and Williams then attacked Harvey for sustaining a 'totalizing' form of theory, which embraces the whole of society and thus has a privileged starting-point (or set of initial assumptions) which are not open to empirical refutation/confirmation:

> It is difficult to see how you can get to the whole by studying the parts and building up from there . . . Totalising discourses are thus always unalterably committed to *a priori* assumptions – Marx's theories of exploitation, class struggle, and historical evolution came, not from studying people in Manchester, but from ideas about society 'as a whole', and these ideas were then mapped onto existing empirical observations (pp. 428–9).

Thus Harvey's marxism is not susceptible to effective falsification; nor, they assert, despite Sayer's arguments, is realism. Both are presented as 'closed' approaches, and although Saunders and Williams share the same goal as Harvey and Sayer, they believe (p. 430) that: 'you will not develop an emancipatory social science before social science itself is opened up'. Hence the debate, which the editor argued was necessary because: capitalist society is changing; the nature of politics is altering; and 'social theory has to take account of these cumulative changes' (Dear, 1987, p. 363).

Liberals and radicals in debate

The material discussed in this chapter illustrates a world view very different from that employed by many human geographers during the 1970s and 1980s. Alongside the debates within the 'radical' approach there have been others which, in the published contributions at least, have been much more heated than that generated by the 'quantitative revolution'. They have also been developed in more depth and detail, using forums, such as the journal *Area*, which were not available twenty years earlier, when publication out lets for 'views and opinions' were few. Whereas the behaviourists caused little real concern within the discipline, and their views were soon coopted within the corpus of acceptable approaches, the so-called radicals of the 1970s and 1980s had much more impact, because they attacked not only the basis of most geographical work but also, in the clear inter-disciplinarity of their approach, both the bureaucratic structure of the discipline and the existence of geography itself (Johnston, 1978c); some, such as Eliot Hurst (1980, 1985), argued explicitly for the abolition of geography as a separate discipline (see also Taylor, 1996). Most of the debate has involved the empiricists/positivists on the one hand and the radicals on the other; relatively little has involved responses to the humanistic approach (but see Billinge, 1983, and Flowerdew, 1989 and the material discussed in the next chapter).

A clear example of the liberal–radical polarization was a debate on the geography of crime initiated by Peet (1975b), who argued that in attempting to make their work relevant geographers avoided asking 'relevant to whom?'; the political consequences of their work were ignored. The studies of crime reviewed (e.g. Harries, 1974) referred only to the surface manifestations of a social problem and could not provide solutions, only ways of ameliorating the problems: 'So it is that "useful" geography comes to be of use only in preserving the existing order of things by diverting attention away from the deepest causes of social problems and towards the details of effect' (Peet, 1975b, p. 277). Furthermore, geographers study only the crimes for which statistics are collected, thereby accepting the definition of crime by the elite; their maps, which are useful to police patrols, can therefore be employed to help maintain the *status quo* of power relations within society. The implicit position of the geographers involved is one of protecting the 'monopoly-capitalist state'.

Harries (1975) responded by attacking the simplistic nature of Peet's arguments, claiming that geographers would have no influence at all if they argued merely that crime is a consequence of monopoly capitalism. He contended that it is best to work within the system, to make the administration of justice more humane and equitable, to protect the potential victims of crime, and to provide employment opportunities for graduate geographers. Approaches based on crime control are likely to be more influential than polemics relating its cause to the mode of production: as Lee (1975, p. 285) also pointed out, Peet 'failed to provide us with any clues as to how he or other radical geographers would study crime' (p. 285).

Peet (1976a) responded to Lee's challenge, outlining a radical theory which would 'contribute directly, through persuasion, to the movement for social revolution' (p. 97). Capitalism harnesses human competitive emotions and generates inequalities in material and power rewards. Aggression is an acceptable part of being competitive, and is often released on the lower classes, who are encouraged to consume but provided with insufficient purchasing power. As the contradictions of this paradox increase, so does the pressure to turn to crime. Thus crime occurs where the lower classes live, and at their spatial interface with the middle class. Harries (1976) replied that being a radical was a luxury few academics could afford, because working at a publicly financed university demanded a pragmatic rather than a revolutionary approach: Peet's theory is overly economic and deterministic; it fails to account for cultural elements, such as the disproportionate criminal involvement of blacks, the sub-culture of violence in the southern United States and other areas, and the fact that all economic systems produce minorities disadvantaged in terms of what they want and what they can get by socially legitimate means. He could offer no alternative theory, however: 'I do not carry in my head a theory of crime causation, and I am quite incapable of synthesizing and attaching value judgements to existing theoretical formulations within a couple of pages of typescript' (p. 102). He encouraged Peet to come off the fence, and to get involved in the production of change within the present system. Wolf (1976), on the other hand, claimed that by concentrating on the traditional concerns of marxism Peet was not radical enough.

Two geographers who were involved in a considerable, often virulent, debate in the 1970s were Brian Berry and David Harvey. Berry (1972b) initiated the exchanges with comments on Harvey's (1972) paper on revolutionary theory and the ghetto: he wondered whether Harvey's rational arguments on the need for a revolution would be accepted: 'because of "commitment", the opposition will quietly drift into corners, the world will welcome the new Messiah, and social change will somehow, magically, transpire' (p. 32). The power to achieve change, he contended, needs more than logical argument – 'nothing less than cudgels has been effective' (p. 32) – and Harvey's belief in logical rationalism will be to no avail. He also argued that Harvey was wrong about the ghetto, for liberal policies were succeeding and the inequalities between blacks and whites were being reduced (Berry, 1974a). Harvey (1974a) responded that scarcity must continue in a capitalist economy, which will leave some people – probably those in the inner city – relatively disadvantaged.

Berry's (1974b) review of *Social Justice and the City* (Harvey, 1973) criticized Harvey's dependence on economic explanations. Based on Daniel Bell's (1973) characterization of post-industrial society, Berry (1974b, p. 144) claimed that the economic function is now subordinate to the political:

> the autonomy of the economic order (and the power of the men who
> run it) is coming to an end, and new and varied, but different, control

systems are emerging. In sum, the control of society is no longer primarily economic but political.

Harvey (1974b) responded that marxism could not be considered as *passé* while the selling of labour power and the collusion between the economically and the politically powerful continue, and that the state has to be considered within a marxist framework too (Harvey, 1976). Berry (1974b, p. 448) retorted:

> I believe that change can be produced *within* 'the system'. Harvey believes that it will come from sources *external* to that system, and then only if enough noise is produced at the wailing wall. ... The choice, after all, is not that hard: between pragmatic pursuit of what is attainable and revolutionary romanticism, between realism and the heady perfumes of flower power.

Harvey's (1975c) response came in a review of Berry's (1973a) *The Human Consequences of Urbanization* – a study of urbanization processes at various times and places and of the planning responses to these – which concluded that the book is 'all fanfare and no substance' (p. 99), revealing that Anglo-American urban theory is 'substantively bankrupt' and that 'it is scholarship of the Brian Berry sort which typically produces such messes' (p. 99):

> It is doubtful if it makes any sense even to consider urbanisation as something isolated from processes of capital formation, foreign and domestic trade, international money flows, and the like, for in a fundamental sense urbanisation is economic growth and capital accumulation – and the latter processes are clearly global in their compass (p. 102).

To Harvey, Berry has nothing to say of any substance, but he recognized that Berry 'is influential and important ... his influence is potentially devastating' (p. 103 – whether he meant this influence as within geography only, or beyond it too, is not clear). Berry's only response was a general comment on the Union of Socialist Geographers (Halvorson and Stave, 1978, p. 233): 'there's no more amusing thing than goading a series of malcontents and kooks and freaks and dropouts and so on, which is after all what that group mainly consists of. There are very few scholars in the group.'

Apart from these very polarized exchanges, several others indicated that whereas some 'liberals' were prepared to consider the radical case seriously, others tended to avoid the issues. Chisholm, for example (1975a, p. 175), claimed that:

> while I am fully sympathetic to the view that the 'scientific' paradigm is not adequate to all our needs, and must be supplemented by other approaches, I am not persuaded that it should be replaced. ...

Harvey wants us to embrace the marxist method of 'dialectic'. This 'method' passes my understanding; so far as it has a value, it seems to be as a metaphysical belief system and not – as its protagonists proclaim – a mode of rational argument.

(See also Chisholm, 1976; Sayer, 1981.) More frequently, reviewers accepted that the radical view added to their appreciation, but argued that is was not, for them at least, tenable in its entirety. Thus Morrill (1974) wrote of *Social Justice and the City* that 'I am pulled most of the way by this revolutionary analysis but I cannot make the final leap that our task is no longer to find truth, but to create and accept a particular truth' (p. 477). King (1976, pp. 294–5) also sought a middle course:

> An economic and urban geography that will be concerned explicitly with social change and policy. . . . Such a middle course will not find favour with the ideologues, who will see it either as another obfuscation favouring only 'status quo' and 'counter-revolutionary' theory, or as a distraction from the immediate task of building elegant quantitative-theoretic structures, but some paths are being cut through the thicket of competing epistemologies, rambling lines of empirical analysis, and gnarled branches of applied studies that now cover the middle ground.

He accepted that much quantitative-cum-theoretical geography had sought mathematical elegance as an end in itself, at the sacrifice of realism; he believed that social science must feed into social policy and generate social change; and he accepted the 'intellectual power' of marxist analysis but believed that prescriptions based on it are acceptable only if the ideological framework is also. His conclusion suggested the need for more quantification, which was operationally useful rather than mathematically elegant (see also Bennett, 1985a, 1985b), and later noted that '*space* . . . should be seen as an element in the political process, an object of competition and conflict between interest groups and different classes' (King and Clark, 1978, p. 12). Finally, Smith (1977, p. 368) concluded that:

> Marx may have been able to dissect the operation of a capitalist economy with particular clarity, and see the essential unity of economy, polity and society that we so often miss today. But Marx does not hold the key to every modern problem in complex, pluralistic society.

The debates continued, with many geographers unable/unwilling to accept both the marxist analysis of society and the marxist programme for action. In part, this reflected a partial reading of marxism – in particular a concentration on strict structural interpretations which neither allow for the activities of knowing individuals nor accept the concept of structuration (see p. 238). Ley (1980, p. 12), for example, saw in marxism not only 'some hidden transcendental phenomenon ... directing the course of human society' – which commits an epistemological error, by denying or at least suppressing the subjective, but also a theoretical error that devalues the power of human action to redirect the course of events and a moral error, which makes humans into puppets and threatens basic freedoms of speech, assembly and worship. Muir (1978) attacked interpretations of marxism, and similarly implied that it threatens individual freedom of the academic 'to pick and choose from among the ... literature' (p. 325). Respondents pointed to the lack of such 'intellectual orthodoxy' and 'sterilized geography' (Manion and Whitelegg, 1979; Duncan, 1979), but Muir (1979, p. 127) remained convinced of the threat implicit in: 'the commands from such little men as Marx, Lenin, Trotsky, Stalin and Mao concerning the primacy of activism, the obligations of party membership and the necessity to subordinate individual judgement to the will of the party'. He claimed that radical geography is contributing to understanding, but to call it marxist is to give it a certain programmatic base. Walmsley and Sorensen (1980), on the other hand, presented marxism as just irrelevant, deflecting attention from 'reformism and relevance'.

Duncan and Ley (1982) published an extensive critique of structural marxism in geography, containing four major themes.

1   Marxist analysis is a form of holism, in which the whole – variously termed capital, the economic structure, economic processes, etc. – is given a life of its own: an abstraction is assumed to exist. This reification (see also Gould, 1988) offended their belief in individuals as conscious, free agents (see Duncan, 1980, on holism in cultural geography, and Agnew and Duncan, 1981) – although they did not proclaim an idealist alternative and accepted that 'individual action cannot be fully explained without reference to the contexts under which individuals act' (p. 32);
2   individuals are represented as agents of the whole – the means of implementing its goals – not as free decision-makers in their own right;
3   the materialist infrastructure of marxism is a form of economism – it presents economic processes as the ultimate cause of all behaviour, which excludes many other influences and leads to the final theme; and
4   'The attempt to cast explanation continually and everywhere in terms of economic imperatives, [leads] ... to a crisis in empirical exposition' (p. 47).

The last comprises what is essentially a positivist critique: 'the form of the explanations is both tautological and empirically untestable. The result is a mystification in explanation of how real processes operate' (p. 55). They concluded that structural marxism in human geography presents a passive view of the individual, offering explanations in terms of abstract wholes which are obfuscatory and not verifiable. Later work by Duncan (1985) showed that he was attracted to the structuration approach, however, which allows for human agency operating within structural constraints – as indeed do most interpretations of marxism.

Others who criticized structural marxism disagreed with its treatment of the individual, and wanted more convincing models of the inter-relationships between infrastructure and superstructure in the process of societal change. Gregory (1981) presented four such models (Figure 7.3):

1 *reification*, in which the individual's actions are entirely determined by the whole – as in a structural marxism that denies human agency;
2 *voluntarism*, in which society has no separate identity but is constructed from the actions of free individuals;
3 *dialectical reproduction*, in which the whole creates the individuals, whose actions then influence the whole – which in turn creates the next generation of individuals; and
4 *structuration*, which starts with individuals rather than with the whole, and portrays them in a continuous dialectic with society from then on.

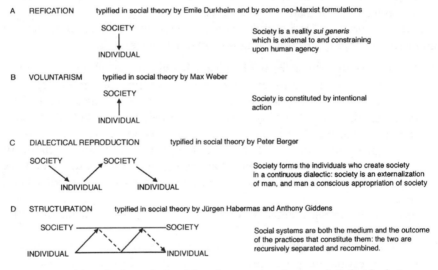

Fig 7.3 Models of the relationships between society and individual (source: Gregory, 1981, p. 11).

These different approaches are based on a variety of conceptions of the nature of the human agent (or 'models of man' as they are frequently termed; for critiques, see Barnes, 1988; Claval, 1983; Harrison and Livingston, 1982 and van der Laan and Piersma, 1982). Gregory opted for the fourth, because it integrated humanist and materialist (or structuralist) perspectives, in a way that critiques such as Duncan and Ley's did not (see also Thrift, 1983a). Such an integration would recognize the materialist base to society (the infrastructure) while accepting the important role of human agency in the superstructure; humanistic geography should not compete with a scientific approach, but should acknowledge: 'the recurrent and recursive relations between the individual and society as being fundamentally implicated in the production and reproduction of both social life and social structure' (p. 15).

Alongside debates on the relevance of marxist ideas, other geographers were defending quantitative spatial science against critiques such as Gregory's (1978a, 1980). Bennett (1981d, p. 24), for example, argued that:

> much of the critique of quantitative geography as 'positivism' has been misplaced and has accepted uncritically the representation of science as positivism given by Harvey in 1969. This has had a pernicious and destructive effect on the subject in three main ways: first, it has suggested that scientific and empirical enquiry is largely socially worthless; second, it has often rejected the links between physical and human geography; and third, it has often rejected the existence of geography as a discipline at all.

(Elsewhere, he characterized the critique as 'at best a mis-representative irrelevance, and at worst a fatuous distraction': Bennett and Wrigley, 1981, p. 10.) Quantification need not be allied with positivism according to Bennett, who identified three messages from Harvey's *Explanation in Geography*: geography is primarily inductive; geography is an objective science; and geography seeks universal laws. But 'each is only a partial representation of the literature and ideas it seeks to describe' (Bennett, 1981d, p. 13). Radical and quantitative approaches must be integrated: 'by empirical analyses, from the integration of environmental, social, political, and economic aspects with space, historical stimuli, and specific modes of thought and their spatial-political manifestations ... by the re-establishment of the geographical subject matter' (p. 24). A separate discipline of geography (human plus physical), in which quantitative geography – 'never ... truly positive' (Bennett and Wrigley, 1981, p. 10) – occupies a central place, would answer 'the fundamental questions of social norms, social distribution, policy impacts and humanistic concerns which the critics rightly emphasize' (p. 10). Many 'radicals' would categorize this as a *status quo* approach (p. 327).

In a later provocative essay, Bennett (1989a) argued that much 'radical geography' was out of touch with what he termed the 'spirit of the times'. He claimed that the welfare state, with its emphasis on rights, had led geographers to concentrate not only on inequalities but also to adopt a welfarist view that '*morally* they should be overcome or at least ameliorated' (p. 279). This leads to the argument that all differences imply relative deprivation and automatic entitlements to state action which would remedy them. But the newly dominant social theories (associated in the UK with 'Thatcherism') stress not 'the negative aspects of capitalism's capacity to create new wants and hence new "relative deprivation" but rather the liberating potential of markets'. According to his argument (see also p. 351 below), geographers should reorient their work to focus on the proper role of the state in such a society, identifying when it should interfere with market forces to ensure basic human rights, rather than promoting a welfare-corporate state that eventually must fail (see Johnston, 1992a, 1993e).

Thrall (1985) reported on a conference on 'scientific geography', which reached the consensus that 'Research in the unified areas of theory and modelling, data measurement and simulation, estimation and verification is central to the discipline of geography' (p. 254); such a perspective, he contended, should be used to promote the image of geography among the sciences and the 'acceptance of scientific approaches within geography'. His equation of science with mathematics, statistics and computer literacy was challenged by Driver and Philo (1986), who argued that 'science is more than a matter of technique' (p. 161) and that a clearly technocratic approach 'effectively marginalizes all other modes of interpretation and explanation' (p. 162). Thrall's (1986) response identified three major geographical traditions: humanistic; scientific; and pure theoretical. Scientific geography, he claimed, is a 'philosophy of research . . . clearly distinct from the earlier quantitative geography movement' (p. 162); he advocated research involving hypothesis-testing leading to theory-creation and verification, rather than either the empiricism of regional geographies or the sterile output of pure theoretical research.

This presentation of geography as an applied, quantitative discipline, but which does not accept the canons of positivism, faced continuing criticism. David Smith (1984, p. 132), for example, concluded an autobiographical essay with the statement that:

> To be quite honest, I no longer care very much about geography, with its smug self-satisfaction and nauseating narrow-minded chauvinism. Who else but geographers would dignify their puerile pursuit of statistics, models and paradigms as a 'revolution' as though it mattered to anyone but themselves? . . . And who else would want to read such trivia? How can we take it all so seriously, when it contributes so little to the improvement of the human condition? Most geography is inconsequential claptrap, and never more so than during the 'quantitative revolution'.

Mercer (1984, p. 194) argued in the same vein, though noting that:

> the last few years have witnessed a small – though perhaps temporary – retreat from the more lunatic excesses of flat earth quantitative geography towards a growing recognition that reality is not in fact beautifully ordered but that it is characterized much more by *contradictions, tension and disharmony*. The daunting . . . task for the

critical geographer – whether 'Marxist' or humanist – [includes] . . . the fight against the hegemony of naive, blinkered, technocratic thinking.

He wanted to unmask 'technocratic geography' and divert attention away from topics such as 'where are the regions of health care need based on access to hospitals?' to more fundamental questions like 'What leads to ill health?'. His basic theme was that selection of a particular research style, with the connotations of how its output may be used, involves an (albeit possibly unconscious) ideological choice: different conceptions of science are based on different views of both its utility and the social order within which that utility is to be employed.

Other writers sought to integrate elements of the positivist, humanistic and structuralist approaches to human geography (e.g. Christensen, 1982; for a fuller review, see Johnston, 1986f: Chapter 8 follows up these themes). Some attempts have been short-lived, such as an essay on the links between catastrophe theory (p. 139) and the discontinuities central to marxist economic theory (Day and Tivers, 1979, pp. 54–8; Alexander, 1979, pp. 228–30). Hay (1979a, p. 22) argued for an empirical, analytical geography that is:

> at the same time a *nomological* geography which seeks, for example, to understand the workings of urban rent theory as positivistically observed, a *hermeneutic* geography which seeks to identify the meaning of the urban rent system for those who are participants (active or passive) within it, and a *critical* geography which points to the extent to which present urban rent systems are themselves transformations of the capitalist system, but which admits that some of its features may indeed be 'invariant regularities'.

Livingstone and Harrison (1981, p. 370) somewhat similarly presented a case for: 'a humanistic geography which is, at the same time, critical, in questioning rather than bracketing our presuppositions, hermeneutic, in interpreting the meanings behind action, and empirical, in examining the subjectively interpreted objective world' but, unlike Hay, they included no nomological – or generalizing – component.

Some, particularly the critics of positivism, argued that such attempts at integration are not feasible because the approaches are incompatible: Gregory (1978a, p. 169) terms it 'inchoate eclecticism' (see also Gregory, 1982b; Eyles and Lee, 1982; Hudson, 1983. Positivist and structuralist approaches sit unhappily together in Rhind and Hudson, 1981.) The biggest problem is with positivism – not quantification (Walker, 1981b; though see Sayer, 1984, 1992a, and the response in Johnston, 1986a, 1993f) – because of its nomological orientation and belief that human geographers can discover 'invariant regularities'. Thus, as illustrated in Chapter 8, most of the links are between the structuralist (radical, realist or marxist) and the humanistic approaches, which treat the knowing actor operating within structural constraints (as in Eyles, 1981). Most of the humanistic approaches, as they have been presented to human geographers, focus entirely on the human as a free decision-maker, with no reference to either structural constraints or to how an individual's decision-making capabilities develop in a societal context. But several authors have indicated (e.g. Duncan, 1981) that human geographers must marry an ability to develop realist theories of the infrastructure with empirical methods of studying human action within the superstructure (Johnston, 1983c).

Increasingly the debates were not directly concerned with different philosophies of geography, but rather with their utility. As a consequence of the demands on education and research from governments of the 'New Right', considerable pressure to advance the study and practice of applied geography built up. (Taylor, 1985c, sets such pressures in historical context; see p. 349 below.) Some, such as Bennett (1985b), Openshaw (1989), and others, saw geographers' technical competence being applied to situations in the public and private sectors without any necessary reference to philosophical issues; applied geographers are thus technicians (as in Gatrell, 1985), who implicitly accept the context within which their skills are used. The philosophical issues remain important to others, however; Golledge *et al.* (1982) countered Hart's (1982, p. 5) case for regional geography that: 'We cannot allow ourselves to be intimidated by those who flaunt the banner of science' with

> We equally cannot allow ourselves to be intimidated by those who flaunt the banner of anti-science, those who would reject all that is scientific about the discipline, and those who would urge a return to the descriptive morass from which we have recently emerged (Golledge *et al.*, 1982, p. 558).

Most arguments for applied geography implied that it should be based in empiricist/positivist philosophies, however; the counter-arguments that humanistic science can be applied to improve self- and mutual awareness and that realistic science can be applied to advance social transformation were rarely presented (Johnston, 1986a).

Others believed that an eclectic pluralism is possible (see the discussion in Johnston, 1986f): some conducted empiricist work (often in the behavioural geography mould) unconcerned with any philosophical ramifications of that practice (see Flowerdew, 1986), however, whereas others generally accepted the realist case and situated their activities within that context. Gould (1985a, p. 296), for example, contended that:

> There is no question in my mind that the appearance of Marxist concern in geography, and its concomitant shaping of the lens through which the world is seen, have greatly enriched our methodological approach. There is an insistence that the things at the surface are not

always what they seem, and that it is crucial to dig down underneath the superficial appearances to get at the deep structures. I think this is quite right.

But he criticized marxists for their claims to truth, their condescending attitudes to 'non-believers', their over-concentration on economic forces, and the 'messianic claims that seem to lead so readily and so often to the sacrifice of human beings today for some promise tomorrow' (p. 300). Finally, there were geographers who, while not convinced of the case for an eclectic pluralism, nevertheless saw the need for a *rapprochement* of the 'radical' and humanistic approaches: their arguments are the focus of the next chapter.

Conclusions

This chapter has illustrated the depth and breadth of the discussions which occupied much of the 1970s and 1980s. An increasing proportion of geographers wanted to be involved in re-shaping societies, either through ameliorative correction of current problems and trends or by designing desirable spatial organizations (Berry, 1973a): some saw this as necessary if human geography was to retain its institutional position. Their motives ranged from 'pure altruism' to 'devoted self-seeking'; their methods varied from those who accept the present mode of production and see humanitarian goals as achievable within its constraints, through those who subscribed to the phenomenological view (Buttimer, 1974, p. 29) that:

> the social scientist's role is neither to choose or decide for people, nor even to formulate the alternatives for choice but rather, through the models of his discipline, to enlarge their horizons of consciousness to the point where both the articulation of alternatives and the choice of direction could be theirs,

to those who believed that a revolution is necessary to remove the causes of society's myriad problems and replace them by an equitable social structure.

Perhaps paradoxically, the 'revolutionaries' in the above classification are not 'activists' in the sense of being deeply involved in contemporary issues. The 'liberals' advanced the strongest arguments for geographical contributions to the solution of societal problems, particularly those which involve public sector intervention (e.g. Bennett, 1983), and pressed for academic engagement in policy-making; many of the 'radicals', on the other hand, argued that their longer-term goals were best served through educational programmes (e.g. Huckle, 1985), although some were involved in policy-making activities for institutions (such as local governments) that promoted alternative (socialist) strategies, especially in the support for employment initiatives (e.g. Duncan and Goodwin, 1985: see, however, Harvey, 1995, on the tensions this can create). The two groups are members of the same discipline, and practise in the same academic environments, yet their goals and methods seem totally incommensurate. Alongside them are many others, who continue to research and teach in an empiricist context, with some recognition of the range of the philosophical arguments but little detailed consideration of their implications for the practice of geography. (Flowerdew, 1986, argues that, on the basis of papers submitted to him as Editor of *Area* over a three-year period, the vast majority of geographical work is empiricist with 'unexamined value positions, no explicit theory and no clear criteria for establishing the truth of statements' – p. 263.) For them, perhaps the majority of human geographers, the task is to assemble material, thereby to sensitize their readers to the world as they perceive it.

By the end of the 1980s, therefore, Anglo-American geography was characterized by a plurality of approaches (Gould and Olsson, 1982), which can be classified into three philosophies, each with a separate epistemology, or theory of knowledge.

1 First was the positivist philosophy, with its belief in the objectivity of scientific description (empiricism) and analysis of the world, its goal of formulating laws about that world, and its assumption that explanation (causal laws) can be derived by studying the outcomes of the laws; the laws of spatial organization and behaviour can be revealed by analysing spatial patterns. At least some of those who promoted the empiricist foundation to this philosophy denied that it necessarily leads to the full positivist commitment. They disagreed over the use of scientific procedures for the evaluation of hypothesis, however, as in the critical rationalism usually associated with Popper (Hay, 1985a; Marshall, 1985; Bird, 1989).

2 Second were the humanistic philosophies based on a belief that people live in subjective worlds of their own creation, within which they act as free agents. Their actions cannot be explained (predicted) as examples of general laws of behaviour, but only understood – or appreciated – through methodologies that pierce their subjectivity.

3 Finally there were various structuralist and realist philosophies which argued that explanations for observed patterns cannot be discovered through analysis of the patterns themselves, but only by the development of theories of the underlying – although inapprehendable – processes that generate the conditions within which human agents can create those patterns. Foremost within this group is marxism, which argues that the processes are themselves changing – and can be altered by concerted political action – so that no laws of spatial organization are possible.

This plurality provided a focus of debate, and even a source of confusion, among human geographers. For some, it presented a polarization that can only be solved when one (paradigm?) is proved triumphant. (Others argue that this cannot be done logically, since there are no common criteria for comparing the approaches.) For many, the plurality offered potential for developing a newer, more robust human geography (e.g. Wilson, 1989a), though the nature of that development is far from clear. (The summaries of the statements made by six geographers – Hall *et al.*, 1987 – to a review by the British Economic and Social Research Council on *Horizons and Opportunities in Social Science* clearly illustrated the pluralism: British geography was warmly commended, however: ESRC, 1988.) For a considerable number, however, it seemed as if the debates were of little relevance. They continued to undertake empirical work, (implicitly) accepting the general structuration–realist thesis but believing that this is of little use in shedding light on the worlds of experiences and events.

But the world in which geographers work, and which they study, continued to change so that, without any resolution of the debates outlined so far, in the mid-1990s they were presented with yet another. Developments outside the discipline began to infiltrate it by the mid-1980s and within a decade these were the source of major shifts in what human geographers did, and why. They are the focus of the next chapter.

8

The cultural turn

The late 1980s saw changes within human geography which were both reactions to preceding disciplinary developments and responses to external stimuli. The majority of human geographers had rejected positivist approaches and no longer considered them viable, especially those associated with spatial science (see Barnes, 1996), although some continued to use the extensive research methodology described by Sayer (p. 219) to explore large data sets in the mould established by behavioural geographers: locational science was a small, relatively isolated sub-set of active geographers, and developments in GIS did not make a substantial impression on much of the discipline (indeed, many were very critical: see Gilbert, 1996).

There was growing disillusion with the radical and marxist approaches, however, despite their continued promotion by Harvey and a small number of others (notably Peet, Sayer, Walker and Neil Smith), because of both their implicit belief in explanation and progress (on which see Barnes, 1996) and their limited treatment of individuals; class was seen as the dominant category within marxist analysis, for example, down-playing (if not largely excluding) others, notably race, gender and sexual orientation, on which attention was increasingly focused (see Sayer, 1992c). Geographers, especially but not exclusively those of younger generations, sought an approach which avoided the implicit narrowly materialist determinism of both positivism and marxism/realism but also was theoretically richer than the voluntarism of the humanistic approaches advanced in the 1970s and 1980s. Some suggested that an accommodation of the two might be sufficient (as in several of the chapters in Kobayashi and Mackenzie, 1989), but others were less convinced.

Most of the critical focus was on the marxist/realist approaches. The rejection of positivism was such that it attracted little further attention, whereas the humanistic efforts were largely ignored and very little referenced. Regarding positivism, for example, Cloke *et al.* (1991) place a brief discussion of spatial science and its critics in the general introduction to their book (subtitled *An Introduction to Contemporary Theoretical Debates*) and begin the substantive chapters with the marxist approaches initiated in the late 1960s. On the humanistic approaches they wrote that (p. 17) 'the adoption and subsequent reformulation of humanist approaches ... [became] a more fragmented process than the evolution of Marxist thought in geography ... in part because the humanist label covers a diversity of different and sometimes incompatible intellectual positions' and concluded (p. 203) that humanistic tools may be valuable in studying 'the everyday mental worlds of everyday people ... with the rider that the theories employed must be prepared to look beyond the transcendental tendencies of much phenomenology and existentialism'. Critiques of the marxist/realist approaches, such as Graham's (1988, 1992; Gibson and Graham, 1992), identified as an important concern the so-called essentialism of the theories employed, especially that derived from Marx, which conflicted with the view that all particular events are 'over-determined'. The essentialist concentration on class interests was difficult to reconcile with a recognition that these interests vary with circumstances, and thus over time and space; class interests are just one of many possible components of an individual's identity, and (quoting Woolf and Resnick, 1987, p. 138) 'since every process exists as the effect of all other processes, each is quite literally a bundle of contradictions'. Everything is related to everything else, in a constant state of flux, so that no single explanation can be identified in any study and no single 'entry point' to their analysis can be privileged over any other. Postmodernism offered such critics an alternative, anti-essentialist perspective that promoted the appreciation of multi-faceted differences, and feminism and a variety of other approaches (several of them derived from the humanities) detailed the major sources of difference.

Beyond the discipline, there were rapid changes in both the economic, social and political context within which academic geography was set and in the practices employed in other social science and humanities disciplines. The former comprised the globalization and flexible accumulation trends within capitalism described earlier, along with major changes in the nature and role of the state apparatus which had profound effects on the funding and expectations of research in higher education. The latter involved attempts to come to terms intellectually with those changes, in the context of widespread disillusion with current academic practices. On both sides of the North Atlantic, the private and public sector structures in place since the 1940s were being dismantled, and replaced by others which appeared much less permanent. In the private sector, for example, many of the large manufacturing industries were very significantly 'downsized', if not totally eliminated, and their replacements were small enterprises, many in service industries, offering less job security and fewer fringe benefits, and with much weaker trades unions representing worker interests. In the public sphere, the major political parties lost much of their core support: the Democratic party's hegemony in the American southern states rapidly dwindled away, for example, and in the UK a majority of skilled working-class trades unionists voted for the Conservative party at the 1983 and 1987 general elections. Change was rapid and was everywhere; understanding why it was happening and what was replacing it provided human geographers with major challenges (as suggested by the essays in Johnston, 1993d) – but the changes were also affecting the institutions in which they worked, thereby increasing their insecurity.

The new project which developed within human geography lacked a core and a focused drive. Instead it brought together groups, sometimes in alliance but occasionally in opposition and often separate, stimulated by a variety of sources and approaches. What they had in common was a distrust of 'grand theory' – a belief that there was a single 'right' answer to research problems – and a celebration of difference within society. Hence use of the term 'cultural turn', which represents, according to Ley and Duncan (1993, p. 331):

> a premise that has been growing in strength in human geography over the past twenty years. To assert that reality is socially constructed is to interrogate the taken-for-granted categories of everyday life and of intellectual endeavours alike. It is a perspective that is vitally concerned with the positioning of the author, with such contingencies as class, race, gender, nationality and political persuasion that shape an outlook and colour an interpretation. The constructionist perspective is concerned not only with the author but also with his or her authority, the power which is always conferred upon the person with the authority to categorize. This power shapes the character of the 'Other'.

That premise appealed to some geographers because place was portrayed as one of the major positional sources within which people defined themselves, and hence the 'Others' who were not of that place.

This use of the term 'cultural turn' suggests a link with the sub-discipline of cultural geography, as traditionally practised (see p. 97). Price and Lewis (1993a), however, suggested that beyond the use of the adjective 'cultural' the 'old' and the 'new' 'share precious little' (p. 2). They argued that the 'new cultural geographers' (they cite Cosgrove, Duncan and Jackson as the 'standard bearers') have 'reinvented' the nature of 'old' cultural geography, in order to criticize it, with the Berkeley school (see p. 48) as 'its currently most misinterpreted form' (p. 3). In return, Cosgrove (1993) accused them of inventing a conspiracy, claimed that he had not wholeheartedly embraced the term 'new cultural geography', and welcomed the growing accommodation between the 'theory-free' Berkeley school and the theoretically grounded approaches. Jackson (1993b) was also accommodating, while pointing to a wide range of other traditions.

More stridently, Duncan (1993b, p. 518) responded that 'I cannot accept that it is possible to employ the term culture in one's work and yet employ no theory of culture', to which Price and Lewis (1993b) responded by re-stressing their case that Duncan misrepresented the Berkeley canon (as in Duncan, 1980), both in its orientation and in its implicit acceptance of 'the tiresome superorganic debate' (p. 521).

The 'cultural turn' involves new developments rather than a substantial building on old foundations, therefore, but its lack of a core, being rather a series of sometimes inter-woven strands coming out of previous approaches, means that this chapter is slightly more disjointed than its pre-decessors. Those different strands provide the successive foci of the major sections that follow.

Postmodernism

Postmodernism has become increasingly popular in the social sciences and humanities over the last decade and has attracted considerable attention from geographers. Its rise, according to Soja (1989), represented an attack on the predominance of historicism in modern thought, with its emphasis on biography (both individual and collective) and the consequent neglect of spatiality. He defined historicism as (p. 15):

> an overdeveloped historical contextualization of social life and social theory that actively submerges and peripheralizes the geographical or spatial imagination . . . [and which produces] an implicit subordina-tion of space to time that obscures geographical interpretations of the changeability of the social world.

Because of this, he claimed that geographers had failed to attain their 'rightful position' within the social sciences for much of the twentieth cen-tury: a 'superficially similar historical rhythm' (p. 33) was assumed to occur across all places, rather than a realization – as stressed in the localities and new regional geography projects discussed above (p. 241) – that social processes are differently constituted in different places, so that the histori-cal flow is not the same everywhere. (Not surprisingly, some sought to link the localities project to postmodernism: see Cooke, 1990.) Thus, for exam-ple, postmodern novels have an apparently chaotic structure when they try to represent different things happening simultaneously in different places, and postmodern architecture lacks a clear, functional structure (Knox, 1987).

This problem of synchronicity has long been recognized by geographers, as Massey (1992) observed with reference to Darby's (1962) paper on 'The problem of geographic description'. Darby pointed out that (p. 2):

> A series of geographical facts is much more difficult to present than a sequence of historical facts. Events follow one another in time in an

inherently dramatic fashion that makes juxtaposition in time easier to convey through the written word than juxtaposition in space. Geographical description is inevitably more difficult to achieve than is historical narrative.

But histories have geographies, and relating concurrent changes over both time and space simultaneously provides particular problems, especially since all mapped distributions are simply static snapshots of continuing processes (Blaut, 1962).

Postmodernism has a wide range of explicit and implicit meanings, and its core is hard to identify; Cloke *et al.* (1991, p. 19) write of it as 'infuriatingly difficult to define'. According to Dear (1994, p. 3):

> Postmodernity is everywhere, from literature, design and philosophy, to MTV [Music Television], ice cream and underwear. This seeming ubiquity only aggravates the problem in grasping its meaning. Postmodern discourse seems capable of instant adaptation in response to context and choice of interlocutors.

Nevertheless, he believes 'we can cut to the heart of the matter by identifying three principal constructs in postmodernism: style, epoch and method' with the first having provided the source of the initial explosion of interest.

1 *Postmodernism as style* originated in literature and literary criticism, and spread to other artistic fields such as design, film, art, photography and architecture; the general trend involved the promotion of difference and the lack of conformity to over-riding structural imperatives. Dear found trends within architecture especially revealing:

> the search for the new was associated with a revolt against the formalism and austerity of the modern style epitomized by the unadorned office tower ... The burgeoning postmodern architecture was disturbingly divorced from any broad philosophical underpinnings, taking the form of an apparently-random cannibalizing of existing architectural archetypes, and combining them into an ironic collage of (or pastiche) of previous styles.

2 *Postmodernism as epoch* portrays current developments within society as a major radical break with the past – hence use of the term 'postmodernity' to contrast it with the modernity of the previous epoch. These 'new times' are characterized by difference, so that study of the postmodern epoch involves

> grappling with the fundamental problem of theorizing contemporaneity, i.e. the task of making sense out of an infinity of concurrent social realities. Any landscape is simultaneously

composed of obsolete, current, and emergent artifacts; but how
do we begin to codify and understand this variety?

3 *Postmodernism as method* is, according to Dear, likely to be the most
enduring of the three main trends. It eschews the notions of universal
truths and meta-theories which can account for 'the Meaning of
Everything'. No portrayal can claim dominance over another; separate
theories are incommensurable and so cannot be evaluated: 'even the
attempt to reconcile or resolve the tensions among competing theories
should *a priori* be resisted'.

This emphasis on 'heterogeneity, particularity and uniqueness' (Gregory,
1989a, p. 70) undoubtedly attracted some human geographers to postmodernism – or, as Dear (1994, p. 3) expressed it in a quote from the *New York
Times*, 'the great lesson of the twentieth century is that all the great truths
are false'. Human geographers under the sway of modernism emphasized
order in their promotion of spatial science, when their empirical observations (as their critics pointed out: see p. 184) could really only identify disorder, which suggested the absence of generally applicable theories and
universal truths (Barnes, 1996). Postmodernism gave them a philosophical
hanger, recognizing (Gregory, 1989a, pp. 91–2) that:

> there is more *disorder* in the world than appears at first sight is
> not discovered until that order is looked for ... we need, in
> part, to go *back* to the question of areal differentiation: but
> armed with a new theoretical sensitivity towards the world in
> which we live and to ways in which we represent it.

(This statement is very similar to that regarding regional geography which
came at the end of his seminal 1978 volume: see p. 247.)

With regard to theory, Dear (1988) observed in his paper introducing
postmodernism to a wide geographical audience that one theoretical viewpoint cannot be promoted over another, because their relative merits are
'undecidable' (p. 266); geographers cannot aspire to 'grand theory' but can
only interpret 'the contemporaneity in social process in time and space'
(p. 272). Recognition that all knowledge is not only time-space specific but
also expressed in language that reflects such specificity (see below) meant,
to him, that geography can 'claim its place alongside history as one of two
key disciplines concerned with the time-space reconstruction of human
knowledge' (p. 272).

Understanding the attractions of this postmodern stance to geographers
requires prior appreciation of why modernist spatial science came to dominate their discipline. According to Gregory (1989b), throughout much of
the nineteenth and twentieth centuries geographical practice has been influenced by particular aspects of, first, anthropology, then sociology, and
finally economics. Two features dominated the adopted modernist
paradigm:

1   its firm base in *naturalism,* which was likely to have 'special significance
    in a discipline like ours, where human geography is yoked to physical
    geography' (p. 352) and the consequent reliance on the aims and proce-
    dures of the natural sciences as relevant to the study of humans and
    their societies; and
2   a *totalization* conception of space, which involves the search for a 'sys-
    tematic order whose internal logic imposes a fundamental coherence on
    the chaos of our immediate impressions', hence the dominance of spatial
    science within that project.

That totalizing, naturalistic approach was being attacked as irrelevant to
understanding the contemporary world; marxism, for example, apparently could not cope with the development of what Lash and Urry
(1987) termed 'disorganised capitalism', and no 'grand' social theories
(including marxism) could comprehend the time-space variety made clear
in Giddens' notions of structuration. Thus for Gregory (1989b, p. 379),
postmodernism   provided   'a   particularly   vibrant   statement   of   the
polyphony that characterizes (and ought to characterize) contemporary
social theory'. It offered a framework within which he could situate both
structuration theory and Habermas's critical theory, but should be
approached with caution and not as a universal panacea (p. 379): 'it
needs to be saved from both its antagonists and its advocates. Its claims
need to be approached openly, scrupulously, and vigilantly.'

Postmodernism presents a substantial critique to the approaches which
dominated much of geography in the period from the 1950s through the
1980s, with their emphases on order and 'grand theory'. As Cloke *et al.*
(1991, p. 200) express it:

> human geographers will increasingly come to recognise the gravity of
> the challenge that postmodernism as attitude poses to the most con-
> ventional theorisations of the human world, and will begin to appre-
> ciate that a sensitivity to the geography of this world – to its
> fragmentation across multiple spaces, places, environments and land-
> scapes – is itself very much bound up with (and an impetus for) a
> postmodern suspicion of modernist 'grand theories' and 'metanarra-
> tives'. . . . These manoeuvres will doubtless cause much unease and
> controversy, however, given that they cast doubt on the stability of the
> foundations from which most human geography had proceeded over
> the last thirty years or so.

Furthermore, that unease will not only erode confidence in what geography
is and how it is done, but also in views of the future; if, as postmodernists
proclaim, there is no 'grand theory', no 'truth' that can eventually be discovered, on what basis can a better future be planned for all? (As Dear,
1994, p. 8, expressed it, some argue that: 'postmodernism's extreme

relativism renders it politically incoherent, and hence useless as a guide for social action': see also Duncan, 1996, p. 453.)

As with other approaches presented to them, such as structuration, for example, one of the problems for many human geographers is how the adoption of postmodernism will influence how they 'do geography', how they conduct their individual empirical inquiries. There are relatively few exemplars. One pressed by some is Soja's (1989) work on the Los Angeles metropolitan area where, he claims, 'it all comes together'. The metropolitan area is, he argued, difficult to represent in a narrative because its many images seem 'to stretch laterally instead of unfolding sequentially' (p. 222), and because 'it too seems limitless and constantly in motion, never still enough to encompass, [it seems] too filled with "other spaces" to be informatively described'. All he says he can offer is 'a succession of fragmentary glimpses' into a 'particularly restless geographical landscape' (p. 223) – what he terms an 'interjacent medley' (p. 247). Its environment is too multi-layered, created by too many authors to be identified: there are too many 'discordant symbols drawing out the underlying themes'.

Soja's work was subjected to a substantial critique by Gregory (1990; see also Chapter 4 of Gregory, 1994). He argued that modernist theories of the capitalist city stressed (Soja, 1989, p. 50) 'geographically uneven development via simultaneous tendencies towards homogenization, fragmentation and hierachization'. He accepted that Los Angeles' landscape contains (p. 246) 'an economic order, an instrumental nodal structure, an essentially exploitative spatial division of labor' but claimed that these cannot be summarized into a 'totalising description'; all that can be offered is a 'series of fragmentary glimpses'. And yet, Gregory stressed, Soja advised those wishing to explore the metropolis that 'it must be reduced to a more familiar and localized geometry to be seen' (Soja, 1989, p. 224) and its emphases are (Gregory, 1994, pp. 300–1):

> Towers and freeways, sites and districts, zones and areas, enclaves and pockets, gradients and wedges: a landscape without figures ... Soja's essay becomes a morphology of landscape that, like Sauer's original, is rarely disturbed by the human form ... Soja's essay [is] so astonishingly univocal. We never hear the multiple voices of those who *live* in Los Angeles – other than Soja himself – and who presumably learn rather different things from it.

To Gregory, Soja excluded much of the difference which is supposedly at the heart of a postmodern approach; he ignored the various social struggles that underpin the making and remaking of the Los Angeles landscape, plus the distinctive urban cultures of ordinary people's everyday lives. Thus whereas Gregory (p. 304) applauded Soja's thesis that 'it is impossible to recover human geographies from a contemplation of their abstract geometries', nevertheless he concluded that Soja 'renders the landscape of Los Angeles as a still life': Soja demonstrated that 'postmodernism can have an insistently critical edge' (p. 312) but there is a clear hermeneutic problem in his work: 'his master-narrative is sometimes so authoritarian that it drowns out the voices of other people engaged in making their own human geographies'. (Indeed, according to Duncan – 1996, p. 443 – Soja's 'goal of reconstructing geography along postmodernist lines . . . belies the fact that Soja is not really a postmodernist!')

A further, wide-ranging critique was launched by Harvey (1989a) in his book *The Condition of Postmodernity*. Its basic thesis is clearly expressed in his brief summary page, entitled 'The argument' (p. vii), which is reproduced here in full:

> There has been a sea-change in cultural as well as in political-economic practices since around 1972.
>
> This sea-change is bound up with the emergence of new dominant ways in which we experience space and time.
>
> While simultaneity in the shifting dimensions of time and space is no proof of necessary or causal connection, strong *a priori* grounds can be adduced for the proposition that there is some kind of necessary relation between the rise of postmodernist cultural forms, the emergence of more flexible modes of capital accumulation, and a new round of 'time-space compression' in the organization of capitalism.
>
> But these changes, when set against the basic rules of capitalistic accumulation, appear more as shifts in surface appearance rather than as signs of the emergence of some entirely new postcapitalist or even postindustrial society.

The observed changes, especially but not only in modes of economic organization and much facilitated by technological changes (notably but not only in information technology), are the outcomes of the processes of continual restructuring by which capitalism overcomes its crisis tendencies (on which see Harvey, 1982; Harvey and Scott, 1989) and within which cultural forms are produced and reproduced (as set out in Harvey, 1985c). His assessment of postmodernism identified positive elements (pp. 113–15):

> in its concern for difference, for the difficulties of communication, for the complexity and nuances of interests, cultures, places and the like, it exercises a positive influence. The meta-languages, meta-theories, and meta-narratives of modernism . . . did tend to gloss over important differences, and failed to pay attention to important disjunctions and details.

But he also rejected what he identified as the negative elements, with postmodernism presented as 'a wilful and rather chaotic movement to overcome all the supposed ills of modernism'. Postmodernism, to him (p. 116), reflects

a particular kind of crisis within ... [modernism], one that emphasizes the fragmentary, the ephemeral, and the chaotic side ... (that
side which Marx so admirably dissects as integral to the capitalist
mode of production) while expressing a deep scepticism as to any particular prescriptions as to how the eternal and immutable should be
conceived of, represented, and expressed.

And so it gets the balance wrong, because its adherents are unwilling to
grapple with the need for a theory which promotes appreciation of the
nature of capitalism and its reproducing tendencies, especially, for geographers, those relating to time-space compression through changed transport
and other technologies and the political strategies which promote the fortunes of people in different places. The shifts identified in the first sentence
of his summary require a dynamic theory of historical materialism, within
which he identified four main agenda items (p. 355):

1 the treatment of difference and 'otherness' (see below) within the dialectics of social change, thereby 'recuperating such aspects of social organization as race, gender, religion, within the overall frame of historical
   materialist enquiry (with its emphasis on the power of money and capital circulation) and class politics (with its emphasis upon the unity of the
   emancipatory struggle);
2 recognition of the importance of cultural practices, including the production of images and discourses, on the reproduction of the social
   order;
3 recognition of the importance of space and time in the geopolitics of
   capitalism, now that 'Historical materialism is finally beginning to take
   its geography seriously'; and
4 acceptance that

   Historical-geographical materialism is an open-ended and dialectical
   mode of enquiry rather than a closed and fixed body of understandings. Meta-theory is not a statement of total truth but an
   attempt to come to terms with the historical and geographical
   truths that characterize capitalism both in general as well as in its
   present phase.

Harvey, in turn, has come under attack. Deutsch (1991) criticized his 'totalizing vision' of society, with aspects of difference merely appended to marxist theory – which she claims is male-dominated – and Massey (1991)
similarly denounced the book because neither women nor the possibility of
a feminist interpretation of postmodernism figure in it. To her, 'gender is a
determining factor in cultural production' (p. 51), but Harvey promotes
marxism, within which, but only within which, gender differences can be
subsumed. Likewise, Dear (1991b) claimed that Harvey 'seems incapable
of tolerating difference' (p. 536) so that 'The catalogue of different voices

that are consequently denied relevance in Harvey's discourse is long and depressing' and his goal appears to be to '*dissolve differences*' (p. 537). Much of the postmodern literature is ignored, he argued – 'Harvey is a much better political economist than he is cultural critic.' He summarized Harvey's position as follows:

1 all aspects of social processes can be encompassed within historical materialism; so that
2 no alternatives need be addressed;
3 flexible accumulation is the right model to apply when considering recent social, economic and political changes; and
4 there is no need for any other theory beyond historical materialism.

He then concluded that 'Perhaps it is time that Harvey tried to transform his Marxism, rather than obliging the world to fit into it': Harvey recognizes difference, but refuses to incorporate it to his thinking.

In response Harvey (1992) noted that he was influenced by some feminist writers, and that incorporation of their insights would have strengthened his book; the work of Young (1990) in particular had considerably influenced his later work on justice (Harvey, 1993a). He accepted the role of situation and position in the production of knowledge (see below, p. 367) but found some of their applications in feminist work 'rather more vulgar' because of their focus on individual biographies not dialectically placed in an evolving mode of production and their denial of the veracity of other accounts. Thus his book was written from a particular (privileged, white male) position and (p. 304):

> emphasized the commonality of our condition as users of commodities and money and as participants in labor markets and the circulation processes of capital. But emphasizing commonality does not deny difference. Properly done, it can enhance the understanding of differences at the same time as it provides a critical basis from which to evaluate the work of those who purport to write critical theory outside the confines of what capitalism as a social system is all about.

In this context, he found the critiques by Deutsch and Massey 'unnecessarily personalized, hurtful and sometimes abusive' (p. 308), part of a strategy of creating a distinctive feminist identity which he claims shows that 'they exhibit not the slightest hint of concern to grapple with the deeper problematic' involved with dissolving gender differences. He wants debate to focus on these and finishes by pointing to a major paradox regarding truth and generalization in all postmodern writing (p. 322):

> postmodernists ... cannot criticise *The Condition* as wrong, misguided, or fundamentally misconceived without deploying truth terms of their own which presuppose they have an ultimate line on a truth they theoretically claim cannot exist. I am reminded here of

William Blake's great aphorism: 'to generalise is to be an idiot; to par-
ticularise is to achieve the greatest distinction of merit', which sounds
great ... until it is recognized as a generalization and thereby self-
condemned as idiotic.

Sayer (1994b) made a similar argument, pointing out that if all knowledge
is fallible, then situated knowledge is as fallible as are grand narratives: just
as Harvey promoted a marxist approach for understanding difference
within systemic wholes, so Sayer advocated realism.

Elsewhere, Dear (1991a, p. 549) criticized both Harvey and Soja, claiming that each rejects the pluralism and celebration of difference that characterizes postmodernism:

> By insisting on their totalizing and reductionist visions, Soja and
> Harvey squander the insights from different voices and alternative
> subjectivities. Difference is relegated to the status of an obstacle hin-
> dering 'our' view of a coherent theoretical and political praxis. ...
> Moreover, for two so overtly committed to progressive politics, it is
> surprising that neither recognizes that the Left, by insisting on unity
> and ignoring divisions in the last, has itself contributed to its dimin-
> ished political efficacy.

Barnes, too, concluded that Harvey's commitment to what he terms the
Enlightenment project – 'a belief in rational progress, the individual subject, a monolithic order, and universal truth' (Barnes, 1996, p. 3) – runs
counter to prevailing trends (as also, he claims, does Sayer's critical rationalism): 'Although [Harvey] wants to recognize that truths are socially
constructed at a given moment in time and space, he also wants to claim
that his theory is a valid and accurate representation of the capitalist
world.'

Dear's evaluation of postmodernism's impact on human geography is
entirely positive (1994, p. 9):

> Simply stated, we live in an era of postmodern consciousness; there is
> no choice in this matter, unless we are prepared to declare in favor of
> ignorance or the *status quo*. I believe that a revolution of sorts is
> occurring in geographical thinking.

He supported this last claim by arguing that since 1984 we have witnessed:

(a) a truly unprecedented increase in quality scholarship devoted to the
relationship between space and society;
(b) a reassertion of the significance and role of space in social theory and
social process;
(c) an effective reintegration of human geography with mainstream social
science and philosophy;
(d) the establishment of theory and philosophy as the *sine qua non* for the
discipline's identity and survival;

(e) a new appreciation of diversity and difference, and a consequent diversification of theoretical and empirical interests; and

(f) a self-conscious questioning of the relationship between geographical knowledge and social action.

He further claimed that a large number of publications which appeared in 1989 or later 'reveals a significant postmodern consciousness' (p. 7) in seven areas:

1 cultural landscapes and 'place-making';
2 the economic landscapes of flexible accumulation and post-Fordism;
3 philosophical and theoretical disputes about space and language;
4 problems of representation in geographical writing and image-making;
5 the politics of postmodernity and difference;
6 the construction of the individual; and
7 a realization of the importance of nature and the environmental question.

This leads him to an optimistic conclusion, provided that geographers are prepared to grapple with the postmodern challenge (p. 9):

> Postmodernism places the construction of meaning at the core of geography's problematic. The key issue here is authority; and postmodernism has served notice on all those who seek to assert or preserve their authority in the academic or everyday world. And yet I understand that geographers, like everyone else, cling tenaciously to their beliefs. Knowledge is, after all, power, and we are all loathe to relinquish the basis for our claims to legitimacy.

Thus,

> To ignore the postmodern challenge is to risk disengaging geography once more from the mainstream. To accept it is to encourage new ways of seeing, to relish participating at the cutting edge of social and philosophical inquiry, to convince our peers of the significance of space in contemporary social thought and social process, and to help forge a new politics for the twenty-first century.

Many human geographers remained concerned as to how these could be achieved, as Dear (1995) recognized in responding to an evaluation of developments since his 1988 paper. Graham (1995, p. 175) accepted that in evaluating Dear's (1994) claim that postmodernism has flourished:

> If judged by the volume of literature which mentions postmodernism then this is certainly true. Postmodernism has become the *lingua franca* of intellectual discourse (Ley, 1993). As yet, however, few are fluent and communication easily breaks down, inevitably producing a disturbingly disorganised discussion (Warf, 1993).

She also noted (p. 176), however, that:

> Postmodernism is emancipatory for it frees us from the dominance of any exclusive paradigm or meta-narrative (Folch-Serra, 1990) yet it is also dis(en)abling because it provides us with no tools for taking yes/no positions (McDowell, 1991). This is one of the dangers of postmodernism, creating an atmosphere in which 'anything goes'.

And she stressed that Dear's (1988) emphasis on separate economic, social and political geographies, set within a postmodern agenda (what he later referred to – 1995, p. 181 – as 'a multi-foundational basis for the discipline'), is an argument against diversity and plurality, privileging some approaches over others. Dear (1995, p. 179) responded that: 'I have concluded that nothing human can be characterized as an absolute truth; our understanding is always and necessarily contingent, ambiguous and context-bound – this last condition being all that is needed by way of a rationalization for a geography.' Thus the indeterminacy of alternative accounts cannot readily be resolved, since there is no 'higher authority' to appeal to – even in the case of the ethics of the Holocaust ('I will readily concede that a moral and ethical case *can* be made against the Holocaust . . . but I also insist that such resolutions are bought only at the cost of silencing the Other, and that knowledge remains conditional. Thus such resolutions should not be uncritically purchased': pp. 179–80). So any approach to problem-resolution must be relative:

> assuming that all truths are local and contingent, and that any conclusions represent no more than a temporary hegemony of a favoured belief. Such studied relativism makes it a condition of knowing and practice that we *make explicit the conditions and criteria under which knowing is to occur and decisions to be deliberated.*

Postmodernism emerges as (p. 180): 'an ontological stance that engages the multiplicities in our own ways of seeing, refuses hegemony and permanent closure in intellectual debate, and seeks deliberately to confront the contested nature of understanding and choice', as a 'sophisticated practice' that as yet 'we are far from attaining'. (Note that Duncan, 1996, p. 447, doubts Dear's commitment to postmodernism; in Barnes's, 1996, terminology, she sees him as committed to the Enlightenment goals of 'progress' and 'better (as in closer to the truth) explanations'.)

Feminism

Gender is one of the main sources of 'difference' identified within the postmodern literature, alongside race. Feminist geography (according to McDowell, 1986a, p. 151) 'emphasizes questions of gender inequality and the oppression of women in virtually all spheres of life', and its goal includes uncovering and countering such inequality and discrimination within the geographical profession itself (on which see the early essays by: Zelinsky, 1973a;. Zelinsky, Monk and Hanson, 1982; Jackson, Smith and Johnston, 1988; Johnson, 1989). That is part of a much larger task (McDowell, 1989, p. 137):

> to demonstrate that women do matter in geography, and to argue that the failure to take gender differences into account impoverishes both geographical teaching and scholarship ... [in addition] it is not enough just to add women in as an additional category. Feminist geography, as opposed to a geography or geographies of women, entails a new look at our discipline. It poses several awkward questions about how we currently divide the subject matter into convenient academic parcels and also challenges current practice in teaching and research.

The goal is not just to add a further fragment to the academic discipline, with the potential for its ghettoization, but rather to ensure that a feminist perspective informs all work within human geography.

Feminist geography, according to Johnson (1989), involves recognizing women's common experience of, and resistance to, oppression by men, and a commitment to end it 'so that women can define and control themselves' (p. 85). Evaluation of geographical practice will demonstrate that it is 'sexist, patriarchal and phallocentric' and will open the way to emancipation, by providing a guide to political practice (Bowlby *et al.*, 1989). The fullest such evaluation has been undertaken by Rose. In a critique of histories of the discipline, for example, she argued that (Rose, 1995, p. 414):

> The writing of certain kinds of pasts is legitimated by, and legitimates, only certain kinds of presents. . . . Traditions are constructed: written, spoken, visualized, taught, lived . . . certain people or kinds of people are included as relevant to the tradition under construction and others are deemed as irrelevant.

This is a procedure which, she claimed, allowed Stoddart (1991) to dismiss the writings of Victorian lady travellers (Domosh, 1991a) as outside what he interpreted as the geographical tradition, and thereby irrelevant to any history he might write. Women have been written out of geography's history, and not only by Stoddart, so that: 'even if we can no longer be certain exactly what geography was in the past, in virtually all histories of geographical knowledges one apparently incontrovertible fact remains: geography, whatever it was, was almost always done by men'.

Rose's (1993) major contribution to the debate on the role (or lack of it) of women within geography involves the argument that:

1 'the academic discipline of geography has historically been dominated by men' (p. 1);
2 within the profession, women have been patronized, harassed and marginalized;
3 feminism remains 'outside the project' of geography (p. 3); and
4 'the domination of the discipline by men has serious consequences both for what counts as legitimate geographical knowledge and who can produce such knowledge. [men] . . . have insisted that geography holds a series of unstated assumptions about what men and women do, and that the discipline concentrates on spaces, places and landscapes that it sees as men's' (p. 2).

This leads her to conclude that the social construction of the discipline of geography is such that it is now necessarily 'masculinist' – it forgets women and concerns itself only with the issues of interest to men (p. 4). This is not presented as a conscious male plot, however: rather, geography has adopted a particular set of positions and practices which attracts both men and women who subscribe, subconsciously, to masculinist positions. Most of her book details the four points of her main argument: the final chapter sets out the basis for a feminist approach to geography which represents women as both prisoners and exiles, and provides a foundation on which resistance to the masculinist hegemony can be based.

Walby (1986) identified five approaches to the study of gender inequality:

1 the demonstration that it is either theoretically insignificant or non-existent;
2 indications that it is derivative of capitalist relations, and has no independent status;
3 suggestions that it results from an autonomous patriarchal system which is the primary form of social inequality;
4 demonstrations that it results from patriarchal relations that are so intertwined with capitalist social relations that they comprise a single system of capitalist patriarchy; and
5 arguments that it results from the interaction of autonomous systems of patriarchy and capitalism.

Bowlby *et al.* (1989) showed that the second and fourth of these have been more commonly adopted among geographers, thus focusing on ways in which patriarchy is structured by capitalism's imperatives. Foord and Gregson (1986) adopted the fifth strategy, however, arguing that although patriarchy and capitalism are empirically linked they are theoretically separate. Just as capitalism can be presented as a particular example of a necessary condition (a mode of production within which economic and social life is structured), with unique individual instances at separate times and places, so, they argued, patriarchy can be represented as a particular example of a general condition (gender relations) which also has its time- and space-specific realizations.

Foord and Gregson (1986, p. 199) argued that the necessary inter-relationships between humans and nature call for a form of social organisation to ensure human survival and social reproduction, so that '*all* social relations must involve gender, and gender . . . relations will be embedded in all forms of social relations'. They could not conceive of a society lacking gender relations, therefore, so that the latter are independent of any mode of production, let alone any particular instance of it. Patriarchy is a partic- ular form of gender relations, in which men dominate the processes of species reproduction; empirical work focuses on the nature of that domina- tion 'in particular periods and places' (p. 206): 'Just as other relations vary and combine differently over time and space, so too must the practices of which comprise these relations' hence the professed importance of studying them within the context of the localities programme (see p. 241), as illus- trated by McDowell and Massey (1984).

Because gender and social relations are separate spheres of activity, Foord and Gregson contended that marxist and feminist theories could not be integrated. Against this view, McDowell (1986b) advanced a capitalist class analysis and rejected the notion of a universal female experience. Biological reproduction is part of the process of capitalist reproduction and (p. 313): 'Patriarchal social relations are further strengthened by the politi- cal and ideological functions of the state that has a vested interest in sup- porting the domination of individual women in the exploited class by individual men in that class.' The contradiction underpinning the need to create surplus value within capitalism is thus the source of women's oppres- sion, not any necessary gender relations (p. 317): 'the social construction of male sexuality and the dominance of family forms based on sexuality and kinship networks to class societies are historical resolutions of the contra- diction'. Gier and Walton (1987, pp. 56–7) took a slightly different approach, disagreeing with Foord and Gregson's contention that gender relations are necessary to all social relations:

> Evidence from anthropology and history as well as other disciplines indicates that gender has not always been used to identify male and female sexual difference and its attendant physical and psychological archetypes . . . [so that] the very identification of the concept is the product of human consciousness and human society.

Gender differences are human creations, according to this view, and need not always be present (see also Knopp and Lauria, 1987). Gregson and Foord (1987, pp. 373–4) responded by defending their position that mode of pro- duction and gender relations are 'distinct and separate objects of analysis which interlock as particular forms (capitalism and patriarchal gender relations) but not as conceptual categories', but were unprepared to admit that this was a preface to the creation of a universal theory of women's oppression.

Although some of the early work presented as feminist geography involved demonstrating that women are discriminated against and oppressed by men in contemporary societies (e.g. Women and Geography Study Group, 1984; Little, Peake and Richardson, 1988), wider issues were soon introduced. Pratt and Hanson (1994), for example, moved from demonstrating differences between men's and women's commuting patterns to analyses of how those patterns were exploited by employers (mainly men) to reproduce segmented labour markets (see Hanson and Pratt, 1995).

The variety of perspectives adopted is illustrated in McDowell's (1993a, 1993b) major review articles. During the 1980s and early 1990s feminist geography, while addressing the discipline's three main concepts of space, place and nature, shifted from analyses of gender differences to concerns over the social creation of gendered beings in particular places, which brings *feminist* geography (as distinct from feminist *geography*) closer to the wider feminist project – 'the study of the lives, experiences and behaviour of women' (1993a, p. 161). Three main themes are identified in the early work:

1 *spatial differences in women's status* – demonstrating 'man's inhumanity to women' (p. 163) – a largely empirical task which emphasized western experience and was increasingly criticized for its ethnocentrism;
2 *gender and place: women and the urban environment,* which stressed that most women were excluded from analyses of urban areas that focused on public rather than private activities, at scales which were larger than the individual household, and so ignored the gender relations that underpinned the home, childbirth and the unpaid labour involved in social reproduction (pp. 165–6):

   in common with the other social sciences, geography takes for granted the Enlightenment distinction between the public and the private, and, implicitly, the gendered associations of those spheres ... [which involves the patriarchal assumptions that] women are the angels of the hearth, to be confined to domesticity preferably in sylvan and suburban surroundings, while men join the fray of public life in the bustling city centres, returning home for emotional and sexual solace, and hot meals to fit them for their continuing labours;

3 *patriarchal power*, which illustrated the 'blindness' of (urban and other) geographers to the 'embodiment of conventional gender divisions' (p. 167) in the built environment on both large (the structuring of urban land-use patterns) and small (the design and layout of buildings) scales.

Thus gender inequalities were added to the others identified by those involved in portraying 'unfairly structured cities' (e.g. Badcock, 1984) as major elements in the reproduction processes of 'patriarchal capitalism'. This 'feminist empiricism' (p. 174) paralleled concerns of other (largely male) geographers for social justice, and stimulated a 'common focus on excluded or oppressed groups – be they women, the working class or ethnic minorities – [that] united "radical" geographers'. These concerns needed to be linked to wider theoretical considerations, however. McDowell (like Sheppard, 1995) saw a bifurcation within human geography (McDowell, 1993a, p. 174):

> It is perhaps not too much of a caricature to argue that throughout the 1970s and early 1980s human geography consisted of two phalanxes going off in sharply different directions. In the early part of the period, the opposing forces consisted of those who held on to mathematically modelled, rational choice, context-free notions facing the grand theoreticians of structuralism whose forward march in a single direction admitted women only in the tail position. Despite their relegation, it was in this camp that feminists found common interests, not least because of its intellectual and political attraction . . . [nevertheless] Despite an enormous shift over the last 15 years or so within economic and urban geography from a structuralist Marxism influenced by Althusser towards a geohistorical materialism alive to specificity, complexity and local struggles . . . gender issues remained marginal to the concerns of 'left' geographers. Thus in 1989 Susan Christopherson was moved to publish a bitter reflection on the ways in which feminism and feminists still remained outside what she saw as the geographical project.

Feminist empiricism was a foundation for a challenge to the masculinist domination, even epistemological foundations, of human geography (p. 175):

> this diverse set of theoretical and empirical books and papers published throughout the 1980s succeeded in placing gender divisions on the geographical agenda. It made women *and men* visible as both academic authors and subjects, thus challenging the implicit masculinity of the discipline. It also succeeded as geography, documenting some of the range of variation in women's social position and circumstances in different parts of the world and raising questions about gender identity and place.

It also laid the groundwork for advance on other fronts:

> this work has enabled more complex theoretical questions about the extent of variety in the spatial constitution of gender and the specific ways in which the characteristics of masculinity and femininity vary between spaces, classes and ethnicities to begin to be raised. Thus by the end of the decade the particular geographic interest in variety coincided with a more widespread interest in the deconstructionism in social theory to enlarge the scope of the agenda for feminist geography in the forthcoming years

in which it was joined by those interested in other sources of difference, such as sexual orientation (see p. 289) and race.

Building on that foundation involved two other feminist perspectives emerging alongside the 'rationalist or empiricist feminism' (McDowell, 1993b):

1 *Anti-rational, or feminist standpoint theory* celebrates gender differ-ences and, rather than present them as unfair or unjust, promotes them and has as its goal the elimination of the traditional allocation of superi-ority to everything associated with masculinity. Knowledge is socially created, in context, so that women's experiences of, for example, men-struation, childbirth and lactation, lead to the construction of different self-identities from those involved in masculine experience, which also vary between places, because of their separately constructed gender rela-tions. Such different standpoints provide a basis not only for under-standing but also for practice, hence the development of various forms of radical feminism, including ecofeminism.

2 *Postrational or postmodern feminism* argues that treatment of women as a single category involves linking together very different groups with separate experiences and needs (involving what Sayer – see p. 217 – terms a chaotic conception). Many feminists initially identified this as a threat to their project, but McDowell argues that the encounter with postmodernism stimulated a debate about 'situated knowledge' which has led to the situation (McDowell, 1993b, p. 310) whereby:

> geographers sympathetic to the postmodern deconstructive project are ... reluctant to abandon gender as *a* difference that makes a dif-ference, if no longer *the* difference. The current aim within feminist geography is a move towards ... 'partial' or 'situated knowledges' that recognize that the positionings of white British women in the academy, to take but one example, are not the same as those of other women, women from different ethnic or class backgrounds, and that this makes a difference to knowledge construction.

For geographers, this emphasis on difference within the 'oppressed' group raises important issues regarding the role of place in identity-creation. Knowledge is both local and gendered – and also linked to other socially constructed categories. Feminists argue that *'some differences are more sig-nificant than others'* (McDowell, 1993b, p. 315); their work

> demands a theoretical analysis of differences, building up an under-standing of the mutual interrelations of gender ... of sexuality, household and family structures and the political economy of domes-tic and workplace relations within and between places. It is a femi-nism located in a theoretical understanding of differences between women, rejecting both the transhistorical and cross-class search for the origins of patriarchy that concerned us at an earlier moment ...

and notions of a cultural feminist essentialism that denies the structure of power relations between women. It is a feminism that recognizes the existence of a material world, or women living in different
social formations, engaged in struggles in which their interests may
converge or diverge.

This presents a major challenge to all geographers, who, as this chapter
illustrates, are having to come to terms with a wide range of situated
knowledges. Bondi and Domosh (1992, p. 210) have argued, however, that
although postmodernism attacks universal truth claims and so should sustain feminist arguments, its application has failed to do so because 'there is
a major impulse within postmodernism that continues a tradition of masculine discourse in which the stereotyping of women is intrinsic to its operation'.

Stoddart (1991) argued against the perceived need for a 'feminist historiography' of geography expressed in Domosh's (1991a) critique of the treatment of women in his *On Geography* (Stoddart, 1986). He argued that
such an enterprise is unnecessary for an understanding of the work of the
small number of women who were influential on nineteenth-century geography: 'No feminist historiography is required to analyse their contributions: they looked after themselves, their careers and their scholarship
perfectly well without such assistance' (p. 485), and to do other is to
impose 'the explanatory whims of an evanescent present on an increasingly
distant past' (p. 486). Domosh (1991b) responded that women should not
be 'tacked on to the list of heroes'; their particular viewpoint and experiences should be explored through a feminist approach. Gould (1994, p.
197), too, challenged some of the challenges, especially those which focus
on language, claiming to have encountered 'not just anger, which is perfectly understandable, but something close to hate'. This was countered by
Peake (1994, p. 205), who characterized Gould's position as:

> woe betide any feminist who adopts a mode of expression that
> offends his notions of clearness, effectiveness, and beauty . . . this situ-
> ation is fine only if we accept the male prerogative to legislate lan-
> guage . . . a proprietary attitude that could be espoused only by
> someone who has appointed himself as the arbiter of language. . . .
> He is refusing women the right to make linguistic changes because, in
> his dominant subject position, he cannot admit to change, to the
> recognition of difference.

Feminist analyses stress that the different experiences which structure
women's views of their worlds are created in contexts that have a clear
power gradient between the oppressors (men) and the oppressed (women):
the differences between the two are separate and unequal. Thus the feminist project is necessarily a political one, seeking to remove the power gradient through emancipatory and other processes. Just as Marx argued that

his project involved the emancipation of the working classes, providing them with the means to control their own destinies, so the feminist project is the basis for 'identity politics' – 'an emancipatory politics of opposition ... resisting and challenging the fraudulent claims of dominant groups' (Bondi, 1993, pp. 86-7). This calls for 'consciousness raising', out of which identities can be constructed and reconstructed (in places) as bases for action.

Positionality, difference and identity politics

Feminism has been a major foundation on which a wider geographical focus on difference has been built since the late 1980s (as illustrated by Jackson's, 1991, essay on the social construction of masculinism, Valentine's, 1993, on sexual identities, and Forrest's, 1995, on the role of place in the construction of gay identity). Pratt (1992), for example, promoted a postcolonial feminism which:

1 no longer focuses solely on male–female differences but incorporates other sources such as race, class and sexual orientation;
2 employs social theory to appreciate how differences are created socially, and what power relations they incorporate; and
3 argues, therefore, that all knowledge is 'situated'.

Situation, or position, is thus the source of understandings of the world and power to act within it – and such positions, to use Thrift's (1983a) terminology, can be either *compositional* (what sort of person you are – as with class and gender), *contextual* (where you come from), or both (working-class females occupy different positions in contemporary 'western' and 'Arabic' societies, for example).

Jackson (1993a) developed this argument with reference to positionality and 'ways of seeing', arguing that powerful groups within societies have imposed their interpretations, of landscape and nature, for example. These interpretations can be challenged by other groups, however, derived from their particular, usually marginalized, standpoints within society; their development is usually the outcome of struggle since it is opposed to the hegemony of powerful individuals and groups at society's core. Postmodern cultural studies emphasize the existence of the 'other' in society, groups whose identity is defined by the dominant groups. Feminism is one of the movements enabling those identified as 'other' to promote their views, as created from their positions within society, and which the postmodern emphasis on difference stresses should be heard. Furthermore, the recognition of these multiple voices within society has two major implications. First, it identifies separate bases for political programmes (Jackson, 1993a, p. 210): 'Contemporary society is characterized by multiple forms of exploitation and oppression, which suggests that our politics should also be increasingly positional'. Second, geographers themselves have to recognize that they too occupy a position within society, so that (p. 211):

> Those of us who wish to change the discipline and have ambitions to change the world should start from modest beginnings, recognizing our own *positionality* with respect to the fundamental inequalities of gender, 'race' and class. For if those dimensions are a source of power to us, they are as surely a source of oppression to those around us. If our geographical imagination is to develop in ways that are genuinely and constructively oppositional, we should begin by changing ourselves.

Gregory (1994, p. ix) claims that he recognized this need when he moved from the University of Cambridge to the University of British Columbia while writing *Geographical Imaginations*. He had to come to terms with what he calls his own 'situatedness', clarified to him by three features of his new home:

1 Canada had a double colonial legacy, linked to both England and France, plus close ties to the USA;
2 Canada is an 'avowedly multicultural society' and many of his students had cultural roots very different from his own; and
3 the University of British Columbia was taking issues of gender and sexuality much more seriously than he had encountered elsewhere.

As a consequence he abandoned his first draft of the book and started again, although Guelke (1995, p. 185) argued that:

> in the text there is scarcely a hint of Canadian culture or intellectual life. This work could have been written anywhere but Canada ... Gregory has a desk in Canada, but he has ignored its artists and intellectuals as thoroughly as any British colonial official might have kept his focus on London in the heart of Africa. Canada for Gregory is an invisible country and its peoples have evidently produced little of importance for him.

This ignores the possibility that Gregory may have been awakened to issues of positionality, such as colonial status and equal rights, in North America, which altered his stance on the issues and literature that he addressed without requiring a particular 'Canadian position': his position was as much compositional as contextual.

A major source for geographers' appreciation of their own positionality and hence partial representation of the world (in every sense of the term partial) has been the writing of Edward Said, especially his classic *Orientalism* (Said, 1978) which argued that 'Westerners' created an 'imaginative geography' of the Orient as part of their imperialist project. Orientals were presented by 'Westerners' as 'others', with different (implicitly if not explicitly inferior) cultures that contrasted with themselves (for examples, see Taylor, 1993). The Orient did not exist, therefore; it was ethnocentrically created by the British and French in the nineteenth century (an activity promoted by the various geographical societies which transmitted this view of the Oriental 'other' to home audiences: (see Bell, Butlin and Heffernan, 1995), in part to promote its colonial dispossession and exploitation (Gregory, 1994, p. 171): 'Orientalism was an active process of othering, the exhibiting of "the" Oriental in a profoundly worldly set of texts which, in a quite fundamental sense, made colonization and dispossession possible' (see also Gregory, 1995b).

The 'othering' of the Orient provided its residents with a position from which they could later challenge the imperial and colonial hegemonies, not only politically but also culturally. This includes, for example, the field of 'subaltern studies' whose goal is to allow the oppressed colonial subjects (in particular women) to express themselves in a non-distorted way. They are countering their representation in the 'imaginative geographies' of those who created the Orient, which structured their perspective on the world, the position from which they are now challenging the (masculine) hegemony of 'the West'. That 'Western' hegemony is based on partial views of the rest of the world, often promoted for political and other reasons (see Sharp, 1993). Emancipation requires that such partiality be removed, but how? Gregory (1994, p. 205) argued that:

> To assume that we are entitled to speak only of what we know by virtue of our own experience is not only to reinstate an empiricism; it is to institutionalize parochialism. Many of us have not been very good at listening to others and learning from them, but the present challenge is surely to find ways of comprehending those other worlds – including our relations with them and our responsibilities toward them – without being invasive, colonizing and violent. If we are to free ourselves from universalizing our own parochialisms, we need to learn how to reach beyond particularities, to speak to the larger questions without diminishing the significance of the places and the people to which they are accountable.

'Ideas about "race" and nation are among the most powerful sources of human identity and division within the contemporary world', according to Jackson and Penrose (1993, p.1), but although they are usually taken for granted they are social constructions rather than natural categories, 'the product of specific historical and geographical forces, rather than biologically given ideas whose meaning is dictated by nature'. Thus those who speak from a racial or national position are doing so within categories defined by others – in many cases (notably with regard to race) from a position of assumed superiority. Such is the power of these constructions that 'resistance is often couched in terms which do not challenge such dominant modes of representation' (Penrose and Jackson, 1993, p. 203). Only when the social construction of identity is recognized does a politics of identity emerge, whereby individuals and groups 'attach significance to certain dimensions and contest the relevance of other designations'.

The politics of identity, like all politics, is a struggle for power and, as Penrose and Jackson (1993, p. 207) argued, 'takes place within a hegemonic system of social relations':

> at any given place and time positions of hegemony are being employed to exercise and preserve power [which] ... includes the capacity to set the parameters for negotiation within any given society. It also gives the freedom to define 'difference' and to enforce this vision through hegemonic institutions of government, law and education. In a system where 'sameness' and conformity are rewarded, the power to define 'difference' becomes the power to disadvantage and disempower.

The disadvantaged and disempowered, who have been marginalized, perhaps even rejected, may then challenge the *status quo*, either directly or subversively – and such challenge is most likely if stimulated by some form of positionality crisis (Bonnett, 1993): the legitimacy of their challenge is also determined by those with power within society, however. Place may be used in the definition of position, and thus provide a territorial focus to the politics of identity – as illustrated in very many nationalist and self-determination movements and in the frequent use of spatial metaphors to represent marginality (Pratt, 1992). Spatial metaphors are also used by a number of social theorists, notably Foucault (Philo, 1992), whose classic work on surveillance focuses on the use of spatial processes in the exercise of power but whose wide use of the metaphor leads Philo to identify 'a danger of his geometric turn effectively elevating an abstract sense of space above a concrete sense of place' (p. 157: surveillance is also the focus of Giddens', 1985, work on the nation-state).

The creation and re-creation of difference is part of a political strategy, therefore – the age-old one of 'divide and rule'. As Soja and Hooper (1993, pp. 184–5) describe it:

> Hegemonic power does not simply manipulate naively given differences between individuals and social groups, it actively *produces and reproduces difference* as a key strategy to create and maintain modes of social and spatial division that are advantageous to its continued empowerment.

The reaction to this is that (eventually) 'those subjected, dominated, or exploited by the workings of hegemonic power and mobilized to resist by their putative positioning, their assigned "otherness", struggle against differentiation and division'. This creation of positions as the basis for the political struggles of identity is part of the process of geographically uneven development. Thus space and place are necessarily implicated in what Soja and Hooper call modernist identity politics, which in each instance mobilizes (p. 186):

> its version of radical subjectivity around a fundamentally epistemological critique of the binary ordering of difference that is particular to it: capital/labour, Self/other, subject/object, colonizer/colonized, white/black, man/woman, majority/minority, heterosexual/homosexual. The critique is aimed at 'denaturalizing' the origins of the binary ordering to reveal its social and spatial construction as a means of producing and reproducing systematic patterns of domination, exploitation and subjection.

To a considerable extent, therefore, all identity politics is geopolitics – 'us here against them there'. It is also fragmented. This is illustrated in Harvey's recent work on workplace conflicts in the UK and the USA (Harvey 1993a, 1993b, 1995). Such conflicts seek justice in the workplace but, Harvey (1993a, p. 48) argues, to contend that conditions in one workplace are socially unjust 'supposes that there are some universally agreed norms as to what we mean or ought to mean by the concept of social justice'. The postmodern critique makes such a conception problematic (p. 50):

> there can be no universal conception of justice to which we can appeal as a normative concept to evaluate some event ... There are only particular, competing, fragmented and heterogeneous conceptions of and discourses about justice, which arise out of the particular situations of those involved.

Thus, for example, business interests which stimulated the 'New Right revolution' of the 1980s (often termed Thatcherism and/or Reaganism) argued that government regulation of the economy and society through the welfare-corporate state was an unjust treatment of personal property rights; free markets, by ensuring that each factor of production (land, labour and capital) received its 'just reward', offered a better conception of social justice, it was contended, than that which argued for 'workers' rights' based on economic security, safety and adequate remuneration.

So what is justice? According to Harvey (pp. 52–3):

> working-class rhetoric on rights and justice is as open to criticism and deconstruction as its capitalist equivalent. Concentration on class alone is seen to hide, marginalize, disempower, repress and perhaps even oppress all kinds of 'others' precisely because it cannot and does not acknowledge explicitly the existence of heterogeneities and differences based on, for example, race, gender, sexuality, age, ability, culture, locality, ethnicity, religion, community, consumer preferences, group affiliation, and the like.

Following Young (1990), Harvey identifies a 'double meaning of universality': the first involves everybody in the moral and social life (i.e.

nobody is excluded from a 'general point of view that leaves behind particular affiliations, feelings, commitments and desires': p. 57); the second involves locally determined, and thus continually renegotiated, determinations of just treatment, which reflect 'particularity, positionality and group difference'. Like all other knowledge (on which see p. 366), conceptions of justice are situated in time and place. But what time and, for the geographer, what place? The latter problem was raised for Harvey by his participation in a research project concerned with the impacts of closure of an Oxford car factory (Hayter and Harvey, 1993) which, as he pointed out, brought together 'radically different positionalities', ranging from those of shop stewards, factory workers and local residents through academics and planners to 'independent leftists' (Harvey, 1995, p. 71). He was challenged, when drafting a conclusion to the project, to identify his loyalties – in both time and place. On the former, was his commitment to the maintenance of the current jobs at the factory (which he described as 'shit jobs'), because there were no alternatives locally available, or was it to the longer-term goal of improved working conditions and security for a wider community? He readily appreciated 'how difficult it is to move on a long-term trajectory when short-term exigencies demand something quite different' (p. 72). On the latter, given his concern about over-capacity in the British car industry, and the ecological problems created by the current Oxford factory, he was faced with the tension between a local and a general solution:

> I could not abandon my own loyalty to the belief that the politics of a supposedly unproblematic extension outwards from the plant of a prospective model of total social transformation is fundamentally flawed. The view that what is right and good from the standpoint of the militant shop stewards at Cowley is right and good for the city and, by extension, for society at large is far too simplistic. Other levels and kinds of abstraction have to be deployed if socialism is to break out of its local bonds and become a viable alternative to capitalism as a working mode of production and social relations. But there is something equally problematic about imposing a politics guided by abstractions upon people who have given their lives and labor over many years in a particular way in a particular place.

In this specific context, therefore, Harvey encountered his general problems with the relativism of postmodernism and related approaches, which he grounds in the space-time differentiation of uneven development.

D. M. Smith (1994) reviewed various approaches to justice, and similarly concluded not only that they were incommensurate but also that postmodernism suggests an ethical relativism – 'we simply choose our favourite theory according to our own moral judgements' (p. 117). But he too was unwilling to deny any universal principles, arguing that a common theme in all of the work surveyed was 'that of equality in some respect ... each attempts to define the social, economic and political conditions under which members of a community or society will be treated as equals'. This leads him to restate his earlier axiom (*'the more equal the better'*: Smith, 1977, p. 152) as *'social justice is manifest in reductions in inequality: in a process of returning to equality'* (Smith, 1994, p. 118), which is:

- 'grounded in the here and now';
- concerned with change; and
- committed to a movement towards greater equality, even if perfect equality can never be achieved.

His case studies lead him to the conclusion that 'social justice should not be left to market forces' (p. 279: see p. 351 for a discussion of Bennett's views).

Language, texts and discourse

One aspect of difference at the forefront of work involved in the cultural turn is language. As indicated earlier (p. 200), language both enables and constrains: without it, meanings cannot be transmitted between people, either orally or in writing; but with it, the normal meanings ascribed to words limit what information can be transmitted. As Barnes and Duncan (1992, p. 2) indicate, writing was for long not considered a problematic issue within geography; the task was just

> the mechanical one of bolting words together in the right order so that the final construction represents the thought or object modelled. In this sense 'earth writing', rather than 'writing about the earth', was a good description of what geographers thought they were doing. Earth came with its own labels, and provided that they were in the correct sequence one's written account was always a mirror representation. Anyone could do it as long as there were a dictionary and style manual at hand.

Writing is now viewed as a highly problematic task:

> Pieces of the world ... do not come with their own labels, and thus representing 'out there' to an audience must involve more than just lining up pieces of language in the right order. Instead it is humans that decide how to represent things, and not the things themselves.

Thus writing is part of the social construction and transmission of knowledge.

If the written word is not a mirror of the world it seeks to represent, but rather a social creation, several consequences follow (Gregory and Walford, 1989).

1 Severing the assumed mirror link between 'reality' and 'text' means that reality does not exist outside the language used to describe it: the world is what we write about it. The texts that we produce draw on other texts, which are the sources of the images that we are trying to convey, such as the metaphors that we use to describe the 'unknown' (Barnes, 1996). This is what philosophers term intertextuality: meanings are created in a continual transition process from text to text (Barnes and Duncan, 1992, p. 3):

> new worlds are made out of old texts, and old worlds are the basis of new texts. In this world of one text careening off another, we cannot appeal to any epistemological bedrocks in privileging one text over another. For what is true is made inside texts, not outside them.

2 Writing reveals as much about the writer, and his or her position, as it does about what is being written about. We write, Barnes and Duncan argue, from our own local setting (where local implies spatial scale, but need not necessarily do so); thus

> the worlds we represent are inevitably stamped with our own particular set of local interests, views, standards, and so on. To understand critically our own representations, and also those of others, we must therefore know the kinds of factors bearing upon an author that makes an account come out the way that it does.

3 All writing involves the use of literary apparatus, such as metaphors and other rhetorical devices, whose use we must appreciate since they are central to the transmission of meanings.

These are illustrated by Barnes and Duncan, and in the essays in their edited volume, with reference to landscape descriptions. Landscapes, following Cosgrove and Daniels (1988, p. 1), may be 'represented in a variety of materials and on many surfaces – in paint on canvas, in writing on paper, in earth, stone, water, and vegetation on the ground. A landscape park is more palpable but no more real, nor less imaginary, than a landscape painting or poem.' Thus the 'texts' to be appreciated may be produced through the medium of words, drawings, paintings or other media, or may be inscribed in the landscape itself. (This is illustrated by: Harvey's, 1979, paper on the Sacré Cœur monument in Paris; by Daniels', 1991, interpretation of 'Constable country'; by work on film, produced both commercially – e.g. Aitken, 1991 – and in communities – e.g. Rose, 1994; and by studies of music – Leyshon *et al.*, 1995.)

Barnes and Duncan identify three major concepts as central to the study of representation: text, discourse and metaphor. *Text* embraces a wide range of cultural products (see also p. 303) that involves the author rewriting what has been 'read' in a hermeneutic exercise (p. 34). Texts give fixity, or concrete expression, to many aspects of social life; they represent the author's intentions but also illustrate the wider context in which they were composed and their author(s) socialized. They are also open to a range of interpretations (p. 6):

> the meaning of a text is unstable, dependent upon the wide range of interpretations brought to bear upon it by various different readers. Similarly, social productions and institutions also address a wide range of possible interpreters. But those interpreters are not free to make of the text what they like, but are subject to discursive practices of specific textual communities.

Thus both how we produce a text and how we interpret one depends upon our textual community – on the language (even the particular form of a language) that we use, reflecting our individual compositional and contextual positions.

Discourses are the larger structures from within which texts are constructed and within which others are read. They comprise (p. 8)

> frameworks that embrace particular combinations of narratives, concepts, ideologies and signifying practices, each relevant to a particular realm of social action. Between discourses words may have different connotations, causing people who ostensibly speak the same language to talk past one another, often without realizing it.

The relationship between a component of the text, such as a word (the signifier), and that which it refers to (the signified) is a social construction, undertaken within the specific textual community. Thus

> discourses are practices of signification, thereby providing a framework for understanding the world. As such, discourses are both enabling as well as constraining: they determine answers to questions, as well as the questions that can be asked. More generally, a discourse constitutes the limits within which ideas and practices are considered to be natural ... These limits are by no means fixed, however. This is because discourses are not unified, but are subject to negotiation, challenge and transformation.

Our language is constantly changing, and with it the worlds that we are representing and reading.

Metaphors are major devices for representing meanings. The world is apprehended and known through study, which requires a language, or some other form of textual representation, for the transmission of meanings. Metaphors are extremely valuable, since they provide a means of describing the unknown using the vocabulary of the known; the unfamiliar is illuminated by comparing it to the familiar. In much science, for example, the use of metaphor steers the study of the unknown, providing a framework for its investigation – as was the case with applications of the gravity model in spatial science (p. 111: Barnes, 1996). Understanding a discourse therefore involves appreciating its metaphors.

If metaphors dominate textual discourse, however, they also ensure instability in the ongoing transmission of meanings, or the reproduction of knowledge. As Barnes (1996, p. 166) argues, drawing on the work of Derrida, a leading deconstructionist:

> the meaning of words and concepts (signifiers) can never be directly tied to particular things (signified). For meaning is derived from a signifier's position with respect to *all* other signifiers in the system [i.e. a metaphor can only be understood in the context in which it was developed]. According to Harland (1987, p. 135), 'In Derrida's conception, one signifier points away to another signifier, which in turn points away to another signifier, which in turn points away to another signifier, and so on *ad infinitum*.' There is no anchor of some final presence or some ultimate origin point of meaning. Meaning, rather, is always produced through displacement and deferral, shaped as much by what is absent as by what is present.

If meaning can only be conveyed through 'an orchestration of signifiers' (metaphors), therefore, then we are led (Barnes, 1996, p. 166) 'to deny identification of an unimpeachable presence. For, if there is no ultimate signified and only a shifting system of signifiers . . . there can only ever be the flux of meaning and no constant presence.' All meaning is relative, therefore, and, as Derrida (1976, p. 158) puts it 'There is nothing outside of the text.'

Metaphor is not the only device used in geographical writing, however. According to Jonathan Smith (1996), geographers use four different modes in their story-telling:

1 *romance* – as in biographical narratives of individual and group struggle, especially where it leads to radical change;
2 *tragedy* – as in representations of deterministic systems, many of which involve prognostications of doom;
3 *comedy* – which represents harmony and reconciled conflict; and
4 *irony* – in which the detached observer occupies a superior position.

These represent release, resignation, reconciliation and removal respectively, and in them the author employs one or more tropes, or figures of speech:

1 *metaphor* involves the use of comparisons to introduce concepts, describing 'the remote in terms of the immediate, the exotic in terms of the domestic, the abstract in terms of the concrete, and the complex in terms of the simple' (p. 12);
2 *metonymy* uses technical terms to provide (accurate) descriptions for specialized audiences;

3 *synecdoche* promotes understanding through synthesizing the general and the particular to impart meanings; and
4 *irony* suggests that representation and understanding are futile, and that general apprehensions can never be produced.

Thus the rhetorical styles chosen by geographers reflect both the general approach to their subject matter and the audiences they are writing for: irony is characteristic of postmodernists, for example, metonymy and tragedy of many analysts of spatial systems, and metaphor and romance of the educator seeking to open eyes to the world.

Language is crucial in the transmission of meanings, for, as Barnes (1996, p. 167) explains Derrida's position:

> we can never have direct access to things in and of themselves. This is because in order to understand those terms, they must already be expressed in language. But if they are part of language, then their meaningfulness only becomes about through a play of difference among signifiers, which ... exclude any fundamental signifieds or presences.

There can never be an empirical world, therefore, only a myriad worlds of meanings: there can be no universal truths. Appreciation of those meanings involves understanding the metaphors employed (as Olsson, 1980, did when exploring the gravity model metaphor which underpinned his early spatial science). This involves appreciating the translation process, represented by Olsson (1992, p. 86) as distortion, and therefore an exercise in the use of power. Doel (1993), for example, uses irony in his characterization of the futility of much geographical writing's failure to represent the world the author wishes to portray, whereas Pred (1989, 1990) has explored a variety of linguistic repertoires ('words, variable meanings, pronunciation, grammar, sign-tactical arrangements, rules of interpretation and expressive bodily gestures at one's command': 1990, p. 33) to show how their production and reproduction are inherent to local struggles. Further, like Olsson, he has employed various writing strategies in order to 'subvert the taken-for-granted (and thereby ideology-riddled and power-laden) nature of the academic printed word' (Pred, 1990, p. 48). This

> seeks to make the taken-for-granted format of representation appear strange and yet comprehensible, that seeks to make the reader understand and mentally see what she otherwise might not understand or mentally see, that seeks, somehow, to push through the filter of pre-conceptions and interpretative predispositions deeply inset in the reader's social, biographical and disciplinary past. Thus, I occasionally resort to chameleon like (mis)spellings, hyphenations and word-couplings that are de- or re-signed either so as to trigger previously unmade associations, or so as to convey the ambiguity, the shifting

subtleties, the multiplicity of meanings characteristic of on-the-ground practices and social relations in any place.

In a critique of such work, Curry (1991) argued that postmodernists' concern with language resulted from two new sets of beliefs:

1 'some combination of modern communication, as advertising and corporate culture has rendered the meanings of words much more slippery than they once were' (p. 215); and
2 'we were always mistaken in thinking that language is "connected" with the world in a simple, unproblematic way' (p. 216).

Thus Pred and others have identified a 'crisis in representation' because the 'modernist conception of a language as a symbol system that describes a fixed reality has failed', which is linked to the rejection of 'authorial figures' in any discourse. Despite their strategies, however, postmodern accounts are (pp. 222–3):

> written in ways that seem little different from modernists. Behind the new typographical trappings there lies that authorial presence, where the author claims ultimate control over the text, which for that reason lies outside of time, as a timeless commentary on a society that is itself ensnared in the logic of epochal change.

Furthermore, postmodernists continue to work (and struggle) within the power and authorial structures of universities and academic disciplines: they fail to practise what they preach! In response, Pred (1992) claimed that he had never labelled himself as postmodern, did not accept either that 'everything turns on language' or that 'there is no reality beyond language' (p. 305), and so could not be associated with postmodernism: he has been influenced by a wide range of 'non-postmodernists', including Marx, and has a wide range of concerns into the experience of modernity within capitalism and the related cultural politics. Curry (1992) replied by reiterating his case that much that is presented as postmodern with regard to language (including those like Pred who deny association with that description) remains firmly modernist, a case for which he was congratulated by Hannah and Strohmayer (1992, p. 309), for showing that 'one cannot consistently *be* a postmodernist'.

Turning to the analysis of texts, the practice of discourse analysis is well illustrated by much of the work in what has become known as 'critical geopolitics' (Dalby, 1991: see also p. 235). This has the goal of 'challenging the commonsense understandings' of (geo)political writing, in which (p. 274):

> the essential moment . . . is the division of space into 'our' place and 'their' place; its political function being to incorporate and regulate 'us' or 'the same' by distinguishing 'us' from 'them', the same from 'the other'. . . . Obvious in appeals to nationalism . . . this ideological

representation of identity and difference is a powerful discursive move widely used in many political situations.

This is illustrated by Dodds' (1994b) analysis of changing representations of Argentina by British Foreign Office officials, illustrated by stereotypical representations such as 'all the vigour, speed and brashness traditionally associated with North America' being found to the east of the Andes, contrasted to a 'more subdued, more European tempo' to the west.

Understanding a text means both reading it critically and setting it in context – appreciating the author's position. Peet (1996) illustrates this with the example of a plaque recording the routing of a rebellion in a New England small town in 1787: explication of the context indicates why it was the success of the routers which was recorded, rather than the gallantry of the rebels fighting for the 'common people', as represented by an alternative plaque temporarily erected in 1987. As he puts it (p. 37): 'the icon [the permanent plaque] would not be where it is, nor would it carry the same message, except for unlikely agents, local contingencies, and spontaneous actions'. Thus any approach to appreciating the landscape and its contents must embrace a range of methodologies, from the structuralist – which identifies the conditions of its production – to what he terms 'the empirics of spontaneity' – which appreciate how individuals have acted within that context.

One of the clearest examples of positionality in geographical writing is given by the separate perspectives of Arab and Jewish authors on Israel–Palestine, who come from different sides of a conflict in which each 'sees itself as being physically threatened with extinction' (Newman, 1996, p. 1). Falah (1994) argued that Israeli geographers have not engaged in critical analyses and that their responses to non-Zionist critiques have taken one of three forms:

1 'take and give' which is charateristic of all academic debate;
2 'take and attack' which seeks to negate the critique by attacking the position from which it is written; and
3 'confuse and attack', which involves unscholarly attempts to discredit critics, often through appeal to an outside 'expert'.

Falah argued that the first of these is little used, and even so is more confrontational than is the case in most academic situations, so that the discourse between the Israeli and Palestinian positions (represented by Falah, 1989, on the one hand and Waterman and Kliot, 1990, on the other) is presented as (Newman, 1996, p. 3): 'competitive rather than collaborative, it seeks to discredit and/or respond to the "other" rather than find common ground for constructive, albeit alternative, debate'. Newman and Falah tried to counter this with joint research on a 'two-state solution' to the Israel–Palestine problem (Falah and Newman, 1995) but encountered severe problems in co-authorship because of the 'mutual suspicion' engendered by their very different socialization processes (Newman, 1996, pp. 4–5). Each was seen by the other as seeking to rewrite history and to alter the accepted terminology – which also potentially alienated each from his established academic peer group. Thus (p. 7): 'When the co-authorship does touch upon the competing interpretations and semantics of historical events, it is often unresolvable. Language implicitly defines the perceptions of the user, enabling him/her to classify and interpret elements of their environment.' They may then get involved in 'bargaining language', involving one or more of: compromise; clarification; exchange of terms; and mutual censorship. Their example is in many ways an extreme one, involving deep-seated categorizations of 'threatening others', but Newman (1996, p. 10) generalizes from the experience regarding other positions and their relationship to power:

> While academics argue for the high ground of tolerance to alternative views, in reality we have become increasingly intolerant in an attempt to re-focus the politico-academic agenda according to what is perceived by certain groups as being 'correct'. This is particularly noticeable in the post-colonial and the feminist agendas. In their respective attempts to highlight processes of discrimination and geographies of plunder, they have opted for a form of academic affirmative action, in which compensation is sought for past injustices, rather than a meeting of minds and conflicting viewpoints . . . If it is acceptable that men be excluded from much of the gender debate, that citizens of past colonial powers be excluded from the decolonisation debate, and that Israelis or Palestinians be excluded from the respective debates on Palestine and/or Zionism, then we have little to contribute to the real world.

Most of the work on texts and positionality relates to the production process, with very little on consumption – on reading as opposed to writing. McDowell (1994) raises this issue with regard to teaching: how do geographers assist students in determining what to read, and how? Postmodernist views are contrary to the notion of 'authorities' in a field whose works should be on all reading lists (in literature this is known as the 'canon'). For teachers educated in previous generations (p. 242):

> the certainties of our own undergraduate training have been overturned. The very ground on which we stood – what we thought we knew, how we knew it and what we in turn decided to teach our students – has been cut from under us. The disembodied rational subject that both stood before us and appeared in the pages of the recommended texts has been revealed as a mere mortal; and not only mortal, but also gendered, classed and 'raced' as a white, western, male bourgeois subject.

Teaching now has to represent 'a fragmented polyphony of voices' and 'we place far more responsibility on the reader to sort through the various truth claims and interpretations' (p. 245). To do that can involve abdicating pedagogic authority, which may empower students (or other readers), or it may disable them (and this will only be partly assisted by hypermedia: see p. 306). McDowell argues against such abdication, however, using feminist perspectives ('the voices of the oppressed are more likely to be true' – p. 246) for the task:

> If the purpose of geographic teaching is, as I believe it to be, to prepare our students to be active citizens, able to think critically about and struggle against social injustices in various locations, then the ability to see through the eyes of others, distant in place, gender, race or class, developed through critical analysis of a variety of texts, be they mono- or poly-phonic, as well as by experiences and political involvement, is surely a necessary basis for constructing a multiple definition of social justice and for social struggles to achieve it.

Such an emancipatory role is the foundation of critical social science for many geographers, who face the problems McDowell identifies.

The essays in Barnes and Duncan's collection illustrate the wide range of applications of devices derived from literary theory, aesthetics and other related fields, with their emphasis on texts, discourses and metaphors. They cover not only literary and artistic representations of landscape, but also the metaphors used in geographical writing and the language of geopolitical practice: O Tuathail (1992), for example, shows that US foreign policy discourse regarding South Africa represented it as both a morally repugnant place, whose practices should be changed, and a vital region within the world, and therefore a focus of American strategic concern – it is presented through two scripts, the one portraying 'tragedy' and the other 'a prize in the global strategic game'; and Hepple (1992) explores the use of organism as a metaphor in discourse regarding the state, initiated by Ratzel in the late nineteenth century and currently applied by South American geopoliticians.

Visual texts are among the most effective for transmitting ideas in many circumstances, and include the geographer's commonly used device – the map. Pickles (1992) illustrates this with one particular category – the propaganda map – whose producer differs from the 'good cartographer' (who seeks to portray 'reality' accurately and comprehensibly) because he/she wishes to send a message which is both believable and convincing. Whereas propaganda maps are extreme examples of bias in representation, however, all cartographic texts are elements of interpretative strategies, a point stressed in a series of papers by a cartographic historian, Brian Harley. His starting-point was that (Harley, 1989, p.1).

> we still accept uncritically the broad consensus, with relatively few dissenting voices, of what *cartographers* tell us maps are supposed to

be. In particular, we often tend to work from the premise that map-
pers engage in an unquestionably 'scientific' or 'objective' form of
knowledge creation. Of course, cartographers believe that they have
to say this to remain credible but historians do not have that obligation. It is better for us to begin from the premise that cartography is
seldom what cartographers say it is.

Cartographers' 'scientistic rhetoric' was becoming more strident with the
development of computer-assisted map-making, he claimed, but he used
deconstruction procedures 'to break the assumed link between reality and
representation which has dominated cartographic thinking' (p. 2); by 'reading between the lines' of maps, he seeks to identify their 'silences and contradictions', their metaphors and rhetorical flourishes.

Cartographers have traditionally assumed that the objects they wish to
represent on their maps are 'real and objective'. Their goal is to display
them accurately, hence their search for ever greater scientific precision; they
can come closer to an exact mirror in their representation and can reject
maps (especially old ones) which fail to conform to their canons of conforming to the rules. But, Harley argued, the production of maps is gov-
erned by cultural as well as scientific rules. Thus, for example, most
societies are ethnocentric in producing maps which have their territories at
the centre of their world, thereby helping to promote geopolitical world
views. In selecting what to show, and what prominence to give it, cartographers frequently employ a 'hierarchicalization of space' (p. 7):

> it is taken for granted in a society that the place of the king is more
> important than the place of a lesser baron, that a castle is more
> important than a peasant's house, that the town of an archbishop is
> more important than that of a minor prelate, or that the estate of a
> landed gentleman is more worthy of emphasis than that of a plain
> farmer. Cartography deploys its vocabulary accordingly so that it
> embodies a systematic social inequality. The distinctions of class and
> power are engineered, reified and legitimated in the map by means of
> cartographic signs.

Thus partial representations are undertaken 'behind a mask of a seemingly
neutral science', and yet the 'rules of society will surface. They have
ensured that maps are at least as much an image of the social order as they
are a measurement of the phenomenal world of objects.'

If we accept maps as cultural texts, Harley contends, we can interrogate
them and come to learn of their functions within the society for whom they
were created. Map-making involves a series of steps, he argues – selection,
omission, simplification, classification, the creation of hierarchies, and
'symbolization' – all of which are rhetorical devices; cartographic rhetoric
is involved in the production of all maps, and is implicated in the exercise
of power:

Power is exerted *on* cartography. Behind most cartographers there is a patron; in innumerable instances the makers of cartographic texts were responding to external needs. Monarchs, ministers, state institutions, the Church, have all initiated programs of mapping for their own ends. In modern Western society maps quickly became crucial to the maintenance of state power – to its boundaries, to its commerce, to its internal administration, to control of populations, and to its military strength.

Thus mapping became a state business, and the publication of maps became subject to laws regarding state security.

The map becomes 'juridical territory': it facilitates surveillance and control. Maps are still used to control our lives in innumerable ways. A mapless society, though we may take the map for granted, would now be politically unimaginable. All this is power *with* the help of maps. It is an external power, often centralized and exercised bureaucratically, imposed from above, and manifest in particular acts or phases of deliberate policy.

As a consequence, cartography (and cartographers) are not just one element in a power structure (p. 13): 'Cartographers manufacture power: they create a spatial panopticon. It is a power embedded in the map text.' Their power is not exercised over people directly, but rather over the knowledge made available to them; 'maps, by articulating the world in mass-produced and stereotyped images, express an embedded social vision' (p. 14; Harley, 1992, exemplified this in his essay on the role of maps in the Columbian encounter with the 'new world').

Deconstructing maps serves three functions (Harley, 1989, p. 15):

1 it challenges the myth that technological improvements 'always produce better delineations of reality';
2 it allows appreciation of the role of maps in the historical processes of creating a socially constructed order to the world; and
3 it promotes the understanding of other texts as the meaning of maps is discovered.

He later extended this argument, asserting that (Harley, 1990, p. 1): 'As a discourse created and received by human agents, maps represent the world through a veil of ideology, are fraught with internal tensions, provide classic examples of power-knowledge, and are always caught up in wider political contexts.' This is illustrated by a range of examples including, for example, cartographic complicity in racial stereotyping through place-name labelling, and the exclusion of 'places to avoid' (such as the *barrios/favelas* which are absent from the maps of most South American cities). Cartographers' interests, he claimed, are dominated by mechanical issues related to 'efficiencies' associated with new technologies, so that the ethical issues of what is and is not depicted go largely unconsidered.

Harley concluded his 1990 paper by noting that (p. 18): 'The challenge to and continual crisis of representation is universal and not peculiar to cartography.' Earlier, he had referred to GIS as extending the crisis of representation to the crisis of the machine. GIS are new ways of presenting the world, new texts to be deconstructed to illustrate meanings and power relationships. This argument has been taken up by Pickles (1995), who identified as a central characteristic of all GIS that they involve 'the production of electronic spatial representations' (p. 3) of data. Those media not only:

1   facilitate data capture, entry and reproduction; and
2   speed up operations on the data; but also
3   allow new forms of representation.

Thus, like maps and other texts, their use relies on signs and representations (signifiers and signified) which call for deconstruction (p. 5): 'We are ... entering a potential new phase of ways of *worldmaking* for which we desperately need new ways of *wordmaking*' because GIS are much more than counting machines with greatly increased efficiency; they are enabling new ways of representing the world, and hence new 'realities', and their use by the technically skilled involves new power relationships. Gilbert (1996), however, encourages geographers associated with the cultural turn not to reject all developments in modern information technology because of the association of GIS with statistics, spatial analysis and positivism. Wider developments of hypermedia (within which GIS can be situated) allow exactly the sorts of explorations of textual and other media that postmodernists and others promote (p. 7):

> In a computer hypertext the reader can use a mouse to point and click on a word, and be instantly taken to related ideas elsewhere in the text. The freedom created for the reader to jump from place to place, to compare, contrast, or simply juxtapose different elements radically changes the nature of writing (or 'authoring' ...) and of reading. ... the hypertext form marks a move away from a modernist concern for 'objects, positions, order and stability' towards a postmodern emphasis on 'processes, relations, chaos, and instability'.

Images (still and moving), diagrams, sounds and other media can be incorporated, allowing a 'polyphonic' (Crang, 1992) approach to geographical writing and study similar to the postmodern novel: the 'reader' becomes an 'active co-author'.

Images, consumption and cultural geography

The role of representation, or image-making, is central to much of economic and social life in (post)modern society, and thus attracts attention from those seeking to appreciate contemporary culture. The images are many and varied, and involve a wide range of representations of the world. As May (1996a, p. 57) indicates, drawing on Said (1978), this involves creating: 'an imaginative geography, a geography that overlays a more tangible geography and helps shape our attitudes to other places and people'. He illustrates this with the growing consumption of the 'exotic' in modern life, such as food, clothing and holidays, which involves representations of the places and peoples involved (and which may lead to racial stereotyping).

Duncan and Ley (1993, pp. 2–3) identified four major modes of representation within Anglo-American human geography:

1 *description of observations obtained through fieldwork*, which dominated cultural geography until relatively recently: its underpinning assumption was that 'trained observation transcribed into clear prose and unencumbered by abstract theorizing produces an accurate understanding of the world';
2 *mimesis*, whereby the world is reflected in media other than words, as with the mathematical modelling of spatial science that had little impact upon cultural geography;
3 *postmodernism*, which 'distrusts and interrogates all meta-narratives including those of the researcher' and, as indicated earlier, rejects the search for 'universal truths'; and
4 *hermeneutic interpretation*, which acknowledges the role of the interpreter, and therefore rules out mimesis – 'reality' cannot be faithfully reproduced:

> Rather than setting up a model of a universal, value-neutral researcher whose task is to proceed in such a manner that s/he is converted into a cipher, this approach recognizes that interpretation is a dialogue between one's data – other places and other people – and the researcher who is embedded within a particular intellectual and institutional context

thereby illustrating Barnes's (1996) argument that all knowledge-production is 'local' (or 'situated').

Duncan and Ley's preference is for the hermeneutic approach, favouring it over the postmodern because it (p. 9): 'allows dialogue between the researcher and his or her subject and yet does not misrepresent the power relations that are structured into the Western academy'. Academics applying such an approach produce a text, using both *an extra-textual field of*

reference (the 'data' employed in the production process) and *an inter-
textual field of reference*, comprising materials culled from other texts that
are drawn upon during the production process: data are interrogated in the
context of other textual materials. Thus the process of producing a text:

> is not a mirroring of the extra-textual . . . but rather *re-presentation*,
> the production of something which did not exist before outside the
> text. This process of academic production is essentially disruptive of
> the extra-textual world. The text disrupts the extra-textual field of
> reference, by highlighting some elements within the field and deleting
> others. Elements are reshuffled within the text, thereby splitting up
> the fields of reference through an act of selection. The basis for this
> selection and reshuffling lies largely within the inter-textual field of
> reference. Both the inter-textual and extra-textual fields serve as con-
> texts for each other within the text. Each plays off the other and helps
> define the possibilities of interpretation. As such, the world within the
> text is a partial truth, a transformation of the extra-textual world,
> rather than something wholly different from it.

Texts of any sort thus transform that which they are representing, the
nature of the transformation process reflecting their author's own context.

A text reflects the relationship between its author and the extra-textual
world, set in the context of his/her inter-textual field. But when it is read, it
becomes part of a new extra-textual world; it is interpreted in the context
of the reader's inter-textual field of reference. A reader may disagree with
what is presented in the text, finding the material presented unsuitable in
some way; or the reader may not share the author's frame of reference (or
theoretical position), so that (Duncan and Ley, 1993, p. 10):

> In either case the reader (by reordering the relationship between the
> text, the extra-textual and the inter-textual) will produce a different
> interpretation of the text than that which the author intends, thereby
> extending the hermeneutic cycle. As long as there are readers for a
> text its reproduction will continue. In this sense representation is not
> only a collective but an iterative process.

By implication, the more iterations there are, the further the reproduced
interpretation may get from the original extra-textual world.

Most work adopting this approach focuses on the production rather than
the reproduction of texts, on the authors rather than their audiences.
Duncan and Ley's book contained four essays on the production of residential landscapes and four on the institutional contexts within which landscapes are produced. Duncan (1993a), for example, writes of a
city-building programme undertaken by a king in Kandya, Sri Lanka, 'both
to celebrate his army's victory over an invading British force, and to
heighten his charismatic appeal' (pp. 232–3), and Kariya (1993) examined
the role of Canada's Department of Indian Affairs and Northern

Development in the definition of who is an Indian. Agnew (1993) focused on the nature and role of space in the representation process. At the national scale, for example, he emphasized the role of territory in what he terms the 'nationalization of social life' (p. 254): in France, for example,

> From diversity and particularity, not to say barbarism, emerged homogeneity and civilisation. This occurred because as France grew more prosperous, roads, railways, markets, schools, national news-papers and military conscription penetrated the countryside. These rationalizing and nationalizing trends undermined rural particular-ism, opened the countryside to new ideas, goods and practices and tied previously isolated communities into the national culture and social life.

A uniform inter-textual field was being created over the national territory, as part of the nation-building process. Other spatial constructions include the core–periphery metaphor; all share what Agnew calls an 'abstract atti-tude', a use of spatial metaphors by social science authors to categorize and classify observations.

Image-making and representation are central to one of the main substan-tial concerns of what is occasionally termed the 'new cultural geography', which focuses on consumption in modern society. Whereas until the mid-1980s little had been written by geographers on the subject, within a decade it had become a major topic, for three reasons (Glennie and Thrift, 1992):

1 commodification is extending further into social life, bringing into the market society a range of activities and forms of consumption which were formerly outside the cash nexus;
2 society is becoming increasingly divided, and new 'structures of taste' are being produced and reproduced for each segment; and
3 new forms of everyday life are being created.

Markets are thus proliferating, along with the means of encouraging con-sumption within them, through advertising, fashion and design. There has been a major shift between the mass market commodities of the modern era and the multiple markets of postmodern flexible accumulation, described by Glennie and Thrift (1992, p. 424) as follows:

> In the preindustrial world . . . commodification was limited by the scope and scale of industrial processes and by the fragmented nature of markets. However, with the advent of industrial capitalism a mass market for commodities formed and the tendrils of commodification could reach into every home and hearth. Styled objects became the province of every class. Desire for these objects was boosted by the related proliferation of advertising, by the advent of designs aimed at mass markets, and by the increasing prevalence of the image, based especially on the photograph. The result was what Barthes called

'neomania'. Individuation based on consumer goods became an important fact of life.

This story then moves on to the postmodern world. Here production and consumption have become segmented, rather than fragmented. This is a world of niche markets tailored to suit specific socioeconomic groups, through electronically mediated signs and images, spectacles and simulations. The result of the confluence of neomania with these electronically mediated signs and images, spectacles and simulations, is individuals who are increasingly fragmented, following particular life-styles that may or may not chime with one another, but which require the adoption of particular personas with which particular commodities are associated.

Our worlds of consumption, like all other worlds, are thus socially created, and in those affluent societies with high levels of consumption needed to sustain the capitalist mode of production, those processes of socially creating 'needs' come to the fore, as Thrift and Olds (1996) illustrate with the modern western practice of celebrating Christmas.

None of this is new to the postmodern era, according to Glennie and Thrift. There have always been tensions between wants and needs that have been mediated by advertising, associated with a further tension between mass-produced and individually tailored goods. Some people at least have for several centuries created their identities around the goods and services that they consume. But the focus on consumption has increased, and the scope and scale of commodification have expanded: spending power is greater and more widely dispersed; segments of society (such as those associated with the ecology movement) are committed to particular forms of consumption; and individuality has been 'marketed' though a whole range of commodities, such as clothing, food and wine, and holidays. These trends have been enhanced by a variety of novel practices, including the availability of electronic means of communication: events are more public and more widely viewed, for example, though not necessarily in consistent ways, and new 'demands' are created:

> The media define certain practices and their associated commodity markets (for example, aerobics and aerobics video tapes), redefine what commodities can be (video tapes are themselves commodities), and reinscribe the relation of people to commodities by creating modes of life (healthy life-styles associated with aerobics) to which, it is often implied, everyone should aspire (Glennie and Thrift, 1992, p. 437).

Individuality, according to Glennie and Thrift, is 'both broadening *and* deepening', and is being developed dialectically by those industries which depend on modern patterns of consumption. These embrace not only the productive processes themselves (the focus of much of the early work on flexible specialization: p. 222) but also the creation of the images used to 'sell' the new goods and services, both directly and indirectly (the role of film, for example), and the contexts in which they are sold – the shopping malls, specialist retail centres and car boot sales, for example, along with the new forms of cinema, outdoor leisure pursuits and tourism. These are all summarized by Pred (1996, p. 11) as:

> the connections between either consumption or identity and the full range of practices, power relations, and experiences that women and men encounter as they conduct their daily lives under socially, spatially, and historically specific circumstances within contemporary commodity societies, as they conduct their daily lives under conditions best characterized as hypermodern, as capitalist modernity accentuated and sped up.

Place is crucial to all of this (p. 13):

> Advertising is rarely the sole or even the most important source of prepurchase knowledge about the existence and qualities of a particular good or service, seldom the single simulator of want and desire, only exceptionally the primary means through which awareness of need arises in everyday life.
> It is through situated practice,
> through social interactions at sites of work, education,
>   and other institutionally embedded activities,
> through formal and informal conversations
>   participated in the conduct of daily life,
> through everyday discourses and representations
>   encountered in public spaces, private spaces and the mass media,
> through visual and aural observations
>   made in the course of site-to-site movements,
> that consumer knowledge is accumulated,
> that the desire to possess is aroused,
> that wants are constructed,
> that requirements become apparent,
> that needs take shape.

The power relations encountered during everyday life in a circumscribed area (a place) are thus crucial to understanding this aspect of postmodern society.

The new versus the old

Each of the various strands within 'the cultural turn' has stimulated debate with supporters of other opinions, as exemplified by Harvey's critique of postmodernism and the responses which it generated (p. 276). Similarly, there has been debate between adherents of different conceptions of cultural geography, as identified by Duncan (1994; see p. 270). Such debates have largely involved people classified as 'social theorists' (Sheppard, 1995) rather than being between social theorists on the one hand and spatial analysts on the other.

As identified above, the relative absence of debate between 'new' cultural geographers and spatial scientists can be attributed in large part to the relative peripheralization of the latter within the geographical enterprise in recent years; they no longer dominate the discipline as they did as little as two decades ago, and are not seen as major contestants for disciplinary hegemony (which, in any case, should run counter to the pluralism promoted by many postmodernist adherents!). One area of quite substantial debate, however, has been over the role of GIS within the discipline.

Although geographers have been very active in the development of GIS – it is taught in most university departments of geography (and increasingly so, in North America at least; see p. 372) – and much basic and applied research is undertaken by geographers either in those departments or in associated research centres, as apparently with remote sensing before it, many of those developments have been undertaken *within* geography rather than *for* it. University departments of geography have provided academic homes for researchers, but from very early on GIS was institutionalized as a quasi-independent professional field with its own conferences, journals and learned societies, and with strong links to user communities in the public and private sectors (Longley and Clarke, 1995). Nevertheless, its presence within those departments of geography is seen as both a competitor for scarce resources and a challenge to the discipline's nature, and whereas some identify the togetherness (if not a symbiosis) not as a threat but rather as a lifeline for the discipline as a whole in increasingly materialist higher education systems (it was described by one as the export staple which sustains the rest of the discipline!), others are concerned at the challenge that it poses, because of the views about geography that some GIS specialists are propounding.

One of the first pieces in a developing debate was a brief editorial by Taylor, who suggested that GIS enthusiasts, working from a naive empiricist base, had stimulated a retreat from knowledge to information, reversing the work of earlier generations of quantifiers (Taylor, 1990a, p. 212):

> Knowledge is about ideas, about putting ideas together into integrated systems of thought we call disciplines. Information is about facts, about separating out a particular feature of a situation and recording it as an autonomous observation. Hence disciplines are defined by the knowledge they produce and not by facts.

Thus unless GIS are transformed into GKS (Geographical Knowledge Systems), they will 'leave geography intellectually sterile – high-tech trivial pursuit'. Goodchild (1990) responded by pointing out that whereas much of the technological research associated with the development of GIS is computer science, the information produced and the methods by which it can be manipulated are more than simply another technical tool for geographers. Just as maps for long were the foundation of much geographical thinking, 'then a major change in the technology of mapping might be expected to provoke reasonably profound thoughts from geographers' (p. 336).

Openshaw's (1991) response was much more strident, not only arguing that without information there can be no knowledge but also claiming that GIS could resolve an identity crisis within geography (p. 622):

> Only rarely does [the] process of geographic fragmentation discover a 'new' geography that is of sufficient stature and importance to have a far-reaching, long-term, and fundamental impact on the nature of geography itself. GIS is probably one of those because it can be regarded as offering the prospect of reversing the disciplinary fission-ing process and replacing it by a fusion; a drawing together of virtu-ally all the subdisciplinary products with their multitude of conflicting paradigms created over the last thirty years within a single (philoso-phy-free or philosophy-invariant or even philosophy-ignorant) inte-grating framework.

But Openshaw's offer was followed by an attack on many of those frag-ments which he sought to integrate: he refers to 'pseudoscience', 'soft and the so-called intensive and squelchy-soft qualitative research paradigms', the 'rambling and ranting era of so-called radical but probably second-rate philosophical thinking' (p. 623), 'technical cripples . . . [and] this sur-vival-by-ignoring-computers strategy' (p. 624), geographers having 'lost their ability to communicate with the public and create any kind of attractive aura' (p. 626), and the need for 'hard empirical proof' to back up the 'realistic but not necessarily nonfictional fairytales' (p. 628): with-out such computing power and the ability to manipulate spatial (or geo-coded) data, all geographers have been doing is to produce 'what are no more than plausible works of fiction and then attribute to them academic credibility by developing philosophical stances that render normal scien-tific proof unnecessary'. (See also Openshaw, 1994, 1995, for his views on using massive computer power and large data sets to search for spa-tial order, which he considers to be the core of scientific human geogra-phy.)

Taylor and Overton (1991) responded by pointing out that Openshaw's polemic was largely assertion rather than argument, and by deconstructing his language in four ways:

1 the use of medical metaphors (terms such as 'infect' and 'like a parasite') is often offensive;

2 there is unfortunate 'psychological analysis' of his opponents' 'fear and anxiety';

3 motivations (such as envy and shame) are attributed to opponents; and

4 confrontational terms are used, such as 'squelchy-soft, and 'poor fools'. (Openshaw – 1996, p. 761 – continued to use the contrast between 'hard' and 'soft', and referred to 'the numerate scientific human geographer [who] can readily deride his or her nonnumerate colleagues as producing rubbish, unverifiable stories: the so-called "Catherine Cookson" approach to reconstructing plausible and knowledgeable geographical fairy-tales that fit the scraps of evidence'.)

They then counter Openshaw's case by arguing that data (the foundations for information) do not just 'exist', they have to be created – and are often done so for political and/or commercial rather than intellectual ends. Further, there is a geography of information as a consequence, notably the inequality between rich and poor countries: they suggest a 'first law of geographical information . . . where the need for information is greatest the amount of information is least', and this unequal access to information can be (is) used to promote power relationships. Finally, they point out that geography, like other disciplines, is a continually contested enterprise, forever redefining itself and seeking new alliances. GIS may generate exhilaration among its proponents for the links it offers with other disciplines, but

> Similar exhilaration can be found in the 'social theory movement' in human geography – both GISers and social-theory geographers have become excited by the recognition of their discipline in other sciences which they regard as crucial, namely information science and social science, respectively. This is a good time for geography, to be sure, when such diverse groups proclaim the centrality of the discipline within important contemporary trends in academia and beyond . . . Let us agree that both are right within their own frame of reference, but neither can ever 'encompass almost everything that geographers do, have done, or may want to do' (Openshaw, 1991, p. 626).

Openshaw (1992, p. 463) responded by claiming that his piece was supposed to be 'relaxed, sarcastic and even whimsical' and not to be taken 'too seriously' (because Humpty-Dumpty was in the title!), and then claiming that Taylor and Overton's piece amounted to 'little more than exaggerating, fabricating, misunderstanding, and changing the original message so that they can create something tangible to criticise'. He represented his case as:

- 'GIS provides an enabling information framework with a core tool-kit that can be used as a platform for all manner of existing, and not yet discovered, geographies' (p. 464);
- in many cases where it is claimed there are no data this is not so, and – although he apologised for using terms such as 'story-telling' and 'fairy

tales' – there are many unexplored potential links between soft under-standing and data-orientated approaches; which lead to

• 'the urgent need to interface the hard and soft parts of geography';

all of which involves the same metaphors of 'soft' and 'hard'!

This debate brought out some extreme views, which clarify many of the differences between 'social theory geographers' and 'GIS-quantifiers'. Although many of the former doubt the value of much quantitative work – whether done using a GIS or not – their concerns are more about the use of GIS in society than about the nature of geography – hence the discussions of GIS in image-making (see p. 306 above) and those about its use in warfare (Smith, 1992). The debate will no doubt continue, not least because of the competition for resources both within universities generally and within geography in particular. Spatial analysis will undoubtedly continue to attract geographers, both for itself and for its potential applied value (as Abler, 1993; Fotheringham, 1993; Clark, 1993 and Gould, 1993 illustrate).

In summary

One of the most (mis)quoted of Marx's aphorisms is that 'men make their own history, but not in circumstances of their own choosing'. That position is largely accepted by adherents to the streams of work identified here within the umbrella term 'the cultural turn', but they reject what they see as Marx's narrow limitation of those circumstances to materialist concerns. Ley and Duncan (1993, p. 329) conclude their book by expressing its message as:

> landscapes and places are constructed by knowledgeable agents who find themselves inevitably caught up in a web of circumstances – economic, social, cultural and political – usually not of their own choosing. Every landscape is thereby a synthesis of charisma and context, a text which may be read to reveal the force of dominant ideas and prevailing practices, as well as the idiosyncrasies of a particular author. ... We are arguing here for a charting of the middle ground between the poles of collectivism and individualism, whereby neither individual nor context is privileged but both are dialectically related in the making of geographies.

Landscapes (and other texts) are read as well as made,

> and the act of reading is not an unproblematic act either. Each reader, like each author, brings a past biography and present intentions to a text, so that the meaning of a place or a landscape may well be unstable, a multiple reality for the diverse groups who produce readings of it.

Thus textual representation is

> a construction that is contingent, partial and unfinished. It is. . . . in the very literal sense a fiction, a fabrication that depends in part upon the position of the interpreter. The insights to be gained from interrogating the positional categories that shape a text [and its reading] are too rich to be ignored.

This concern with relativism was reflected in an initiative taken by the Social and Cultural Geography Study Group of the IBG in the early 1990s (having recently introduced 'and cultural' to its title), entitled *New Words, New Worlds: Reconceptualising Social and Cultural Geography* (Philo, 1991).The group's committee produced a discussion document (entitled 'De-limiting human geographies') which outlined the main themes being explored as:

1 *moral philosophy, moral geographies and the geographer's morality* – stressing the need to downplay the dominant economic focus of much geography and replace it by the moral frameworks which inform life;
2 *processes of social differentiation* – involving greater appreciation than heretofore of the various categories within society – class, gender, race, ethnicity, sexuality, age, health, etc. – which have been largely taken for granted in discussions of spatial differentiation;
3 *constructions and boundaries of the self* – how individuals define themselves and relate to others within the contexts of the various categories used within society, which involves interrogating psychoanalytic literature, something not previously undertaken by geographers (Pile, 1991, 1993);
4 *globality and territoriality* – the location of individuals and groups in spaces and places, and the cultural practices involved; and
5 *society, culture and the natural environment* – addressing the social construction of 'nature' and 'environment' and their importance to approaches to the resolution of 'environmental problems'.

They conclude the chapter (p. 26):

> the varying arguments pursued above may seem rather strange to many human geographers not just because they are sketches of debates rather than careful and fully-documented accounts of research, but because they appear to skirt around the sorts of materials that we would usually expect to find when discussing 'the social' and 'the cultural' in human-geographical inquiries. . . . this is certainly not because we regard existing geographical work . . . as misguided or irrelevant, far from it . . . [but] what we would insist is that there are numerous other conceptual and empirical questions now in need of answers – questions to do with the 'moral' well-springs of differentiation amongst social groups; to do with the connections

between emotion, passion and 'morality' that become tangled up as selves develop in relation to others; to do with the complex clash of 'moralities' currently shaping the actions of nations, religions and other social groupings in the modern world-system – and what we would then insist is that all of these 'moral' questions are ones demanding a geographical sensitivity to how 'moralities' are made and remade across space.

Such a 'manifesto' may not have appeared when other 'turns' were being promoted within human geography even only a decade earlier – perhaps because the discipline was less democratic then and power was more concentrated in a few hands; perhaps because new publishing technologies allow easier preparation and distribution of such documents – but there is little to differentiate these recent developments within human geography from those charted earlier in the present book. They are based on dissatisfaction with what has been, and is being, done in the name of the discipline, and a wish to change it, at least by adding new perspectives if not by also expunging some of their predecessors.

9

Applied geography and the relevance debate

The period since the late 1960s has contained traumatic years in the countries being studied here. The underlying problem has been economic uncertainty: after two decades of relatively high rates of economic growth and prosperity, the American and British economies began to experience serious difficulties. Further, it became increasingly clear that the prosperity of the previous decades had not been shared by all; this was highlighted in the United States by the growing tempo of the civil rights movement and in the United Kingdom by the unrest in Northern Ireland, which began as a civil rights movement, and a series of inner-city riots, most of them linked to racial discrimination issues. Student protest erupted in several countries during 1968, much though not all of it related to the Vietnam War, and increasing concern was being expressed about human-induced degradation of the physical environment. Discrimination against women in all parts western societies was also brought to the forefront of attention (including the geographical profession: Zelinsky, 1973a).

   Much of the protest of the late 1960s centred on particular issues and was relatively ephemeral. For many of those involved, the aim was to win reforms within society, in the classical liberal manner, while leaving its major structure untouched. For some, however, disillusion stimulated what Peet (1977) termed a 'breaking-off' from liberalism and a move to more radical political stances:

> The starting point was the liberal political social scientific paradigm, based on the belief that societal problems can be solved, or at least significantly ameliorated, within the context of a modified capitalism. A corollary of this belief is the advocacy of pragmatism – better to be involved in partial solutions than in futile efforts at revolution. Radicalization in the political arena involved, as its first step, rejecting the point of view that one more policy change, one more 'new face', would make any difference (p. 242).

The 'radicals' eventually exploded 'through the thick layers of ideology which in the most dangerous, mass-suicidal way, protect late capitalism' (p. 243) and settled on forms of socialism (see also Blaut, 1979: Peet, 1985b, has updated this history).

The substantive concerns of those termed 'radicals' from the late 1960s on were treated in detail in a previous chapter. This chapter is concerned with general issues of applied geography and its role within the discipline. As noted in Chapter 1, the pressures for more applied work built up in the late 1970s and through the 1980s, as a reaction to the economic recession that afflicted the countries considered here during that period. In addition, government policies towards higher education, especially in the United Kingdom, placed an increased emphasis on applied work, both in terms of providing vocational and professional training for students and in the pressure to cover institutional costs through 'earnings' from research activity. Thus the debate over applied geography concerned pragmatic issues of disciplinary survival as well as concerns with regard to relevance.

Tracking the development of applied geography is not as easy as study of other aspects of the discipline, because relatively little of its output is published in academic journals. This is partly because of the nature of the research contracts and the associated funding for researchers. Some contracts make publication in academic journals difficult, because of commercial sensitivity of the research results, and even where this is not a particular problem the time pressures on those involved – to complete one project and then go on to the next, if for no other reason than to secure a continuing salary for the contract researchers – militate against preparation of academic papers and mature reflection on the theoretical and other implications of particular (usually empirical) research findings. As universities have been pressed to do more of this type of work, however, and as geographers have identified commercial niches for themselves, so the practice of applied geography has somewhat disappeared from general view. Nevertheless, its nature and validity have been debated, and that forms the focus of the present discussion.

Disenchantment and disillusion in academic geography

A forthcoming revolution in human geography, against the innate conservatism of behavioural studies, was foreseen by Kasperson (1971, p. 13):

> The shift in the objects of study in geography from supermarkets and highways to poverty and racism has already begun, and we can expect it to continue, for the goals of geography are changing. The

new men see the objective of geography as the same as that for medicine – to postpone death and reduce suffering.

It can be illustrated by one senior geographer's writings in the 1970s. Wilbur Zelinsky was president of the Association of American Geographers (AAG) in 1973; his views may not be entirely typical, and are certainly more firmly stated than those of his peers, but they reflect both growing disillusion within geographical circles with past achievements (see also Cooke and Robson, 1976) and a wondering about future directions.

Zelinsky's (1970, p. 499) first statement began:

> This is a tract ... The reader is asked to consider what I have come to regard as the most timely and momentous item on the agenda of the human geographer: the study of the implications of a continuing growth in human numbers in the advanced countries, acceleration in their production and consumption of commodities, the misapplication of old and new technologies, and of the feasible responses to the resultant difficulties.

He developed three basic arguments:

1 that people are inducing for themselves a state of acute frustration and a crisis of survival;
2 that these conditions originate, and can only be solved, in the 'advanced' nations; and
3 that the current 'growth syndrome' has profound geographical implications.

Material accumulation can no longer be considered progress, he argued, because it is unsustainable; effort is currently misallocated on a massive scale, and there is a major geographical task involved in its sensible reallocation.

Zelinsky identified five typical academic reactions towards the problems of the growth syndrome:

1 ignore them;
2 accept that major consequences will occur, but only eventually;
3 admit that problems exist, but argue that they are easily solved by the free market, perhaps with state guidance;
4 claim that nothing can be done, but that in any case we will survive; and
5 realize the potential for immediate, unprecedented trouble.

His own reaction clearly fell into the fifth type, but many others have suggested that it was (and remains) an over-reaction, as illustrated by Beckerman's (1995) coruscation of the 'sustainable development' thesis. Zelinsky identified three roles which geographers could play in facing up to the perceived oncoming disasters. The first – which involves a minimal political commitment and 'should not offend even the most rock-ribbed conservative scholar' (p. 518) – is the *geographer as diagnostician*, applying 'the geographic stethoscope to a stressful demography' (p. 519), mapping what he calls geodemographic load, environmental contamination, crowd-ing and stress. The second involves the *geographer as prophet*, projecting and forecasting likely futures. Finally, there is the *geographer as architect of utopia*, educating with regard to problems and possible solutions and pro-viding support for the unknown leaders who have the political will to guide society through the coming 'Great Transition'.

Ending his 'declaration of conscience' on a pessimistic note, Zelinsky argued regarding geographers:

> how woefully deficient we are in terms of practitioners, in terms of both quantity and quality, how we are still lacking in relevant tech-niques, but most of all that we are totally at sea in terms of ideology, theory and proper institutional arrangements (p. 529).

Those criticisms were not confined to geographers, however, and he applied them to scientists *en masse* in his AAG presidential address (Zelinsky, 1975). Science, he contended, is the twentieth-century religion, which has failed to avert the oncoming crisis. (See also Harvey, 1973, 1989a, 1994, discussed on pp. 327 and 330.) Its disciplinary specialisms and separatism 'fog perception of larger social realities' (p. 128), while 'fresher, keener insights, along with much better prose' (p. 129) are produced by the brighter contemporary journalists.

Zelinsky identified five crucial axioms as the foundations of science:

1 that the principle of causality is valid for studying all phenomena;
2 that all problems are soluble (see Johnston's, 1990b, response to Pacione, 1990a);
3 that there is a final state of perfect knowledge;
4 that findings have universal validity; and
5 that total scientific objectivity is possible.

The social sciences have failed to live up to these, he claims, for several rea-sons: their immaturity; their use as a refuge for mediocre personnel; the dif-ficulty of their subject matter concerning interpersonal relationships; their problems of observation and experimentation; and the political and other problems involved in applying their proposed solutions. Furthermore, and the major cause of their failure, the natural science model is irrelevant to the study of society:

> If we are in pursuit of nothing more than information or knowledge, then there is some value in copying the standard formula of a research paper in the so called hard sciences. ... But if we are in pursuit of something more difficult and precious than just knowledge, namely understanding, then this simple didactic pattern has very limited value (p. 141).

The natural science approach adopted by positivist spatial scientists helps to describe the world, but not to understand it.

Zelinsky's views were echoed in another AAG presidential address, with Ginsburg (1973, p. 2) writing that:

> Much so called theory in geography . . . is so abstracted from reality that we hardly recognize reality when we see it. . . . The increasing demand for rigor to cast light on trivia has come to plague all of the social sciences . . . the most important questions tend not to be asked because they are the most difficult to answer.

At about that time, the Association established a Standing Committee on Society and Public Policy (Ginsburg, 1972), which White (1973, p. 103) hoped would 'be alert to distinguishing the fatuous problems and the activities that are pedestrian fire-fighting or flabby reform'. This statement was made at a session on geography and public policy at the Association's 1970 conference; a similar theme was chosen for the 1974 conference of the Institute of British Geographers (IBG). Not all geographers accepted White's aim – using learned societies to influence public policy – whilst not denying the value of geographic method in social engineering. Trewartha (1973, p. 79) stated that:

> I must demur when he proposes that it should be a corporate responsibility of our professional society to become an instrument for social change. . . . From the beginning, the unique purpose of the Association of American Geographers has been to advance the cause of geography and geographers; it was never intended to be a social-action organization. . . . All kinds of research, pure as well as applied, should be equally approved and supported by the AAG.

There are two major components to Zelinsky's case: geographical research should be relevant to the solution of major societal problems; and the positivist-based spatial-science methodology may be inappropriate for such a task. Several were quick to point out that neither was particularly new, especially the first: House (1973), for example, reviewed the tradition of involvement in public policy by British geographers, and Stoddart (1975b, p. 190) identified the late nineteenth-century views of Reclus and Kropotkin as 'the origins of a socially relevant geography' – the latter was later rediscovered by the 'radicals' (Peet, 1978). Nor was the more 'revolutionary' approach of those who had 'broken off' from liberalism particularly novel: Santos (1974) reminded English-speaking geographers of the marxist-inspired works of Jean Dresch on capital flows in Africa and Jean Tricart on class conflict and human ecology, both prior to the Second World War, for example, and Keith Buchanan's presidential address to the New Zealand Geographical Society on the need for studying 'the absolute geographical primacy of the state; especially in the non-Western world' (Buchanan, 1962) produced an acid response from Spate (1963).

Relevance, to what and for whom?

Claims that geographical work should be more relevant to major societal problems raised queries about the nature of that relevance, and it soon became apparent that there was no consensus on what should be done, and why. The ensuing debate is illustrated by a number of contributions to the British journal *Area* during the early 1970s.

Chisholm (1971b) opened it, identifying differences between governments, with their interests in cost-effective research and their primacy in decision-making, and academics, some of whom are concerned to protect their academic freedom and their right to be the sole judges of what they study and publish. Traditionally, geographers had advised governments in the roles of information-gatherers and 'masterful synthesizers' and they had not been involved in the final stages of policy-making: they had been delvers and dovetailers, but not deciders. On the latter role, unfortunately:

> The magic of quantification is apt to seem rather less exciting when the specifications of the goods it can deliver are inspected at close quarters (p. 66). . . . The danger with empirical science is the absence of guidance at the normative level as to which of various options one should take (pp. 67–8).

The challenge to human geography, according to Chisholm, was to define such norms. Chisholm himself later worked for different governments in a role which was neither delver nor dovetailer: as a member of independent commissions (the Local Government Boundary Commission in the 1970s and the Local Government Commission in the 1990s) he was involved in advising the British government on the best way to redraw important components of England's administrative map (Chisholm, 1975b, 1995). Eyles (1971) responded that the focus of relevant research should be 'some of the social and spatial inequities in society' (p. 242), and the first challenge is to study the distribution of power in society, thereby identifying the mechanisms for allocating scarce resources. Research could then isolate the disadvantages of relative powerlessness and provide the basis for policy aimed at redistributing resources.

British readers were introduced to the ongoing debates in American geography in two reports on the 1971 AAG conference. (This was not the first at which major social issues had been raised. The 1969 meeting should have been in Chicago but was transferred to Ann Arbor as a protest over events at the 1968 Democrat Party Convention in Chicago; the radical journal *Antipode* was launched there, and those attending were confronted by the problems of the inhabitants of Detroit's black ghetto, small groups of whom 'invaded' several of the sessions to present their concerns at the academics' 'unworldliness'.) By 1971, 'many geographers were deeply frustrated by a sense of failure' to deal with major social issues (Prince, 1971b, p. 152), but although some members were taking notice of 'the sufferings of the outside world' (p. 152) other scholars remained 'locked in private debates, preoccupied with trivia, mending and qualifying accepted ideas' (p. 153).

Smith's (1971) report suggested that American geography was about to undergo another revolution, to counter a situation in which 'geography is overpreoccupied with the study of the production of goods and the exploitation of natural resources, while ignoring important conditions of human welfare and social justice' (p. 154). This forthcoming revolution would involve a fundamental re-evaluation of research, teaching activities and basic social philosophies, and was represented at the conference by activities such as the foundation of SERGE (the Socially and Ecologically Responsible Geographer) by Zelinsky and others and by a motion at the Annual General Meeting condemning United States' involvement in Vietnam. Smith was unsure whether this revolution, with its emphasis on social as against economic concerns, would spread to Britain, however:

> The conditions which have helped to spawn radical geography in the United States include the existence of large oppressed racial minorities, inequalities between rich and poor with respect to social justice, a power structure and value system largely unresponsive to the needs of the underprivileged, and an unpopular war which is sapping national economic and moral strength. These conditions do not exist in Britain or exist in a less severe form, and the stimulus for social activism in geography is thus considerably less than in America (pp. 156–7).

Smith would almost certainly have changed his mind a decade later: by the mid-1990s he was writing of the need 'to place social justice at the heart of human geography' (Smith, 1994, p. 1). Dickenson and Clarke (1972) responded to his 1991 claim by arguing that British geographers had long been concerned with 'relevant' issues, with particular respect to the Third World.

Another commentary on AAG's 1971 conference came from Berry (1972c), who felt that he had observed just a new fad involving 'new entrants to the field seeking their "turf"'. He could identify no real commitment:

> The majority of the new revolutionaries, it seems, are essentially white liberals – quick to lament the supposed ills of society and to wear their bleeding hearts like emblems or old school ties – and quicker to avoid the hard work that diagnosis and action demand. A smaller group of hard-line marxists keeps bubbling the potage of liberal laments. *In neither group is there any profound commitment to producing constructive change by democratic means. . . . If either of*

these will be the 'new geography' of the 1970s, count me out (pp. 77–8).

To him, academic geographers should provide a knowledge base on which policy can be built, which implies close involvement with the education of future generations of policy-makers. To Blowers (1972), however, 'The issue is not how we can cooperate with policy-makers, but whether and in what sense we should do so. It is a question of values' (p. 291); he argued that the sort of activities proposed by Berry would be strongly supportive of the *status quo*, and unlikely to produce fundamental social reform.

Smith (1973a, p. 1) responded to Berry that: 'bleeding hearts sometimes help to draw attention to important issues, and marxists can make valuable contributions in the search for alternatives to existing institutions and policies', and he pointed out that the current 'fad' was no more pronounced than that of the quantifiers a decade earlier. The results of their 'revolution' offered little for the solution of social problems, however, and Smith doubted the value of the large projects established by the AAG as part of its geography and public policy drive. (One of those projects produced a major series of books on Metropolitan America: Adams, 1976.) Research should highlight particular problems, and teaching should place emphases on (p. 3): 'a man in harmony with nature rather than master of it, on social health rather than economic health, on equity rather than efficiency, and on the quality of life rather than the quantity of goods'. Chisholm (1973) advocated caution in the corridors of power, because geographers had done insufficient substantive research to back up a 'hard sell' (a point made three decades earlier by Ackerman: see p. 54); Eyles (1973, p. 155) argued that 'any entry to those corridors assumes that the structure underlying policy alternatives is basically sound'; and Blowers (1974) wrote that in the corridors one can only influence, not decide, and that for the latter task geographers must develop their political convictions and act accordingly. (Blowers has been very active in English local government for some decades, focusing on environmental issues, on which he has also written major academic works: Blowers, 1984.)

This debate on if, and how, geographers should contribute to the solution of societal problems was a major issue at the 1974 annual conference of the IBG. Coppock's (1974) presidential address presented the challenges, opportunities and implications of geographical involvement in public policy, an involvement which he felt the current generation of students welcomed. He argued that policy-makers were largely ignorant of potential geographical contributions, while at the same time geographers seemed unaware that 'there is virtually no aspect of contemporary geography which is not affected to some degree by public policy' (p. 5). Coppock wanted to change this, to have geographers identify the contributions that they could make, to encourage research relevant to those contributions, and to enter à dialogue with those who advise on and implement public policy.

Other conference contributors were neither as optimistic nor as committed as Coppock. Hare (1974), himself an adviser to the Canadian government, reacted to cries that geographers were not consulted enough with the reply 'Thank goodness': geography, as a discipline, is irrelevant to the separate domain of public policy-making, although geographers, as individuals, because of the breadth of their training, could offer much that was valuable, so his conclusion was 'Geography no, geographers yes' (p. 26). His was a different response to mounting social concerns from Steel's (1974, p. 200) who told the British Geographical Association:

> As geographers we often get hot under the collar over the number of theoretical economists who are called on to advise the governments of developing countries. We comment on how much better World Bank surveys of countries would be if they were prepared, at least in part, by geographers. ... We wonder why university departments of geography are not engaged on a consultancy basis more often than they are, and we marvel that the Overseas Development Administration in London has only a handful of geographers on its staff where, we feel, an army would be more appropriate.

Hare (1977) subsequently argued that a major reason for a lack of geographical contributions to public policy may be the poverty of their training: in recent years we have 'swept geography departments into the social–science divisions of faculties of arts and sciences where, from playing second fiddle to geologists or literary critics, we learned to play second fiddle to economists and sociologists' (p. 263). Geographers would have to rebuild their discipline based on the centrality of society–environment interactions, he argued, with a new brand of physical geography that leans heavily on biological ideas and sources. Thus: 'We must reassert the old, essential truth that geography is the study of the earth as the habitat of man, and not some small sub-set of that gigantic theme' (p. 266). Regional syntheses should be stressed, because 'Realizing that regional perspectives are necessary in politics is to get one's manhood back' (p. 269). (See also Steel, 1982; Hart, 1982.) Hare's views on the current irrelevance of geography to public policy were supported by Hall (1974, p. 49) who argued that:

> geography, most clearly of all the social sciences, has neither an explicit nor an implicit normative base ... spatial efficiency ... is rather a description of what men seek to do in actuality ... not ... an objective to be achieved or objective function to be maximized.

Policy-makers must seek their norms elsewhere; geographers, meanwhile, must develop a new political geography which will aid them in understanding the crucial role of political decisions in structuring spatial systems (Johnston, 1978b).

Two other papers given at the 1974 IBG conference argued against Coppock's programme. Leach (1974), for example, claimed that governments, as paymasters, already constrained what geographers could do research on, and as a result geographers were being used; their only alternative was political action. Harvey's (1974c) contribution was entitled 'What kind of geography for what kind of public policy?'. Individuals wishing to become involved in policy-making were, he argued, stimulated by motives such as personal ambition, disciplinary imperialism, social necessity and moral obligation: at the level of the whole discipline, on the other hand, geography had been coopted, through the universities, by the growing corporate state, and geographers had been given some illusion of power within a decision-making process designed to maintain the *status quo*; he portrayed the corporate state as 'proto-fascist' (p. 23), a transitional step on the path to the barbarism of Orwell's *1984*. The function of academics, he claimed, was to counter such trends, to expunge the racism, ethnocentrism and condescending paternalism from within their own discipline and to build a humanistic subject, thereby assisting all human beings 'to control and enhance the conditions of our own existence' (p. 24).

Harvey (1973, p. 129) had previously argued that the current mode of analysis in geography offered little for the solution of pressing societal concerns:

> There is an ecological problem, an urban problem, an international trade problem, and yet we seem incapable of saying anything of depth or profundity about any of them. When we do say anything, it appears trite and rather ludicrous. ... It is the emerging objective social conditions and our patent inability to cope with them which essentially explain the necessity for a revolution in geographic thought.

He recognized three types of theory:

1 *status quo*, which represents reality accurately but only in terms of static patterns, and therefore cannot make predictions which will lead to fundamental social change;
2 *counter-revolutionary*, which also represents reality, but obfuscates the real issues because it ignores (either deliberately or accidentally) the important causative factors and so can be used to promote changes that will not bring about significant alterations to the operation of those factors. It is 'a perfect device for non-decision making, for it diverts attention from fundamental issues to superficial or non-existent issues' (p. 151); and
3 *revolutionary*, which is grounded in the reality it seeks to represent, and

is formulated so as to encompass the contradictions and conflicts which produce social change.

Harvey wished to write revolutionary theory, thereby overthrowing the current paradigm; his blueprint for geography

> does not entail yet another empirical investigation. . . . In fact, mapping even more evidence of man's patent inhumanity to man is counter-revolutionary in the sense that it allows the bleeding-heart liberal in us to pretend we are contributing to a solution when in fact we are not. This kind of empiricism is irrelevant. There is already enough information. . . . Our task does not lie here. Nor does it lie in what can only be termed 'moral masturbation' of the sort which accompanies the masochistic assemblage of some huge dossier of the daily injustices. . . . This, too, is counter-revolutionary for it merely serves to expiate guilt without our ever being forced to face the fundamental issues, let alone do anything about them. Nor is it a solution to indulge in that emotional tourism which attracts us to live and work with the poor for a while. . . . These . . . paths . . . merely serve to divert us from the essential task at hand.
>
> This immediate task is nothing more nor less than the self-conscious and aware construction of a new paradigm for social geographic thought through a deep and profound critique of our existing analytical constructs. This is what we are best equipped to do. We are academies, after all, working with the tools of the academic trade . . . our task is to mobilize our power of thought, which we can apply to the task of bringing about a humanizing social change (pp. 144–5).

To Harvey, then, relevant geography involves building new geographic theory on a marxist base, and its dissemination will achieve social reform through the education process. (This gradualist view, promoting reform through education, is only implicit in Harvey's work; Johnston, 1974, p. 189; 1986g.) Blaut (1979) contended that marxist theory offered two benefits because it can handle two crucial issues which positivist theory cannot: increased injustice and heightened economic and social instability.

Harvey (1984) presented a forceful elaboration of his views a decade later in a paper subtitled 'an historical materialist manifesto'. Geography, he contended, not only records, analyses and stores information about society but also 'promotes conscious awareness of how such conditions are subject to continuous transformation through human action' (p. 1). The nature of the knowledge that it produces and propagates reflects the social context of place and time, hence the role of geography in the 'Bourgeois era' as an 'active vehicle for the transmission of doctrines of racial, cultural, sexual, or national superiority' (p. 3). The positivist movement had sought to establish a universal science of spatial relations, which was countered by both humanistic and marxist critiques. Together the latter suggested the development of a revitalized geography, but they lacked:

> a clear context, a theoretical frame of reference, a language which can simultaneously capture global processes restructuring social, economic and political life in the contemporary era and the specifics of what is happening to individuals, groups, classes, and communities at particular places at certain times (p. 6).

Historical materialism provides that framework, he argues (though see Eliot Hurst's, 1985, argument that, if Harvey accepted his own (1972) contention that disciplinary boundaries are counter-revolutionary, then it is hard to understand how he can later promote a disciplinary-based manifesto).

Harvey argued that geographers cannot be neutral; their work must be of value to some special interest group within society. For him, that group should not be 'generals, politicians, and corporate chiefs' (p. 7) but the disenfranchised. A people's geography, 'threaded into the fabric of daily life with deep taproots into the well-springs of popular consciousness', 'must also open channels of communication, undermine parochialist world views, and confront or subvert the power of the dominant classes or the state. It must penetrate the barriers to common understanding by identifying the material base to common interests' (p. 7). It would not only reveal to the disenfranchised how societies are structured and restructured so that 'centers exploit peripheries, the first world subjugates the third, and capitalist powers compete for domination of protected space (markets, labor power, raw materials). People in one place exploit and struggle against those in another place' (p. 9), but would also help them to

> Define a political project that sees the transition from capitalism to socialism in historico-geographical terms . . . we must define, also, a radical guiding vision: one that explores the realms of freedom beyond material necessity, that opens the way to the creation of new forms of society in which common people have the power to create their own geography and history in the image of liberty and mutual respect of opposed interests. The only other course . . . is to sustain a present geography founded on class oppression, state domination, unnecessary material deprivation, war, and human denial (p. 10).

Harvey returned to this theme in 1989, in a retrospective volume reporting on a conference held to mark the twentieth anniversary of the publication of *Models in Geography* (see p. 93). He set out a brief critique of the scientific approach adopted in the earlier book, claiming that the type of modelling adopted could not be used to tackle what he identified as the major geographical questions (such as the 'evolutionary path of capitalism itself' – Harvey, 1989b, p. 212: in this context, he portrays 'all geography as historical geography'). Indeed: 'when put in the context of these grander questions, the modelling effort appeared both puny and not particularly revealing'. Many geographers retained that modelling approach, however, by fragmenting the discipline and focusing on narrow questions. They had restricted the nature of the questions asked, and as a consequence (pp. 212–13):

> we can now model spatial behaviour like journey-to-work, retail activity, the spread of measles epidemics, the atmospheric dispersion of pollutants, and the like, with much greater security and precision than once was the case. And I accept that this represents no mean achievement. But what can we say about the sudden explosion of third world debt in the 1970s, the remarkable push into new and seemingly quite different modes of flexible accumulation, the rise of geopolitical tensions, even the definition of key ecological problems? What more do we know about major historical-geographical transitions (the rise of capitalism, world wars, socialist revolutions and the like)? Furthermore, pursuit of knowledge by the positivist route did not necessarily generate usable configurations of concepts and theories. There must be thousands of hypotheses proven correct at some appropriate significance level in the geographical literature by now, and I am left with the impression that *in toto* this adds up to little more than the proverbial hill of beans.

He found it hard to understand why marxism had had such little influence on geography until the 1970s, and noted that when it did 'geographers who turned to Marx were swept up in that, and many were so submerged in it that they entirely forgot their own disciplinary identity' (p. 215). The submersion was necessary, but so was the identity: the production of knowledge is a political project, and the history and future of the discipline have to be appreciated in that context. Thus (p. 216):

> any project to 'remodel' contemporary geography must take the achievements of the Marxist thrust thoroughly into account at the same time as it recognizes the limits of positivism and the restricted domain of the modelling endeavours that derived therefrom. This is not to rule all forms of mathematical representation, data analysis and experimental design out of order, but to insist that those batteries of techniques and scientific languages be deployed within a much more powerful framework of historical-materialist analysis.

This, he accepts, is necessarily ideological, 'because it is necessarily political and built upon some conception of our collective agency in history'.

Smith's (1994) *Geography and Social Justice* took forward his concerns with inequality and welfare, reflecting changing circumstances (p. xiii). He had been involved [in] a variety of engagements with geographical aspects of inequality and human welfare. Reflecting the dominant mood of the times, explicit concern for social justice remained muted, for the most part. But the theme was still there, and increasingly required reassertion as social change on the world stage – east, west and south – began to resurrect some basic questions concerning the distribution of benefits and burdens under alternative economic and political arrangements.

He determined to explore those aspects of moral philosophy related to the issue of distributive justice, to 'see what can be made of this in the geographical context'. This involved a return to normative thinking: 'with how we conceive of what is right or wrong, better or worse, in human affairs lived out in geographical space'. He concluded that 'social justice should not be left to market forces' because 'to commend observed market outcomes on the grounds that they are the results of a just process is not credible' (p. 279), and he argues instead for a foundation in egalitarianism which, despite postmodern critiques, should embrace certain universal needs (see also p. 277). Harvey (1993a, 1993b, 1996) also addressed issues of justice, focusing on the interplay between universal conceptions and local circumstances (see p. 278).

The liberal contribution

Liberalism is defined as here the idea 'shared by many modern American liberals who, characteristically, combine a belief in democratic capitalism with a strong commitment to executive and legislative action in order to alleviate social ills' (Bullock, 1977, p. 347). Liberals are concerned that all members of society do not fall below certain minimum levels of well-being (variously defined), and are prepared for state action within the capitalist structure in order that this can be achieved. Within geography, much of the work conducted in this ethos focused on description rather than theory-construction: Chisholm's (1971a) investigation of the potential of welfare economics as a basis for normative theory which does not involve profit-maximization goals was one of the few exceptions (see also Wilson, 1976b).

There is a long tradition of liberal contributions by 'applied geographers', both in their research and associated activities and in their teaching (see House, 1973, and Hall, 1981b, for reviews). In Britain, applied geography as information-gathering and synthesizing has a substantial record, starting with Stamp's Land Utilization Survey of Britain in the 1930s, and his subsequent involvement in the preparation for post-war land-use planning (Stamp, 1946). Other aspects of land-use planning were of interest to geographers in the 1930s and 1940s, and were discussed several times at the Royal Geographical Society (Freeman, 1980b); similar geographical involvement occurred in the United States (Kollmorgen, 1979). After the Second World War – during which geographers made many contributions as information synthesizers and gatherers, the latter including the development of air-photo interpretation – land-use planning was established on a large scale, and trained geographers provided a large proportion of its personnel. Many academic geographers remained active in applied work. Technical developments in cartography and data handling saw them involved in redistricting for Congressional elections (Morrill, 1981), a wide range of mapping activities (Rhind and Adams, 1980), and developing regional bases for the presentation of census statistics (Coombes *et al.*, 1982), for example, and developments in GIS during the 1980s and 1990s stimulated very substantial increases in mapping and enumeration (see Charlton, Rao and Carver, 1995). Alongside this, various policies have been evaluated, such as those aimed at changing the distribution of industrial activity, and much of the entropy-maximizing systems modelling (p. 136) was intended to provide procedures for the joint activities of land-use and transport planning (Wilson, 1974; on whether it did, see Batty, 1989).

Most of this work used an empiricist and (usually implicit) positivist framework. Geographers are perceived as having valuable skills in the collection and ordering of data, as in land-use surveys, for example, and the presentation of such data frequently assumed the existence, and desirability of maintaining, certain causal relationships; planning agricultural land use, for example, often assumed a clear causal link between the physical environment and agricultural productivity and that of industrial location assumed the need for efficiency via the minimization of total travel costs. (Interestingly, some critics pointed out that the greatest use of such optimizing models occurred not in the 'capitalist west' but rather in the 'socialist east'.) As positivist work on the allocation of land uses and traffic flows increased, so the potential for geographic inputs to spatial planning was promoted (and many people initially trained as geographers became professional planners). Most of this was pragmatic application of technical skills, though there were some attempts to evaluate policy impacts (e.g. Hall *et al.*, 1973) and to develop a theory of decision-making in this context (Hall, 1981a, 1982).

Geographers' empiricist role continued to be advanced, in response to economic crises and the perceived need for valid data. In Britain, for example, the government introduced a three-year programme in 1983 to provide 'new blood' for university research efforts; 792 lectureships were available, in open competition between universities, with a further 146 in the fields of information technology. Geography departments received 11.5 of these posts (1.1 per cent of the total; the staff of geography departments comprised 2 per cent of all university posts in 1982–3). D. M. Smith (1985) argued not only that geography lost out relatively in that contest (the 'winners' were engineering and technology and the physical and biological sciences) but also that within geography the posts allocated selectively focused on certain aspects of the discipline: five were for research in remote sensing/digital mapping, for example, and three more were for various aspects of mathematical modelling (two in physical geography); only one reflected 'place-specific' issues, a post in historical–cultural geography. Smith interprets this as follows:

> The predominance of remote sensing (and associated digital mapping) reflects a view of geography as a technologically sophisticated means of gathering and displaying information, in the tradition of the geographer as map-maker linked to the contemporary preoccupation with information technology . . . [that] appears to conflate the needs of the discipline with a conception of the needs of society in which the emphasis is much more economic than social. It is hard to see more than one or two of the posts contributing much to the solution of *social* problems. The predominant impression is of an a-social view of the world, in which social relations, class structure and political power seem strangely absent (p. 2).

Clayton (1985a) responded that this was unlikely to be the only example of much greater state interference with the direction of academic research (see also his analysis of how geographers should react to that: Clayton, 1985b). In 1986 the University Grants Committee produced its first rating of the research record of every university department – on criteria that were not clear but within which externally earned research income was clearly important – and began a selective allocation of funds to universities that reflected those ratings (see Smith, 1986). Similar rating exercises were undertaken in 1989, 1992 and 1996 (each using peer review, and more transparent than the first: Thorne, 1993); virtually all of the research money distributed to the universities by the funding councils (some £600 million annually in the early 1990s) is now allocated by a formula linked to those ratings, with most money going to the largest departments whose research records and plans win the highest ratings (on the funding implications, see Johnston, 1993b, 1995b; on the ethics of such a distribution, see Smith, 1995, 1996; Rhind, 1996; Curran, 1996).

Alongside those who argued for a greater commitment to applied geography in the empiricist/positivist mould, and therefore for an implicit acceptance of a particular ideology (Johnston, 1981a), others challenged this as the best way to respond to pressing societal problems. They argued for a reappraisal of how geographers could assist in understanding the genesis of those problems rather than in the suggestion of solutions which rarely tackled the root causes. The major contributions of the two groups are the subject of the next sections.

Mapping welfare

A lot of research was reported in the 1960s and 1970s, under the general title of factorial ecologies (p. 90), applications of multivariate statistical procedures to large data matrices as a means of representing spatial variations in population characteristics. These, according to Smith (1973b, p. 43), were over-reliant on certain types of census data and therefore provided little information on social conditions. Earlier attempts had been made to structure analyses of such data towards particular ends, as in the work of rural sociologists on farmers' levels of living (Hagood, 1943) – a concept introduced to the geographical literature by Lewis (1968), and Thompson *et al.*'s (1962) investigation of variations in levels of economic health among different parts of New York State.

Factorial ecology procedures were adapted to the task of mapping social welfare in the 1970s, led by two workers. Knox (1975) promoted the mapping of social and spatial variations in the quality of life as a fundamental objective for geography, to provide both an input to planning procedures and a means of monitoring policies aimed at improving welfare. He divided the concept of 'level-of-living' into three sets of variables – physical needs (nutrition, shelter and health); cultural needs (education, leisure and recreation and security); and higher needs (to be purchased with surplus income) – and used statistical procedures to produce accurate portrayals of spatial variations in meeting these needs. With the resultant maps, geographers must then decide whether they are playing a sufficient role in awakening human awareness of the extent of the disparities or are 'under an obligation to help society improve the situation' (p. 53).

Smith's (1973b) very similar work was set in the context of the American social indicators movement and the growing belief there that GNP and national income 'are not necessarily direct measures of the quality of life in its broadest sense' (p. 1). He initiated the collation and dissemination of territorial social indicators, in order to illustrate the extent of discrimination by place of residence which occurs in the United States; multivariate statistical procedures generated the maps, on inter-state, inter-city and intra-urban scales. (Cox, 1979, extended the treatment to the international scale.) Smith's (1979, p. 11) goal was to present 'the basis for a better understanding of the origins of inequality as a geographical condition and of the difficulties in the way of plans to promote greater equality in human life chances'.

In Chisholm's terms, these two works represented the geographer as delver and dovetailer, producing information on which more equitable social planning could be based. Other studies performed similar roles, and also suggested spatial policies which could lead to improvements. Harries (1974), for example, studied spatial variations in crime rates and the administration of justice, and argued that positivist predictive models of criminal patterns could assist the organization of police work; Shannon and Dever (1974) and Phillips and Joseph (1984) investigated variations in the provision of health-care facilities and argued for spatial planning which would improve the services offered to the sick (which is different from a geography of prophylaxis: Fuller, 1971); and Morrill and Wohlenberg (1971) studied the geography of poverty in the United States, proposing both social policies – higher minimum wages, guaranteed incomes, guaranteed jobs and stronger anti-discrimination laws – and spatial policies (such as an extensive programme of economic decentralization to a network of regional growth centres) which would alleviate this major social problem.

An alternative, highly personal, programme of mapping variations in human welfare was advanced by Bunge (1971), who prepared a 'geobiography' of his home area in Detroit's black ghetto. His deeply humanitarian concern for the future was interpreted as a need to ensure a healthy existence for children; he wanted a 'dictatorship of the children' (Bunge, 1973b, p. 329) with regions – 'may the world be full of happy regions' (p. 331) – designed for them. This requires a reduction in the worship of machines, which are inimical to children's health (Bunge, 1973c), and a mapping of the sorts of variables never collected by external agencies and therefore requiring the development of geographical expeditions within the world's large cities; these maps would include roach regions, parkless spaces, toyless regions, and rat-bitten-children regions (Bunge, 1973d), and some were prepared for Detroit and for Toronto (Bunge and Bordessa, 1975). This work was later described by Merrifield (1995, p. 57) as involving excursions 'beyond the cloisters of the academy' and 'a redefinition of the research problematic and intellectual commitment of the researcher away from a smug campus career to one incorporating a dedicated community perspective'. The task of empathizing with and situating oneself authentically within an impoverished community is difficult, and the researcher's own biography is likely to influence the process of learning about and becoming committed to such a community. This raises the issues of voice that have increasingly concerned those studying various forms of representation (see p. 297). By becoming involved, the geographer will become 'a person of action, a radical problem-raiser, [and] a responsible critical analyst participating *with* the oppressed' (p. 63), but:

> there is always an immanent hazard that the voice heard in the supposed symbiosis between academic geographer and folk geographer is skewed towards the overzealous – though well-meaning – academic geographer. As the voice of the oppressed is muted, the expedition program degenerates into a paternalism reminiscent of 19th century Western missionaries and settlement houses.

Attempts at understanding

The mapping investigations just discussed were very largely descriptive, and any prescriptions offered based on limited theoretical foundations. Attempts were made to develop the necessary theoretical understanding, however. Cox (1973), for example, looked at the urban crisis in the United States – the racial tensions and riots, municipal bankruptcies, and the role of the government in the urban economy – presenting his analysis in terms of conflict over access to sources of power. This was intended as part of an educational exercise, for:

> It would be utopian to think that we can propose solutions on the basis of our analysis. The locational problems and locational conse-
> quences of policies weave too intricate a web for that to be possible.
> All we can hope to do is inform. To be aware of the problems and
> of their complexity may induce some sensitivity in a citizenry which
> has shown as yet precious little tolerance for the other point of view
> (p. xii).

Nevertheless, his final chapter, entitled 'Policy implications', discussed two imperatives towards greater equity in the provision of public services – the moral imperative and the efficiency imperative (the latter applies to the total level of welfare in society as well). The policies presented involve spatial reorganization to achieve the desired equity (including metropolitan government integration), community control, population redistribution and transport improvements. Massam's (1976) review of geographical contributions to social administration also focused on spatial reorganization: he evaluated service provision using the spatial variables of distance and accessibility, with major chapters on the size and shape of administrative districts and on the efficient allocation of facilities within such districts (see also Hodgart, 1978).

Cox's work heralded an increased geographical interest in a much neglected field, the role of the state in capitalist society (see also Cox, Reynolds and Rokkan, 1974; Dear and Clark, 1978; Johnston, 1982a; Cox, 1979). Political geography had traditionally been concerned with the state at the macro-scale only, dealing with political regions and boundaries and with the operations of the international political system (e.g. Muir, 1975). The similarly under-developed field of electoral geography had highlighted spatial variations in voting, but there had been little work on either the geographical inputs to voting or the geographical consequences of the translation of votes into political power (Taylor, 1978, 1985a; Taylor and Johnston, 1979). The state is involved in many aspects of economic and social geography, however, as both Buchanan (1962) and Coppock (1974) had stressed, but few geographers had investigated this involvement in any detail, or the electoral base on which it is founded (Brunn, 1974; Johnston, 1978b).

A framework for understanding spatial variations in well-being was presented by Coates, Johnston and Knox (1977). They defined the components of well-being, whose variations were mapped at three scales (international, intra-national and intra-urban), and argued that those variations reflected three sets of causes: the spatial division of labour; accessibility to goods and facilities; and the political manipulation of territories. (See also Cox, 1979.) Spatial policies aimed at the reduction of spatial inequalities were evaluated, such as various forms of positive discrimination by areas, leading to the conclusion that the division of labour is the primary determinant of levels of social well-being. Creation of this division is a social and not a spatial process, though it has clear spatial consequences, so that:

> The root causes of spatial inequalities cannot be tackled by spatial policies alone, therefore. Inequalities are products of social and economic structures, of which capitalism in its many guises is the predominant example. Certainly inequalities can be alleviated by spatial policies ... but alleviation is not cure: whilst capitalism reigns, however, remedial social action may be the best that is possible ... the solution of inequalities must be sought in the restructuring of societies (pp. 256–7).

(See also Johnston, 1986h.)

Smith (1977) essayed a more ambitious attempt at explaining spatial variations in well-being, arguing that: 'the well-being of society as a spatially variable condition should be the focal point of geographical enquiry ... if human beings are the object of our curiosity in human geography, then the quality of their lives is of paramount interest' (pp. 362–3) which was the foundation of his case for 'a restructuring of human geography around the theme of *welfare* ... to provide both positive knowledge and guidance in the normative realm of evaluation and policy formulation' (p. ix). His book proceeded from theory through measurement to application. The theoretical section is an amalgam of normative welfare economics with marxian perspectives on the creation of value, plus the political conflict for power.

> The analysis will inevitably reveal certain fundamental weaknesses of the contemporary capitalist-competitive-materialistic society, but the temptation to offer a more radical critique of existing structures has been resisted, in favour of an approach that builds on the discipline's established intellectual tradition (p. xi).

Two types of solution to perceived spatial inequalities are identified – liberal intervention, and radical, structural reform – with the former emphasized because 'most social change is incremental rather than revolutionary' (p. xii). Thus one concluding chapter – entitled 'Spatial Reorganization and Social Reform' – emphasized reorganization of administrative areas, although geographers are later urged to be 'open-minded enough to see that there may be greater merit to some of the development strategies practised with evident if not painless success in the USSR, China, Cuba and so on' (p. 358). Counter-revolutionary possibilities were foreseen, however – 'overpreoccupation with spatial reorganization may serve the interests of the existing rich and powerful by helping to obscure the more fundamental issues' (p. 359) – leading to socialist arguments 'That spaces and natural resources should be privately owned, with their use subject to the chances of individual avarice, altruism or whim, is increasingly an anachronism' (p. 360). He concluded that:

> As geographers we have a special role – a truly creative and revolutionary one – that of helping to reveal the *spatial* malfunctionings and injustices, and contributing to the design of a spatial form of society in which people can be really free to fulfil themselves. This, surely, would be progress in geography (p. 373).

Such design, he argued in a later book on the same theme (Smith, 1979), would require control of multi-national corporations, with greater public participation involving a decentralization of power, but he was not optimistic that this would lead to the equality hoped for in socialist arguments because 'Striving to become unequal seems almost to be an imperative of human existence' (p. 363).

Environmentalism

The late 1960s saw a rapid increase in concern about environmental problems; in the United States (Mikesell, 1974, p. 1):

> Towards the end of the 1960s the American public was overwhelmed with declarations of an impending environmental crisis. . . . Since that time, crisis rhetoric and a yearning for simple answers to complicated questions have given way to a more sophisticated and deliberate search for environmental understanding. Ecology has been institutionalized.

Two of the leaders of the public debate disagreed over the cause of the problems (O'Riordan, 1976, pp. 65–80). Ehrlich argued for the primacy of population increase, and popularized the concept of zero population growth; Commoner claimed that technological advances and the consequent rapid depletion of resources plus deposition of pollutants created the major problems. Both arguments have clear geographical components, and it was stressed that geographers have a strong record of activity in resource conservation: George Perkins Marsh had written on the topic in 1864 (Lowenthal, 1965) and the climatologist Warren Thornthwaite had been closely involved in the 1930s' soil conservation movement established as a consequence of the Dust Bowl phenomenon. The interest in landscape modification was advanced by Sauer and his followers, and reflected in the symposium *Man's Role in Changing the Face of the Earth*; similar interest elsewhere was exemplified by Cumberland's (1947) pioneering classic on soil erosion in New Zealand. Nevertheless, Mikesell argued that 'developments in geography have been such that the several phases of national preoccupation with environmental problems have not produced a general awareness of our interests and skills' (p. 2).

As part of the AAG's increased commitment to public affairs in the early 1970s, its Commission on College Geography established a Panel on Environmental Education and sponsored a Task Force on Environmental Quality. The latter reported (Lowenthal *et al.*, 1973) that geographers would make excellent leaders for the educational tasks in hand because of:

1 the breadth of their training and their ability to handle and synthesize material from a range of sources;
2 their acceptance of the complexity of causation;
3 the range of information which they are trained to tap;
4 their interest in distributions; and
5 their long tradition of study in this area.

All had fostered expertise in work on environmental perception, on vegetation succession, and on relationships between land use and soil erosion, which could be used as the bases for environmental impact statements, the elaboration of environmental choices, and international research collaboration.

Two types of work characterized geographers' activities in the area of society–environment inter-relationships. The first involved the traditional geographical concern with description and analysis. Review volumes such as *Perspectives on Environment* (Manners and Mikesell, 1974) were prepared, and a particular interest in problems of the physical environment of urban areas was generated (Detwyler and Marcus, 1972; Berry and Horton, 1974; Berry *et al.*, 1974; a later, British, addition was Douglas, 1983). The second type of effort focused on issues of environmental management (O'Riordan, 1971a, 1971b), with particular emphasis on its economic aspects and on societal response to environmental hazards (Hewitt, 1963): as Kates (1972, p. 519) pointed out, economics provided the theories and prescriptions of the 1960s (and later; see Rees, 1985). A topic of special interest was the study of leisure, of the growing demand for recreational facilities, and the impact of recreational and tourism activities on the environment (Patmore, 1970, 1983; see the critique of much of that work in Owens, 1984).

Despite such activity, Mikesell concluded that geographical contributions to environmentalism had not been great up to 1974. He commented regarding the prognostications of *The Limits to Growth* (Meadows *et al.*, 1972), for example, that 'the debate on this most relevant of all issues has attracted remarkably little attention from geographers' (Mikesell, 1974, p. 19) – though see Eyre (1978) – and he concluded more generally that: 'one must add hastily that many of the environmental problems exposed in recent years and also many of the social and philosophical issues debated during the environmental crusade have not been given adequate attention by geographers'. This conclusion was supported by analysis of the contents of recent geographical journals and O'Riordan's (1976) lengthy bibliography, and sustained by Goudie (1993) nearly two decades later.

A powerfully argued case not only for more work on the society–environment interface but also for its centrality to the whole of geographical activity has been presented by Stoddart (1987), who contended that instead of celebrating the achievements of a century of professional geography, many of his colleagues are 'despondent, morose, disillusioned, almost literally devoid of hope, not only about Geography as it is today but as it might be in the future' (p. 328). This, he believed, was because so many of them 'have either abandoned or failed ever to recognize what I take to be our subject's central intent and indeed self-evident role in the community of knowledge' (p. 329). For him, geography had become diffuse and lacked a central focus, which should be (p. 331): 'Earth's diversity, its resources, man's survival on the planet'. This calls for a unified discipline, human and physical, in which: 'The task is to identify geographical problems, issues of man and environment within regions – problems not of geomorphology or history or economics or sociology, but geographical problems: and to use our skills to work to alleviate them, perhaps to solve them.' Focusing on 'the big questions, about man, land, resources, human potential' would involve geographers reclaiming 'the high ground' (p. 334) and abandoning much that is currently done:

> Quite frankly I have little patience with so-called geographers who ignore these challenges. I cannot take seriously those who promote as topics worthy of research subjects like geographical influences in the Canadian cinema, or the distribution of fast-food outlets in Tel Aviv. Nor have I a great deal more time for what I can only call the chauvinist self-indulgence of our contemporary obsession with the minutiae of our own affluent and urbanized society ... We cannot afford the luxury of putting so much energy into peripheral things. Fiddle if you will, but at least be aware that Rome is burning all the while.

Bird (1989, p. 212) pointed out that although Stoddart called for a geography that is 'real, united and committed ... we are not told exactly what it is, though for Stoddart such a geography obviously exists'.

Similar, though less strident, calls have been made by others (e.g. Douglas, 1986; Goudie, 1986b; see also Cosgrove and Daniels, 1989); in response, the relevance of much that Stoddart would disregard in contemporary social science has been promoted as necessary to an appreciation of society–environment relations (Blaikie, 1985; Blaikie and Brookfield, 1987; Johnston, 1989b – the latter was written to counter the relative naiveté of much writing by physical geographers regarding the potential role of the state in the resolution of environmental problems: Trudgill, 1990; see also Johnston, 1996c). The nature of risk has been closely scrutinized by Adams (1995), for example, as have varying perceptions of nature (Simmons, 1993) and political approaches to environmental problems (Pepper, 1996); a successor to the 1954 Wenner-Gren Symposium explored human exploitation of the environment in considerable depth (Turner *et al.*, 1990) as have others on specific issues such as desertification (Granger, 1990; Middleton and Thomas, 1994).

Stoddart's argument, in a paper which was first given as the ninth Carl O. Sauer Memorial Lecture at the University of California, Berkeley, was clearly set in the cultural ecology mould. A similar case was made in the same year by Kates (1987, p. 526), who regretted the dominance of spatial science within geography and argued that when environmental issues became important on the public and political agenda in the early 1970s:

> No discipline was better situated than was geography to provide intel-
> lectual and scientific leadership. The natural science for the environ-
> mental revolution should have been the science of the human
> environment. Instead, intellectual leadership was split among biology,
> economics, and engineering, each of which transferred onto the
> human environmental realm their own theories of nature, of econ-
> omy, or of technology, but none of these offered a truly integrated
> view . . . The theory of the human environment, then, was the theory
> of plant or animal ecosystems, or of pervasive externalities, or of
> technological and managerial fixes.

For geographers, a perceived inability to respond to the demand for environmental scientists (producing 'geographers who can sit astride the natural and social science boundary to provide analysis, integration and leadership') was an opportunity lost, a 'road not taken'. But the Malthusian dilemma remains, posing great questions for society to which geographers can bring their special disciplinary advantages to in the search for answers (p. 532):

> We possess more than passing knowledge of both the natural and
> social sciences . . . We have some useful tools for organizing data and
> information . . . We possess a strong tradition of empirical field
> research . . . And perhaps most important we have and we teach a
> respect for other peoples' theories. Our answer, then, as to why geo-
> graphy . . . is that we are needed and that we are useful. When they go
> forth, our students understand the nature of the great questions, have
> more than a passing knowledge of natural and social science, have

been in the field, have collected and organized new data, and have placed these data into a theoretical perspective.

To be sure that they will be called upon to perform in this way, geographers must put their house in order and some university departments should rebuild centres of expertise in the human environment tradition. (The University of Oxford established an Environmental Change Centre along-side its School of Geography, for example, and its first director had an established international reputation in the study of the interactions between climatic change and human activity: Parry and Duncan, 1995.)

One of the major concepts to emerge from the increased concern over environmental problems in the late 1980s was sustainable development, a considerably ambiguous term which is generally taken to imply continual increases in material living standards without any diminution in environmental capacity to meet the needs of future generations (Turner, 1993). This was a focal concern of the 1992 Earth Summit at Rio de Janeiro, but the difficulties of implementing global environmental policies have been frequently stressed (see the essays in Johnston, Taylor and Watts, 1995), although others believe that the problems have been over-stated and can readily be addressed, within a satisfactory time scale, using instruments of economic policy that promote continued economic development (Beckerman, 1995; see also Kates, 1995).

Whilst these broader concerns were being aired, geographers were continuing to be active in the tradition established by White in the 1940s (see p. 149). In 1937, for example, White's memorandum to President Roosevelt had helped to convince him not, as Congress had proposed, to make the Secretary of War responsible for flood control projects, thereby bypassing local and regional planning bodies, and in 1965 he prepared a paper for the Bureau of the Budget on national policy options for dealing with floods and similar hazards. The result was the National Flood Insurance Program, after which White held annual seminars in Boulder to bring public-sector officials and academics together to discuss disaster response and mitigation policies (Platt, 1986). In this, he and the many others who have studied environmental policy and contributed to its implementation have adopted what Feldman (1986, p.189) termed the citizen-scholar model:

> The scholar chooses his or her research concerns on the basis of perceived social need but attempts to conduct the inquiry free of influence from outside the inquiry itself. The scholar is committed to social and political action as follow-up to the research, based on the findings and quality of the inquiry. But political activity is, during the course of the scholarship, kept separate . . . [In addition] the citizen-scholar teaches public affairs both within a curriculum and by example, letting her or his own activities demonstrate good citizenship and good scholarship without being didactic.

Earlier, White (1972, p. 322) had asked 'What shall it profit a profession if it fabricate a nifty discipline about the world while that world and the human spirit are degraded?', and in seeking to avoid that problem he developed an approach which Wescoat (1992) compares to that of the philosopher of pragmatism, John Dewey. It had four components, he argued:

1 recognition of the precariousness of existence;
2 a pragmatic conception of the nature of inquiry – problems arise and are tackled within situations, for there are no absolute truths; hence
3 a tradition of learning from experience; and
4 a belief in public discourse and democracy

all of which fitted White's deep involvement with the work of the Society of Friends.

Geographers have also turned their attention to other 'great questions' within contemporary society, such as nuclear weapons and nuclear power. They have criticized attitudes to civil defence policies and likely deaths from nuclear blast and fall-out (Openshaw, Steadman and Green, 1983), for example, with others setting that concern within a developing geographical contribution to peace studies (Pepper and Jenkins, 1985), and have addressed issues relating to the siting of nuclear power stations (Openshaw, 1986) and the transport of nuclear waste. (Openshaw's work on these topics involved innovative prototypes of GIS technology.) Arguments for geographical contributions to other aspects of peace studies, such as international relations (van der Wusten and O'Loughlin, 1986), were advanced (Pepper and Jenkins, 1983), none more forcibly than Gilbert White's (1985, p. 14), who saw the human family 'tortured and driven by its newfound capacity to throw the whole set of processes out of kilter by more violent action'. The two editions of *A World in Crisis?* (Johnston and Taylor, 1986a, 1989; see also Johnston, Taylor and Watts, 1995) focused on many of these issues, ranging from the macro-scale in the study of geopolitics to the investigation of individual human rights.

The breadth of study in modern environmentalism is illustrated by O'Riordan's (1981) work, much of it set in the liberal humanitarian tradition already illustrated here, and from which he drew four conclusions:

1 modern environmentalism challenges many aspects of Western capitalism;
2 it points out paradoxes rather than clear solutions;
3 it involves a conviction that better modes of existence are possible; and
4 it is a politicizing and reformist movement, based on a realization of the need for action in the face of impending scarcity and a lack of faith in the western democracies (pp. 300–1).

A new social, environmental order is required and O'Riordan identified three possibilities: centralized; authoritarian; and anarchist. He chose a middle of the road, liberal option:

we must individually and collectively seize the opportunities of the present situation to end the era of exploitation and enter a new age of humanitarian concern and cooperative endeavour with a driving desire to re-establish the old values of comfortable frugality and cheerful sharing (p. 310).

Such a new era, involving a new political order based on a combination of local self-determination and supra-nationalism, can be achieved through education, so environmental education will form a preparation for citizenship. Others are more doubtful: Pepper (1996, pp. 324–5), for example, is sceptical about the liberal arguments that: 'a basic unselfishness and communalism which is human nature would come to the fore in ecotopia's unalienated society, as it has never been allowed to do in industrial capitalism' (arguments that he terms 'simply hard to believe') and promotes instead the view 'that by making humanist, egalitarian and socialist aspirations a prerequisite for an ecological society, rather than something that is supposed automatically to follow, we can avoid making ecological society a repressive dystopia'. Social change must precede the resolution of environmental problems.

Some writers have argued that the revival of interest in environmental issues – mainly through the study of resources and their management – provides a contemporary linking of human and physical geography. Relatively little of the research and textbook writing indicates any integration of the two, however (Johnston, 1983a, 1989b), because the focus is almost invariably on the processes studied in one of the sub-disciplines only. Physical geographers portray trends such as demographic growth, technological sophistication, urbanization, and demands for resources as catalysts for environmental processes and changes, but they take those trends for granted and do not address what processes generate them. Similarly, human geographers take the resources of the physical environment as given and do not inquire about the physical and chemical laws underpinning their genesis. Thus while connections have been made across the human–physical geography interface, there has been no integration of the study of physical and societal processes; for human geographers their links with other social scientists are very much stronger than those with environmental scientists (as illustrated, for example, by the lack of references to the physical geography literature in O'Riordan, 1976, and Pepper, 1996).

Geographers and policy

Many of the studies referred to here have been concerned with identifying problems and suggesting solutions. Underlying their varied approaches has been the basic thesis that geographers should be much more involved in the creation and monitoring of policies. But what sort of policies, and what sort of involvement?

Berry (1973a) categorized planning policies into four types:

1 *Ameliorative problem solving* involves identifying problems and proposing immediate solutions, as with the removal of a traffic bottleneck. Such solutions are likely to stimulate further problems in the future, since it is only the proximate cause that is tackled (the features of the bottleneck) rather than the real cause (the growth of traffic).
2 *Allocative trend-modifying: planning towards the future* involves identifying trends, evaluating what is likely to be the best outcome of the several which they imply, and then allocating resources to steer the system being planned towards that end.
3 *Exploitative opportunity-seeking: planning with the future* identifies trends and then seeks to gain the maximum benefit from them, irrespective of the possible long-term consequences. Compared to the previous category, this one has a dominantly short-term focus.
4 *Normative goal-oriented planning for the future* begins with a statement of goals, a vision of the future, and then prepares a strategy which will ensure that they are achieved.

Relatively little policy-making is of the fourth type; most involves elements of the other three, with general statements about goals but no clear strategy regarding a foreseeable future. Some critics interpret this as meaning that geographers who become involved in policy-making and evaluation are likely to be uncritically accepting the dominant forces in society, which leads to arguments that their claimed scientific objectivity and neutrality are an (often unrealized) cloak for ideological political judgements about the nature of society. Whatever their individual motives, such geographers are acting on behalf of interest groups (private and public) whose sustenance depends on maintaining an unjust and unequal structure to society.

Clark (1982, p. 43) accepted that 'the academic community is not independent: there are no objective standards between competing explanations and thus policy advice' but nevertheless agreed 'with the principle of policy analysis and the involvement of academics in policy-making' (p. 48). That involvement cannot be presented as neutral and objective, however, since all social science 'explanations' are incomplete and compete with others to provide plausible accounts and prescriptions. He suggested that academic contributions to policy analysis should be guided by four propositions (pp. 55–9):

1 academics must acknowledge their own values and beliefs in presenting policy alternatives and impact assessments;
2 policy analysts must be advocates for particular causes rather than supposedly independent and objective adjudicators of knowledge;
3 policy science should be critical of the *status quo*; and

4 sponsoring institutions must encourage advocate briefs and make those
   briefs accessible to the public.

In accepting these, academics who become involved in policy analysis
should be considered 'part of the political process' (p. 57) whose role
would be to ensure that 'Choices would be brought squarely into the open
and be dependent upon the political, as opposed to expert, process' (p. 59).
The present situation idolizes experts, and promotes a myth of social science as objective, neutral knowledge. Clark argues that social scientists
cannot be neutral experts, because their 'values, interests and normative
views of the world' mean that their presentations, although scholarly and
rigorous, are necessarily partial. Sayer (1981) made a similar case in a
slightly different context, arguing that any rigorous social scientific inquiry
must be based on rationally defended value judgements; objectivity does
not require neutrality. (Clark, 1991, has illustrated this from his experience
as an expert witness in US court cases.)

   In an essay on the role of urban geographers in applied work, Pacione
(1990a) argued that practitioners have paid insufficient attention to conceptual issues underlying what they do. He derived the following 'principles
or guidelines' to remedy those failings:

1 the notion of value-free research is an illusion;
2 towns, as examples of places, are meaningful entities on which to work;
3 a spatial perspective is of substantial value;
4 the main emphasis of applied urban geography is on problem-
   solving;
5 a realist position provides the context for such work (see above,
   p. 216);
6 analysis must integrate various spatial scales;
7 a wide methodological tool-kit of quantitative and qualitative procedures must be employed; and
8 geography must integrate the findings of many disciplines.

Johnston (1990b) responded by posing six questions to Pacione: what is a
problem?; are problems always soluble?; what is science?; what is a geographic perspective?; who solves?; and what sort of society? He concluded
that some of the principles/guidelines are trivial and/or irrelevant, some are
unsupported, and some are wrong; unless Pacione:

   is prepared to tackle the fundamental issues of what problems are and
   how they can be tackled, he is unlikely to help those of us who want
   both an end to the problems that we currently perceive *and* a society
   which would no longer produce such problems.

Few problems are soluble, which points to the need for resolution between
opposing points of view, none of which have any claim to absolute truth.
(This point is made in another way in a debate over how different

geographers perceive the geography of Israel: Waterman, 1985; Falah, 1989; Waterman and Kliot, 1990: see p. 301.)

Undertaking research characterized as 'relevant' – pertinent to tackling societal problems – raises major issues for the individuals concerned. Mitchell and Draper (1982, p. 2) argued that 'when functioning as an advocate or consultant, the geographer must consciously decide how to resolve a conflict which may arise regarding the promotion of one perspective versus critical assessment and balanced judgement about all viewpoints'. There are also issues of *ethics*: 'when functioning as a pure researcher, the geographer must balance a concern for obtaining necessary information against a concern for respecting the dignity and integrity of those people or things being studied' (p. 3) which also applies to 'relevant' research. They claimed that geographers have largely ignored these ethical issues, and their professional bodies, unlike those of other disciplines, have promulgated no codes of conduct. Conflict is frequently likely between striving to 'discover truth' and respecting the rights of those being studied, however, and they advocated individual, institutional and external controls.

An example of the problem was raised in 1996 within the merged RGS-IBG (see p. 41). The Shell oil company was one of the society's commercial sponsors – an arrangement concluded with the RGS prior to the merger of the two societies. It donated £40,000 per annum towards the costs of the RGS's Expeditions Advisory Service, and was presented as an example of the sort of commercial sponsorship for geographical work which academic geographers could benefit from within the merged society. In late 1995, however, several members of the Ogoni tribe in the Biafra region of Nigeria, including a well-known author Ken Saro-Wiwa, were sentenced to death by a military tribunal (in a trial which western observers claimed violated their basic human rights) and were subsequently executed, despite political pressure on the Nigerian government which culminated in its exclusion from the Commonwealth heads of government conference held in New Zealand in November 1995. Their 'crimes' were related to protests over Shell's treatment of the Ogoniland environment, where it was involved in exploiting major oil and gas fields. A number of academic geographers argued that Shell's corporate patronage of the RGS-IBG should be ended because of its environmental record there and its alleged complicity in the trial of the Ogoni leaders and the arming of the Nigerian police. A motion to that effect was passed by a very substantial majority at the Research Section (formerly the IBG) conference in January 1996 (an event which attracted considerable media attention worldwide) and the RGS-IBG as a consequence set up a working party to consider all aspects of corporate sponsorship, which proposed a code of practice and ethical guidelines.

Most of the critiques of the arguments for greater involvement by geographers in policy analysis are entirely sympathetic. They make a case for *sensitive* geographical involvement; others question the grounds for such involvement and instead focus on the development of revolutionary theory (see p. 327). The latter focus almost entirely on applied work in the tradition of positivist, empirical science, in which the goal is perceived to be to improve well-being through contributing to one or more of:

1   the preparation of public (i.e. state) policies;
2   the development of commercial (i.e. profit-making) strategies; and
3   the attack on environmental problems.

Virtually none of this work is concerned with the applied 'arms' of the two other types of science identified in Chapter 1, leading to either greater self- and mutual awareness (see Chapter 6) or emancipation (Chapter 7). To that extent, most applied geography in the English-speaking world accepts, and seeks to serve, the ruling ideas.

The nature of applied geography has been restricted by some to work at the 'society–environment interface'. When launching the journal *Applied Geography* in 1981, for example, its founding editor, following Stamp (1960, p. 10), defined it as addressing 'some of the great world problems – the increasing pressure of population on space, the development of under-developed areas, or the attempt to improve living conditions'. Two decades later (Briggs, 1981, p. 2): 'These problems remain. Indeed, they are growing; environmental problems such as pollution, damage to wildlife, destruction of habitats, soil erosion and resource depletion; the problems of human deprivation and inequality'. He set out the fundamental basis of applied geography as use of resources:

> The exploitation of scarce resources represents a dominating theme to human existence. It is from the pursuit of these resources and from the attempt to decide between alternative policies of exploitation, that not only environmental damage, but also the greater part of political, social and economic problems emerge; they can be seen as expressions of man's inability to organize himself and his world to his best, long-term advantage.

To overcome that inability, the applied geographer must be brave (p. 6): 'He needs to commit himself before he knows all the answers. He needs to be able to make public mistakes. But he must also be prepared to learn from them.' Most of the applied work undertaken by academic geographers has been commissioned by one or more arms of the public sector and most geographers who have applied their skills in non-academic careers have probably entered the public sector too. Given the importance of the welfare state in the first four post-Second World War decades, this has been not been surprising. Nevertheless much work has always been done for the private sector too: in economic geography, for example, geographers such as Applebaum (1954) were involved in work for supermarket and similar companies in the United States during the 1950s and geographers were central to the establishment of an Institute for Retail Studies at the University of Stirling in the mid-1980s. Others, building on their expertise in spatial data handling, have moved into information management, including some who work in prestigious university business schools.

The pressure on university academics to obtain more external financial support for research activities in the 1980s and 1990s has stimulated much more work for the private sector – a trend accentuated by the attempts to reduce the size of the public sector and to increase competitiveness and efficiency in the provision of public goods (including, of course, higher education). Some of this work has involved spatial analysis in its various forms – as by the GMap company established by the school of geography at the University of Leeds, which has applied and developed Alan Wilson's pioneering work on location-allocation models to a wide range of problems in, for example, operations research for health-care providers and the location of franchises for major car firms (Birkin, Clarke, Clarke and Wilson, 1990; Birkin, Clarke and George, 1995). Much more of it has used geographers' skills in spatial data handling (what Openshaw and others refer to as 'adding value' to geocoded data) in the growing field of geodemographics, whereby marketing and other campaigns are spatially targeted to people living in areas where demand for particular goods and services is most likely to be generated (Batey and Brown, 1995; Birkin, 1995). Other work still is in the areas of resource and environmental management.

How much of this work has been done is difficult to appreciate because, unlike 'pure' research, little of it is reported in the discipline's academic literature – in part because of confidentiality constraints imposed by the research 'purchasers'; in part by timetables which press researchers to obtain further grants (to keep their staff in employment, for example) as soon as, if not before, one project is completed and the reports delivered; and in part because much of the work done involves routine application of standard procedures ('normal science') which makes no original contributions to the corpus of disciplinary knowledge. But more is certainly being done, as illustrated by the increased volume of research money being obtained by academic geographers from commercial and other purchasers.

Changing contexts and applied geography

The demands for greater involvement in what is generally termed applied geography have grown in recent decades, largely as a response to changes in the societal contexts within which Anglo-American human geographers work. This is not a surprising trend, according to Taylor's (1985c) analysis of the history of geography. In periods of economic recession, cutbacks in public funding for higher education and research can be expected, and to counter the loss of support geographers are forced to seek financial backing elsewhere (including arms of the state which contract for research); that backing is only likely to come if individuals within the discipline, and even organizations representing it, can convince potential sponsors of the value (i.e. potential 'profitability') of investing in geography and geographers. From this observation, and following Grano (1981), Taylor identified two external influences on disciplinary developments: 'Within academia geographers had to be given an intellectual foundation to satisfy intellectual peers, and outside in the wider world geography had to be justified as a useful activity on which to spend public money' (p. 100). These produce 'pure' and 'applied' geography respectively. Both are necessary to a discipline's future, but the relative emphasis placed on each will vary over time and space:

> Outside pressures will be particularly acute in periods of economic recession when all public expenditure has to prove its worth. All disciplines will tend to emphasize their problem-solving capacity and we can expect applied geography to be in the ascendancy. . . . In contrast in a period of expanding economies and social optimism outside pressures will diminish and academia can be expected to be under less external pressure. Geographers will thus be able to contemplate their discipline and feel much less guilty about this activity. We can expect bursts of pure geography to occur in these periods.

Taylor identified cycles of pure and applied geography corresponding to cycles in economic prosperity, and Hagerstrand (1977, p. 329) reached a similar conclusion: 'When the world is stable and/or unhampered liberalism prevails, then there is probably not much to do for geographers except surviving in academic departments trying to keep up competence and train schoolteachers in how wisely arranged the world is.'

The period since the late 1970s should demonstrate applied geography in the ascendancy, according to Taylor's analysis, an expectation generally supported by experience of those years. Pressure to justify the discipline in terms of utility to economic goals was strong, and the search for research contracts of all types became much more determined (so that learned journals and newsletters began to list new grants and contracts won by geographers and institutions such as the IBG prepared documents to promote the discipline which emphasized its utilitarian aspects, including one which graduates could show to potential employers). That pressure for applied geography has not been released, largely because the attitude within the state apparatus, particularly but not exclusively in the United Kingdom, did not favour a return to the type of funding of higher education which would sustain a further period of pure geography. Instead, academic departments were continually required to canvass for research funds from outside sponsors (which they were apparently very successful at: Johnston, 1995b).

Changing the focus

The desirability of a continued period within which applied geography is in the ascendancy has been the focus of further debate of the type discussed above. Some clearly accepted and argued for that ethos, but their main concern was either with geography using its perceived particular skills in what Harvey (1973; see p. 327) termed the application of *status quo* theory (as in Pacione, 1990a and b) or with exploring how geographers could make a committed contribution towards the achievement of change (Clark, 1982).

Bennett (1989a) based a case for a reorientation of what geographers do on a belief that there has been a major shift in the 'culture of the times', away from 'welfarism' – a consensus ideology aimed at 'improving the quality of life and provision of needs through collective and governmental intervention' (p. 273) – towards what he terms 'post-welfarism', which emphasizes individual rather than collective decision-making, and the role of markets rather than states (p. 286):

> the emergent 'culture of the times' has been happier to see the market as both the creator and the provider of new wants. Rather than markets being seen as an inhumane and exploitative system, socialism and even corporatist social democracy have come to be associated with the odious and paternalistic treatment of individuals.

This led him to criticize geographers who work with 'social theory', and who accept the welfarist ethos, as being necessarily of the 'political left', and he challenged their 'core concept of relative deprivation and its consequential focus on relative inequality and hence "social" and "spatial" justice' (p. 285). Rather than a society in which 'every difference in outcome is translated into an entitlement to state intervention', which inevitably leads to a situation of 'total state intervention in everything', Bennett (p. 286) argued that:

> Where the Thatcher era has heralded consumer choice and economic change, social theory and socialist politics have sought to defend the mode of production and to trap people in labour-intensive work practices and unattractive jobs vulnerable to technological change. The spirit of market freedom of individuals has heralded a consumer and service economy which has offered the release from the least attractive toils and labours, and has seemed to offer the potential to satisfy many of people's most avaricious dreams.

This market-oriented society poses an important challenge to geographers' applied role. It has been embraced by a number of geographers, including some who work with GIS (see p. 315). Longley (1995, p. 127), for example, argues that spatial analysis using GIS technology can enhance business management systems enabling users 'to gain competitive advantage in sophisticated consumer-led markets'. He criticizes the 'deskilling of geography and planning' associated with arguments such as Harvey's (1989b; see p. 329 above) that quantitative research has produced little relevant output. Not only does the development and teaching of GIS equip students with 'a range of flexible skills for use in continually restructuring labour markets' (p. 129), but in addition the range of applications:

> provides clear testimony that quantitative spatial analysis is most certainly not preoccupied with techniques that do not work to analyse problems that do not matter. Social science that does not show interest in real world issues of popular concern is doomed to remain on the sidelines of academic respectability and perceived social relevance, and reinvigorated spatial analysis is central to the measurement and modelling of economic and social aspects of human behaviour.

Bennett portrayed the geographical research he criticized as focused on aiding government intervention in economy and society, without questioning its validity. He believed such intervention to be no longer viable, and he also rejected its underpinning critical stance, based on a critique of capitalism. He identified the intellectual challenge for geographers to involve dismissing notions of a welfare state founded on a social theory which emphasizes rights and relative needs, because policies based on such theory 'cannot be proved to work ... Even social democracy offers no easy solutions' (p. 287) . But markets do fail, so:

> The two key questions for a post-welfarist society are: first how can support be improved by practical policies that can be demonstrated to work and are reasonably cost-effective; and second at what point does governmental action end ... Hence what is needed is a better definition of what is 'socially' possible through collective action and what is not. Social theory and social democracy have, perhaps, promised too much and hence led to disillusionment in its own promises (pp. 287–8).

Answering those questions is central to the role he casts for geographers: 'The grand objective of the discipline should be to contribute to the debate around these issues. But it must be a contribution to practice.' This implies a discipline in which practical concern with means is more important than theoretical debate about ends. In a post-welfarist society:

> The welfare state, and its associated public decision-making, no longer have the privileged status that it can be justified by mere statements of belief in public or governmental goods: public goods and the policy that provides them have to be demonstrated to be *effective* in

meeting social needs; policies have to work and be more effective than alternatives. (pp. 288–9).

Bennett provided no detail regarding what such a post-welfarist geography would contain, though he referred approvingly to Openshaw's (1989) arguments (see p. 349 above). He offered two brief suggestions, however. The first is a criticism of marxist and related work (p. 289):

> The key aspect . . . is a better understanding of the structure of economic incentives and rights, rather than class . . . We need to identify a new language not of class but of 'rights' or nature. By which I mean choice conceptions of rights which promote autonomy, freedom, self-determination and human development, and not 'interest' conceptions of rights which make people passive beneficiaries of the services of others.

The other is a recommendation that:

> For the academic discipline of geography this means adaptation of its frameworks of teaching and research. I would argue that one aspect of this requires more intensive training in analytical methods including model-based approaches, information systems and elementary analytical skills.

Thus geographers are called upon to participate in a re-evaluation of the welfare state, focusing attention on the limits to both individual choice and collective action, and to develop the analytical skills which will advance policy appraisal and so contribute 'useful knowledge to the research process, to policy debate and to practice' (p. 290). The extent to which the call will be followed, in the creation of a new form of applied geography, waits to be seen: in his commentary at the end of the book in which Bennett's chapter appears, Macmillan (1989b, p. 306) comments that 'the idea that the welfarist tradition is in terminal decline seems highly debatable'.

As already noted, this position locates applied geography within the spatial–science approach to the discipline; that which is applicable is that which is based on empiricism. But each of the three types of science identified in Chapter 1 has its 'applied arm' (see p. 33): hermeneutic sciences have as their goal the development of self- and mutual understanding, a project shared with the critical sciences, whose emancipatory aspirations seek similar understanding within the wider context of appreciation of the economic, social, cultural and political contexts within which individuals and groups are embedded. (Whereas radical geography sought emancipation largely in the context of class interests, therefore, that imbued with the cultural turn promotes a wide range of identity politics, within which class is only of the foundations: May, 1996b.)

Such applied goals are political and ideological to those opposed to them – but no more so than those promoted within the constraints of empirical science: the difference is that whereas the latter are largely supportive of the *status quo* – in particular its reliance on market mechanisms founded on the ideology of private property – the former challenge it, and argue for altered economic, social and political arrangements. As Harvey argued in 1974 (1974c), the crucial questions to be asked about any work claiming to be 'relevant' are 'relevant to whom?' and 'for what?'.

|10|

A changing discipline?

The reasonable man adapts himself to the world: the unreasonable man persists in trying to adapt the world to himself. Therefore all progress depends on the unreasonable man.

George Bernard Shaw

I was to learn later in life that we tend to meet a new situation by reorganising, and a wonderful method it can be for creating the illusion of progress while producing confusion, inefficiency and demoralisation.

Petronius Arbiter

It must be considered that there is nothing more difficult to carry out, nor more doubtful of success, nor more dangerous to handle, than to initiate a new order of things. For the reformer has enemies in all those who profit by the old order, and only lukewarm defenders in all those who would profit by the new order.

Niccolò Machiavelli

The previous eight chapters have outlined the major debates among Anglo-American human geographers since 1945 regarding their discipline's contents – what it studies, how, and why. The chapter titles themselves indicate that several very different approaches to the discipline have been advocated; their contents suggest a large number of unreasonable academics promoting new arrangements, but being resisted by defenders of the old order! The purpose of this final chapter is not to assess progress within those approaches (Lowe and Short, 1990), let alone to establish whether they have contributed towards the attainment of 'higher levels of intellectual, social and physical well-being for [our] fellow men' (Wise, 1977,

p. 10); indeed, according to some interpretations, progress in the Enlightenment sense of that term is not feasible (Barnes, 1996). Rather, the evaluation here returns to the issues raised in Chapter 1, where several different models of the development of scientific disciplines were presented. No formal testing of those models is presented, for no methodology has been outlined that would allow such a task. Instead the general relevance of the ideas presented there is assessed against the material outlined above.

Human geographers and models of disciplinary progress

Along with members of almost every other academic discipline, human geographers have been attracted to the ideas and language of Kuhn's paradigm model (Harvey and Holly, 1981, p. 11):

> the use of the word paradigm has become fashionable in geography as well as having become a pivotal concept for courses in geographic thought on both sides of the Atlantic. Thomas Kuhn has become as familiar to students of geography as Hartshorne or Humboldt.

Kuhnian ideas were applied in the 1960s and 1970s with relatively little reference to the major debates they had stimulated throughout Anglo-American academia, however. Most human geographers relied on the first (1962) edition of *The Structure of Scientific Revolutions*; they seemed unaware either that 'The loose use of "paradigm" in his book has made [it] amenable to a wide variety of incompatible interpretations' (Suppe, 1977a, p. 137) or that 'Kuhn's views have undergone a sharply declining influence on contemporary philosophy of science' (1977c p. 647). Furthermore, Kuhn (1977) substantially revised his ideas later. (See also the exegesis of his work in Barnes, 1982.) Geographers at that time displayed a general tendency to adopt ideas from other disciplines rather uncritically. Agnew and Duncan (1981, p. 42), for example, argued that:

> Recent reviews and programmatic statements concerning trends in Anglo-American human geography leave the impression that little attention has been given by geographers to the philosophical compatibility of borrowed ideas . . . that the political implications of different ideas have largely been ignored . . . and that controversy on source disciplines or literatures has not excited much interest.

Their examples do not include the import of Kuhnian ideas into geography, but their conclusions certainly hold in this case. (For a critical discussion of that importing of Kuhnian ideas by geographers, see Mair, 1986.)

Kuhnian concepts were first used in the geographical literature by Haggett and Chorley (1967), as part of a normative argument for a revolution in geographical method. They argued that the then dominant paradigm, as they defined it (discussed in Chapter 2 of the present book), could not handle either the explosion of relevant data for geographical research or the increasing fragmentation and compartmentalization of the sciences. They proposed a new 'model-based' paradigm:

> able to rise above this flood-tide of information and push out confidently and rapidly into new data-territories. It must possess the scientific habit of seeking for relevant pattern and order in information, and the related ability to rapidly discard irrelevant information (p. 38).

The new paradigm had been launched more than a decade previously (see Chapter 3 above); Haggett and Chorley's goal was to spread the new ideas, and to win British converts to a new orientation of work within geography. (In the same book, Stoddart, 1967b, used the paradigm model to promote an 'organic paradigm' – as a 'general conceptual model': p. 512. When that essay was reprinted in 1986, he added a footnote 'that I would now take a less enthusiastic view of Kuhn's analysis' – p. 231.)

Haggett and Chorley were attracted by the concept of scientific revolutions. A similar concept was argued by Burton (1963), whose paper appeared almost contemporaneously with Kuhn's book, and so contained no reference to it. Burton introduced the term 'quantitative and theoretical revolution', and argued that:

> *An intellectual revolution* is over when accepted ideas have been overthrown or have been modified to include new ideas. *An intellectual revolution* is over when the revolutionary ideas themselves become part of the conventional wisdom. When Ackerman, Hartshorne and Spate are in substantial agreement about something, then we are talking about the conventional wisdom. Hence, my belief that the quantitative revolution is over and has been for some time. Further evidence may be found in the rate at which schools of geography in North America are adding courses in quantitative methods to their requirements for graduate degrees (Burton, 1963, p. 153).

Davies (1972b) also used Kuhnian terminology in the title of his book, *The Conceptual Revolution in Geography*. He identified a change from contemplation of the unique to adoption of 'the more rational scientific methodology' (p. 9) as a revolution: none of the contributions reprinted in his collection refer to Kuhn, but their context is clearly influenced by Kuhnian ideas. A year later, Harvey (1973) used the term, like Haggett and Chorley, in a normative sense in his search for a new world view.

The authors cited so far used the paradigm concept at either the macro-scale of a world view or the meso-scale of a disciplinary matrix (see above, p. 16, and Mair, 1986). Others employed it as a general descriptive tool (e.g. Buttimer, 1978b, 1981; Holt-Jensen, 1981, 1988 – see also Asheim, 1990; it has the largest number of entries in the index to James and Martin, 1981), and as a framework for summarizing sub-disciplinary changes (Herbert and Johnston, 1978). Kuhn's micro-scale concept of a paradigm as an exemplar has also been employed: Taylor (1976), for example, identified seven separate revolutions during the preceding decades and Webber (1977) proposed an entropy-based paradigm (see p. 136). Harvey and Holly (1981, p. 31) identified five paradigms within geography during the last century, associating each with an individual scholar: 'we can tentatively assign paradigmatic status to . . . Ratzel with the paradigm of determinism, Vidal with that of possibilism, Sauer with the landscape paradigm, Hartshorne with the chronological paradigm and Schaefer with the spatial organization paradigm'. They were focusing on 'schools of thought' associated with particular scholars, some of which existed concurrently rather than consecutively. They claimed that a single paradigm dominated during the 1960s. (Whether it was stimulated by Schaefer is open to question, despite the views of Bunge, 1979, and others; Cox, 1995, p. 306, claimed that 'Schaefer arguably was rescued from intellectual oblivion only in order to provide some philosophical justification for what was happening in human geography at that time', though Getis, 1993, reported that Schaefer's paper was widely read at the University of Washington in the late 1950s: see also Berry, 1993.) The 1970s were characterized by a 'diversity of viewpoints' (Harvey and Holly, 1981, p. 37), within which spatial organization remained important. Zelinsky (1978, p. 8) called that decade one 'of confused calm, or rather of pluralistic stalemate, as geographers explore a multiplicity of philosophical avenues and research strategies . . . without that single firm conviction as to destination that guided most of us in the past'. The discipline was technically more capable, substantively more catholic, philosophically more mature, internationally more merged, socially more relevant, and academically more linked to other disciplines, he claimed, and this would probably lead to a continued plurality of approaches: 'I happen to believe that, more than anything else, this philosophical coming of age, this rising above the superficiality and tunnel vision that blemished so much geographical work earlier in this century, justifies' (p. 10) the title of his edited collection – 'Human geography: coming of age'.

Although some were uncertain about the relevance of Kuhn's concepts to changes in the 1970s and 1980s, in comparison with the 1950s and 1960s, others were less equivocal. In his introduction to a collection of essays on *The Nature of Change in Geographical Ideas*, Berry (1978a, pp. vii, ix) claimed that: 'The changes in geographical ideas that we have discussed are distinctly Kuhnian . . . What, then, is progress in geography? The perspective provided by the essays in this book is distinctly Kuhnian.' The other contributors' essays provide little supporting evidence for this claim, however, and Berry's own contribution suggested that pluralism and inter-paradigm conflict were much more common than periods of normal science. Writing of geographical theories of social change, he noted (Berry, 1978b, pp. 19–22):

> a diversity arising from the mosaic quality of modern geography ... they have ... moved from one paradigm to another, and in the last decade they have been extremely dynamic. ... With multiple ideas and multiple origins, modern geography could rightly be characterized as a mosaic within a mosaic (Mikesell, 1969).

Human geographers and a critique of Kuhnian applications

Kuhnian concepts and terminology have been widely used by human geographers, therefore, although some treatments of recent disciplinary history entirely ignore this literature (Freeman, 1980a, 1980b) and others make little use of it: Buttimer (1993, p. 70) has a single, brief reference to Kuhn, for example, and Livingstone (1992), after noting that (p. 14) 'In the wake of Kuhn's treatise, a batch of historians working in various disciplines set out on a paradigm hunt, looking for paradigms, paradigm-shifts, and what not. Geographers were no exception', concluded that some, like Haggett and Chorley, Berry, and Harvey, used the paradigm as 'little more than a flag for rallying the troops' (p. 24) to a new cause. Gould (1994, p. 196) argued that 'They challenge, sometimes with deliberate overstatement, using polemic to grab the ear of an established coterie, and rhetoric to persuade it of its folly', and Berry (1993, p. 438) claimed that this was necessary: in the context of Hartshorne's responses to Schaefer (p. 57) those promoting the new approach in the late 1950s were engaged in a debate with the 'discipline's luminaries':

> Mainline geographers were suspicious, threatened, antagonistic; and we reciprocated. We felt we had to fight and fight we did, earning reputations for brashness and abrasiveness. So be it: if we had not been aggressive, geography would have rolled over us. Instead, we tried to roll geography.

In that fight they sought support from the emerging field of regional science (see p. 83), and as a consequence of their victory, according to Morrill (1993, p. 442):

> I believe we saved geography from extinction as a serious university discipline, by attracting and training good students, by writing articles and books that developed theory and method, by gaining a foothold in science at large, and by applying these methods and theories to contemporary social problems.

Nevertheless, he still finds geography's position 'weak, fragile and almost invisible' (p. 443): 'we survived but did not succeed in the goal of propelling geography into the mainstream' because 'much of geographic work ... was not good enough' and some of the best spatial scientists 'defected'.

Livingstone's attitude to discussions of Kuhn and the revolutionary fervour of some spatial scientists was that (1992, p. 24) 'The details of these (and other similar cases) need not be reviewed here. Suffice to say that their revolutionary gung ho spirit of triumphalism was scarcely what Kuhn had in mind as he portrayed the mega-level Gestalt-shifts in the history of science.' Many of the presentations were superficial, however (see Graves, 1981), both cavalier and uncritical in basing their descriptions on Kuhnian foundations. (Mair, 1986, includes previous editions of this book in that category.) Livingstone (1992, p. 25) found that 'it became plain that much of this writing amounted to misdirected effort', but agreed with Mair that: 'even among [the] ... critics the flavour of Kuhn's work still lingers. The Kuhnian ghost, it seems, is proving rather hard to exorcize from the history of geography.' Nevertheless, he welcomed the sociological approach to disciplinary history which the adaptation of Kuhn's ideas heralded; it countered the belief in 'conceptual cumulation, disciplinary progress, and internal chronology', and encouraged contextual readings in its stead.

Other geographers have been even less convinced of the value of a Kuhnian interpretation, with two claiming that it has 'distorted even perverted the development of geography' (Haines-Young and Petch, 1978, p. 1). The concepts of revolutions and normal science have been criticized as providing poor descriptions of geography in recent decades (e.g. Holt-Jensen, 1988). Individual geographers may have experienced personal revolutions and rapid shifts from one paradigm to another (at the meso- if not the macro-scale) (Harvey, 1973, indicates this for himself: see the essays in Billinge, Gregory and Martin, 1984; the editors – p. 11 – quote Cox, Gould, Olsson, Scott and D. M. Smith as additional examples). Sheppard (1995, p. 287), employing citation data, contrasts a drastic shift in the discipline's 'master weavers' (the term is drawn from Bodman's, 1991, 1992, depiction of the most-cited human geographers) from spatial science to social theory during the 1980s, and also argues that:

> It is of interest to note that 8 of the 12 social theorists [all are classified as either spatial scientists or social theorists], as well as a large number of other influential individuals in the social-theory group, began their careers within the research traditions of spatial analysis.

Among them, he identified three cohorts.

1 those who wrote the classic spatial analysis papers of the 1960s (he cites Bunge, Cox, Harvey, Johnston, Olsson, Pred, Scott, Soja and Webber);

2 those 'whose early work was solidly within spatial analysis, and recognized as such, but for whom this work did not represent a major period in research careers that shifted rapidly from spatial analysis to Marxism and social theory' (he lists Dear, Massey, Peet, Taylor and Thrift); and

3 those whose careers began in the late 1970s with contributions to spatial analysis when it was already under heavy criticism, 'only to re-identify themselves prominently with the concerns of social theory'.

For Sheppard, therefore, there have been only two major competing world views – the spatial scientists' versus the social theorists': most of the debates discussed in this book can presumably either be encapsulated within that competition or involve controversy within a world view (over disciplinary matrices and/or exemplars, but not basic orientation, which Barnes – 1996 – portrays as for or against the Enlightenment project with its belief in progress and universal truths).

To claim that such individual experiences can be amalgamated into disciplinary revolutions strikes some observers as both inapt and inconsistent with the evidence (Bird, 1977, p. 105): 'Perhaps so many revolutions in so short a time indicate in themselves either a continuously rolling programme, or something basically wrong with the overturning metaphor.' Bird (1978, p. 134) suggested that mono-paradigm dominance of a discipline is inconsistent with 'the fact that society itself is organized around more than one major principle'. The focus of Bird's criticism is not clear, however: it is much more convincing on the world-view scale than at that of the disciplinary matrix, for example, and even less so on the scale of the exemplar.

Stoddart's criticism was even more pointed; he initially saw some value in the paradigm model (Stoddart, 1967b) but later argued (Stoddart, 1977, p. 1) that:

the concept sheds no light on the processes of scientific change, and readily becomes caricature. I suggest that as more is understood of the complexities of change in geography over the last hundred years, and especially of the subtle interrelationships of geographers themselves, the less appropriate the concept of the paradigm becomes.

His analysis showed both the absence of consensus (normal science) and the slow pace of change (which is more readily represented in Lakatos's schema). In his view the paradigm concept became part of the 'boosterism' image with which geographers conducted debates (Stoddart, 1981a):

the paradigm terminology has been used to illuminate either the establishment of views of which a commentator approved, or to advocate the rejection of those he did not (p. 72) . . . the concept of revolution bolsters the heroic self-image of those who see themselves as innovators and who use the term paradigm in a polemical manner

... those who propound the Kuhnian interpretation have done so in ways which tend to make it self-fulfilling (p. 78).

He argued (Stoddart, 1977, p. 2) 'There is scope for sociological enquiry into the extent to which the concept has been used in recent years as a slogan in interactions between different age groups, schools of thought, and centres of learning.' Billinge, Gregory and Martin (1984, p. 6) claimed that Chorley and Haggett's initial use of Kuhn's terminology was 'In some measure ... only gestural' because although they clearly distinguished between normal science and extraordinary research: 'the "anomalies" within the traditional paradigm which were central to Kuhn's thesis were never identified in any detail'. Chorley and Haggett were pressing for a revolution in the nature of geography, but it was not a revolution generated by the failure of the previous paradigm (using failure in a Kuhnian context). Billinge *et al.* pointed out that the term 'paradigm' was being used more for propaganda – thereby gaining prestige by locating their would-be revolution alongside those in physics described by Kuhn (Taylor, 1976) – than for historiography. With regard to the latter, they accepted that the Kuhnian model (as presented in earlier editions of this book) should be rejected, but argued that since Kuhn himself did not expect the model to fit the social sciences, such a conclusion is hardly surprising. Mair (1986, p. 359) took the criticism further, contending that, with the single exception of Billinge, Gregory and Martin:

> geographers have entirely misinterpreted Kuhn's contributions. They have been wrong on Kuhn in the simple sense that the skeletal Kuhnian model is misrepresented as Kuhn's major contribution. More fundamentally, however, they have been wrong on Kuhn in misconceiving his entire project.

For him, the main values of Kuhn's work are:

- the concept of the exemplar as an analogy to be used in the creative process;
- the notion of incommensurability in the comparison on competing paradigms; and
- the clear need to study the sociology of scientific communities.

Despite such criticisms, relatively little had been done until recently to suggest an alternative sociology, either one based in another model of the history of science or one developed specifically to represent the situation in human geography. Wheeler (1982) found Lakatos's work attractive, arguing that several separate programme cores could be identified in contemporary human geography (he cites areal differentiation, spatial science, cognitive-behavioural approaches and marxist structuralism), in each of which the operation of positive heuristics (see p. 18) can be identified. These research programmes are in continuing competition, whose nature will change over time (Wheeler, 1982, p. 4):

Given the deficiencies in Kuhn's scheme and the rather ill-defined nature of geography, it seems unlikely that the future of the discipline will be characterized by sequences of revolutionary change interspersed with efficient problem solving. Moreover expectations of revolutionary progress appear to be unjustified. Instead it seems that a variety of approaches, which will undoubtedly wax and wane in popularity, will continue to be employed.

Mohan (1994), drawing on the notion of a product cycle that has informed much work in economic geography, suggests that there has been a rapid turnover of theories within geography in recent years, as a consequence of pressures within the academic profession: he identified 'four "big" approaches over the past decade hailed as capable of explaining social reality' alone (p. 387: the four were critical realism, structuration, postmodernism, and postcolonialism). He associates this with the growing 'commodification of knowledge' and the competition for status within academia, so that:

> these ideas were not paradigmatic and their diffusion into geography uneven. Likewise some of these intellectuals [those promoting the 'big approaches'] have persisted with their theoretical frameworks and have not continually experimented with new ones ... [nevertheless] Many academics fight for currency and exploit 'new markets' in knowledge. This increased competition has, in the spirit of flexible specialisation, resulted in the 'niching' of academic thought despite the recent emphasis on trans-disciplinary pursuits.

Some of the forces impelling these trends are general within academia, but one at least is specific to geographers (p. 389):

> As a trans-disciplinary subject geography has often lacked kudos in wider social theory. The result is that geography has tended to borrow and incorporate any social theory that appears spatial or contains spatial metaphors. The increased commodification of knowledge has served to heighten this tendency. For example, we have had Giddens' *locales*, Foucault's *disciplinary spaces* and Mohanty's *contested cartographies*. It seems that few have time to actually apply any of these potentially useful ideas before the next theoretical innovation superseded them.

Disciplinary matrices and exemplars come and go at an alarming rate according to this view, with geography equated to a supermarket in its concern for 'turnover time' and short shelf lives as its practitioners seek status through the novelty of their approaches.

Geography and its environment

Some geographers have developed the thesis that the major influence on the discipline's content and approaches comes from its environment, particularly its economic, social and political milieux (and also from other subjects: in general, it is believed that geography is more likely to change because of developments in other disciplines than vice versa). Stoddart (1981b, p. 1), for example, introduced a book of essays as demonstrating 'that both the ideas and the structure of the subject have developed in response to complex social, economic, ideological and intellectual stimuli'. Evidence for this was presented, as a second theme of the essays, in the reciprocal relationship between geographers and their milieux: 'throughout its recent history geographers have been not only concerned with narrowly academic issues, but have also been deeply involved with matters of social concern'.

Many geographers who employ a contextual approach stress that any discipline's contents must be linked to its milieux, with disciplinary changes (revolutionary or not) associated with significant events there. Thus Berdoulay (1981, p. 10) wrote of the role of the *Zeitgeist* (or spirit of the age) influencing what geographers do. Grano (1981) took this much further (Figure 10.1); geographers are a sub-set within the community of scientists, which is itself a sub-set of wider society, whose culture includes a scientific component that strongly influences the content of geography. The community of geographers is an 'institutionalizing social group' (p. 26), providing the context within which individual geographers are socialized and defining their disciplinary goals within the constraining and enabling characteristics of the external structures. A major goal when geography was established as an academic discipline was to create an identity, to 'establish an object of study that could be regarded as geography's own and that differed from that of other disciplines' (p. 30); this, he claimed, was a 'passive education

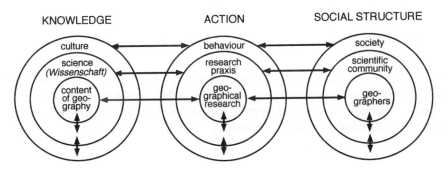

Fig 10.1 Geography's context (source: Grano, 1981, p. 19).

role' (p. 32) disseminating knowledge of society–environment inter-relationships, but it was replaced in the 1970s when: 'geography began to contribute actively to a transformed and replanned world. Applied geography was created and geography became a *profession* outside the small world of the university' (p. 32). (See also Goodson, 1981.)

Geographers made and continue to remake geography (Grano, 1981, p. 30): 'It was the external goals of society that brought the establishment of geography as an academic discipline. This took place without any noticeable contribution from any other scientists.' (See, however, Stoddart, 1986.) The initial period of institutionalization involved geography operating largely as a pedagogic subject, meeting the needs for training teachers whose activities would promote the interests of the expanding 'nation-states' (Capel, 1981; Taylor, 1985c). Capel (1981, p. 36) referred to the creation of geography as a discipline resulting from:

> the presence of geography in primary and secondary education at the time when the European countries began the rapid process of diffu-sion of elementary education; the necessity to train geography teach-ers for primary and middle schools was the essential factor which led to the institutionalization of geography in the university and the appearance of the scientific community of geographers (p. 36).

(See also Freeman 1961, 1980a, and, for a more critical account, Rieser, 1973.) As the context changed, however, so did geography for (Capel, 1981, p. 66):

> The established community employs strategies tending to reproduce and amplify itself. Never will it opt for self-liquidation; the commu-nity will defend its survival, even if other communities of scientists investigate similar problems with like methods, or if the logical inco-herence of the conceptions that they defend is revealed ... Everything will be sacrificed for the reproduction and growth of the community, including the coherence of the very conception of the discipline: dif-ferent conceptions can defend themselves in distinct moments or even simultaneously, without putting into doubt the continuity of the sci-ence practised.

The promotion of national interests, especially national commercial inter-ests, by nineteenth-century geographical societies (see the essays in Bell, Butlin and Heffernan, 1995) provided the context for the creation of geography as an academic discipline, and subsequent changes in economic, social and polit-ical structures and needs generated new demands, to which geographers responded. (See, for example, Taylor's, 1993, evaluation of different geo-graphical perspectives on 'the global' during the twentieth century.)

Scott's (1982) answer to the question 'why do geographers, regional sci-entists, urban economists and others study the spatial patterning of social events?' (p. 141) provides an example of disciplinary reconstruction in the context of changed circumstances. He concentrated on the ways in which late capitalism is organized, not through market relations but rather by bureaucratic intervention by an all-embracing state. (Note that this was written in 1982 in North America; a decade later, the context was very different.) Because 'the geography of late capitalist society is shot through with problems and predicaments' (p. 145: Scott uses geography there in the vernacular sense – Johnston, 1986b – rather than the professional) state action is required in 'more subtle forms of social, cultural, and psychological management' (p. 146) on which science thrives. By participating in those managerial tasks, however, scientists contribute to the creation of countervailing forces, with which they may also become involved:

> the endemic crisis of economic production and growth in late capitalist society creates the need for specific problematics and policy discourses out of which technical control may be accomplished. But technical control creates an advanced set of social conditions in which a countervailing set of human predicaments makes its appearance – alienation, the destruction of affective human relations, the repoliticization of human and regional planning, and so on (p. 152).

Hence the radical and humanistic responses to spatial science.

Geographers make and remake geography in context. There is no necessity for geography, no 'specific necessities in scientific knowledge' (Grano, 1981, p. 65). Rather, according to Taylor (1985c, p. 93):

> Geography is a social institution. Like all such institutions its value to society varies over time and place. The creation of any social institution is a result of a group of people who identify a particular need and are able to find the resources to meet that need. As needs change the institution has to adopt to survive.

It may fail; the forms of scientific discourse created and practised may not command sponsorship and resources. Scott (1982, p. 151) argued strongly that success requires identification of clear social needs: 'only discourses that are posited upon existing problems of social life and practice, and upon existing political interests, stand any likelihood of commanding a significant consensus of scholars and scientists'.

There is nothing deterministic in these arguments, for they depend on geographers identifying what is and is not sustainable in a particular context and successfully promoting themselves as able to meet the perceived needs. How they do this will reflect what Berdoulay (1981) calls their individual 'circles of affinity', social networks extending outside their disciplinary base. To appreciate those one must appreciate biographies, as Buttimer (1981) argued and two volumes of autobiographical recollections illustrate (Buttimer, 1983; Billinge, Gregory and Martin, 1984). Buttimer (1983, p. 3) defends such an approach because 'each person's life echoes the drama of his or her times and milieu; in all, to varying degrees, the propensity to submit or rebel. Through our own biographies we reach toward understanding, being and becoming.' Autobiography provides what she terms 'choreographic awareness', the 'moral, esthetic, and emotional commitments which are related to lived experience and which underpin a scholar's eventual choice of epistemological presentation and style of practice' (p. 12), and can indicate aspects of disciplinary history that are absent from the written record, the influence of 'pioneers of geographic thought whose inspiration flowed through their teaching and field experiences, through their counselling and listening' (Buttimer, 1981, p. 88). This theme was taken up explicitly by Pred (1979, 1984a), using the language of time geography (p. 158). A particular feature of a scientist's autobiography, however, is that the encounters which influence career development and change need not be inter-personal. Bird (1975, 1985) used Popper's concept of World Three (the world of recorded knowledge) to show how geographers can be strongly influenced by reading both ancient and contemporary thoughts, so that our milieux are not as bounded (in time and space) as are those of people who rely much more on inter-personal transmission of information (see Johnston, 1984e) – though, of course, we are reliant on what materials are, or can be, made available to us locally, which is increasing very rapidly through developments in information technology (Adams, 1995). (Biographies offer an alternative source, but there are few on geographers who have contributed significantly to the contemporary period – Paterson, 1985; Johnston, 1986g – and the only available biographical dictionaries are slight in their treatment of geographers' intellectual contributions relative to their career paths: Larkin and Peters, 1993.)

More recently, geographers have extended the approach to contextual influences on their discipline's content and direction by drawing on the ideas of sociologists of science regarding 'situated, or local, knowledge'. In presenting an 'alternative history' of economic geography, for example, Barnes (1996, p. 105) argued that understanding geographers' practices requires knowing 'something about the local context and not overarching principles of rationality'. Thus (p. 124)

> Economic geographers will continue to do economic geography in the way that they have always done, that is, by creatively drawing upon existing ideas and beliefs and responding – sometimes with great originality and imagination, other times with less of each – to the concerns of their own local context. . . . this is all there is to practice. There is no foolproof method of directing inquiry.

The nature of the 'local' is not addressed in depth, however, so that the interaction of 'place as context' with 'World Three as context' remains unexplored. (For Barnes himself, for example, the key local context was the department of geography at the University of Minnesota in the late 1970s when he was a graduate student there, working with Fred Lukermann and Eric Sheppard and alongside Michael Curry. A decade later, he was working in Vancouver and still writing with Curry and Sheppard, who were in Los Angeles and Minneapolis respectively.)

The case for understanding the changing nature of geography contextually closely parallels that promoted by realists and structurationists for appreciating all human activity; the operation of human agency must be analysed within the constraining and enabling conditions provided by its environment. Thus Johnston (1983d, p. 4) enlisted structuration theory as a framework for analysing the changes in human geography that are the concern of this book, arguing that 'the content of a discipline at any one time and place reflects the response of the individuals involved to external circumstances and influences, within the context of their intellectual socialization'. Mikesell (1981) took a similar position, though without the language of structuration, describing the history of geography in the United States as a sequence of 'temporary enthusiasms or episodes' (p. 9), which he claimed were responses to contemporary stimuli – hence the popularity of urban studies in recent decades, the decline in foreign-area studies after the Vietnam withdrawal, and the impact of national concern about the physical environment in the 1970s. He also suggested that physical milieux influenced the culture of individual geographers:

> The first generation of American geographers grew up in a country that was still strongly influenced by the *mores* of small towns. Most of the students now attracted to geography are products not only of an urban but increasingly of a suburban environment ... the geographical profession has changed and is changing as a consequence of the suburban, middle-class origin of most of its current members (p. 12).

Porter (1978) advanced similar arguments, identifying two types of American geography: a midwest version that 'was a characteristically optimistic, action-oriented, "can-do" kind' (p. 17); and a Californian geography, based on Sauer, 'historical, uncompromisingly academic, speculative, suspicious of government, keenly interested in cultures other than the dominant Anglo culture of the United States' (p. 18).

This environmental influence thesis should not become environmental determinism, however: Mikesell noted that urban geography flowered in the 1950s and 1960s not only in Chicago but also in Iowa. But it stresses the importance of studying human geography and human geographers in context and, as Capel (1981) noted, the disciplinary community (or major elements within it) will seek to maintain its identity by bending to perceived shifts in their milieux (as clearly demonstrated for anthropologists by Patterson, 1986). Mikesell (1981, p. 13; see also Johnston, 1996d) also argued – following Harvey (1973) and others – that reactions to environmental shifts involve individual scholars seeking not only to defend and promote their chosen discipline but also to defend and promote their own status and careers within it:

> innovation will continue to be regarded as a virtue. Much of the development that has already taken place in American geography is a consequence of the attempt of individual scholars to stake a claim for themselves, to be or at least to seem to be different from their rivals. The fact that most academics see virtue in innovation means that there is reward for innovation.

Mikesell did not discuss the scale of the innovation, however; does it have to involve developments within the positive heuristic of a research programme, to use Lakatos's terms, or does it require the launch of a new research programme? Can one bring potentially greater rewards, and disasters, than the other?

One little explored aspect of this sociological approach is that it focuses on the winners rather than the losers, on those whose ideas have a major impact on at least parts of their discipline. Many ideas are not taken up at the time they are published, but are later revived, whereas others are never followed through. Why some individuals and their ideas are much more influential than others is an intriguing aspect of a discipline's history, as illustrated by the belated recognition of Hagerstrand's seminal work (Duncan, 1974b), the slight impact of Wreford Watson's ideas regarding the social geography of cities, which predated the studies referred to above (p. 88) by several years but were almost ignored (Robinson, 1991; Johnston, 1993a), and the focus on only one aspect of Gottmann's voluminous writings (Johnston, 1996b). In general, it seems that the most influential are those who are well-networked with other influential scholars, especially those who have been trained in large graduate schools with a clear agenda for the discipline. Morrill (1993, p. 442) recognized this with regard to the developments he was involved in during the 1950s and 1960s, at both the University of Washington (see p. 66, and also Berry, 1993, on the depth of the resources available there, outwith as well as inside the geography department) and Northwestern University:

> This happened when and where it did because at a few universities there was a fortunate combination of an emphasis on interdisciplinary study, existing development of theory in these related fields, notably regional economics and sociology, some students ready for change, and some faculty ready to encourage them.

Following this, leading proponents of the new approach were able to promote the discipline more widely within the social sciences, thus achieving what King (1993, p. 545) refers to as 'the professionalization of geography as a social science'.

Context is not deterministic, of course: it provides constraints and opportunities, including the opportunities to question and to seek new directions. (Indeed, in academic life perhaps more than in any other, because of the emphasis on the development of a critical intellect and ordered scepticism, the opportunities to question rather than conform are very likely to be taken up.) Thus Hanson's (1993) reflections on graduate school at Northwestern University, whilst stressing the positive aspects of the education provided (such as the concern for theory and explanation, the importance of space, and the emphasis on standards of evidence), also noted that (p. 553):

> Coexisting with these considerable strengths, counterbalancing and – at the time – sometimes overwhelming them, were blind spots and inflexibilities bred in revolutionary fervor and in earlier bruising conflicts with the reigning paradigm. Our mentors exuded a certain intolerance, a certain hubris, a certain arrogance, a certain *certainty* that there was only one way to the Truth, and they knew it. There was a caricaturing and a trivializing of the idiographic/regional approach, much as today critical social theorists caricature and demean spatial analysis and scientific geography.

Hanson and some of her contemporaries were involved in the emergence of behavioural geography as a counter to some of the 'certainties' being advanced by the 'space cadets' (her term for the Seattle-trained spatial scientists then on the staff at Northwestern) (p. 534): 'By dethroning economic man and replacing him with a variety of decision makers, behavioral geography ignited the first glimmer of the notion that each viewpoint is partial, incomplete and dependent on the subject's [i.e. the decision-maker's] location.' She was led, by her questioning of the assumptions which underpinned the then-popular theories, to the realization that (pp. 555–6): 'people's decision making and behavior are not immune to context but are conditioned, *inter alia*, by time, place, class and gender', an argument which applies as much to geography itself as to that which it studies.

The current situation

The impact of external circumstances on geographical practice has been illustrated several times in this book. It has been very noticeable since the late 1970s as a consequence of economic recessions, cut-backs in higher education, attacks on social-science research (in 1996 the US House of Representatives voted to end funding for the social-science divisions of the National Science Foundation) and the policy prescriptions of the 'New Right' designed to achieve recovery from the recession by sustaining what has become widely known as 'flexible accumulation' (Hudson, 1988) through a 'free economy and a strong state' (Gamble, 1988; Johnston and Pattie, 1990). Many have reacted to this by promoting human geography as an 'applied discipline', offering relevant skills for the attack on contemporary problems. In the December 1981 issue of the Association of American Geographers' (AAG) *Newsletter* an article entitled 'A survival package for geography and other endangered disciplines' indicated that (Kish and Ward, 1981, p. 8): 'On 19 June, 1981, the Board of Regents of the University of Michigan decided, by unanimous vote, to terminate the Department of Geography at the end of the 1981–82 academic year.' Their experience of this decision led Kish and Ward to suggest how other departments of geography could counter similar attacks. They emphasized the teaching role, recognizing the need to attract students in a competitive market and present 'our wares in a stimulating and excellent way' involving 'the virtues of applied geography. As students become more conscious of careers, there may be a corresponding need to increase a skill-oriented curriculum. This could demonstrate the relevancy of geography and enhance its appeal to students' (p. 14). Kish and Ward – 'To appease some traditionalists' – did not advocate reorganizing the entire corpus of geography, 'but only [an] attempt to broaden the appeal of geography to the student population'. (See also Ford, 1982; Powell, 1981.) Their prescriptions were written at a time when, according to Haigh (1982, p. 185), geographers' peers saw their discipline as 'small, marginal and perhaps immature': the late 1970s had seen a net loss of 32 university departments, including one-sixth of those in 'private and denominationally funded institutions'. Avoiding the discipline's demise was presented as a crucial task (Wilbanks and Libbee, 1979), a problem stimulated by the absence of geography in most high schools there and the very small number of undergraduate students intending to take geography courses. Despite the vitality of geographical research, therefore, 'Geographical education is still regarded as a marginal activity by American schools and colleges' (Haigh, 1991, p. 189).

Promotion of geography as an applied discipline was heavily emphasized in research as well as teaching. The announcement of a new editorial policy for *The Professional Geographer* in the AAG *Newsletter* for April 1992 reported that (p. 1): '[The new editor] hopes to emphasize work in applied geography and to include information on corporation activities, state and local government projects, Federal government activities, activities by United Nations and other international agencies, and research projects.' The need for more applied work has been widely accepted among American human geographers as a necessary means for the reproduction and even growth of their discipline. Mikesell (1981, p. 14) suggested that:

geographers fortunate enough to have secure teaching positions will worry about what they should do. The best response to this concern could be a decision *to* do what they have been doing, but with a keener appreciation of context and a greater willingness to be influenced by environment.

The response has taken a variety of forms, with many arguing for a greater concentration on cartographic, remote sensing and geographical information systems skills.

The changing context is illustrated by a group of papers published by the AAG in 1995. In 1993, the National Academy of Sciences/National Research Council undertook its first major review of geography for nearly three decades (see p. 85). It established a committee of 16 charged with 'Rediscovering geography: new relevance for the new century' and given five objectives (Wilbanks, 1995):

1 identify critical issues and constraints for the discipline of geography;
2 clarify priorities for teaching and research;
3 link developments in geography as a science with national needs for geography education;
4 increase the appreciation of geography within the scientific community; and
5 communicate with the international scientific community about future directions of the discipline in the United States.

Within this context, a report was commissioned from the AAG on the supply and demand for geographers, which was produced by its Employment Forecasting Committee. Introducing papers based on that report, Wilbanks (1995, p. 316) noted that:

> many geography programs in universities across the United States are already experiencing unprecedented growth in student demand. . . . A major concern for geography should be that, because of supply limitations (e.g. a lack of additional faculty positions in geography departments) the demand will be artificially and arbitrarily truncated by limitations on course enrollments.

This is described as a 'very welcome but very difficult disciplinary transition from relative penury to relative abundance'.

The committee's report on the supply side of the equation (Gober *et al.*, 1995a) referred to an earlier study (Goodchild and Janelle, 1988) on the internal structure of the discipline which found that (Gober *et al.*, 1995a, p. 317): 'Technical expertise and interest in geographical information systems (GIS) were burgeoning, especially among young geographers, while regionally oriented specialities were shrinking.' (Turner and Varlyguin's, 1995, report on foreign-area dissertations in the same issue of *The Professional Geographer* shows that the latter trend was relative, not absolute.) To identify which skills and interests university departments of geography were focusing on, department chairs were asked about specializations and the occupations students were being prepared for. Of the 212 respondents, 139 identified at least one specialization, and the full list was:

> led by programs in environmental/resource management, techniques (GIS, cartography, and remote sensing), and urban planning. These tracks appear to be designed to prepare students for the occupations in which geographers traditionally have found work rather than to develop their interests in regional geography or the systematic specialties like urban, economic or physical geography that have traditionally formed the core of the academic discipline.

More specifically

> We also asked chairs to indicate the specific occupations for which students were being prepared. The occupations indicated by the highest numbers of departments were (1) GIS/remote sensing specialist, (2) secondary school teacher, (3) cartographer, (4) environmental manager/technician, and (5) urban/regional planner. ... GIS was by far the most popular occupational trajectory (involving almost 11% of students enrolled in programs offering GIS training).

The committee also surveyed the labour market experiences of recent graduates (Gober *et al.*, 1995b, p. 331), finding that: 'Among those who were employed and listed occupations closely related to geography, respondents clustered into five predictable occupations: teacher (15.6%), environmental manager/technician (12.9%), GIS/remote sensing specialist (10.5%), cartographer (8.2%), and planner (6.7%).' They also studied the employers and job seekers using the placement service at the AAG's 1994 annual meeting: 45 per cent of the employers were looking for candidates with technical qualifications, and they comprised 80 per cent of all non-academic employers using the service.

Finally, the committee surveyed the future labour market by asking department chairs the areas of expertise of staff who would be retiring over the next decade, and which they would be recruiting in (Gober *et al.*, 1995c). Technical areas of the discipline predominated in the latter category, with GIS specified in 17 per cent of the cases – nearly 12 times more frequently than it was among the retirements. As the discipline reproduces itself over the next decade, according to these figures, the relative number of technical specialists will increase substantially whereas traditional areas, such as agricultural, historical, political, cultural and economic geography, will decline substantially. The committee's conclusions, based on interviews with the AAG's corporate sponsors, implied a major shift in the nature of the discipline, as taught to undergraduates (p. 346):

> The debate over geography as a broad-based liberal arts discipline or as a technical, semiprofessional field ignores the realities of the

current labor market. Sponsors told us they want employees who can combine technical skills with a broad-based background. Geography's comparative advantage over other social sciences lies in its ability to combine technical skills with a more traditional liberal arts perspective. Successful geography programs will be those that are able to find the appropriate balance of field-based technical skills like GIS, cartography and air-photo interpretation with competence in literacy, numeracy, decision making, problem solving, and critical thinking.

(See also Miyares and McGlade, 1994.)

The external pressures became just as great in the UK during the 1980s, where there has been substantial expression of the need for applied geography and the development of what are known as 'transferable skills' (problem-solving, group-working, IT awareness, etc.) within undergraduate curricula (see Matthews and Livingstone, 1996). Bennett (1982, p. 69), for example, found at the 1982 conference of the Institute of British Geographers, that it was:

> possible to discern a strong and growing set of foci which, if they do not yet demark a new core, at least show an emerging commonality of interest. For this writer these foci were a widespread assertion of 'relevant' research, the reassertion of quantitative and analytical methods, and the rejection of recent anti-empirical movements.

He welcomed these, noting that 'hot [i.e. relevant] issues are not ones which British geographers are particularly noted for tackling *en masse* – and, as a result, the discipline has suffered a lack of public exposure, and a marked inability to influence public and private decisions'. And

> At a time when higher education as a whole is under considerable challenge, and when geography as a discipline may suffer particularly severe pressure in some institutions it is heartening to see the emergence of concern with the hot issues expressed at this meeting (p. 71).

Beaumont (1987, p. 172) echoed this a few years later:

> the issues raised for the next twenty years are a practical and developmental, rather than a research, orientation. The future is unknown, but it could be exciting, if geographers are prepared to become involved (probably with new partners) in doing geography.

Against this, others saw societal changes presenting an alternative challenge: Harvey (1989a, p. 16), for example, sought alliances that would 'mitigate if not challenge the hegemonic dynamic of capitalist accumulation to dominate the historical geography of social life'. His 'applied geography' was neither narrowly technical nor materialist, therefore: like many others who can be classified as social theorists rather than spatial scientists, his goals are emancipation and long-term change.

Richards and Wrigley (1996, p. 41) presented the early 1990s as a period of even greater and faster change, arguing that in contrast earlier periods

> seem in retrospect almost to have been years of relative stability and calm, prefacing the maelstrom of change which swept through the whole British education system in the early to mid-1990s. ... This period has been one of system-wide expansion, of curriculum and quality control through external review ... and of institutional change. ... These have combined to alter, quite fundamentally, the size, nature and structure of British geography.

The assessment of teaching in university departments of geography was set in the context of a national agenda concerned not only with the quality of the education provided (which some believe conflicts with an over-emphasis on research: Gibbs, 1995; Johnston, 1996e) but also with the development of 'transferable skills' rather than subject-based knowledge and a critical intellect (Johnston, 1996b). The assessment of research has produced not only a grading of all departments but also substantial differentials in funding (Johnston, 1993b), while government directives on research funding make it (Richards and Wrigley, 1996, p. 47):

> likely that future research will be constrained to ask specific questions, will be increasingly 'applied', and will be manipulated by government, business and industry for ends which are unlikely to assist the less material aspects of the quality of civilized life; and critical (social or environmental) science will be marginalized.

On the last point, however, it is important to note a considerable 'culture of resistance' among many geographers against the material imperative, as exemplified by the burgeoning work discussed in Chapter 8 which does not fit the materialist agenda.

A generational model?

The importance of context as an influence on the nature of the practice of human geography suggests that the paradigm model might be rephrased as a generational model (Johnston, 1978c, 1979b). Changes in the external environment provide necessary, but not sufficient, stimuli to changes within the discipline which may be interpreted as attempts to develop new research programmes, if not to launch a revolution and create a new normal science.

Associated with the external changes must be a set of conditions within the discipline itself which is sympathetic to the new demands of the milieux. In most cases, these conditions are best met by younger members of the discipline. Stegmuller (1976, p. 148) argues that 'it is mostly young people who bring new paradigms into the world. And it is young people who are most inclined to champion new causes with religious fervour, to thump the propaganda drums.' To win influence, however, especially in times of resource shortage within higher education, younger workers need the patronage of some established members of the discipline (see Chapter 1). According to Lemaine *et al.* (1976, p. 5), for example:

> Mendel's work, and that of his successors, was a response to scientific problems. But the scientific implications of their results were not pursued until there existed a strong group of scientists who, owing to their academic background and their position in the research community, were willing to abandon established conceptions.

Such responses to environmental shifts may involve attempts to create a new paradigm or new research programme, or they may only require new branches of an existing paradigm/programme. Whichever it is, even when establishment support has been obtained, success is more likely when certain criteria are met, including (van den Daele and Weingart, 1976):

1 an autonomous system of evaluation and reputation;
2 an autonomous communication system;
3 acknowledgement of the new ability to solve puzzles within the confines of the disciplinary matrix;
4 a formal organization providing training programmes which allow reproduction and expansion of the new group's membership;
5 an informal structure with leaders; and
6 resources for research.

These rarely create problems if resources are available; Capel, Mikesell and Taylor all indicated that innovation is encouraged if it brings status, charisma and resources to the discipline. During periods of stagnation and retrenchment, however, conditions are less favourable and major shifts are more likely to be achieved by revolutions among existing members of the discipline. The generational model suggests that the latter is rare. With regard to the research record, Law (1976, p. 228) has counselled that:

> it may well be the case that scientists do lay special emphasis on the accounts in scientific papers, but my hunch is that there is immense (and non-trivial) variation between scientists on this count. For some, science is something you do in the laboratory, something you talk about, and something you get excited about. For others, science is what they write and what they read in the journals. I would even hypothesize (in conformity with the invisible college notion) that those who are generally felt to be of higher status locate science less in their journals than in their own and other people's heads.

Many academics are not particularly active as researchers, and some get 'left behind' as changes proceed. The 'normal science' that they continue to teach is probably based on the world view, disciplinary matrix and exemplars into which they were socialized. But their colleagues, socialized later and influenced by subsequent environmental conditions, operate in different ways. The academic career cycle, in combination with a changing milieu, can produce a multi-paradigm teaching, if not research, discipline.

The key elements in the generational model (modified after Johnston, 1978c) are:

1   The external environment is a major influence on a discipline's contents, especially in the social sciences that are closely linked with that environment.
2   At times, the nature of this environment may change significantly, provoking a reaction among a minority of members of the discipline who try to stimulate change in disciplinary practice by its established members and to generate interest in that change among the youngest generation of research workers – the latter is usually much more successful than the former.
3   Together, this grouping presents a new 'school of thought', although in some cases opposing new schools may be stimulated.
4   The new school is coopted into the disciplinary career structure.
5   The publications of the new school come to dominate the disciplinary research output, as the productivity of the earlier generation declines.
6   Students face two or more separate generational schools in a department's teaching syllabus.
7   Over time, members of the new school attain seniority and political influence within the discipline.

Several consequences may follow. If a discipline fails to react to a changing milieu it could stagnate, and so innovators are encouraged. Another is that some potential innovators may be unable to influence their discipline, either because conditions are not conducive or the 'establishment' does not react positively to their suggestions, or they cannot obtain permanent positions within the academic career structure. Duncan (1974b, p. 109) illustrated these 'processes of resistance' using the example of Hagerstrand's (1968) ideas on spatial diffusion which were originally published in 1953; citation analysis shows that widespread recognition was much delayed, compared to the average for all publications in geography (Stoddart, 1967a). This was not, Duncan claims, because of either language or Hagerstrand's relative isolation in Sweden (though see Getis, 1993, p. 519, on the 'excitement' generated by Hagerstrand's visit to Seattle in 1960, which led to Morrill doing his Ph.D. work in Sweden), but the apparent irrelevance of his work to those steeped in another paradigm. Only when spatial science had been established was Hagerstrand's seminal contribution recognized:

Hagerstrand's own attempts to disseminate his work met rejection from adherents to orthodoxy, but later enthusiasm from those pioneering spatial science. The eventual relay of information to this community owed more to dogged personal effort than to the formal communication system of normal science, and general recognition was not achieved until professional allegiances were reorganized (p. 130: see also Johnston, 1993c, 1996b).

Even when a new idea has been recognized, however, several virtually independent groups may be involved in its development (Gatrell, 1982).

Human geography: paradigms or research programmes or . . .?

Several models of scientific progress have been applied to the task of interpreting changes within human geography over the last five decades, therefore. This final discussion promotes the general relevance of Kuhn's concept of a paradigm (at all three scales of definition), without necessarily suggesting either substantial periods of mono-paradigm dominance and normal science for the discipline as a whole or major revolutionary events that involved large numbers of geographers switching from one paradigm to another. (In some ways, paradigms as disciplinary matrices can be equated with research programmes in Lakatos's terminology; the Kuhnian term is preferred here.) In seeking to understand the relative popularity of various paradigms at different times, however, it is necessary to employ the contextual approach largely ignored by Kuhn's disciples.

The 1950s and 1960s

Regionalism, with its empiricist and implicit exceptionalist philosophy, dominated the discipline in the 1940s, though in several versions: in the United States, for example, areal differentiation was stressed by geographers emanating from the midwest, notably Chicago and Madison, whereas those from Berkeley focused on the evolving cultural landscape, with a third distinct school based in Clark University (Bushong, 1981; Prunty, 1979). The regional theme dominated in Britain, too, though with a greater emphasis on physical geography and less evidence of distinct 'schools' (Freeman, 1979; Johnston and Gregory, 1984). Systematic studies increasingly replaced regional descriptions through the 1950s; their stated (though often unrealized) aims were to increase the content of regional descriptions, advancing the understanding of particular places through knowledge of the general processes which interact to produce unique characteristics.

Regionalism was a disciplinary matrix with several different exemplars, therefore, but regional synthesis as their disciplinary *raison d'être* was the focus of most geographers' orientation to their work. In the 1950s and 1960s, they experienced what Entrikin (1981, p. 1) terms 'transition between reigning orthodoxies . . . in which the spatial theme superseded the regional theme'. This was marked by the growth of systematic studies, the distancing of many geographers from the core belief in regional synthesis, and an increasing emphasis on finding laws of spatial organization, involving distance as a basic influence on human behaviour. As Guelke (1977a, 1977b, 1978) and Entrikin (1981) have indicated, the distancing was not rapid – in part no doubt because of the political need to sustain the unity and identity of the discipline – and some of those promoting the new paradigm only argued that quantitative analysis and spatial science offered better and more rigorous procedures for identifying and understanding regions (as in Berry, 1964b). There was a methodological shift (Chisholm, 1975a), which involved new exemplars but not a new disciplinary matrix, let alone a new world view.

The methodological shift alone was increasingly portrayed as insufficient, however, and a philosophical shift slowly emerged (elsewhere termed a 'quiet revolution': Johnston, 1978c, 1979b, 1981b). Regional synthesis as the disciplinary core was unattractive to many younger geographers (e.g. Gould, 1979), for whom contact with other social scientists introduced the excitements of systematic specialisms practised according to the positivist model of science, which implied the search for laws and applicable research findings: the links with the regional core were severed and a new disciplinary matrix of spatial science established. (Batty, 1989, argues that by the time its models were well developed and applicable the demand for them had gone: 'It is an irony of history that such good models finally exist which could well have produced excellent advice in their day had they been available. But that day has passed' – p. 156.) The term 'region' took on a very different meaning (Johnston, 1984d). The shared values of the new disciplinary matrix varied from those of regionalism, and to the extent that the adherents of spatial science, most of them from the new generation of geographers who were part of the educational boom of those decades, came to dominate the discipline (see Mikesell, 1984, on whether they did) so a revolution can be said to have occurred. Whether it was a revolution in Kuhnian terms is doubtful, however, since, as noted above (p. 361), the shift hardly fits into the 'response to anomalies' component of Kuhn's model. That would involve not only a shift in disciplinary matrix but also a shift in world view, in the conception of the nature of science but, as Hartshorne makes clear (p. 59), traditional regional geography as he conceived it did not deny the possible relevance of generalizations about processes for the understanding of places. Bird (1989) identified a 'one and only' revolution, however, dating it to June 1966 with the publication of Bunge's (1966) note which criticized those who argued that the uniqueness of location meant that positivism was inapplicable in human geography.

The new disciplinary matrix of spatial science was firmly established in Anglo-American human geography by the end of the 1960s, and has been sustained since. Over nearly 30 years it has seen many shifts in exemplars, a lot of them linked to methodological – especially technical – developments in data collection, collation, analysis and display. (Compare, for example, Haggett, 1965c with Haggett, Cliff and Frey, 1977, for shifts in the general orientation; Cliff and Ord, 1973 with Cliff and Ord, 1981, for a single methodological issue; and Hagerstrand, 1968, with Cliff *et al.,* 1987, for a particular substantive topic.) Such paradigm shifts at the level of exemplar are closer to Kuhn's original presentation, with new ways of doing research accepted as superior to those previously used. They brought greater substantive success, too: Haggett (1978, p. 161), writing on 'The spatial economy', claimed that it:

> is more carefully defined than before, we know a little more about its organization, the ways it responds to shocks, and the way some regional sections are tied to others. There now exist theoretical bridges, albeit incomplete and shaky, which span from pure, spaceless economics through to a more spatially disaggregated reality.

In addition to the shifts among methodological exemplars there were also changes within the spatial science disciplinary matrix which had a wider import. The first was the reorientation away from normative modelling, which involved testing observed spatial organization against *a priori* models, and towards behavioural studies; as Cox (1981) and Hanson (1993) indicate, the disciplinary matrix was not queried but there were major shifts in emphasis and style. The second was the advocacy for 'welfare geography' (p. 337), described by Eyles and Smith (1978) as a response to social conditions and a desire to make contemporary human geography more relevant to them.

The 1950s and 1960s saw the establishment of a new disciplinary matrix for human geographers, therefore, with twin concentrations on the spatial organization of society and human spatial behaviour. During those decades, and much more so since, particular exemplars have waxed and waned. They have introduced new methodological procedures and new substantive foci, in part as reactions to anomalies thrown up within the discipline (the failure of certain normative models, for example, and the shortcomings of certain quantitative procedures) and in part as responses to trends in society, as illustrated by the many branches and sub-branches of the disciplinary matrix, reflecting the substantive interests of geographers (urban social, agricultural, etc.) and the technical arsenal that they deploy. A review of American geography published at the end of the 1980s is almost entirely structured within those systematic specialisms (Gaile and Willmott, 1989).

The 1970s and 1980s

Although the spatial-science disciplinary matrix expanded rapidly in the 1960s and early 1970s, it never entirely dominated human geography then, although it was relatively unchallenged until the mid-1970s (Taylor, 1976). Responses to the output from that disciplinary matrix and to events and issues elsewhere in society then stimulated two major challenges, contesting not just the exemplars, nor even the disciplinary matrix, but the world view implicit in the disciplinary matrix of spatial science.

The first of these challenges – termed humanistic here – was a response both to the nature of spatial science and to the ideology of society that it reflected. Spatial science is technocratic in its orientation and application; it tends to reduce people to terms in equations, and thereby ignores their individuality and freedom of action (Ley, 1981), and to ignore the immense variety among places in favour of a universalistic view of how people think and act. As Barnes (1996, p. 6ff.) describes it, spatial scientists adhered (implicitly in most cases) to the 'Enlightenment project' launched in seventeenth-century Europe, which emphasized:

1 the *notion of progress* in the development of scientific understanding, through the application of the power of reason and the rationality of scientific methods;
2 a belief in *autonomous, sovereign, self-consciously directed individuals*, so that all are in essence the same, able to make the same moral and other judgements;
3 a belief that *'the world had order and humans could find it'* through application of their rational methods, notably those based on logic and mathematics; and
4 an acceptance that there are *universal truths*, which hold at all times and in all places, and whose discovery is the goal of all scientific endeavour.

These tenets are rejected by humanistic geographers, whose reading of the other social sciences and philosophy led them to argue for a focus on subjectivity, which was clearly incommensurable with positivistic spatial science, so that the choice of which to practise was ideological (Johnston, 1986a). According to Barnes (pp. 8–10), they had their own core beliefs (shared with other counter-Enlightenment scholars) which:

1 rejected 'both the epic of progress and the power of rationality and reason' (p. 8);
2 contended that individuals are 'shaped from the outside rather than from the inside' (p. 9; i.e. by their context);
3 rejected any notions of a 'monolithic order'; and
4 argued that science advances not through the use of universal constructs but rather through 'a set of local social practices of a given time and place'.

Humanistic geography was introduced as an alternative science of geography, therefore, and not as a reorientation of the existing way, though some proponents claimed roots in earlier geographical practices such as those of the French geographer Vidal de la Blache (Buttimer, 1978a). It had its own disciplinary matrix, and its own variety of exemplars – in terms of both their philosophy and their subject matter, as set out in Chapter 6. Its introduction did not generate a revolution, however, for there was no major switch in the contents of the discipline as a whole. Rather, it offered an alternative conception of human geography and competed (ultimately unsuccessfully) with the others on that basis.

The second challenge – summarized here as radical – was also a response to the contents of the spatial-science disciplinary matrix, but much more influenced by the external environment than was the humanistic. As detailed above (p. 210), those who launched what was initially known as radical geography were concerned with the failure of positivist spatial science to tackle and solve pressing societal problems. They advocated applied geography, but defined relevance in a very different way (Harvey, 1974c; Johnston, 1981a). As with humanistic geography, this (eventually, though not initially among those advocating 'welfare geography') involved promoting a scientific revolution at the level of the world view; the realist science that they advanced was incommensurable with both positivist and humanistic science.

Once established, realist-marxist science also became a disciplinary matrix with a variety of exemplars. The core of its programme is the desire to uncover the mechanisms that drive society, providing accounts of how people act and how the empirical world is organized. How such understanding can be achieved, and how the knowledge should be used, has been the subject of much debate – both between marxists and non-marxists and among various types of marxist – and exemplars have waxed and waned rapidly in recent years as a consequence. The debates within the realist conception of science are about how to achieve agreed scientific goals; the debates with positivist and humanistic geographers are about the nature of science. The former produce internal revolutions; the latter seek to promote major disciplinary revolutions. (Note, however, Barnes' critique of both Harvey – the leading advocate of marxism within geography – and Sayer – the leading advocate for realism – as both implicit part-adherents to the Enlightenment project, and hence to the same world view as the spatial scientists: 'my argument is that as with Harvey, Sayer totters between Enlightenment and anti-Enlightenment views' – p. 23.)

These decades were turbulent for human geography, which Ley (1981) found both exciting and confusing, in part because of the rapidity with which human geographers explored new ideas. He believed that such exploration was often superficial, with ideas discarded almost as soon as they had been adopted – 'In true North American fashion, obsolescence is setting in more and more speedily' (p. 209). Others argued that ideas were often divorced from their original context, and that geographers who imported them may have been unaware of the controversies surrounding their use (Agnew and Duncan, 1981; Duncan, 1980; Mohan, 1994). Some sought to reconcile the various world views (Harrison and Livingstone, 1982; Hay, 1979a; Johnston, 1980a, 1982c; Livingstone and Harrison, 1981); some argued that is impossible (Eyles and Lee, 1982); and others still contended that each of the various world views is informed by the others (e.g. Thrift, 1987, p. 401 and his claim that marxist political economy now 'forms a vital subtext to most theorising'), so that geographers should use their synthesizing powers to integrate the various perspectives (Brookfield, 1989, p. 314).

The 1990s

The current decade has seen yet more turbulence, again reflecting both dissatisfaction with the various world views and a rapidly changing external environment – economically, socially, politically, technically and culturally. Within it, the challenge of humanistic geography very substantially withered, in that very few followed the postulates of idealism, phenomenology and the other approaches discussed in Chapter 6 although, as stressed in Chapter 8, rejection of the Enlightenment model of the individual was at the core of the various critiques of marxism-realism. Spatial science did not wither, however: its world view remained strong, although often on the margins of the discipline, the majority of whose members (in part on political-ideological grounds) rejected not only the implicit positivism but also the emphasis on technical rigour in data analysis. The 'new generation' emphasized difference, although in a rather different way from that promulgated by Hartshorne some half a century earlier (Campbell, 1994).

By the 1990s, the discipline was substantially fragmented, not only in its world view and disciplinary matrices but also in its substantive concerns. It comprised a variety of approaches held together as much by the *realpolitik* of university politics and funding as by any adherence to core beliefs, and relatively few of its practitioners interact across sub-disciplinary frontiers (Johnston, 1991, 1996a). There are signs that some of those frontiers are increasingly being crossed by the latest generation of human geographers (as illustrated in Chapter 8), but only some of them (Johnston, 1996d): the relative peripheralization of much of the technical work in GIS and spatial statistics remains. This view has been challenged by Gould (1994, p. 194): 'I find the idea of fragmentation absurd, and too frequently raised by people who long for others to conform to their monolithic ideological positions.' He argued that geographers with different specialist interests (as reflected, for example, by their affiliations to institutional study groups) nevertheless are following his advice of 'don't specialize ... read eclectically' and:

Despite some contrary opinion, [there] is no evidence that they are disconnecting from the rich body of ideas that constitute the geographic 'way of looking' today. For their research, they may be using remotely sensed images, geographic information systems, structural-Marxist concepts, hydrological science, socially constructed ideas of the environment, Michel Foucault and the power of surveillance, deconstructions of official government reports, goal programming, a heightened awareness of gender issues . . . I mean, you name it! . . . to illuminate the spatio-temporal human condition at a place . . . embedded in a larger regional and national space. With all that conceptual and methodological integration, where, literally on earth, is the fragmentation?

Others were much more concerned about fragmentation, however, When the AAG Council decided to initiate a follow-up volume to that edited by James and Jones (1954) 30 years earlier, it rejected a proposal from Gaile and Willmott (subsequently published independently: Gaile and Willmott, 1989) because it drew largely on the AAG's specialty groups and so reflected a fragmented view of the discipline rather than focus on 'cross-cutting and unifying concepts and methods' (Abler, Marcus and Olson, 1992a p. xvii). The book which they edited for the AAG (*Geography's Inner Worlds: Pervasive Themes in Contemporary American Geography*) had four main sections:

1 *what geography is about* – 'geography's worlds': 'places and regions': 'representations of the world';
2 *what geographers do* – 'observation': 'visualization': 'analysis': 'modelling': 'communication';
3 *how geographers think* – 'location, place, region and space': 'movements, cycles and systems': 'the local-global continuum': 'scale in space and time'; and
4 *why geographers think that way* – 'paradigms for inquiry?': 'humanism and science in geography': 'applications of geographic concepts and methods': 'the peopling of American geography'.

This was done to counter the perceived fragmentation (Abler, Marcus and Olson, 1992b, p. 2):

Specialization and specialty groups foster better communication in a multi-faceted discipline. They permit scholars and practitioners with common interests to achieve identity without forming independent associations . . . [but they also] foster intellectual isolation by retarding the cross-fertilization that occurs when geographers encounter unexpected ideas. Despite the fact that the fission evident within geography is common in the physical and social sciences, specialization may have gone as far as it can or should go in American geography.

Hence their book was structured to bring geographers together

> to highlight the common elements within a discipline whose practi-
> tioners are in danger of forgetting their shared heritage and ideals. In
> their preoccupation with the great diversity of geographic problems
> with which they deal on a day to day basis, geographers are prone to
> overlook how much they share emotionally and intellectually with
> other geographers that they do not share with colleagues in other dis-
> ciplines.

A few of the chapters are multi-authored, presumably as a way of bringing
individuals with diverse interests together to identify common elements,
and the outcome led the editors to conclude that the discipline is shown to
be 'simultaneously less fragmented than we had feared it would be and less
coherent than we had hoped' (Abler, Marcus and Olson, 1992c, p. 391),
but nevertheless 'wonderfully diverse in substance, method and philosophy'
(p. 395):

> Compared with other disciplines, geography has always been tolerant
> of variety, and often enthusiastic about multiplicity. With the possible
> exception of the period before 1920 when environmental determinism
> achieved a measure of intellectual hegemony, American geography
> has never been a normal, paradigmatic science of the kind postulated
> by Thomas Kuhn.

Thus although concerned about fragmentation they also promote it as a
'geographical good thing':

> American geography was postmodern long before the term was
> invented. It has historically been eclectic and self-contradictory in
> many respects. It has often playfully delighted in juxtaposing disso-
> nant substantive interests and intellectual traditions. American geo-
> graphy will prosper in the 1990s not so much because geography will
> change radically, but rather because the discipline's social and intellec-
> tual environments have now evolved to where geography has long
> been. Geographers are at last professing and practicing a postmodern
> discipline in a postmodern age.

Nevertheless, if it is to prosper, the basis for coherence (necessary in a
postmodern discipline?) has to be realized, which they argued involves:
overcoming the indifference to physical geography among some human
geographers; finding 'the language and narrative forms that will engage
the attention of American society' (p. 398); and ensuring mutual recognition between the two main 'camps' within human geography – termed
spatial scientists and social theorists elsewhere in this chapter ('The critical matter is that all parties avoid the wholesale rejection of each others'
viewpoints that occurred in geography after 1920 and again after 1955':
p. 400).

An abundance of turbulence

Evaluating these turbulent years, when the turbulence shows little sign of abating, is not easy. Kuhn's model, sometimes presented to geographers as periods of normal scientific progress punctuated by major revolutions, is not relevant to what has happened within their discipline since 1945. But each of the basic Kuhnian definitions of a paradigm (world view, disciplinary matrix, and exemplar) is relevant to appreciating what has occurred and is occurring within geography, as Mair (1986) argued for the concept of the paradigm as a scientific community (or *disciplinary matrix*) with shared values.

At the lowest definitional level of a paradigm, the *exemplar*, human geographers socialized within any of the available disciplinary matrices have shifted the orientation of their work as 'better' ways have been suggested to them; minor revolutions have occurred, and frequently. Such shifts have gone in a variety of directions, presenting an apparent anarchy – not chaos, which is the vernacular use of that term, but 'free and voluntary cooperation of individuals and groups' (Labedz, 1977, p. 22). But the core values hold.

There is also considerable evidence of the relevance of the largest scale definition of a paradigm, the *world view*, to recent changes in human geography. Very different and entirely incommensurable conceptions of the nature of science are competing for geographers' attention. They differ in their scientific and their societal goals, and demand choice. Some have made a very clear choice, switching from one conception of science to another because, as Harvey (1973) expressed it with regard to marxism, 'I can find no other way of accomplishing what I set out to do or of understanding what has to be understood' (p. 17). Some argue that the choice is really between two world views only (spatial science and social theory: Sheppard, 1995): others identify three and, according to Duncan and Barnes (1993, p. 248), one of them has triumphed:

> R. J. Johnston's *Geography and Geographers since 1945* [*sic*] . . . could have been subtitled 'Modernism and its Discontents'. For its narrative structure is compellingly organized around the contest for intellectual supremacy among the triad of approaches: empiricist, positivist and modernist social theory. Although from Johnston's perspectives the social theorists win the day, and the others are relegated to the dustbin of geographical history, it is only modernism that really triumphed, because it is the only game in town.

Each of those positions, they contend, seeks to provide the 'one best method to explain geographical phenomena' and 'None has seriously entertained the notion that there is no best method' (p. 249). The differences among the three approaches are, they claim, 'relatively minor when compared to their shared assumptions', and they fear that the most radical aspects of postmodern epistemology have been rejected by geographers who 'wish to incorporate certain of its important insights into a reconstructed modernist [i.e. Enlightenment] project'. If the postmodern project was fully embraced, they argue, then the discipline would become much more variable since it would no longer be characterized by the views of 'trained academics':

> The problem lies in our conceptualization of difference. Difference for geographers is other spaces, other places, other regions or other landscapes. To embrace difference is in Gregory's ... words to embrace 'areal differentiation'. It is instructive, however, to compare the postmodernist geographers' conception of difference to that of postmodernist ethnographers. For the latter, difference is other people and, to be more precise, other peoples' voices. Their plea is for an end to monovocality and authorial authority. Let us hear direct, they say, from those from whom we as academics have for so long spoken (p. 253).

Such a shift in the nature of geography is politically threatening because of its challenge to the academics' authority. Furthermore:

> Multivocality (taking others' voices seriously) is not something that can be easily accomplished in geography, for we have no tradition of ethnography. Unless we begin to focus more of our energies on developing our techniques for listening to others our calls for difference are highly suspect, for we will continue systematically to silence difference. Those outside the academy have no voice in our work other than the one we choose to give them ... to speak for another is not a politically innocent act. We have appropriated their voice – colonized their perspective.

To adopt their agenda would be to increase the turbulence exponentially!

Buttimer (1993), drawing on Pepper (1942), identified four 'root metaphors', or world views, underpinning the practice of geography, each of which 'projects a distinct interpretation of reality':

1 the *world as a mosaic of patterns and forms*, which she sees as the 'central root metaphor for geography's chorological tradition, the most practiced of all four' (p. 22);
2 the *world as a mechanism* of causally integrated interacting systems;
3 the *world as an organism*, as a whole which comprises unity in diversity; and
4 the *world as arena*, as the context within which 'spontaneous and possibly unique events may occur' (p. 23).

The four are not presented as exclusive, nor are they treated as such: 'Most creative scholars ... avail themselves of more than one root metaphor in

the course of a career. Only the rare dogmatist clings to one throughout.'
Nevertheless, she sees them as 'vectors of distinct *thought styles* whose
appeal has varied through different moments in Western social history'
(p. 24). Which styles are dominant when depends on the interaction of
people and context:

> Acceptance or rejection of a particular paradigm, model, or method
> within the discipline of geography has as much to do with the aes-
> thetic, emotional or moral connotations of a root metaphor as it does
> with purely epistemological reasoning. The succession of metaphors
> within any tradition raises questions about the interplay of internal
> and external circumstances. Career stories of geographers reveal
> important clues about their succession and relative appeal. At any
> moment of disciplinary history, all four root metaphors ... may be
> simultaneously co-present, although one or more may appear domi-
> nant within particular periods.

Her descriptions of the four metaphors, based to a considerable extent on
autobiographical data, does not lead to a detailed appreciation of why
one or more of the metaphors has been more visible in some periods than
others. Instead, she identifies a major tension between, on the one hand
(p. 212), 'the integrated approaches of *organism* and *mechanism* [which]
have invited research of wider scale and have apparently enhanced the
status of geography within the academy' which is desirable given 'the
political expediency of the times' and the 'political economy of research
grantsmanship', and, on the other, 'The dispersed approaches of *mosaic*
and *arena* have yielded more sensitive accounts of life and landscape at
local and regional scales of inquiry.' Her own preference is clearly for the
latter pair, for 'the humanist's emancipatory hope' and 'communication
and understanding' (p. 219). Humanism must be 'the leaven in the dough
and not a separate loaf in the smorgasbord of geographic endeavor'
(p. 220), but she does not seek to impose (p. 212): 'The integrity of disciplinary practice ... demands a flexibility to changing educational needs
and to new substantive research challenges. Ideally each individual,
department or research team, given its resources, aims and context,
should assume responsibility for designing and adapting its agenda to
such changing demands.'

Within each world view the *disciplinary matrix* identifies the framework
within which research is conducted. The shift from regionalism to spatial
science involved a change of disciplinary matrix, for example; whilst the
world view remained constant – that of positivist science with its foundation in empiricism (although it was hidden until relatively late in the shift) –
the shared goals altered very considerably. The relative popularity of marxism, realism, structuration, and postmodernism at different times reflects
changes in the disciplinary matrix of the 'geography as social theory' world
view.

Kuhn's model as usually presented does not fit the experience of human geography since 1945, therefore. But its major components are valuable in appreciating much of what has happened within the discipline over that period. In summary, the salient elements of the period have been:

1 Mono-paradigm dominance at the level of the world view until the late 1960s, with empiricism as the foundation and an increasingly explicit acceptance of positivism.

2 A steady switch within that period from one disciplinary matrix (regionalism) to another (spatial science), but with the former never entirely ousted by the latter.

3 A substantial number of exemplars provided the framework for conducting research within the spatial science disciplinary matrix, which remains buoyant, reflecting both technical developments and the systematic sub-division of the discipline.

4 Two further world views were promoted from the late 1960s on representing two very different views of science, both from each other and from spatial science. One ('radical' or 'social theory') is now well established within the discipline, with substantial numbers of adherents, but has not achieved disciplinary hegemony: the other ('humanistic') informed the shift away from the search for 'universal truths', but its particular research routes are rarely travelled now (and a case could be made that its various components – idealism, phenomenology, etc. – are now no more than exemplars within the social theory world view).

5 Different disciplinary matrices have been advocated within each of these world views: none has dominated humanistic geography; realism (incorporating marxism) increasingly prevailed within the 'radical structural' world view.

6 During the 1990s a new group of disciplinary matrices has developed within the social theory world view, each emphasizing 'difference' within structural constraints.

Thus human geography is currently characterized by a multi-paradigm situation at the world-view level, by competition between disciplinary matrices within at least two of those world views, and by a wealth of exemplars on which research is based in all three. It is because of the last element, and because many human geographers have not been deeply schooled into any one disciplinary matrix, let alone the use of any one exemplar, that the discipline appears replete with fickle allegiances, as individuals explore various ways of practising geography (Johnston, 1981b, pp. 313–14):

Much geographical work is exploratory, and is conducted by individuals who operate independently. Indeed, many, although influenced by what they read, are in no sense socialized into a particular matrix or set of exemplars which might be associated with a 'research school' and its leader. The charisma of certain individuals and their published works may occasionally produce the prophet and disciple situation. Much more usual, however, is a situation of fickle allegiances.

Such fickleness suggests anarchy, in the sense used by Feyerabend (1975; Johnston, 1976b, 1978c). Some human geographers shift frequently among exemplars, sometimes between disciplinary matrices, and just occasionally between world views.

The last five decades have seen individual human geographers occasionally promoting a particular paradigm (world view, disciplinary matrix or exemplar) in opposition to prevailing practice. That has rarely been done in isolation; either contacts with others or reading (or both) will suggest the argument, and support is then sought within the discipline, and perhaps resources and sponsorship from within society too. Why are some people better able to promote paradigm change than others? Little work has been done on the sociology of geography as a discipline, especially with regard to power over the acceptance/rejection of proposed paradigm shifts and the ability to convince others of the 'rightness' of any approach. Work has been done on the influential figures in the discipline's early development (e.g. in Blouet, 1981, and in the biobibliography series edited by Freeman, 1977) and the nature of the profession in Britain has been explored (Johnston and Brack, 1983), but the dynamics of the discipline in recent years have not been the subject of detailed analysis; nor, interestingly, has the impact of the 'commercialisation of knowledge' by publishing houses been scrutinized (Barnett and Low, 1996), although the increased stress on research publications in the UK from the mid-1980s on led to problems for textbook publishers (Davey *et al.*, 1995).

Within the discipline, analysis of citations (mostly from the data base constructed by the Institute of Scientific Information in Philadelphia) has been used:

1　to identify frequently referenced works that can be categorized as exemplars (e.g. Whitehand, 1985; Wrigley and Matthews, 1986; Bodman, 1991, 1992);

2　to discover research communities, groups of scholars who refer to each other's writings on a particular topic and so occupy a disciplinary matrix (Gatrell, 1984a, 1984b);

3　to rate journals and chart their interdependence (e.g. Whitehand, 1984; Gatrell and Smith, 1984);

4　to rank departments in the United States and the United Kingdom in terms of publications produced, publications cited and peer evaluation, for example (see Morrill, 1980; Jones, Lindsey and Coggeshall, 1982; Turner and Meyer, 1985 on the US; on the UK, see Bentham, 1987; Smith, 1988b); and

5 to conduct exploratory studies of particular sub-disciplines and the
   main exemplars which have stimulated how individuals practise (e.g.
   Phillips and Unwin, 1985).

All describe aspects of the discipline's structure but tell us little of its
processes; like spatial science they provide valuable descriptive data, but
not necessarily much understanding.

For that understanding, it is necessary not only to appreciate the contexts in which people worked but also to realize that those contexts are
not determinate. A change in the level of economic prosperity in a country will not necessarily bring forth a certain reaction from geographers;
individuals may respond to their interpretations of that change, and their
responses may stimulate others to follow them. This is the most likely
reason for shifts at the level of the world view and, probably, the disciplinary matrix too, as exemplified by the introduction of a radical/struc-
turalist world view and by the shift within the empiricist world view
from regionalism to spatial science. But such shifts can stimulate
counter-reactions, as with the advancement of humanistic geography as a
counter to spatial science; those promoting that particular cause were
influenced both by currents of thinking outwith geography and by their
explorations of World Three. At the level of the exemplar, shifts are
most likely to occur as a response to events within the discipline, as
Kuhn suggests; this occurred within the spatial-science disciplinary
matrix with the movement from normative to behavioural analyses, but
the later move towards welfare geography was strongly influenced by
external factors.

No simple model can be applied to the changes within human geography to provide an explanation of why the discipline has altered in the
ways described here. Kuhn's work provides a valuable vocabulary and
organizing framework within which to describe the alterations, but what
has occurred reflects the perceptions and actions of individual human
geographers. Like all other aspects of society, geography is a discipline
created by and for geographers, and is continually fought over and recreated by them, at a variety of scales and in social, economic and political
contexts that vary over space and through time. The last point is crucial,
for – as the structuration approach stresses – people are socialized in particular contexts and they then create part of the milieux within which
others are socialized. Those places are not isolated, and changes in one
can influence changes in others. But, as illustrated here, there have been
substantial differences from one place to another within Anglo-American
human geography regarding how the discipline is practised, reflecting
both the nature of those places and the people in them; on the international scale, the differences are even greater (Johnston and Claval, 1984).

And the future?

In the early 1970s, several leading geographers were asked their views of the discipline's future (Chorley, 1973a). Little of what they suggested is reflected in what happened; the contents of a similar book produced twelve years later (Johnston, 1985c) are very different. Debates about how geography should be practised and taught continue, reflecting not only internal divisions but also trends in the societies to which human geographers belong. The contents and contexts of those future debates cannot be predicted; one could suggest with some certainty, however, that human geography will not be characterized by mono-paradigm dominance in the next decade.

In the introduction to his autobiographical essay, David Smith (1984, p. 118) wrote that:

> my own professional activities seem to have been a continual struggle to come to terms with (or keep up with) the rapidly shifting focus of human geography. The struggle arises in large measure from the difficulty of breaking free from one's own intellectual heritage. . . . If my own struggles represent anything more than one half-life experience, it may well be the theme of the geographer or social scientist as creature of his or her times. . . . If anything is to be learned from the instant replay of such recent events . . . it is that 'scientific' advance is not conducted in a social vacuum but as an integral part of human history, within which the element of chance arising from individual personality and creativity plays an important part. So let us proceed with the recollection of one of the random variables.

We are all random variables, it seems. (Robson, 1984, p. 104, used a similar turn of phrase: 'it is clear how small a part in my own development seems to have been played by clearly guided aims and how much has been contributed by the collage of rather random influences and serendipitous events to which I both responded and contributed'.) Hence our individual projects and life-paths can be appreciated and set in context, but no more. We are very uncertain where we, as individuals, are going, let alone where geography as a set of linked yet anarchistic communities is going. We are making the future of geography as we practise it, just as I have been remaking its past by writing this book.

Bibliography

ABLER, R. F. 1971: Distance, intercommunications, and geography. *Proceedings, Association of American Geographers* 3, 1–5.

_____ 1993: Everything in its place: GPS, GIS, and geography in the 1990s. *The Professional Geographer* 45, 131–9.

ABLER, R. F., ADAMS, J. S. and GOULD, P. R. 1971: *Spatial organization: the geographer's view of the world.* Englewood Cliffs, NJ: Prentice-Hall.

ABLER, R. F., MARCUS, M. G. and OLSON, J. M. 1992a: Preface. In R. F. Abler, M. G. Marcus and J. M. Olson (eds) *Geography's inner worlds: pervasive themes in contemporary American geography.* New Brunswick, NJ: Rutgers University Press, xv–xx.

_____ 1992b: Contemporary American geography. In R. F. Abler, M. G. Marcus and J. M. Olson (eds) *Geography's inner worlds: pervasive themes in contemporary American geography.* New Brunswick, NJ: Rutgers University Press, 1–8.

_____ 1992c: Afterword. In R. F. Abler, M. G. Marcus and J. M. Olson (eds) *Geography's inner worlds: pervasive themes in contemporary American geography.* New Brunswick, NJ: Rutgers University Press, 391–402.

ACKERMAN, E. A. 1945: Geographic training, wartime research, and immediate professional objectives. *Annals of the Association of American Geographers* 35, 121–43.

_____ 1958: *Geography as a fundamental research discipline.* Chicago: University of Chicago, Department of Geography Research Paper 53.

_____ 1963: Where is a research frontier? *Annals of the Association of American Geographers* 53, 429–40.

ADAMS, J. 1994: *Risk.* London: Routledge.

ADAMS, J. S. 1969: Directional bias in intra-urban migration. *Economic Geography* 45, 302–23.

_____ (ed.) 1976: *Urban policymaking and metropolitan dynamics: a comparative geographical analysis.* Cambridge MA: Ballinger.

ADAMS, P. C. 1995: A reconsideration of personal boundaries in space-time. *Annals of the Association of American Geographers* 85, 267–85.

AGNEW, J. A. 1984: Place and political behaviour: the geography of Scottish nationalism. *Political Geography Quarterly* 3, 191–202.

_____ 1987a: *The United States in the world-economy: a regional geography.* Cambridge: Cambridge University Press.

_____ 1987b: *Place and politics: the geographical mediation of state and society.* Boston: Allen and Unwin.

_____ 1989: The devaluation of place in social science. In J. A. Agnew and J. S. Duncan (eds) *The power of place*. Boston: Unwin Hyman, 9–29.

_____ 1990: Sameness and difference: Hartshorne's *The Nature of Geography* and geography as areal variation. In J. N. Entrikin and S. D. Brunn (eds) *Reflections on Richard Hartshorne's The Nature of Geography*. Washington: Association of American Geographers, 121–40.

_____ 1993: Representing space: space, scale and culture in social science. In J. S. Duncan and D. Ley (eds) *Place/culture/representation*. London: Routledge, 251–71.

AGNEW, J. A. and DUNCAN, J. S. 1981: The transfer of ideas into Anglo-American human geography. *Progress in Human Geography* 5, 42–57.

_____ 1989: Introduction. In J. A. Agnew and J. S. Duncan (eds) *The power of place*. Boston: Unwin Hyman, 1–8.

AITKEN, S. 1991: A transactional geography of the image-event: the films of Scottish director, Bill Forsyth. *Transactions, Institute of British Geographers* NS16, 105–18.

AITKEN, S. C., CUTTER, S. L., FOOTE, K. E. and SELL, J. S. 1989: Environmental perception and behavioral geography. In G. L. Gaile and C. J. Willmott (eds) *Geography in America*. Merrill, Columbus, 218–38.

ALEXANDER, D. 1979: Catastrophic misconception? *Area* 11, 228–30.

ALEXANDER, J. W. and ZAHORCHAK, G. A. 1943: Population-density maps of the United States: techniques and patterns. *Geographical Review* 33, 457–66.

AMEDEO, D. and GOLLEDGE, R. G. 1975: *An introduction to scientific reasoning in geography*. New York: John Wiley.

ANDERSON, J. 1973: Ideology in geography: an introduction. *Antipode* 5(3), 1–6.

ANON, 1968: A preliminary contribution to the geographical analysis of a Poohscape. *IBG Newsletter* 6, 54–63.

APPLEBAUM, W. 1954: Marketing geography. In P. E. James and C. F. Jones (eds) *American geography: inventory and prospect*. Syracuse: Syracuse University Press, 245–51.

APPLETON, J. 1975: *The experience of landscape*. London: John Wiley.

_____ 1994: *How I made the world: shaping a view of landscape*. Hull: University of Hull Press.

ARCHER, J. C. and TAYLOR. P. J. 1981: *Section and party*. Chichester: John Wiley.

ARMSTRONG, M. P. 1993: On automated geography. *The Professional Geographer* 45, 440–2.

ASHEIM, B. T. 1990: How to confuse rather than guide students: a review of Holt-Jensen's *Geography – history and concepts*. *Progress in Human Geography* 14, 281–92.

BADCOCK, B. A. 1970: Central-place evolution and network development in south Auckland, 1840–1968: a systems analytic approach. *New Zealand Geographer* 26, 109–35.

_____ 1984: *Unfairly structured cities*. Oxford: Basil Blackwell.

BAHRENBERG, G., FISCHER, M. M. and NIJKAMP, P. (eds) 1984: *Recent developments in spatial data analysis: methodology, measurement, models*. Aldershot: Gower Press.

BAKER, A. R. H. 1972: Rethinking historical geography. In A. R. H. Baker (ed.), *Progress in historical geography*. Newton Abbott: David & Charles, 11–28.

_____ 1979: Historical geography: a new beginning? *Progress in Human Geography* 3, 560–70.

_____ 1981: An historico-geographical perspective on time and space and on period and place. *Progress in Human Geography* 5, 439–43.

_____ 1984: Reflections on the relations of historical geography and the *Annales*

school of history. In A. R. H. Baker and D. Gregory (eds). *Explorations in historical geography*. Cambridge: Cambridge University Press, 1–27.
BAKER, A. R. H. and GREGORY, D. 1984: Some terrae incognitae in historical geography: an exploratory discussion. In A. R. H. Baker and D. Gregory (eds) *Explorations in historical geography*. Cambridge: Cambridge University Press, 180–94.
BALL, M. 1987: Harvey's Marxism. *Environment and Planning D: Society and Space* 5, 393–4.
BALLABON, M. B. 1957: Putting the economic into economic geography *Economic Geography* 33, 217–23.
BARNES, B. 1974: *Scientific knowledge and sociological theory*. London: Routledge & Kegan Paul.
―――― 1982: *T. S. Kuhn and social science*. London: Macmillan.
BARNES, T. J. 1985: Theories of international trade and theories of value. *Environment and Planning A* 17, 729–46.
―――― 1988: Rationality and relativism in economic geography: an interpretive review of the *homo economicus* assumption. *Progress in Human Geography* 12, 473–96.
―――― 1989a. Place, space and theories of economic value: contextualism and essentialism in economic geography. *Transactions, Institute of British Geographers* NS14, 299–316.
―――― 1989b: Structure and agency in economic geography and theories of economic value. In A. Kobayashi and S. Mackenzie (eds) *Remaking human geography*. Boston: Unwin Hyman, 134–48.
―――― 1996: *Logics of dislocation: models, metaphors and meaning of economic space*. New York: The Guilford Press.
BARNES, T. J. and DUNCAN, J. S. 1992: Introduction: writing worlds. In T. J. Barnes and J. S. Duncan (eds) *Writing worlds: discourse, text and metaphor in the representation of landscape*. London: Routledge, 1–17.
BARNETT, C. 1995: Awakening the dead: who needs the history of geography? *Transactions, Institute of British Geographers* NS20, 417–19.
BARNETT, C. and LOW, M. 1996: Speculating on theory: towards a political economy of academic publishing. *Area* 28, 13–24.
BARROWS, H. H. 1923: Geography as human ecology. *Annals of the Association of American Geographers* 13, 1–14.
BASSETT, K. and SHORT, J. R. 1980: *Housing and residential structure: alternative approaches*. London: Routledge & Kegan Paul.
BATEY, P. and BROWN, P. 1995: From human ecology to customer targeting: the evolution of geodemographics. In P. Longley and G. Clarke (eds) *GIS for business and service planning*. Cambridge: GeoInformation International, 73–103.
BATTY, M. 1976: *Urban modelling: algorithms, calibrations, predictions*. London: Cambridge University Press.
―――― 1978: Urban models in the planning process. In D. T. Herbert and R. J. Johnston (eds) *Geography and the urban environment*, vol. 1. Chichester: John Wiley, 63–134.
―――― 1989: Urban modelling and planning: reflections, retrodictions and prescriptions. In B. Macmillan (ed.) *Remodelling geography*. Oxford: Basil Blackwell, 147.
BEAUMONT, J. R. 1987: Quantitative methods in the real world: a consultant's view of practice. *Environment and Planning A* 19, 1441–8.
BEAUMONT, J. R. and GATRELL, A. C. 1982: *An introduction to Q-analysis*. CATMOG 34, Norwich: Geo Books.
BEAUREGARD, R. A. 1988: In the absence of practice: the locality research debate. *Antipode* 20, 52–9.

BECKERMAN, W. 1995: *Small is stupid: blowing the whistle on the greens.* London: Duckworth.

BELL, C. and NEWBY, H. 1976: Community, communion, class and community action: the social sources of new urban politics. In D. T. Herbert and R. J. Johnston (eds) *Social areas in cities,* volume 2: *Spatial perspectives on problems and policies.* Chichester: John Wiley, 189–208.

BELL, D. 1973: *The coming of post-industrial society.* New York: Basic Books.

BELL, M., BUTLIN, R. J. and HEFFERNAN, M. (eds) 1995: *Geography and imperialism: 1820–1940.* Manchester: Manchester University Press.

BENNETT, R. J. 1974: Process identification for time-series modelling in urban and regional planning. *Regional Studies* 8, 157–74.

——— 1975: Dynamic systems modelling of the Northwest region: 1. Spatio-temporal representation and identification. 2. Estimation of the spatio-temporal policy model. 3. Adaptive parameter policy model. 4. Adaptive spatio-temporal forecasts. *Environment and Planning A* 7, 525–38, 539–66, 617–36, 887–98.

——— 1978a: Forecasting in urban and regional planning closed loops: the examples of road and air traffic forecasts. *Environment and Planning A* 10, 145–62.

——— 1978b: *Spatial time series: analysis, forecasting and control.* London: Pion.

——— 1979: Space-time models and urban geographical research. In D. T. Herbert and R. J. Johnston (eds) *Geography and the urban environment: progress in research and application,* vol. 2. London: John Wiley, 27–58.

——— 1981a: Quantitative geography and public policy. In N. Wrigley and R. J. Bennett (eds) *Quantitative geography.* London: Routledge & Kegan Paul, 387–96.

——— 1981b: A hierarchical control solution to allocation of the British Rate Support Grant. *Geographical Analysis* 13, 300–14.

——— (ed.) 1981c: *European progress in spatial analysis.* London: Pion.

——— 1981d: Quantitative and theoretical geography in Western Europe. In R. J. Bennett (ed.) *European progress in spatial analysis.* London: Pion, 1–32.

——— 1982: Geography, relevance and the role of the Institute. *Area* 14, 69–71.

——— 1983: Individual and territorial equity. *Geographical Analysis* 15, 50–87.

——— 1985a: A reappraisal of the role of spatial science and statistical inference in geography in Britain. *L'Espace Geographique* 14, 23–8.

——— 1985b: Quantification and relevance. In R. J. Johnston (ed.) *The future of geography.* London: Methuen, 211–24.

——— 1989a: Whither models and geography in a post-welfarist world? In B. Macmillan (ed.) *Remodelling geography.* Oxford: Basil Blackwell, 273–90.

——— 1989b: Demography and budgetary influence on the geography of the poll tax: alarm or false alarm? *Transactions, Institute of British Geographers* NS14, 400–17.

BENNETT, R. J. and CHORLEY, R. J. 1978: *Environmental systems: philosophy, analysis and control.* London: Methuen.

BENNETT, R. J. and HAINING, R. P. 1985: Spatial structure and spatial interaction: modelling approaches to the statistical analysis of geographical data. *Journal of the Royal Statistical Society A* 148, 1–36.

BENNETT, R. J., HAINING, R. P. and WILSON, A. G. 1985: Spatial structure, spatial interaction and their integration: a review of alternative models. *Environment and Planning A* 17, 625–46.

BENNETT, R. J. and THORNES, J. B. 1988: Geography in the United Kingdom, 1984–1988. *The Geographical Journal* 154, 23–48.

BENNETT, R. J. and WRIGLEY, N. 1981: Introduction. In N. Wrigley and R. J. Bennett (eds), *Quantitative geography.* London: Routledge & Kegan Paul, 3–11.

BENTHAM, G. 1987: An evaluation of the UGC's rating of the research of British university geography departments. *Area* 19, 147–54.

BERDOULAY, V. 1981: The contextual approach. In D. R. Stoddart (ed.) *Geography, ideology and social concern.* Oxford: Blackwell, 8–16.

BERRY, B. J. L. 1958: A critique of contemporary planning for business centers. *Land Economics* 25, 306–12.

_____ 1959a: Ribbon developments in the urban business pattern. *Annals of the Association of American Geographers* 49, 145–55.

_____ 1959b: Further comments 'geographic' and 'economic' economic geography. *The Professional Geographer* 11(1), 11–12.

_____ 1964a: Cities as systems within systems of cities. *Papers, Regional Science Association* 13, 147–63.

_____ 1964b: Approaches to regional analysis: a synthesis. *Annals of the Association of American Geographers* 54, 2–11.

_____ 1965: Research frontiers in urban geography. In P. M. Hauser and L. F. Schnore (eds) *The study of urbanization.* New York: John Wiley, 403–30.

_____ 1966: *Essays on commodity flows and the spatial structure of the Indian economy.* Chicago: University of Chicago, Department of Geography, Research Paper 111.

_____ 1967: *The geography of market centers and retail distribution.* Englewood Cliffs, NJ: Prentice-Hall.

_____ 1968: A synthesis of formal and functional regions using a general field theory of spatial behavior. In B. J. L. Berry and D. F. Marble (eds) *Spatial analysis.* Englewood Cliffs, NJ: Prentice-Hall, 419–28.

_____ 1969: Review of B. M. Russett *International regions and the international system. The Geographical Review* 59, 450–1.

_____ (ed.) 1971: Comparative factorial ecology. *Economic Geography* 47, 209–367.

_____ 1972a: Hierarchical diffusion: the basis of development filtering and spread in a system of growth centers. In N. M. Hansen (ed.) *Growth centers in regional economic development.* New York: The Free Press, 108–38.

_____ 1972b: 'Revolutionary and counter-revolutionary theory in geography' – a ghetto commentary. *Antipode* 4(2), 31–3.

_____ 1972c: More on relevance and policy analysis. *Area* 4, 77–80.

_____ 1973a: *The human consequences of urbanization.* London: Macmillan.

_____ 1973b: A paradigm for modern geography. In R. J. Chorley (ed.) *Directions in geography.* London: Methuen, 3–22.

_____ 1974a: Review of H. M. Rose (ed.) *Perspectives in geography 2. Geography of the ghetto, perceptions, problems and alternatives. Annals of the Association of American Geographers* 64, 342–5.

_____ 1974b: Review of David Harvey, *Social Justice and the City. Antipode* 6(2), 142–5, 448.

_____ 1978a: Introduction: a Kuhnian perspective. In B. J. L. Berry (ed.) *The nature of change in geographical ideas.* de Kalb: Northern Illinois University Press, vii–x.

_____ 1978b: Geographical theories of social change. In B. J. L. Berry (ed.) *The nature of change in geographical ideas.* de Kalb: Northern Illinois University Press, 17–36.

_____ 1993: Geography's quantitative revolution: initial conditions, 1954–1960: a personal memoir. *Urban Geography* 14, 434–41.

_____ 1995: Whither regional science? *International Regional Science Review* 17, 297–306.

BERRY, B. J. L. and BAKER, A. M. 1968: Geographic sampling. In B. J. L. Berry and D. F. Marble (eds) *Spatial analysis.* Englewood Cliffs, NJ: Prentice-Hall, 91–100.

BERRY, B. J. L. and GARRISON, W. L. 1958a: The functional bases of the central place hierarchy. *Economic Geography* 34, 145–54.

_____ 1958b: Recent developments in central place theory. *Papers and Proceedings, Regional Science Association* 4, 107–20.

BERRY, B. J. L. and HORTON, F. E. (eds) 1974: *Urban environmental management: planning for pollution control.* Englewood Cliffs, NJ: Prentice-Hall.

BERRY, B. J. L. *et al.* 1974: *Land use, urban form and environmental quality.* Chicago: Department of Geography, Research Paper 155, University of Chicago.

BHASKAR, R. 1978: A *realist theory of science.* Brighton: Harvester Press.

BILLINGE, M. 1977: In search of negativism: phenomenology and historical geography. *Journal of Historical Geography* 3, 55–68.

_____ 1983: The mandarin dialect: an essay on style in contemporary geographical writing. *Transactions, Institute of British Geographers* NS8, 400–20.

BILLINGE, M., GREGORY, D. and MARTIN, R. L. (eds) 1984: *Recollections of a revolution: geography as spatial science.* London: Macmillan.

BIRD, J. H. 1975: Methodological implications for geography from the philosophy of K. R. Popper. *Scottish Geographical Magazine* 91, 153–63.

_____ 1977: Methodology and philosophy. *Progress in Human Geography* 1, 104–10.

_____ 1978: Methodology and philosophy. *Progress in Human Geography* 2, 133–40.

_____ 1985: Geography in three worlds: how Popper's system can help elucidate dichotomies and changes in the discipline. *The Professional Geographer* 37, 403–9.

_____ 1989: *The changing worlds of geography: a critical guide to concepts and methods.* Oxford: Clarendon Press.

BIRKIN, M. 1995: Customer targeting, geodemographics and lifestyle approaches. In P. Longley and G. Clarke (eds) *GIS for business and service planning.* Cambridge: GeoInformation International, 104–49.

BIRKIN, M. and CLARKE, M. 1988: SYNTHESIS: a synthetic spatial information system: methods and examples. *Environment and Planning A* 20, 645–71.

_____ 1989: The generation of individual and household incomes at the small area level using SYNTHESIS. *Regional Studies* 23, 535–48.

BIRKIN, M., CLARKE, G., CLARKE, M. and WILSON, A. G. 1990: Elements of a model-based GIS for evaluation of urban policy. In L. Worrall (ed.) *Geographic information systems: development and applications.* London: Belhaven Press, 131–62.

BIRKIN, M., CLARKE, M. and GEORGE, F. 1995: The use of parallel computers to solve nonlinear spatial optimisation problems: an application to network planning. *Environment and Planning A* 27, 1049–68.

BIRKIN, M. and WILSON, A. G. 1986: Industrial location models 1. a review and integrating framework and 2. Weber, Palander, Hotelling and extensions within a new framework. *Environment and Planning A* 18, 175–206 and 293–306.

BLAIKIE, P. M. 1978: The theory of the spatial diffusion of innovations: a spacious cul-de-sac. *Progress in Human Geography* 2, 268–95.

_____ 1985: *The political economy of soil erosion.* London: Longman.

_____ 1986: Natural resource use in developing countries. In R. J. Johnston and P. J. Taylor (eds) *A world in crisis? Geographical perspectives.* Oxford: Basil Blackwell, 107–26.

BLAIKIE, P. M. and BROOKFIELD, H. C. (eds) 1987: *Land degradation and society.* London: Methuen.

BLAUG, M. 1975: Kuhn versus Lakatos, or paradigms versus research programmes in the history of economics. *History of Political Economy* 7, 399–419.

BLAUT, J. M. 1962: Object and relationship. *The Professional Geographer* 14, 1–6.

——— 1979: The dissenting tradition. *Annals of the Association of America Geographers* 69, 157–64.

——— 1987: Diffusionism: a uniformitarian critique. *Annals of the Association of American Geographers* 77, 30–47.

BLOUET, B. W. (ed.) 1981: *The origins of academic geography in the United States.* Hamden CT: Archon Books.

BLOWERS, A. T. 1972: Bleeding hearts and open values. *Area* 4, 290–2.

——— 1974: Relevance, research and the political process. *Area* 6, 32–6.

——— 1984: *Something in the air: corporate power and the environment.* London: Harper and Row.

BLUMENSTOCK, D. I. 1953: The reliability factor in the drawing of isarithms. *Annals of the Association of American Geographers* 43, 289–304.

BOAL, F. W. and LIVINGSTONE, D. N. 1989: The behavioural environment: worlds of meaning in a world of facts. In F. W. Boal and D. N. Livingstone (eds) *The behavioural environment.* London: Routledge, 3–17.

BODDY, M. J. 1976: The structure of mortgage finance: building societies and the British social formation. *Transactions, Institute of British Geographers* NS1, 58–71.

BODMAN, A. R. 1991: Weavers of influence: the structure of contemporary geographic research. *Transactions, Institute of British Geographers* NS16, 21–7.

——— 1992: Holes in the fabric: more on the master weavers in human geography. *Transactions, Institute of British Geographers* NS17, 108–9.

BONDI, L. 1993: Locating identity politics. In M. Keith and S. Pile (eds) *Place and the politics of identity.* London: Routledge, 84–101.

BONDI, L. and DOMOSH, M. 1992: Other figures in other places: on feminism, postmodernism and geography. *Environment and Planning D: Society and Space* 10, 199–213.

BONNET, A. 1993: Contours of crisis: anti-racism and reflexivity. In P. Jackson and J. Penrose (eds) *Constructions of race, place and nation.* London: UCL Press, 163–80.

BOOTS, B. N. and GETIS, A. 1978: *Models of spatial processes.* Cambridge: Cambridge University Press.

BOWLBY, S. R., FOORD, J. and MCDOWELL, L. 1986: The place of gender in locality studies. *Area* 18, 327–31.

BOWLBY, S. R., LEWIS, J., McDOWELL, L. and FOORD, J. 1989: The geography of gender. In Peet, R. and Thrift, N. J. (eds) *New models in geography,* vol. 2. London: Unwin Hyman, 157–75.

BOYLE, M. J. and ROBINSON, M. E. 1979: Cognitive mapping and understanding. In D. T. Herbert and R. J. Johnston (eds) *Geography and the urban environment,* vol. 2. Chichester: John Wiley, 59–82.

BRACKEN, I., HIGGS, G., MARTIN, D. and WEBSTER, C. 1990: *A classification of geographical information systems literature and applications.* CATMOG 52, Norwich: Environmental Publications.

BRADLEY, P. N. 1986: Food production and distribution – and hunger. In R. J. Johnston and P. J. Taylor (eds) *A world in crisis? Geographical perspectives.* Oxford: Basil Blackwell, 89–106.

BREITBART, M. M. 1981: Peter Kropotkin, the anarchist geographer. In D. R. Stoddart (ed.) *Geography, ideology and social concern.* Oxford: Blackwell, 134–53.

BRIGGS, D. J. 1981: The principles and practice of applied geography. *Applied Geography* 1, 1–8.

BRITTAN, S. 1977: Economic liberalism. In A. Bullock and O. Stallybrass (eds) *The Fontana dictionary of modern thought.* London: William Collins, 188–9.

BROOKFIELD, H. C. 1962: Local study and comparative method: an example from Central New Guinea. *Annals of the Association of American Geographers* 52, 242–54.

_____ 1964: Questions on the human frontiers of geography. *Economic Geography* 40, 283–303.

_____ 1969: On the environment as perceived. In C. Board *et al.* (eds) *Progress in Geography 1*. London: Edward Arnold, 51–80.

_____ 1973: On one geography and a Third World. *Transactions, Institute of British Geographers* 58, 1–20.

_____ 1975: *Interdependent development*. London: Methuen.

_____ 1989: The behavioural environment: how, what for, and whose? In F. W. Boal and D. N. Livingstone (eds) *The behavioural environment*. London: Routledge, 311.

BROWETT, J. 1984: On the necessity and inevitability of uneven spatial development under capitalism. *International Journal of Urban and Regional Research* 8, 155–76.

BROWN, L. A. 1968: *Diffusion processes and location: a conceptual framework and bibliography*. Regional Science Research Institute, Bibliography Series 3, Philadelphia.

_____ 1975: The market and infrastructure context of adoption: a spatial perspective on the diffusion of innovation. *Economic Geography* 51, 185–216.

_____ 1981: *Innovation diffusion: a new perspective*. London: Methuen.

BROWN, L. A. and MOORE, E. G. 1970: The intra-urban migration process: a perspective. *Geografiska Annaler* 52B, 1–13.

BROWN, R. H. 1943: *Mirror for Americans: likenesses of the eastern seaboard 1810*. New York: American Geographical Society.

BROWN, S. E. 1978: Guy-Harold Smith, 1895–1976. *Annals of the Association of American Geographers* 68, 115–18.

BROWNING, C. 1982: *Conversations with geographers: career pathways and research styles*. University of North Carolina at Chapel Hill, Department of Geography, Occasional Paper 16.

BRUNN, S. D. 1974: *Geography and politics in America*. New York: Harper & Row.

BRUSH, J. E. 1953: The hierarchy of central places in southwestern Wisconsin. *Geographical Review* 43, 380–402.

BUCHANAN, K. 1962: West wind, east wind. *Geography* 47, 333–46.

_____ 1973: The white north and the population explosion. *Antipode* 5(3), 7–15.

BULLOCK, A. 1977: Liberalism. In A. Bullock and O. Stallybrass (eds) *The Fontana dictionary of modern thought*. London: William Collins, 347.

BUNGE, W. 1962: second edition, 1966. *Theoretical geography*. Lund Studies in Geography, Series C 1, Lund: C. W. K. Gleerup.

_____ 1966: Locations are not unique. *Annals of the Association of America Geographers* 56, 375–6.

_____ 1968: Fred K. Schaefer and the science of geography. *Harvard Papers in Theoretical Geography*, Special Papers Series, Paper A, Laboratory for Computer Graphics and Spatial Analysis, Harvard University, Cambridge, MA.

_____ 1971: *Fitzgerald: geography of a revolution*. Cambridge, MA: Schlenkman.

_____ 1973a: Spatial prediction. *Annals of the Association of American Geographers* 63, 566–8.

_____ 1973b: Ethics and logic in geography. In R. J. Chorley (ed.) *Directions in geography*. London: Methuen, 317–31.

_____ 1973c: The geography of human survival. *Annals of the Association of American Geographers* 63, 275–95.

_____ 1973d: The geography. *The Professional Geographer* 25, 331–7.

_____ 1979: Fred K. Schaefer and the science of geography. *Annals of the Association of American Geographers* **69**, 128–33.

BUNGE, W. and BORDESSA, R. 1975: *The Canadian alternative: survival, expeditions and urban change*. Geographical Monographs, Atkinson College, York University, Downsview, Ontario.

BUNTING, T. E. and GUELKE, L. 1979: Behavioral and perception geography: a critical appraisal. *Annals of the Association of American Geographers* **69**, 448–62, 471–4.

BURNETT, K. P. (ed.) 1981: Studies in choice, constraints, and human spatial behavior. *Economic Geography* **57**, 291–383.

BURTON, I. 1963: The quantitative revolution and theoretical geography. *The Canadian Geographer* **7**, 151–62.

BURTON, I., KATES, R. W. and WHITE, G. F. 1978: *The environment as hazard*. New York: Oxford University Press.

BUSHONG, A. D. 1981: Geographers and their mentors: a genealogical view of American academic geography. In B. W. Blouet (ed.) *The origins of academic geography in the United States*. Hamden, CT: Archon Books, 193–220.

BUTLIN, R. A. 1982: *The transformation of rural England c. 1580–1800*. Oxford: Oxford University Press.

BUTTIMER, A. 1971: *Society and milieu in the French geographical tradition*. Chicago: Rand McNally.

_____ 1974: *Values in geography*. Commission on College Geography, Resource Paper 24, Association of American Geographers, Washington.

_____ 1976: Grasping the dynamism of lifeworld. *Annals of the Association of American Geographers* **66**, 277–92.

_____ 1978a: Charism and context: the challenge of *La Geographie Humaine*. In D. Ley and M. S. Samuels (eds) *Humanistic geography: prospects and problems*. Chicago: Maaroufa Press, 58–76.

_____ 1978b: On people, paradigms and progress in geography. Institutionen for Kulturgeografi oeh Eeonomisk Geografi vid Lunds Universitet, *Rapporter och Notiser* 47.

_____ 1979: Erewhon or nowhere land. In S. Gale and G. Olsson (eds) *Philosophy in geography*. Dordrecht: D. Reidel, 9–38.

_____ 1981: On people, paradigms and progress in geography. In D. R. Stoddart (ed.) *Geography, ideology and social concern*. Oxford: Blackwell, 70–80.

_____ 1983: *The practice of geography*. London: Longman.

_____ 1993: *Geography and the human spirit*. Baltimore: Johns Hopkins University Press.

BUTTIMER, A. and HAGERSTRAND, T. 1980: *Invitation to dialogue: a progress report*. DIA Paper 1, Lund: University of Lund.

_____ 1995: Book review. *Annals of the Association of American Geographers*, **85**, 406–7.

BUTZER, K. W. 1989: Cultural ecology. In G. L. Gaile and C. J. Willmot (eds) *Geography in America*. Columbus, OH: Merrill, 192–208.

_____ 1990: Hartshorne, Hettner, and *The Nature of Geography*. In J. N. Entrikin and S. D. Brunn (eds) *Reflections on Richard Hartshorne's The Nature of Geography*. Washington: Association of American Geographers, 35–52.

CADWALLADER, M. 1975: A behavioral model of consumer spatial decision making. *Economic Geography* **51**, 339–49.

_____ 1986: Structural equation models in human geography. *Progress in Human Geography* **10**, 24–47.

CAMERON, I. 1980: *To the farthest ends of the earth*. London: Macdonald.

CAMPBELL, C. S. 1994: The second nature of geography: Hartshorne as humanist. *The Professional Geographer* **46**, 411–17.

CAMPBELL, J. A. 1989: The concept of 'the behavioural environment', and its origins, reconsidered. In F. W. Boal and D. N. Livingstone (eds) *The behavioural environment*. London: Routledge, 33–76.

CAMPBELL, J. A. and LIVINGSTONE, D. N. 1983: Neo-Lamarckism and the development of geography in the United States and Great Britain. *Transactions, Institute of British Geographers* NS8, 267–94.

CAPEL, H. 1981: Institutionalization of geography and strategies of change. In D. R. Stoddart (ed.) *Geography, ideology and social concern*. Oxford: Blackwell, 37–69.

CAREY, H. C. 1858: *Principles of social science*. Philadelphia: J. Lippincott.

CARLSTEIN, T. 1980: *Time, resources, society and ecology*. Lund: Department of Geography, University of Lund.

CARLSTEIN, T., PARKES, D. N. and THRIFT, N. J. (eds) 1978: *Timing space and spacing time* (three volumes). London: Edward Arnold.

CARR, M. 1983: A contribution to the review and critique of behavioural industrial location theory. *Progress in Human Geography* 7, 386–402.

CARROLL, G. R. 1982: National city-size distributions: what do we know after 67 years of research? *Progress in Human Geography* 6, 1–43.

CARROTHERS, G. A. P. 1956: An historical review of the gravity and potential concepts of human interaction. *Journal, American Institute of Planners* 22, 94–102.

CASTELLS, M. 1977: *The urban question*. London: Edward Arnold.

CHAPMAN, G. P. 1977: *Human and environmental systems: a geographer's appraisal*. London: Academic Press.

CHAPPELL, J. E., Jr. 1975: The ecological dimension: Russian and American views. *Annals of the Association of American Geographers* 65, 144–62.

_____ 1976: Comment in reply. *Annals of the Association of American Geographers* 66, 169–73.

CHAPPELL, J. M. A. and WEBBER, M. J. 1970: Electrical analogues of spatial diffusion processes. *Regional Studies* 4, 25–39.

CHARLTON, M., RAO, L. and CARVER, S. 1995: GIS and the census. In S. Openshaw (ed.) *Census users' handbook*. Cambridge: GeoInformation International.

CHISHOLM, M. 1962: *Rural settlement and land use*. London: Hutchinson.

_____ 1966: *Geography and economics*. London: G. Bell & Sons.

_____ 1967: General systems theory and geography. *Transactions, Institute of British Geographers* 42, 45–52.

_____ 1971a: In search of a basis for location theory: micro-economics or welfare economics? In C. Board *et al.* (eds) *Progress in Geography 3*, London: Edward Arnold, 111–34.

_____ 1971b: Geography and the question of 'relevance'. *Area* 3, 65–8.

_____ 1973: The corridors of geography. *Area* 5, 43.

_____ 1975a: *Human geography: evolution or revolution?* Harmondsworth: Penguin Books.

_____ 1975b: The reformation of local government in England. In R. Peel, M. Chisholm and P. Haggett (eds) *Processes in physical and human geography: Bristol essays*. London: Heinemann, 305–18.

_____ 1976: Regional policies in an era of slow population growth and higher unemployment. *Regional Studies* 10, 201–13.

_____ 1995: Some lessons from the review of local government in England. *Regional Studies* 29, 563–9.

CHISHOLM, M., FREY, A. E. and HAGGETT, P. (eds) 1971: *Regional forecasting*. London: Butterworth.

CHISHOLM, M. and MANNERS, G. (eds) 1973: *Spatial policy problems of the British economy*. London: Cambridge University Press.

CHISHOLM, M. and O'SULLIVAN, P. 1973: *Freight flows and spatial aspects of the British economy.* Cambridge: Cambridge University Press.

CHORLEY, R. J. 1962: Geomorphology and general systems theory. *Professional Paper 500–B,* United States Geological Survey, Washington.

———— 1964: Geography and analogue theory. *Annals of the Association of American Geographers* 54, 127–37.

———— 1973a: Geography as human ecology. In R. J. Chorley (ed.) *Directions in geography.* London: Methuen, 155–70.

———— (ed.) 1973b: *Directions in geography.* London: Methuen.

CHORLEY, R. J. and BENNETT, R. J. 1981: Optimization: control models. In N. Wrigley and R. J. Bennett (eds) *Quantitative geography.* London: Routledge & Kegan Paul, 219–24.

CHORLEY, R. J. and HAGGETT, P. 1965a: Trend-surface mapping in geographical research. *Transactions and Papers, Institute of British Geographers* 37, 47–67.

———— (eds) 1965b: *Frontiers in geographical teaching.* London: Methuen.

———— (eds) 1967: *Models in geography.* London: Methuen.

CHORLEY, R. J. and KENNEDY, B. A. 1971: *Physical geography: a systems approach.* London: Prentice-Hall International.

CHOUINARD, V., FINCHER, R. and WEBBER, M. 1984: Empirical research in scientific human geography. *Progress in Human Geography* 8, 347–80.

CHRISMAN, N. R., COWEN, D. J., FISHER, P. F., GOODCHILD, M. F. and MARK, D. M. 1989: Geographic information systems. In G. L. Gaile and C. J. Willmott (eds) *Geography in America.* Columbus, OH: Merrill, 776–96.

CHRISTALLER, W. 1966: *Central places in southern Germany* (translated by C. W. Baskin). Englewood Cliffs, NJ: Prentice-Hall.

CHRISTENSEN, K. 1982: Geography as a human science. In P. Gould and G. Olsson (eds) A *search for common ground.* London: Pion, 37–57.

CHRISTOPHERSON, S. 1989: On being outside 'the project'. *Antipode* 21, 83–9.

CLARK, A. H. 1954: Historical geography. In P. E. James and C. F. Jones (eds) *American geography: inventory and prospects.* Syracuse: Syracuse University Press, 70–105.

———— 1977: The whole is greater than the sum of the parts: a humanistic element in human geography. In D. R. Deskins *et al.* (eds) *Geographic humanism, analysis and social action: a half century of geography at Michigan.* Ann Arbor Michigan Geographical Publication No. 17, 3–26.

CLARK, D., DAVIES, W. K. D. and JOHNSTON, R. J. 1974: The application of factor analysis in human geography. *The Statistician* 23, 259–81.

CLARK, G. L. 1982: Instrumental reason and policy analysis. In D. T. Herbert and R. J. Johnston (eds) *Geography and the urban environment,* vol. 5. Chichester: John Wiley, 41–62.

———— 1985: *Judges and the cities.* Chicago: University of Chicago Press.

CLARK, G. L. and DEAR, M. J. 1984: *State apparatus.* Boston: George Allen & Unwin.

CLARK, K. G. T. 1950: Certain underpinnings of our arguments in human geography. *Transactions, Institute of British Geographers* 16, 15–22.

CLARK, W. A. V. 1975: Locational stress and residential mobility in a New Zealand context. *New Zealand Geographer* 31, 67–79.

———— 1981: Residential mobility and behavioral geography: parallelism or interdependence? In K. R. Cox and R. C. Golledge (eds) *Behavioral problems in geography revisited.* London: Methuen, 182–205.

———— 1991: Geography in court: expertise in adversarial settings. *Transactions, Institute of British Geographers* NS16, 5–20.

_____ 1993: Applying our understanding: social science in government and the marketplace. *Environment and Planning A* anniversary issue, 38–47.

CLARKE, M. and HOLM, E. 1987: Towards an applicable human geography: some developments and observations. *Environment and Planning A* **19**, 1525–41.

_____ 1988: Microsimulation methods in spatial analysis and planning. *Geografiska Annaler* **69B**, 145–64.

CLARKE, M. and WILSON, A. G. 1985a: A model-based approach to planning in the National Health Service. *Environment and Planning B: Planning and Design* **12**, 287–302.

_____ 1985b: The dynamics of urban spatial structure: the progress of a research programme. *Transactions, Institute of British Geographers* NS10, 427–51.

_____ 1989: Mathematical models in human geography. In R. Peet and N. J. Thrift (eds) *New models in geography*, vol. 2. London: Unwin Hyman, 30–42.

CLARKSON, J. D. 1970: Ecology and spatial analysis. *Annals of the Association of American Geographers* **60**, 700–16.

CLAVAL, P. 1981: Epistemology and the history of geographical thought. In D. R. Stoddart (ed.) *Geography, ideology and social concern*. Oxford: Blackwell, 227–39.

_____ 1983: *Models of man in geography*. Syracuse: Department of Geography, Syracuse University, Discussion Paper 79.

CLAYTON, K. M. 1985a: New blood by (government) order. *Area* **17**, 321–2.

_____ 1985b: The state of geography. *Transactions, Institute of British Geographers* NS10, 5–16.

CLIFF, A. D. and HAGGETT, P. 1988: *Atlas of disease distributions: analytical approaches to epidemiological data*. Oxford: Blackwell Publishers.

_____ 1989: Spatial aspects of epidemic control. *Progress in Human Geography* **13**, 315–47.

_____ 1995: Disease implications of global change. In R. J. Johnston, P. J. Taylor and M. J. Watts (eds) *Geographies of global change: remapping the world in the late twentieth century*. Oxford: Blackwell Publishers, 206–22.

CLIFF, A. D., HAGGETT, P. and ORD, J. K. 1987: *Spatial aspects of influenza epidemics*. London: Pion.

CLIFF, A. D., HAGGETT, P. and SMALLMAN-RAYNOR, M. 1993: *Measles: an historical geography of a major human viral disease from global expansion to local retreat, 1840–1990*. Cambridge: Cambridge University Press.

CLIFF, A. D. and ORD, J. K. 1973: *Spatial autocorrelation*. London: Pion.

_____ 1981: *Spatial process* London: Pion.

CLIFF, A. D. *et al.* 1975: *Elements of spatial structure: a quantitative approach*. London: Cambridge University Press.

_____ *et al.* 1981: *Spatial diffusion,* Cambridge: Cambridge University Press.

CLOKE, P., PHILO, C. and SADLER, D. 1991: *Approaching human geography: an introduction to contemporary theoretical debates*. London: Paul Chapman.

COATES, B. E., JOHNSTON, R. J. and KNOX, P. L. 1977: *Geography and inequality*. Oxford: Oxford University Press.

COCHRANE, A. 1987: What a difference the place makes: the new structuralism of locality. *Antipode* **19**, 354–63.

COFFEY, W. J. 1981: *Geography: towards a general spatial systems approach*. London: Methuen.

COLE, J. P. 1969: Mathematics and geography. *Geography* **54**, 152–63.

COLE, J. P. and KING C. A. M. 1968: *Quantitative geography*. London: John Wiley.

CONZEN, M. P. 1981: The American urban system in the nineteenth century. In D. T. Herbert and R. J. Johnston (eds) *Geography and the urban environment*, vol. 4. Chichester: John Wiley, 295–348.

COOK, I. and CRANG, M. 1995: *Doing ethnographies. Concepts and techniques in modern geography* 58. Norwich: Geo Books.

COOKE, P. N. 1986: The changing urban and regional system in the United Kingdom. *Regional Studies* 20, 243–52.

_____ 1987a: Clinical inference and geographic theory. *Antipode* 19, 69–78.

_____ 1987b: Individuals, localities and postmodernism. *Environment and Planning D: Society and Space* 5, 408–12.

_____ (ed.) 1989a: *Localities: The changing face of urban Britain.* London: Unwin Hyman.

_____ 1989b: Locality theory and the poverty of 'spatial variation' (A response to Duncan and Savage). *Antipode* 21, 261–73.

_____ 1990: *Back to the future: modernity, postmodernity and locality.* London: Unwin Hyman.

COOKE, R. U. 1985a: Applied geomorphology. In A. Kent (ed.) *Perspectives on a changing geography.* Sheffield: The Geographical Association, 36–47.

_____ 1985b: *Geomorphological hazards in Los Angeles.* London: George Allen & Unwin.

_____ 1992: Common ground, shared inheritance: research imperatives for environmental geography. *Transactions, Institute of British Geographers* NS17, 131 51.

COOKE, R. U. and ROBSON, B. T. 1976: Geography in the United Kingdom, 1972–1976. *Geographical Journal* 142, 3–22.

COOMBES, M.G. *et al.* 1982: Functional regions for the population census of Great Britain. In D. T. Herbert and R. J. Johnston (eds) *Geography and the urban environment,* vol. 5. Chichester: John Wiley, 63–111.

COOPER, W. 1952: *The struggles of Albert Woods.* London: Jonathan Cape.

COPPOCK, J. T. 1974: Geography and public policy: challenges, opportunities and implications. *Transactions, Institute of British Geographers* 63, 1–16.

CORBRIDGE, S. 1986: *Capitalist world development.* London: Macmillan.

_____ 1988: Deconstructing determinism. *Antipode* 20, 239–69.

_____ 1989: Marxism, post-Marxism, and the geography of development. In R. Peet and N. J. Thrift (eds) *New models in geography,* vol. 1. London: Unwin Hyman, 224–54.

_____ 1993: Marxists, modernities, and moralities: development praxis and the claims of distant strangers. *Environment and Planning D: Society and Space* 11, 449–72.

CORBRIDGE, S., MARTIN, R. L. and THRIFT, N. J. (eds) 1994: *Money, space and power.* Oxford: Blackwell Publishers.

COSGROVE, D. E. 1983: Towards a radical cultural geography: problems of theory. *Antipode* 15, 1–11.

_____ 1984: *Social formation and symbolic landscape.* London: Croom Helm.

_____ 1989a: Geography is everywhere: culture and symbolism in human landscapes. In D. Gregory and R. Walford (eds) *Horizons in Human Geography.* London: Macmillan, 118–135.

_____ 1989b: Models, description and imagination in geography. In B. Macmillan (ed.) *Remodelling geography.* Oxford: Blackwell, 23–44.

_____ 1993: Commentary. *Annals of the Association of American Geographers* 83, 515–17.

_____ 1996: Classics in human geography revisited. *Progress in Human Geography* 20, 197–9.

COSGROVE, D. E. and DANIELS, S. J. (eds) 1988: *The iconography of landscape.* Cambridge: Cambridge University Press.

_____ 1989: Fieldwork as theatre: a week's performance in Venice and its region. *Journal of Geography in Higher Education* 13, 169–82.

COSGROVE, D. E. and JACKSON, P. 1987: New directions in cultural geography. *Area* 19, 95–101.

COUCLELIS, H. 1986a: Artificial intelligence in geography: conjectures on the shape of things to come. *The Professional Geographer* 38, 1–10.

_____ 1986b: A theoretical framework for alternative models of spatial decision and behavior. *Annals of the Association of American Geographers* 76, 95–113.

COUCLELIS, H. and GOLLEDGE, R. G. 1983: Analytic research, positivism, and behavioral geography. *Annals of the Association of American Geographers.* 73, 331–9.

COURT, A. 1972: All statistical populations are estimated from samples. *The Professional Geographer* 24, 160–1.

COWEN, D. J. 1983: Automated geography and the DIDS. *The Professional Geographer* 35, 339–40.

COX, K. R. 1969: The voting decision in a spatial context. In C. Board *et al.* (eds) *Progress in Geography 1.* London: Edward Arnold, 81–118.

_____ 1973: *Conflict, power and politics in the city: a geographic view.* New York: McGraw-Hill.

_____ 1976: American geography: social science emergent. *Social Science Quarterly* 57, 182–207.

_____ 1979: *Location and public problems.* Oxford: Basil Blackwell.

_____ 1981: Bourgeois thought and the behavioral geography debate. In K. R. Cox and R. G. Golledge (eds), *Behavioral problems in geography revisited.* London: Methuen, 256–79.

_____ 1989: The politics of turf and the question of class. In J. Wolch and M. Dear (eds) *The power of geography.* Boston: Unwin Hyman, 61–90.

_____ 1995: Concepts of space, understanding in human geography, and spatial analysis. *Urban Geography* 16, 304–26.

COX, K. R. and GOLLEDGE R. G. (eds) 1969: *Behavioral problems in geography: a symposium.* Evanston: Northwestern University Studies in Geography 17.

_____ (eds) 1981: *Behavioural problems in geography revisited.* London: Methuen.

COX, K. R. and McCARTHY, J. J. 1982: Neighbourhood activism as a politics of turf: a critical analysis. In K. R. Cox and R. J. Johnston, (eds) *Conflict, politics and the urban scene.* London: Longman, 196–219.

COX, K. R. and MAIR, A. 1989: Levels of abstraction in locality studies. *Antipode* 21, 121–32.

COX, K. R., REYNOLDS, D. R. and ROKKAN, S. (eds) 1974: *Locational approaches to power and conflict.* New York: Halsted Press.

COX, N. J. 1989: Modelling, data analysis and Pygmalion's problem. In B. Macmillan (ed.) *Remodelling geography.* Oxford: Basil Blackwell, 204–10.

COX, N. J. and JONES, K. 1981: Exploratory data analysis. In N. Wrigley and R. J. Bennett (eds) *Quantitative geography.* London: Routledge & Kegan Paul, 135–43.

CRANE, D. 1972: *Invisible colleges.* Chicago: University of Chicago Press.

CRANG, M. 1994: On the heritage trail: maps of and journeys to olde Englande. *Environment and Planning D: Society and Space* 12, 341–56.

CRANG, P. 1992: The politics of polyphony: reconfigurations in geographical authority. *Environment and Planning D: Society and Space* 10, 527–49.

CROMLEY, R. G. 1993: Automated geography ten years later. *The Professional Geographer* 45, 442–3.

CROWE, P. R. 1936: The rainfall regime of the Western Plains. *Geographical Review* 26, 463–84.

_____ 1938: On progress in geography. *Scottish Geographical Magazine* 54, 1–19.

_____ 1970: Review of *Progress in Geography 1. Geography* 55, 346–7.

CUMBERLAND, K. B. 1947: *Soil erosion in New Zealand.* Wellington: Whitcombe & Tombs.

CURRAN, P. J. 1984: Geographic information systems. *Area* 16, 153–8.

_____ 1996: Differential research funding. *Area* 28.

CURRY, L. 1967: Quantitative geography. *The Canadian Geographer* 11, 265–74.

_____ 1972: A spatial analysis of gravity flows. *Regional Studies* 6, 131–47.

CURRY, M. 1982a: The idealist dispute in Anglo-American geography. *The Canadian Geographer* 26, 37–50.

_____ 1982b: The idealist dispute in Anglo-American geography: a reply. *The Canadian Geographer* 26, 57–9.

_____ 1991: Postmodernism, language and the strains of modernism. *Annals of the Association of American Geographers* 81, 210–28.

_____ 1992: Reply. *Annals of the Association of American Geographers* 82, 310–12.

_____ 1994: Image, practice and the hidden impacts of geographic information systems. *Progress in Human Geography* 18, 460–90.

_____ 1995: GIS and the inevitability of ethical inconsistency. In J. Pickles (ed.) *Ground truth: the social implications of geographical information systems.* New York: Guilford Press, 68–87.

_____ 1996: On space and spatial practice in contemporary geography. In C. Earle, K. Mathewson and M. S. Kenzer (eds) *Concepts in human geography.* Lanham MD: Rowman & Littlefield, 3–32.

DACEY, M.F. 1962: Analysis of central-place and point patterns by a nearest-neighbor method. In K. Norborg (ed.) *Proceedings of the IGU Symposium in Urban Geography, Lund 1960.* Lund: C. W. K. Gleerup, 55–76.

_____ 1968: A review on measures of contiguity for two and k-color maps. In B. J. L. Berry and D. F. Marble (eds) *Spatial analysis.* Englewood Cliffs, NJ: Prentice-Hall, 479–95.

_____ 1973: Some questions about spatial distributions. In R. J. Chorley (ed.) *Directions in geography.* London: Methuen, 127–52.

DALBY, S. 1991: Critical geopolitics: discourse, difference, and dissent. *Environment and Planning D: Society and Space* 9, 261–83.

DALE, A. 1993: Office of Population Censuses and Surveys longitudinal study. *Environment and Planning A* 25, 83–6.

DANIELS, P. W. 1982: *Service industries: growth and location.* Cambridge: Cambridge University Press.

_____ 1985: *Service industries: a geographical appraisal.* London: Methuen.

DANIELS, S. J. 1991: The making of Constable country. *Landscape Research* 16 (2), 9–18.

DARBY, H. C. 1948: The regional geography of Thomas Hardy's Wessex. *The Geographical Review,* 38, 426–43.

_____ 1953: On the relations of geography and history. *Transactions and Papers, Institute of British Geographers* 19, 1–11.

_____ 1962: The problem of geographical description. *Transactions and Papers, Institute of British Geographers* 30, 1–14.

_____ (ed.) 1973: *A new historical geography of England.* London: Cambridge University Press.

_____ 1977: *Domesday England.* London: Cambridge University Press.

_____ 1983a: Historical geography in Britain, 1920–1980: continuity and change. *Transactions, Institute of British Geographers* NS8, 421–8.

_____ 1983b: Academic geography in Britain, 1918–1946. *Transactions, Institute of British Geographers* NS8, 14–26.

DAVEY, J., JONES, R., LAWRENCE, V., STEVENSON, I., JENKINS, A. and

SHEPHERD, I. D. H. 1995: Issues and trends in textbook publishing: the views of geography editors/publishers. *Journal of Geography in Higher Education* 19, 11–28.

DAVIES, R. B. and PICKLES, A. R. 1985: Longitudinal versus cross-sectional methods for behavioural research: a first-round knockout. *Environment and Planning A* 17, 1315–30.

DAVIES, W. K. D. 1972a: Geography and the methods of modern science, In W. K. D. Davies (ed.) *The conceptual revolution in geography*. London: University of London Press, 131–9.

_____ 1972b: Introduction: the conceptual revolution in geography. In W. K. D. Davies (ed.) *The conceptual revolution in geography*. London: University of London Press, 9–18.

DAVIS, W. M. 1906: An inductive study of the content of geography. *Bulletin of the American Geographical Society* 38, 67–84.

DAY, M. and TIVERS, J. 1979: Catastrophe theory and geography: a Marxist critique. *Area* 11, 54–8.

DAYSH, G. H. J. (ed.) 1949: *Studies in regional planning*. London: Philip & Son.

DEAR, M. J. 1987: Society, politics and social theory. *Environment and Planning D: Society and Space* 5, 363–6.

_____ 1988: The postmodern challenge: reconstructing human geography. *Transactions, Institute of British Geographers* NS13, 262–74.

_____ 1991a: The premature demise of postmodern urbanism. *Cultural Anthropology* 6, 538–52.

_____ 1991b: Review of Harvey, The condition of postmodernity. *Annals of the Association of American Geographers* 81, 533–9.

_____ 1994: Postmodern human geography: an assessment. *Erdkunde* 48, 2–13.

_____ 1995: Practising postmodern geography. *Scottish Geographical Magazine* 111, 179–81.

DEAR, M. J. and CLARK, G. L. 1978: The state and geographic process: a critical review. *Environment and Planning A* 10, 173–84.

DEAR, M. J. and MOOS, A. I. 1986: Structuration theory in urban analysis: 2. empirical application. *Environment and Planning A* 18, 351–74.

DEAR, M. J. and SCOTT, A. J. (eds) 1981: *Urbanization and urban planning in capitalist society*. London: Methuen.

DENNIS, R. J. 1984: *English industrial cities in the nineteenth century: a social geography*. Cambridge: Cambridge University Press.

DERRIDA, J. 1976: *Of grammatology*. Baltimore: Johns Hopkins University Press.

DESBARATS, J. 1983: Spatial choice and constraints on behavior. *Annals of the Association of American Geographers* 73, 340–57.

DETWYLER, T. R. and MARCUS, M. G. (eds) 1972: *Urbanization and environment*. Belmont, CA: Duxbury Press.

DEUTSCH, R. 1991: Boy's town. *Environment and Planning D: Society and Space* 9, 5–30.

DICKEN, P. 1986: *Global shift*. London: Harper and Row.

DICKENS P., DUNCAN, S. S., GOODWIN, M. and GRAY, F. 1985: *Housing, states and localities*. London: Methuen.

DICKENSON, J. P. and CLARKE, C. G. 1972: Relevance and the 'newest geography'. *Area* 3, 25–7.

DICKINSON, R. E. 1933: The distribution and functions of smaller urban settlements of East Anglia. *Geography* 18, 19–31.

_____ 1947: *City, region and regionalism*. London: Routledge & Kegan Paul.

DINGEMANS, D. 1979: Redlining and mortgage lending in Sacramento. *Annals of the Association of American Geographers* 69, 225–39.

DOBSON, J. E. 1983a: Automated geography. *The Professional Geographer* 35, 135–43.

———— 1983b: Reply to comments on 'Automated geography'. *The Professional Geographer* 35, 349–53.

———— 1993: The geographic revolution: a retrospective on the age of automated geography. *The Professional Geographer* 45, 431–9.

DODDS, K.-J. 1994a: Geopolitics and foreign policy: recent developments in Anglo American political geography and international relations. *Progress in Human Geography* 18, 186–208.

———— 1994b: Geopolitics in the Foreign Office: British representations of Argentine 1945–1961. *Transactions, Institute of British Geographers* NS19, 273–90.

DODDS, K.-J. and SIDAWAY, J. D. 1994: Locating critical geopolitics. *Environment and Planning D: Society and Space* 12, 515–24.

DOEL, M. A. 1993: Proverbs for paranoids: writing geography on hollowed ground. *Transactions, Institute of British Geographers* NS18, 377–94.

DOMOSH, M. 1991a: Towards a feminist historiography of geography. *Transactions, Institute of British Geographers* NS16, 95–104.

———— 1991b: Beyond the frontiers of geographical knowledge. *Transactions, Institute of British Geographers* NS16, 488–90.

DORLING, D. 1995: *A social atlas of Britain.* Chichester: John Wiley.

DOUGLAS, I. 1983: *The urban environment.* London: Edward Arnold.

———— 1986: The unity of geography is obvious. *Transactions, Institute of British Geographers* NS11, 459–63.

DOWNS, R. M. 1970: Geographic space perception: past approaches and future prospects. In C. Board *et al.* (eds) *Progress in Geography* 2. London: Edward Arnold, 65–108.

———— 1979: Critical appraisal or determined philosophical skepticism? *Annals of the Association of American Geographers* 69, 468–71.

DOWNS, R. M. and MEYER, J. T. 1978: Geography and the mind. *Human geography: coming of age. American Behavioral Scientist* 22, 59–78.

DOWNS, R. M. and STEA, D. (eds) 1973: *Image and environment.* London: Edward Arnold.

———— 1977: *Maps in mind.* New York: Harper & Row.

DRIVER, F. and PHILO, C. 1986: Implications of 'scientific' geography. *Area* 18, 161–2.

DRYSDALE, A. and WATTS, M. 1977: Modernization and social protest movements. *Antipode* 9(1), 40–55.

DUFFY, P. 1995: Literary reflections on Irish migration in the nineteenth and twentieth centuries. In R. King, J. Connell and P. White (eds) *Writing across worlds: literature and migration.* London: Routledge, 20–38.

DUNCAN, J. S. 1980: The superorganic in American cultural geography. *Annals of the Association of American Geographers* 70, 181–98.

———— 1985: Individual action and political power: a structuration perspective. In R. J. Johnston (ed.) *The future of geography.* London: Methuen, 174–89.

———— 1993a: Representing power: the politics and poetics of urban form in the Kandyan Kingdom. In J. S. Duncan and D. Ley (eds) *Place/culture/representation.* London: Routledge, 232–50.

———— 1993b: Commentary. *Annals of the Association of American Geographers* 83, 517–19.

———— 1994: After the civil war: reconstructing cultural geography as heterotopia. In K. E. Foote, P. J. Hugill, K. Mathewson and J. M. Smith (eds) *Re-reading cultural geography.* Austin: University of Texas Press, 401–8.

DUNCAN, J. S. and BARNES, T. S. 1993: Afterword. In T. J. Barnes and J. S.

Duncan (eds) *Writing worlds: discourse, text and metaphor in the representation of landscape*. London: Routledge, 248–53.

DUNCAN, J. S. and LEY, D. 1982: Structural marxism and human geography: a critical assessment. *Annals of the Association of American Geographers* 72, 30–59.

_____ 1993: Introduction: representing the place of culture. In J. S. Duncan and D. Ley (eds) *Place/culture/representation*. London: Routledge, 1–21.

DUNCAN, N. 1996: Postmodernism in human geography. In C. V. Earle, K. Mathewson and M. S. Kenzer (eds) *Concepts in human geography*. Lanham MD: Rowman and Littlefield, 429–58.

DUNCAN, O. D. 1959: Human ecology and population studies. In P. M. Hauser and O. D. Duncan (eds) *The study of population*. Chicago: University of Chicago Press, 678–716.

DUNCAN, O. D., CUZZORT, R. P. and DUNCAN, B. 1961: *Statistical geography*. New York: The Free Press.

DUNCAN, O. D. and SCHNORE, L. F. 1959: Cultural, behavioral and ecological perspectives in the study of social organization. *American Journal of Sociology* 65, 132–46.

DUNCAN, S. S. 1974a: Cosmetic planning or social engineering? Improvement grants and improvement areas in Huddersfield. *Area* 6, 259–70.

_____ 1974b: The isolation of scientific discovery: indifference and resistance to a new idea. *Science Studies* 4, 109–34.

_____ 1975: Research directions in social geography: housing opportunities and constraints. *Transactions, Institute of British Geographers* NS1, 10–19.

_____ 1979: Radical geography and marxism. *Area* 11, 124–6.

_____ 1981: Housing policy, the methodology of levels, and urban research: the case of Castells. *International Journal of Urban and Regional Research* 5, 231–54.

_____ 1989: What is a locality? In R. Peet and N. J. Thrift (eds) *New models in geography*, vol. 2. London: Unwin Hyman, 221–54.

DUNCAN, S. S. and GOODWIN, M. 1985: The local state and local economic policy. *Capital and Class* 27, 14–36.

DUNCAN, S. S. and SAVAGE, M. 1989: Space, scale and locality. *Antipode* 21, 179–206.

DUNFORD, M. F. and PERRONS, D. 1983: *The arena of capital*. London: Macmillan.

EARLE, C. V. 1996: Classics in human geography revisited. *Progress in Human Geography*, 20, 195–7.

EILON, S. 1975: Seven faces of research. *Operational Research Quarterly* 26, 359–67.

ELIOT HURST, M. E. 1972: Establishment geography: or how to be irrelevant in three easy lessons. *Antipode* 5(2), 40–59.

_____ 1980: Geography, social science and society: towards a de-definition. *Australian Geographical Studies* 18, 3–21.

_____ 1985: Geography has neither existence nor future. In R. J. Johnston (ed.) *The future of geography*. London: Methuen, 59–91.

ENTRIKIN, J. N. 1976: Contemporary humanism in geography. *Annals of the Association of American Geographers* 66, 615–32.

_____ 1980: Robert Park's human ecology and human geography. *Annals of the Association of American Geographers* 70, 43–58.

_____ 1981: Philosophical issues in the scientific study of regions. In D. T. Herbert and R. J. Johnston (eds) *Geography and the urban environment*, vol. 4. Chichester: John Wiley, 1–27.

_____ 1989: Place, region, and modernity. In J. A. Agnew and J. S. Duncan (eds) *The power of place*. Boston: Unwin Hyman, 30–43.

_____ 1990: Introduction: *The Nature of Geography* in perspective. In J. N. Entrikin and S. D. Brunn (eds) *Reflections on Richard Hartshorne's The Nature of Geography*. Washington: Association of American Geographers, 1–16.

ESRC 1988: *Horizons and opportunities in social science*. London: ESRC.

EVANS, M. 1988: Participant observation: the researcher as research tool. In J. Eyles and D. M. Smith (eds) *Qualitative methods in human geography*. Cambridge: Polity Press, 197–218.

EYLES, J. 1971: Pouring new sentiments into old theories: how else can we look at behavioural patterns? *Area* 3, 242–50.

_____ 1973: Geography and relevance. *Area* 5, 158–60.

_____ 1974: Social theory and social geography. In C. Board *et al.* (eds) *Progress in Geography* 6. London: Edward Arnold, 27–88.

_____ 1981: Why geography cannot be Marxist: towards an understanding of lived experience. *Environment and Planning A* 13, 1371–88.

_____ 1989: The geography of everyday life. In D. Gregory and R. Walford (eds) *Horizons in human geography*. London: Macmillan, 102–17.

EYLES, J. and LEE, R. 1982: Human geography in explanation. *Transactions, Institute of British Geographers* NS7, 117–21.

EYLES, J. and SMITH, D. M. 1978: Social geography. *Human geography: Coming of Age. American Behavioral Scientist* 22, 41–58.

EYRE, S. R. 1978: *The real wealth of nations*. London: Edward Arnold.

EYRE, S. R. and JONES, G. R. J. (eds) 1966: *Geography as human ecology*. London: Edward Arnold.

FALAH, G. 1989: Israelization of Palestine human geography. *Progress in Human Geography* 13, 535–50.

_____ 1994: The frontier of political criticism in Israeli geographic practice. *Area* 26, 1–12.

FALAH, G. and NEWMAN, D. 1995: The spatial manifestation of threat: Israelis and Palestinians seek a 'good boundary'. *Political Geography* 14, 689–706.

FELDMAN, E. J. 1986: The citizen-scholar: education and public affairs. In R. W. Kates and I. Burton (eds) *Geography, resources and environment*, vol. 2. *Themes from the work of Gilbert F. White*. Chicago: University of Chicago Press, 188–206.

FEYERABEND, P. 1975: *Against method*. London: New Left Books.

FIELDHOUSE, E., PATTIE, C .J. and JOHNSTON, R. J. 1996: Tactical voting and party constituency campaigning at the 1992 British general election in England. *British Journal of Political Science* 26, 403–18.

FINGLETON, B. 1984: *Models of category counts*. Cambridge: Cambridge University Press.

FISCHER, M. M. and GOPAL, S. 1993: Neurocomputing – a new paradigm for geographic information processing. *Environment and Planning A* 25, 757–60.

FISHER, P. F. 1989a: Geographical information system software for teaching. *Journal of Geography in Higher Education* 13, 69–80.

_____ 1989b: Expert system applications in geography. *Area* 21, 279–87.

FITZSIMMONS, M. 1989: The matter of nature. *Antipode* 21, 106–21.

FLEMING, D. K. 1973: The regionalizing ritual. *Scottish Geographical Magazine* 89, 196–207.

FLEURE, H. J. 1919: Human regions. *Scottish Geographical Magazine* 35, 94–105.

FLOWERDEW, R. 1986: Three years in British geography. *Area* 18, 263–4.

_____ 1989: Some critical views of modelling in geography. In B. Macmillan (ed.) *Remodelling geography*. Oxford: Basil Blackwell, 245–54.

FOLCH-SERRA, M. 1990: Place, voice and space: Mikhail Bakhtin's dialogical landscape. *Environment and Planning D: Society and Space* 8, 255–74.

FOLKE, S. 1972: Why a radical geography must be Marxist. *Antipode* 4(2), 13–18.

_____ 1973: First thoughts on the geography of imperialism. *Antipode* 5(3), 16–20.

FOLKE, S. and SAYER, A. 1991: What's left to do? Two views from Europe. *Antipode* 23, 240–8.

FOORD, J. and GREGSON, N. 1986: Patriarchy: towards a reconceptualisation. *Antipode* 18, 186–211.

FOOTE, D. G. and GREER-WOOTTEN, B. 1968: An approach to systems analysis in cultural geography. *The Professional Geographer* 20, 86–90.

FORD, L. R. 1982: Beware of new geographies. *The Professional Geographer* 34, 131–5.

FORER, P. C. 1974: Space through time: a case study with New Zealand airlines. In E. L. Cripps (ed.) *Space-time concepts in urban and regional models*. London: Pion, 22–45.

FORREST, B. 1995: West Hollywood as symbol: the significance of place in the construction of a gay identity. *Environment and Planning D: Society and Space* 13, 133–58.

FORRESTER, J. W. 1969: *Urban dynamics*. Cambridge, MA: MIT Press.

FOTHERINGHAM, A. S. 1981: Spatial structure and distance-decay parameters. *Annals of the Association of American Geographers* 71, 425–36.

_____ 1993: On the future of spatial analysis: the role of GIS. *Environment and Planning A* anniversary issue, 30–4.

FOTHERINGHAM, A. S. and MACKINNON, R. D. 1989: The National Center for Geographic Information and Analysis. *Environment and Planning A* 21, 141–4.

FOUCAULT, M. 1972: *The archaeology of knowledge*. London: Tavistock Publications.

FREEMAN, T. W. 1961: *A hundred years of geography*. London: Gerald Duckworth.

_____ 1977: *Geographers: biobibliographical studies*. London: Mansell.

_____ 1979: The British school of geography. *Organon* 14, 205–16.

_____ 1980a: *A history of modern British geography*. London: Longman.

_____ 1980b: The Royal Geographical Society and the development of geography. In E. H. Brown (ed.) *Geography, yesterday and tomorrow*. Oxford: Oxford University Press, 1–99.

FUKUYAMA, F. 1992: *The end of history and the last man*. London: Penguin Books.

FULLER, G. A. 1971: The geography of prophylaxis: an example of intuitive schemes and spatial competition in Latin America. *Antipode* 3(1), 21–30.

GAILE, G. L. and WILLMOTT, C. J. (eds) 1984: *Spatial statistics and models*. Dordrecht: D. Reidel.

_____ (eds) 1989: *Geography in America*. Columbus, OH: Merrill Publishing.

GALE, N. and GOLLEDGE, R. G. 1982: On the subjective partitioning of space. *Annals of the Association of American Geographers* 72, 60–7.

GAMBLE, A. 1988: *The free economy and the strong state*. London: Macmillan.

GARRISON, W. L. 1953: Remoteness and the passenger utilization of air transportation. *Annals of the Association of American Geographers* 43, 169.

_____ 1956a: Applicability of statistical inference to geographical research. *Geographical Review* 46, 427–9.

_____ 1956b: Some confusing aspects of common measurements. *The Professional Geographer* 8, 4–5.

_____ 1959a: Spatial structure of the economy I. *Annals of the Association of American Geographers* 49, 238–49.

_____ 1959b: Spatial structure of the economy II. *Annals of the Association of American Geographers* 49, 471–82.

_____ 1960a: Spatial structure of the economy III. *Annals of the Association of American Geographers* 50, 357–73.

_____ 1960b: Connectivity of the interstate highway system. *Papers and Proceedings, Regional Science Association* 6, 121–37.

_____ 1962: Simulation models of urban growth and development. In K. Norborg (ed.) *IGU symposium in urban geography,* Lund Studies in Geography B 24. Lund: C. W. K. Gleerup, 91–108.

_____ 1979: Playing with ideas. *Annals of the Association of American Geographers* 69, 118–20.

_____ 1995: Living with and loving a no-win situation. *International Regional Science Review* 17, 327–32.

GARRISON, W. L., BERRY, B. J. L., MARBLE, D. F., NYSTUEN, J. D. and MORRILL, R. L. 1959: *Studies of highway development and geographic change.* Seattle: University of Washington Press.

GARRISON, W. L. and MARBLE, D. F. 1957: The spatial structure of agricultural activities. *Annals of the Association of American Geographers* 47, 137–44.

_____ (eds) 1967a: *Quantitative geography. Part I: economic and cultural topics.* Evanston, ILL Northwestern University Studies in Geography, Number 13.

_____ (eds) 1967b: *Quantitative geography, Part II. physical and cartographic topics.* Evanston, ILL Northwestern University Studies in Geography, Number 14.

GATRELL, A. C. 1982: *Geometry in geography and the geometry of geography.* Discussion Paper 6, Department of Geography, Salford: University of Salford.

_____ 1983: *Distance and space: a geographical perspective.* Oxford: Oxford University Press.

_____ 1984a: The geometry of a research specialty: spatial diffusion modelling. *Annals of the Association of American Geographers* 74, 437–53.

_____ 1984b: Describing the structure of a research literature: spatial diffusion modelling in geography. *Environment and Planning B: Planning and Design* 11, 29–45.

_____ 1985: Any space for spatial analysis? In R. J. Johnston (ed.) *The future of geography.* London: Methuen, 190–208.

GATRELL, A. C. and LOVETT, A. A. 1986: The geography of hazardous waste disposal in England and Wales. *Area* 18, 275–83.

GATRELL, A. C. and SMITH, A. 1984: Networks of relations among a set of geographical journals. *The Professional Geographer* 36, 300–7.

GEARY, R. C. 1954: The contiguity ratio and statistical mapping. *The Incorporated Statistician* 5, 115–41.

GETIS, A. 1963: The determination of the location of retail activities with the use of a map transformation. *Economic Geography* 39, 1–22.

_____ 1993: Scholarship, leadership and quantitative methods. *Urban Geography* 14, 517–25.

GETIS, A. and BOOTS, B. N. 1978: *Models of spatial processes.* Cambridge: Cambridge University Press.

GIBBS, G. 1995: The relationship between quality in research and quality in teaching. *Quality in Higher Education* 1, 147–57.

GIBSON, E. 1978: Understanding the subjective meaning of places. In D. Ley and M. S. Samuels (eds) *Humanistic geography: problems and prospects.* Chicago: Maaroufa Press, 138–54.

GIBSON, K. and GRAHAM, J. 1992: Rethinking class in industrial geography: creating a space for alternative politics of class. *Economic Geography* 68, 109–27.

GIDDENS, A. 1976: *New rules of sociological method.* London: Hutchinson.

_____ 1981: *A critique of contemporary historical materialism.* London: Macmillan.

_____ 1984: *The constitution of society*. Oxford: Polity Press.

_____ 1985: *The nation state and violence*. Oxford: Polity Press.

GIER, J. and WALTON, J. 1987: Some problems with reconceptualising patriarchy. *Antipode* 19, 54–8.

GILBERT, A. 1988: The new regional geography in English- and French-speaking countries. *Progress in Human Geography* 12, 208–28.

GILBERT, D. 1996: Between two cultures: geography, computing and the humanities. *Ecumene* 2, 1–14.

GINSBURG, N. 1972: The mission of a scholarly society. *The Professional Geographer* 24, 1–6.

_____ 1973: From colonialism to national development: geographical perspectives on patterns and policies. *Annals of the Association of American Geographers* 63, 1–21.

GLACKEN, C. J. 1956: Changing ideas of the habitable world. In W. L. Thomas (ed.) *Man's role in changing the face of the earth*. Chicago: University of Chicago Press, 70–92.

_____ 1967: *Traces on the Rhodian shore: nature and culture in western thought from ancient times to the end of the eighteenth century*. Berkeley: University of California Press.

_____ 1983: A late arrival in academia. In A. Buttimer, *The practice of geography*. London: Longman, 20–34.

GLENNIE, P. D. and THRIFT, N. J. 1992: Modernity, urbanism and modern consumption. *Environment and Planning D: Society and Space* 10, 423–44.

GOBER, P. *et al.* 1995a: Employment trends in geography, Part 1: enrollment and degree patterns. *The Professional Geographer* 47, 317–28.

_____ 1995b: Employment trends in geography, Part 2: current demand conditions. *The Professional Geographer* 47, 329–35.

_____ 1995c: Employment trends in geography, Part 3: future demand conditions. *The Professional Geographer* 47, 336–46.

GODDARD, J. B. and ARMSTRONG, P. 1986: The 1986 Domesday project. *Transactions, Institute of British Geographers* NS11, 279–89.

GODLEWSKA, A. 1995: Map, text and image. The mentality of enlightened conquerors: a new look at the *Description de l'Egypte*. *Transactions, Institute of British Geographers* NS20, 5–28.

GOHEEN, P. G. 1970: *Victorian Toronto*. Chicago: University of Chicago, Department of Geography, Research Paper 127.

GOLD, J. R. 1980: *An introduction to behavioural geography*. Oxford: Oxford University Press.

GOLLEDGE, R. G. 1969: The geographical relevance of some learning theories. In K. R. Cox and R. G. Golledge (eds) *Behavioral problems in geography: a symposium*. Evanston: Northwestern University Studies in Geography 17, 101–45.

_____ 1970: Some equilibrium models of consumer behavior. *Economic Geography* 46, 417–24.

_____ 1980: A behavioral view of mobility and migration research. *The Professional Geographer* 32, 14–21.

_____ 1981a: Misconceptions, misinterpretations, and misrepresentations of behavioral approaches in human geography. *Environment and Planning A* 13, 1325–44.

_____ 1981b: A critical response to Guelke's 'Uncritical rhetoric'. *The Professional Geographer* 33, 247–51.

_____ 1983: Models of man, points of view, and theory in social science. *Geographical Analysis* 15, 57–60.

GOLLEDGE, R. G. and AMEDEO, D. 1968: On laws in geography. *Annals of the Association of American Geographers* 58, 760–74.

GOLLEDGE, R. G. and BROWN, L. A. 1967: Search, learning and the market decision process. *Geografiska Annaler* 49B, 116–24.

GOLLEDGE, R. G., BROWN, L. A. and WILLIAMSON, F. 1972: Behavioral approaches in geography: an overview. *The Australian Geographer* 12, 59–79.

GOLLEDGE, R. G. and COUCLELIS, H. 1984: Positivist philosophy and research in human spatial behavior. In T. F. Saarinen, D. Seamon and J. L. Sell (eds) *Environmental perception and behavior: an inventory and prospect.* Chicago: Department of Geography, University of Chicago, Research Paper 209, 179–90.

GOLLEDGE, R. G., COUCLELIS, H. and GOULD, P. R. (eds) 1988: *A search for common ground.* London: Pion.

GOLLEDGE, R. G. and RAYNER, J. N. (eds) 1982: *Proximity and preference: problems in the multidimensional analysis of large data sets.* Minneapolis: University of Minnesota Press.

GOLLEDGE, R. G. and RUSHTON, G. 1984: A review of analytic behavioural research in geography. In D. T. Herbert and R. J. Johnston (eds) *Geography and the urban environment: progress in research and application,* vol. 6. Chichester: John Wiley, 1–44.

GOLLEDGE, R. G. and STIMSON, R. J. 1987: *Analytical behavioural geography.* London: Croom Helm.

GOLLEDGE, R. G. and TIMMERMANS, H. (eds) 1988: *Behavioural modelling in geography and planning.* London: Croom Helm.

_____ 1990: Applications of behavioural research on spatial problems: I. Cognition. *Progress in Human Geography* 14, 57–100.

GOLLEDGE, R. G. *et al.* 1982: Commentary on 'The highest form of the geographer's art'. *Annals of the Association of American Geographers* 72, 557–8.

GOODCHILD, M. F. 1990: Comment: just the facts. *Political Geography Quarterly* 10, 335–7.

_____ 1993: The years ahead: Dobson's automated geography in 1993. *The Professional Geographer* 45, 444–6.

_____ 1995: GIS and geographic research. In J. Pickles (ed.) *Ground truth: the social implications of geographic information systems.* New York: Guilford, 31–50.

GOODCHILD, M. F. and JANELLE, D. F. 1988: Specialization in the structure and organization of geography. *Annals of the Association of American Geographers* 78, 11–28.

GOODSON, I. 1981: Becoming an academic subject: patterns of explanation and evolution. *British Journal of Sociology of Education* 2, 163–79.

GOSS, J. 1995: Marketing the new marketing: the strategic discourse of geodemographic information systems. In J. Pickles (ed.) *Ground truth: the social implications of geographic information systems.* New York: Guilford, 130–70.

GOTTMANN, J. 1951: Geography and international relations. *World Politics* 3, 153–73.

_____ 1952: The political partitioning of our world: an attempt at analysis. *World Politics* 4, 512–19.

GOUDIE, A. S. 1986a: *The human use of the environment.* Oxford: Basil Blackwell.

_____ 1986b: The integration of human and physical geography. *Transactions, Institute of British Geographers* NS11P, 464–7.

_____ 1993: Land transformation. In R. J. Johnston (ed.) *The challenge for geography. A changing world: a changing discipline.* Oxford: Blackwell Publishers, 117–37.

GOULD, P. R. 1963: Man against his environment: a game theoretic framework. *Annals of the Association of American Geographers* 53, 290–7.

_____ 1966: On mental maps. Michigan Inter-University Community of

Mathematical Geographers, Discussion Paper 9. Reprinted in R. M. Downs and D. Stea (eds) 1973, *Image and environment*. London: Edward Arnold, 182–220.

_____ 1969: Methodological developments since the fifties. In C. Board *et al.* (eds) *Progress in Geography 1*, London: Edward Arnold, 1–50.

_____ 1970a: Is *statistix inferens* the geographical name for a wild goose? *Economic Geography* 46, 439–48.

_____ 1970b: Tanzania 1920–63: the spatial impress of the modernization process. *World Politics* 22, 149–70.

_____ 1972: Pedagogic review. *Annals of the Association of American Geographers* 62, 689–700.

_____ 1975: Mathematics in geography: conceptual revolution or new tool? *International Social Science Journal* 27, 303–27.

_____ 1977: What is worth teaching in geography? *Journal of Geography in Higher Education* 1, 20–36.

_____ 1978: Concerning a geographic education. In D. A. Lanegran and R. Palm (eds) *An invitation to geography*. New York: McGraw-Hill, 202–26.

_____ 1979: Geography 1957–1977: the Augean period. *Annals of the Association of American Geographers* 69, 139–51.

_____ 1980: Q-analysis, or a language of structure: an introduction for social scientists, geographers and planners. *International Journal of Man-Machine Studies* 12, 169–99.

_____ 1981a: Letting the data speak for themselves. *Annals of the Association of American Geographers* 71, 166–76.

_____ 1981b: Space and rum: an English note on espacien and rumian meaning. *Geografiska Annaler* 63B, 1–3.

_____ 1985a: *The geographer at work*. London: Routledge & Kegan Paul.

_____ 1985b: Will geographical self-reflection make you blind? In R. J. Johnston (ed.) *The future of geography*. London: Methuen, 276–90.

_____ 1988: The only perspective: a critique of marxist claims to exclusiveness in geographical inquiry. In R. G. Golledge, H. Couclelis and P. R. Gould (eds) *A ground for common search*. Santa Barbara: The Santa Barbara Geographical Press, 1–10.

_____ 1993: Why not? The search for spatiotemporal structure. *Environment and Planning A* anniversary issue, 48–55.

_____ 1994: Sharing a tradition – geographies from the enlightenment. *The Canadian Geographer* 38, 194–202.

GOULD, P. R. and OLSSON, G. (eds) 1982: *A search for common ground*. London: Pion.

GOULD, P. R. and WHITE, R. 1974: *Mental maps*. Harmondsworth: Penguin Books.

_____ 1986: *Mental maps* (second edition). London: George Allen & Unwin.

GRAHAM, E. 1986: The unity of geography: a comment. *Transactions, Institute of British Geographers* NS11, 464–7.

_____ 1995: Postmodernism and the possibility of a new human geography. *Scottish Geographical Magazine* 111, 175–8.

GRAHAM, J. 1988: Postmodernism and marxism. *Antipode* 20, 60–5.

_____ 1992: Post-Fordism as politics: the political consequences of narratives on the left. *Environment and Planning D: Society and Space* 10, 393–420.

GRANGER, A. 1990: *The threatening desert*. London: Earthscan.

GRANO, O. 1981: External influence and internal change in the development of geography. In D. R. Stoddart (ed.) *Geography, ideology and social concern*. Oxford: Blackwell, 17–36.

GRAVES, N. J. 1981: Can geographical studies be subsumed under one paradigm or are a plurality of paradigms inevitable? *Terra* 93, 85–90.

GRAY, F. 1975: Non-explanation in urban geography. *Area* 7, 228–35.
_____ 1976: Selection and allocation in council housing. *Transactions, Institute of British Geographers* NS1, 34–46.
GREEN, N. P., FINCH, S. and WIGGINS, J. 1985: The 'state of the art' in Geographical Information Systems. *Area* 17, 295–301.
GREENBERG, D. 1984: Whodunit? Structure and subjectivity in behavioral geography. In T. F. Saarinen, D. Seamon and J. L. Sell (eds) *Environmental percep tion and behavior: an inventory and prospect*. Chicago: Department of Geography, University of Chicago, Research Paper 209, 191–208.
GREER-WOOTTEN, B. 1972: *The role of general systems theory in geographic research*. Department of Geography, Toronto: York University, Discussion Paper No. 3.
GREGORY, D. 1976: Rethinking historical geography. *Area* 8, 295–9.
_____ 1978a: *Ideology, science and human geography*. London: Hutchinson.
_____ 1978b: The discourse of the past: phenomenology, structuralism, and historical geography. *Journal of Historical Geography* 4, 161–73.
_____ 1980: The ideology of control: systems theory and geography. *Tijdschrift voor Economische en Sociale Geografie* 71, 327–42.
_____ 1981: Human agency and human geography. *Transactions, Institute of British Geographers* NS6, 1–18.
_____ 1982a: *Regional transformation and industrial revolution. a geography of the Yorkshire woollen industry*. London: Macmillan.
_____ 1982b: Solid geometry: notes on the recovery of spatial structure. In P. R. Gould and G. Olsson (eds) *A search for common ground*. London: Pion, 187–222.
_____ 1985a: Suspended animation: the stasis of diffusion theory. In D. Gregory and J. Urry (eds) *Social relations and spatial structures*. London: Macmillan, 296–336.
_____ 1985b: People, places and practices: the future of human geography. In R. King (ed.) *Geographical futures*. Sheffield: The Geographical Association, 56–76.
_____ 1989a: Areal differentiation and post-modern human geography. In D. Gregory and R. Walford (eds) *Horizons in human geography*. London: Macmillan, 67–96.
_____ 1989b: The crisis of modernity? Human geography and critical social theory. In R. Peet and N. J. Thrift (eds) *New models in geography*, vol. 2. London: Unwin Hyman, 348–85.
_____ 1990: *Chinatown*, Part Three? Soja and the missing spaces of social theory. *Strategies: A Journal of Theory, Culture and Politics* 3, 40–104.
_____ 1994: *Geographical imaginations*. Oxford: Blackwell Publishers.
_____ 1995a: Between the book and the lamp: imaginative geographies of Egypt, 1849–50. *Transactions, Institute of British Geographers* NS20, 29–57.
_____ 1995b: Imaginative geographies. *Progress in Human Geography* 19, 447–85.
GREGORY, D. and LEY, D. 1988: Culture's geographies. *Environment and Planning D: Society and Space* 6, 115–16.
GREGORY, D. and URRY, J. 1985: Introduction. In D. Gregory and J. Urry (eds) *Social relations and spatial structures*. London: Macmillan, 1–8.
GREGORY, D. and WALFORD, R. 1989: Introduction: making geography. In D. Gregory and R. Walford (eds) *Horizons in human geography*. London: Macmillan, 1–7.
GREGORY, K. J. 1985: *The nature of physical geography*. London: Edward Arnold.
GREGORY, S. 1963: *Statistical methods and the geographer*. London: Longman.

_____ 1976: On geographical myths and statistical fables. *Transactions, Institute of British Geographers* NS1, 385–400.

GREGSON, N. 1986: On duality and dualism: the case of structuration and time geography. *Progress in Human Geography* 10, 184–205.

_____ 1987a: The CURS initiative: some further comments. *Antipode* 19, 364–70.

_____ 1987b: Structuration theory: some thoughts on the possibilities for empirical research. *Environment and Planning D: Society and Space* 5, 73–91.

GREGSON, N. and FOORD, J. 1987: Patriarchy: comments on critics. *Antipode* 19, 371–5.

GRIFFITHS, M. J. and JOHNSTON, R. J. 1991: What's in a place? An approach to the concept of place as illustrated by the British National Union of Mineworkers' strike, 1984–1985. *Antipode* 23, 185–213.

GRIGG, D. B. 1977: E. G. Ravenstein and the laws of migration. *Journal of Historical Geography* 3, 41–54.

GROSSMAN, L. 1977: Man-environment relationships in anthropology and geography. *Annals of the Association of American Geographers* 67, 126–44.

GUDGIN, G. and TAYLOR, P. J. 1979: *Seats, votes and the spatial organisation of elections.* London: Pion.

GUELKE, L. 1971: Problems of scientific explanation in geography. *The Canadian Geographer* 15, 38–53.

_____ 1974: An idealist alternative in human geography. *Annals of the Association of American Geographers* 14, 193–202.

_____ 1975: On rethinking historical geography. *Area* 7, 135–8.

_____ 1976: The philosophy of idealism. *Annals of the Association of American Geographers* 66, 168–9.

_____ 1977a: The role of laws in human geography. *Progress in Human Geography* 1, 376–86.

_____ 1977b: Regional geography. *The Professional Geographer* 29, 1–7.

_____ 1978: Geography and logical positivism. In D. T. Herbert and R. J. Johnston (eds) *Geography and the urban environment: progress in research and applications,* 1. London: John Wiley, 35–61.

_____ 1981a: Uncritical rhetoric: 'A classic disservice'. *The Professional Geographer* 33, 246–7.

_____ 1981b: Idealism. In M. E. Harvey and B. P. Holly (eds) *Themes in geographic thought.* London: Croom Helm, 133–47.

_____ 1982: The idealist dispute in Anglo-American geography: a comment. *The Canadian Geographer* 26, 51–7.

_____ 1995: Review of D. Gregory *Geographical imaginations. The Canadian Geographer* 39, 184–6.

GUTTING, G. 1980: Introduction. In G. Gutting (ed.) *Paradigms and revolutions.* Notre Dame, IN: University of Notre Dame Press, 1–22.

HABERMAS, J. 1972: *Knowledge and human interests.* London: Heinemann.

HACKING, I. 1983: *Representing and intervening.* Cambridge: Cambridge University Press.

HAGERSTRAND, T. 1968: *Innovation diffusion as a spatial process.* Chicago: University of Chicago Press.

_____ 1975: Space, time and human conditions. In A. Karlquist, L. Lundquist and F. Snickars (eds) *Dynamic allocation of urban space.* Farnborough: Saxon House, 3–12.

_____ 1977: The geographers' contribution to regional policy: the case of Sweden. In D. R. Deskins *et al.* (eds) *Geographic humanism, analysis and social action: a half century of geography at Michigan.* Ann Arbor: Michigan Geographical Publications No. 17, 329–46.

_____ 1982: Diorama, path and project. *Tijdschrift voor Economische en Sociale Geografie* 73, 323–39.

_____ 1984: Presence and absence: a look at conceptual choices and bodily necessities. *Regional Studies* 18, 373–8.

HAGGETT, P. 1964: Regional and local components in the distribution of forested areas in southeast Brazil: a multivariate approach. *Geographical Journal* 130, 365–77.

_____ 1965a: Changing concepts in economic geography. In R. J. Chorley and P. Haggett (eds) *Frontiers in geographical teaching.* London: Methuen, 101–17.

_____ 1965b: Scale components in geographical problems. In R. J. Chorley and P. Haggett (eds) *Frontiers in geographical teaching.* London: Methuen, 164–85.

_____ 1965c: *Locational analysis in human geography.* London: Edward Arnold.

_____ 1967: Network models in geography. In R. J. Chorley and P. Haggett (eds) *Models in geography.* London: Methuen, 609–70.

_____ 1973: Forecasting alternative spatial, ecological and regional futures: problems and possibilities. In R. J. Chorley (ed.) *Directions in geography.* London: Methuen, 217–36.

_____ 1978: The spatial economy. *Human geography: coming of age. America Behavioral Scientist* 22, 151–67.

_____ 1990: *The geographers' art.* Oxford: Blackwell Publishers.

HAGGETT, P. and CHORLEY, R. J. 1965: Frontier movements and the geographical tradition. In R. J. Chorley and P. Haggett (eds) *Frontiers in geographical teaching.* London: Methuen, 358–78.

_____ 1967: Models, paradigms, and the new geography. In R. J. Chorley and P. Haggett (eds), *Models in geography.* London: Methuen, 19–42.

_____ 1969: *Network models in geography.* London: Edward Arnold.

_____ 1989: From Madingley to Oxford. In B. Macmillan (ed.) *Remodelling geography.* Oxford: Basil Blackwell, xv–xx.

HAGGETT, P., CLIFF, A. D. and FREY, A. E. 1977: *Locational analysis in human geography* (second edition). London: Edward Arnold.

HAGOOD, M. J. 1943: Development of a 1940 rural farm level of living index for counties. *Rural Sociology* 8, 171–80.

HAIGH, M. J. 1992: The crisis in American geography. *Area* 14, 185–9.

HAINES-YOUNG, R. 1989: Modelling geographical knowledge. In B. Macmillan (ed.) *Remodelling geography.* Oxford: Basil Blackwell, 22–39.

HAINES-YOUNG, R. and PETCH, J. R. 1978: *The methodological limitations of Kuhn's model of science.* Salford: University of Salford, Department of Geography, Discussion Paper 8.

_____ 1985: *Physical geography: its nature and methods.* London: Harper & Row.

HAINING, R. P. 1980: Spatial autocorrelation problems. In D. T. Herbert and R. J. Johnston (eds) *Geography and the urban environment, 3.* Chichester: John Wiley, 1–44.

_____ 1981: Analysing univariate maps, *Progress in Human Geography* 5, 58–78.

_____ 1989: Geography and spatial statistics: current positions, future developments. In B. Macmillan (ed.) *Remodelling geography.* Oxford: Basil Blackwell, 191–203.

_____ 1990: *Spatial data analysis in the social and environmental sciences.* Cambridge: Cambridge University Press.

HALL, P. 1974: The new political geography. *Transactions, Institute of British Geographers* 63, 48–52.

_____ 1981a: *Great planning disasters.* London: Penguin.

_____ 1981b: The geographer and society. *Geographical Journal* 147, 145–52.

_____ 1982: The new political geography: seven years on. *Political Geography Quarterly* **1**, 65–76.

HALL, P., JACKSON, P., MASSEY, D., ROBSON, B. T., THRIFT, N. J. and WILSON, A. G. 1987: Horizons and opportunities in research. *Area* **19**, 266–72.

HALL, P. *et al.* 1973: *The containment of urban England.* London: George Allen & Unwin.

HALVORSON, P. and STAVE, B. M. 1978: A conversation with Brian J. L. Berry. *Journal of Urban History* **4**, 209–38.

HAMILTON, F. E. J. 1974: A view of spatial behaviour, industrial organizations, and decision-making. In F. E. I. Hamilton (ed.) *Spatial perspectives on industrial organization and decision-making.* London: John Wiley, 3–46.

HAMNETT, C. 1977: Non-explanation in urban geography: throwing the baby out with the bath water. *Area* **9**, 143–5.

HANNAH, M. and STROHMAYER, U. 1992: Postmodernism (s)trained. *Annals of the Association of American Geographers* **82**, 308–10.

HANSON, S. 1993: 'Never question the assumptions' and other scenes from the revolution. *Urban Geography* **14**, 552–6.

HANSON, S. and PRATT, G. 1995: *Gender, work and space.* London: Routledge.

HARE, F. K. 1974: Geography and public policy: a Canadian view. *Transactions, Institute of British Geographers* **63**, 25–8.

_____ 1977: Man's world and geographers: a secular sermon. In D. R. Deskins *et al.* (eds) *Geographic humanism, analysis and social action: a half century of geography at Michigan.* Ann Arbor: Michigan Geographical Publication No. 17, 259–73.

HARLAND, R. 1987: *Superstructuralism: the philosophy of structuralism and post-structuralism.* London: Methuen.

HARLEY, J. B. 1989: Deconstructing the map. *Cartographica* **26**, 1–20.

_____ 1990: Cartography, ethics and social theory. *Cartographica* **27**, 1–23.

_____ 1992: Rereading the maps of the Columbian encounter. *Annals of the Association of American Geographers* **82**, 522–42.

HARRIES, K. D. 1974: *The geography of crime and justice.* New York: McGraw-Hill.

_____ 1975: Rejoinder to Richard Peet: 'The geography of crime: a political critique'. *The Professional Geographer* **27**, 280–2.

_____ 1976: Observations on radical versus liberal theories of crime causation. *The Professional Geographer* **28**, 100–13.

HARRIS, C. D. 1954a: The geography of manufacturing. In P. E. James and C. F. Jones (eds) *American geography: inventory and prospect.* Syracuse: Syracuse University Press, 292–309.

_____ 1954b: The market as a factor in the localization of industry in the United States. *Annals of the Association of American Geographers* **44**, 315–48.

_____ 1977: Edward Louis Ullman, 1912–1976. *Annals of the Association of American Geographers* **67**, 595–600.

HARRIS, C. D. and ULLMAN, E. L. 1945: The nature of cities. *Annals of the American Academy of Political and Social Science* **242**, 7–17.

HARRIS, R. C. 1971: Theory and synthesis in historical geography. *The Canadian Geographer* **15**, 157–72.

_____ 1977: The simplification of Europe overseas. *Annals of the Association of American Geographers* **67**, 469–83.

_____ 1978: The historical mind and the practice of geography. In D. Ley and M. S. Samuels (eds) *Humanistic geography: problems and prospects.* Chicago: Maaroufa Press, 123–37.

HARRISON, R. T. and LIVINGSTONE, D. N. 1982: Understanding in geography:

structuring the subjective. In D. T. Herbert and R. J. Johnston (eds) *Geography and the urban environment*, 5. Chichester: John Wiley, 1–40.

HART, J. F. 1982: The highest form of the geographer's art. *Annals of the Association of American Geographers* 72, 1–29.

———— 1990: Canons of good editorship. *The Professional Geographer* 42, 354–8.

HARTSHORNE, R. 1939: *The nature of geography*. Lancaster, PA: Association of American Geographers.

———— 1948: On the mores of methodological discussion in American geography. *Annals of the Association of American Geographers* 38, 492–504.

———— 1954a: Political geography. In P. E. James and C. F. Jones (eds) *American geography: inventory and prospect*. Syracuse: Syracuse University Press, 167–225.

———— 1954b: Comment on 'Exceptionalism in geography'. *Annals of the Association of American Geographers* 44, 108–9.

———— 1955: 'Exceptionalism in geography' re-examined. *Annals of the Association of American Geographers* 45, 205–44.

———— 1958: The concept of geography as a science of space from Kant and Humboldt to Hettner. *Annals of the Association of American Geographers* 48, 97–108.

———— 1959: *Perspective on the nature of geography*. Chicago: Rand McNally.

———— 1972: Review of *Kant's concept of geography*. *The Canadian Geographer* 16, 77–9.

———— 1979: Notes towards a bibliography of *The Nature of Geography*. *Annals of the Association of American Geographers* 69, 63–76.

———— 1984: In *The Geographical Journal* 150, 429.

HARVEY, D. 1967a: Models of the evolution of spatial patterns in geography. In R. J. Chorley and P. Haggett (eds) *Models in geography*. London: Methuen, 549–608.

———— 1967b: Editorial introduction: the problem of theory construction in geography. *Journal of Regional Science* 7, 211–16.

———— 1969a: *Explanation in geography*. London: Edward Arnold.

———— 1969b: Review of A. Pred, *Behavior and location part I*. *Geographical Review* 59, 312–14.

———— 1969c: Conceptual and measurement problems in the cognitive-behavioral approach to location theory. In K. R. Cox and R. G. Golledge (eds) *Behavioral problems in geography: a symposium*. Northwestern University Studies in Geography 17, 35–68.

———— 1970: Behavioral postulates and the construction of theory in human geography. *Geographica Polonica* 18, 27–46.

———— 1972: Revolutionary and counter-revolutionary theory in geography and the problem of ghetto formation. *Antipode* 4(2), 1–13.

———— 1973: *Social justice and the city*. London: Edward Arnold.

———— 1974a: A commentary on the comments. *Antipode* 4(2), 36–41.

———— 1974b: Discussion with Brian Berry. *Antipode* 6(2), 145–8.

———— 1974c: What kind of geography for what kind of public policy? *Transactions. Institute of British Geographers* 63, 18–24.

———— 1974d: Population, resources and the ideology of science. *Economic Geography* 50, 256–77.

———— 1974e: Class-monopoly rent, finance capital and the urban revolution, *Regional Studies* 8, 239–55.

———— 1975a: Class structure in a capitalist society and the theory of residential differentiation. In R. Peel, M. Chisholm and P. Haggett (eds) *Processes in physical and human geography: Bristol essays*. London: Heinemann, 354–69.

———— 1975b: The political economy of urbanization in advanced capitalist societies: the case of the United States. In G. Gappert and H. M. Rose (eds) *The social economy of cities.* Beverly Hills: Sage Publications, 119–63.

⎯⎯⎯⎯ 1975c: Review of B. J. L. Berry, *The human consequences of urbanization.* *Annals of the Association of American Geographers* **65**, 99–103.

⎯⎯⎯⎯ 1976: The marxist theory of the state. *Antipode* 8(2), 80–9.

⎯⎯⎯⎯ 1978: The urban process under capitalism: a framework for analysis. *International Journal of Urban and Regional Research* **2**, 101–32.

⎯⎯⎯⎯ 1979: Monument and myth. *Annals of the Association of American Geographers* **69**, 362–81.

⎯⎯⎯⎯ 1982: *The limits to capital.* Oxford: Blackwell.

⎯⎯⎯⎯ 1984: On the history and present condition of geography: an historical materialist manifesto. *The Professional Geographer* **36**, 1–11.

⎯⎯⎯⎯ 1985a: *The urbanization of capital.* Oxford: Basil Blackwell.

⎯⎯⎯⎯ 1985b: The geopolitics of capitalism. In D. Gregory and J. Urry (eds) *Social relations and spatial structures.* London: Macmillan, 128–63.

⎯⎯⎯⎯ 1985c: *Consciousness and the urban experience.* Oxford: Basil Blackwell.

⎯⎯⎯⎯ 1987: Three myths in search of a reality in urban studies. *Environment and Planning D: Society and Space* **5**, 367–76.

⎯⎯⎯⎯ 1989a: *The condition of postmodernity.* Oxford: Basil Blackwell.

⎯⎯⎯⎯ 1989b: From models to Marx: notes on the project to 'remodel' contemporary geography. In B. Macmillan (ed.) *Remodelling geography.* Oxford: Basil Blackwell, 211–16.

⎯⎯⎯⎯ 1989c: From managerialism to entrepreneurialism: the transformation of urban governance in late capitalism. *Geografiska Annaler* **71B**, 3–17.

⎯⎯⎯⎯ 1992: Postmodern morality plays. *Antipode* **24**, 300–26.

⎯⎯⎯⎯ 1993a: Class relations, social justice and the politics of difference. In M. Keith and S. Pile (eds) *Place and the politics of identity.* London: Routledge, 41–66.

⎯⎯⎯⎯ 1993b: From space to place and back again: reflections on the condition of postmodernity. In J. Bird *et al.* (eds) *Mapping the futures: local cultures, global change.* London: Routledge, 3–29.

⎯⎯⎯⎯ 1994: The nature of environment: the dialectics of social and environmental change. In R. Miliband and L. Panitch (eds) *Real problems, false solutions: socialist register 1993.* London: The Merlin Press, 1–51.

⎯⎯⎯⎯ 1995: Militant particularism and global ambition: the conceptual politics of place, space, and environment in the work of Raymond Williams. *Social Text* **42**, 69–98.

⎯⎯⎯⎯ 1996a: *Justice, nature and the geography of difference.* Oxford: Blackwell Publishers.

⎯⎯⎯⎯ 1996b: Cities or urbanization. *City* **1–2**, 38–61.

HARVEY, D. and SCOTT, A. J. 1989: The practice of human geography: theory and empirical specificity in the transition from Fordism to flexible accumulation. In B. Macmillan (ed.) *Remodelling geography.* Oxford: Basil Blackwell, 217–29.

HARVEY, M. E. and HOLLY, B. P. 1981: Paradigm, philosophy and geographic thought. In M. E. Harvey and B. P. Holly (eds) *Themes in geographic thought.* London: Croom Helm, 11–37.

HAY, A. M. 1978: Some problems in regional forecasting. In J. I. Clarke and J. Pelletser (eds) *Régions géographique et régions d'aménagement.* Collection les hommes et les lettres, 7. Lyon: Editions Hermes.

⎯⎯⎯⎯ 1979a: Positivism in human geography: response to critics. In D. T. Herbert and R. J. Johnston (eds) *Geography and the urban environment: progress in research and applications,* 2. London: John Wiley, 1–26.

⎯⎯⎯⎯ 1979b: The geographical explanation of commodity flow. *Progress in Human Geography* **3**, 1–12.

_____ 1985a: Scientific method in geography. In R. J. Johnston (ed.) *The future of geography*. London: Methuen, 129–42.

_____ 1985b: Statistical tests in the absence of samples: a comment. *The Professional Geographer* 37, 334–8.

HAY, A. M. and JOHNSTON, R. J. 1983: The study of process in quantitative human geography. *L'Espace Géographique* 12, 69–76.

HAYNES, R. M. 1975: Dimensional analysis: some applications in human geography. *Geographical Analysis* 7, 51–68.

_____ 1978: A note on dimensions and relationships in human geography, *Geographical Analysis* 10, 288–92.

_____ 1982: *An introduction to dimensional analysis for geographers*. CATMOG 33, Norwich: Geo Books.

HAYTER, R. and WATTS, H. D. 1983: The geography of enterprise. *Progress in Human Geography* 7, 157–81.

HAYTER, T. and HARVEY, D. (eds) 1993: *The factory and the city: the story of the Cowley Automobile Workers in Oxford*. Brighton: Mansell.

HELD, D. 1980: *Introduction to critical theory: Horkheimer to Habermas*. London: Hutchinson.

HEPPLE, L. W. 1992: Metaphor, geopolitical discourse and the military in South America. In T. J. Barnes and J. S. Duncan (eds) *Writing worlds: discourse, text and metaphor in the representation of landscape*. London: Routledge, 136–54.

HERBERT, D. T. and JOHNSTON, R. J. 1978: Geography and the urban environment. In D. T. Herbert and R. J. Johnston (eds) *Geography and the urban environment: progress in research and applications*, 1. London: John Wiley, 1–29.

HERBERTSON, A. J. 1905: The major natural regions, *Geographical Journal* 25, 300–10.

HEWITT, K. (ed.) 1983: *Interpretations of calamity*. London: George Allen & Unwin.

HILL, M. R. 1982: Positivism: a 'hidden' philosophy in geography. In M. E. Harvey and B. P. Holly (eds) *Themes in geographic thought*. London: Croom Helm, 38–60.

HODGART, R. L. 1978: Optimizing access to public services:. a review of problems, models, and methods of locating central facilities. *Progress in Human Geography* 2, 17–48.

HOLT-JENSEN, A. 1981: *Geography: its history and concepts*. London: Harper & Row.

_____ 1988: *Geography: its history and concepts*, second edition. London: Harper & Row.

HOOK, J. C. 1955: Areal differentiation of the density of the rural farm population in the northeastern United States. *Annals of the Association of American Geographers* 45, 189–90.

HOOSON, D. J. M. 1981: Carl O. Sauer. In B. W. Blouet (ed.) *The origins of academic geography in the United States*. Hamden, CT: Archon Books, 165–74.

HOUSE, J. W. 1973: Geographers, decision takers and policy makers. In M. Chisholm and B. Rodgers (eds), *Studies in human geography*. London: Heinemann, 272–305.

HUCKLE, J. 1985: Geography and schooling. In R. J. Johnston (ed.) *The future of geography*. London: Methuen, 291–306.

HUDSON, R. 1983: The question of theory in political geography: outlines for a critical theory approach. In N. Kliot and S. Waterman (eds) *Pluralism and political geography*. London: Croom Helm, 39–55.

_____ 1988: Uneven development in capitalist societies. *Transactions, Institute of British Geographers* NS13, 484–96.

HUGGETT, R. J. 1980: *Systems analysis in geography*. Oxford: Oxford University Press.
_____ 1994: *Geoecology: an evolutionary approach*. London: Routledge.
HUGGETT, R. J. and THOMAS, R. W. 1980: *Modelling in geography*. London: Harper & Row.
HUGILL, P. J. and FOOTE, K. E. 1994: Foreword: culture and geography: thirty years of advance. In K. E. Foote, P. J. Hugill, K. Mathewson and J. M. Smith (eds) *Re-reading cultural geography*. Austin: University of Texas Press, 9–26.
ISARD, W. 1956a: *Location and space economy*. New York: John Wiley.
_____ 1956b: Regional science, the concept of region, and regional structure. *Papers and Proceedings, Regional Science Association* 2, 13–39.
_____ 1960: *Methods of regional analysis: an introduction in regional science*. New York: John Wiley.
_____ 1975: *An introduction to regional science*. Englewood Cliffs, NJ: Prentice-Hall.
ISARD, W. *et al.* 1969: *General theory: social, political, economic and regional*. Cambridge, MA: MIT Press.
ISSERMAN, A. M. 1995: The history, status, and future of regional science: an American perspective. *International Regional Science Review* 17, 249–96.
JACKSON, P. 1984: Social disorganization and moral order in the city. *Transactions, Institute of British Geographers* NS9, 168–80.
_____ 1985: Urban ethnography. *Progress in Human Geography* 9, 157–76.
_____ 1988: Definitions of the situation. In J. Eyles and D. M. Smith (eds) *Qualitative methods in human geography*. Cambridge: Polity Press, 49–74.
_____ 1989: *Maps of meaning*. London: Unwin Hyman.
_____ 1991: The cultural politics of masculinity: towards a social geography. *Transactions, Institute of British Geographers* NS16, 199–213.
_____ 1993a: Changing ourselves: a geography of position. In R. J. Johnston (ed.) *The challenge for geography. A changing world: a changing discipline*. Oxford: Blackwell Publishers, 198–214.
_____ 1993b: Berkeley and beyond: broadening the horizons of cultural geography. *Annals of the Association of American Geographers* 83, 519–20.
JACKSON, P. and PENROSE, J. 1993: Introduction: placing 'race' and nation. In P. Jackson and J. Penrose (eds) *Constructions of race, place and nation*. London: UCL Press, 1–26.
JACKSON, P. and SMITH, S. J. 1981: Introduction. In P. Jackson and S. J. Smith (eds) *Social interaction and ethnic segregation*. London: Academic Press, 1–18.
_____ 1984: *Exploring social geography*. London: George Allen & Unwin.
JACKSON, P., SMITH, S. J. and JOHNSTON, R. J. 1988: An equal opportunities policy for the IBG. *Area* 20, 279–80.
JACOBS, J. 1994: Negotiating the heart: heritage, development and identity in postimperial London. *Environment and Planning D: Society and Space* 12, 751–72.
JAMES, P. E. 1942: *Latin America*. London: Cassell.
_____ 1954: Introduction: the field of geography. In P. E. James and C. F. Jones (eds) *American geography: inventory and prospect*. Syracuse: Syracuse University Press, 2–18.
_____ 1965: The President's session. *The Professional Geographer* 17(4), 35–7.
_____ 1972: *All possible worlds: a history of geographical ideas*. Indianapolis: Odyssey Press.
JAMES, P. E. and JONES, C. F. (eds) 1954: *American geography: inventory and prospect*. Syracuse: Syracuse University Press.
JAMES, P. E. and MARTIN, G. J. 1981: *All possible worlds: a history of geographical ideas* (2nd edition). New York: John Wiley.

JANELLE, D. G. 1968: Central-place development in a time-space framework. *The Professional Geographer* 20, 5–10.

_____ 1969: Spatial reorganization: a model and concept. *Annals of the Association of American Geographers* 59, 348–64.

_____ 1973: Measuring human extensibility in a shrinking world. *The Journal of Geography* 72 (5), 8–15.

JENKINS, A. 1995: The impact of Research Assessment Exercises on teaching in selected geography departments in England and Wales. *Geography* 80, 367–74.

JOHNSON, J. H. and POOLEY, C. G. (eds) 1982: *The structure of nineteenth-century cities*. London: Croom Helm.

JOHNSON, L. 1989: Geography, planning and gender. *New Zealand Geographer* 45, 85–91.

JOHNSTON, R. J. 1969: Urban geography in New Zealand 1945–1969. *New Zealand Geographer* 25, 121–35.

_____ 1971: *Urban residential patterns: an introductory review*. London: G. Bell & Sons.

_____ 1974: Continually changing human geography revisited: David Harvey: *Social Justice and the City. New Zealand Geographer* 30, 180–92.

_____ 1976a: *The world trade system: some enquiries into its spatial structure* London: G. Bell & Sons.

_____ 1976b: Anarchy, conspiracy and apathy: the three conditions of geography. *Area* 8, 1–3.

_____ 1976c: Observations on accounting procedures and urban-size policies. *Environment and Planning A* 8, 327–40.

_____ 1978a: *Multivariate statistical analysis in geography: a primer on the general linear model*. London: Longman.

_____ 1978b: *Political, electoral and spatial systems*. London: Oxford University Press.

_____ 1978c: Paradigms and revolutions or evolution: observations on human geography since the Second World War. *Progress in Human Geography* 2, 189–206.

_____ 1979a: Urban geography: city structures. *Progress in Human Geography* 3, 133–8.

_____ 1979b: *Geography and geographers: Anglo-American human geography since 1945* (1st edn), London: Edward Arnold.

_____ 1980a: On the nature of explanation in human geography. *Transactions, Institute of British Geographers* NS5, 402–12.

_____ 1980b: *City and society*. London: Penguin.

_____ 1981a: Applied geography, quantitative analysis and ideology. *Applied Geography* 1, 213–19.

_____ 1981b: Paradigms, revolutions, schools of thought and anarchy: reflections on the recent history of Anglo-American human geography. In B. W. Blouet (ed.) *The origins of academic geography in the United States*. Hamden, CT: Archon Books, 303–18.

_____ 1982a: *Geography and the state*. London: Macmillan.

_____ 1982b: On the nature of human geography. *Transactions, Institute of British Geographers* NS7, 123–5.

_____ 1982c: On ecological analysis and spatial autocorrelation. In L. le Rouzic (ed.), *L'autocorrelation spatiale*. Reims, Travaux de l'Institut de Géographie, 3–16.

_____ 1983a: Resource analysis, resource management and the integration of human and physical geography. *Progress in Physical Geography* 7, 127–46.

_____ 1983b: *Philosophy and human geography: an introduction to contemporary approaches* (1st edn). London: Edward Arnold.

_____ 1983c: Texts, actors, and higher managers: judges, bureaucrats and the political organization of space. *Political Geography Quarterly* 2, 3–20.

_____ 1983d: On geography and the history of geography. *History of Geography Newsletter* 3, 1–7.

_____ 1984a: The political geography of electoral geography. In P. J. Taylor and J. W. House (eds) *Political geography: recent advances and future directions.* London: Croom Helm, 133–48.

_____ 1984b: *Residential segregation, the state and constitutional conflict in American urban areas.* London: Academic Press.

_____ 1984c: The world is our oyster. *Transactions, Institute of British Geographers* NS9, 443–59.

_____ 1984d: The region in twentieth century British geography. *History of Geography Newsletter* 4, 26–35.

_____ 1984e: A foundling floundering in World Three. In M. Billinge, D. Gregory and R. Martin (eds) *Recollections of a revolution.* London: Macmillan, 39–56.

_____ 1984f: Quantitative ecological analysis in human geography: an evaluation of four problem areas. In G. Bahrenberg, M. Fischer and P. Nijkamp (eds) *Recent developments in spatial data analysis.* Aldershot: Gower, 131–44.

_____ 1985a: Places matter. *Irish Geography* 18, 58–63.

_____ 1985b: *The geography of English politics: the 1983 general election.* London: Croom Helm.

_____ (ed.) 1985c: *The future of geography.* London: Methuen.

_____ 1985d: To the ends of the earth. In R. J. Johnston (ed.) *The future of geography.* London: Methuen, 326–38.

_____ 1986a: *On human geography.* Oxford: Basil Blackwell.

_____ 1986b: Four fixations and the quest for unity in geography. *Transactions, Institute of British Geographers* NS11, 449–53.

_____ 1986c: Placing politics. *Political Geography Quarterly* 5, s63–s78.

_____ 1986d: Individual freedom and the world-economy. In R. J. Johnston and P. J. Taylor (eds) *A world in crisis? Geographical perspectives.* Oxford: Basil Blackwell, 173–95.

_____ 1986e: The neighbourhood effect revisited: spatial science or political regionalism. *Environment and Planning D: Society and Space* 4, 41–56.

_____ 1986f: *Philosophy and human geography: an introduction to contemporary approaches* (second edition). London: Edward Arnold.

_____ 1986g: Review of John L. Paterson: *David Harvey's Geography. Antipode* 18, 96–108.

_____ 1986h: Understanding and solving American urban problems: geographical contributions? *The Professional Geographer* 38, 229–33.

_____ 1987: Job markets and housing markets in the 'developed world'. *Tijdschrift voor Economische en Sociale Geografie* 78, 328–35.

_____ 1988: There's a place for us. *New Zealand Geographer* 44, 8–13.

_____ 1989a: Philosophy, ideology and geography. In D. Gregory and R. Walford (eds) *Horizons in human geography.* London: Macmillan, 48–66.

_____ 1989b: *Environmental problems: nature, economy and state.* London: Belhaven Press.

_____ 1990a: The challenge for regional geography: some proposals for research frontiers. In R. J. Johnston, J. Hauer and G.A. Hoekveld (eds) *The challenge of regional geography.* London: Routledge, 124–41.

_____ 1990b: Some misconceptions about conceptual issues. *Tijdschrift voor Economische en Sociale Geografie* 81, 14–18.

_____ 1991: *A question of place: exploring the practice of human geography.* Oxford: Blackwell Publishers.

_____ 1992: The rise and decline of the corporate-welfare state: a comparative analysis in global context. In P. J. Taylor (ed.) *Political geography of the twenti-eth century: a global analysis*. London: Belhaven Press, 115–70.

———— 1993a: The geographer's degrees of freedom: Wreford Watson, postwar progress in human geography and the future of scholarship in UK geography. *Progress in Human Geography* 17, 319–32.

———— 1993b: Removing the blindfold after the game is over: the financial out-comes of the 1992 Research Assessment Exercise. *Journal of Geography in Higher Education* 17, 174–80.

———— 1993c: A voice in the wilderness. *Geography* 78, 204–7.

———— (ed.) 1993d: *The challenge for geography. A changing world: a changing discipline*. Oxford: Blackwell Publishers.

———— 1993e: Meet the challenge: make the change. In R. J. Johnston (ed.) *The challenge for geography. A changing world: a changing discipline*. Oxford: Blackwell Publishers, 151–80.

———— 1993f: Real political geography. *Political Geography* 12, 473–80.

———— 1994: One world, millions of places: the end of History and the ascendancy of Geography. *Political Geography* 13, 111–22.

———— 1995a: Geographical research, geography and geographers in the changing British university system. *Progress in Human Geography* 19, 355–71.

———— 1995b: The business of British geography. In A. D. Cliff, P. R. Gould, A. G. Hoare and N. J. Thrift (eds) *Diffusing geography: essays for Peter Haggett*. Oxford: Blackwell Publishers, 317–41.

———— 1995c: Territoriality and the state. In G. B. Benko and U. Strohmeyer (eds) *Territoriality and the social sciences*. Ottawa: University of Ottawa Press.

———— 1996a: The expansion and fragmentation of geography in higher educa-tion. In R. J. Huggett, M. Robinson and Douglas, I. (eds) *Companion encyclope-dia of geography*. London: Routledge, 794–817.

———— 1996b: Jean Gottmann: French regional and political geographer *extraordi-naire*. *Progress in Human Geography* 20, 183–93.

———— 1996c: *Nature, state and economy: the political economy of environmental problems*. Chichester: John Wiley.

———— 1996d: Academic tribes and territories: the *realpolitik* of opening up the social sciences. *Environment and Planning A* 28.

———— 1996e: Quality in research, quality in teaching and quality in debate: a response to Graham Gibbs. *Quality in Higher Education* 2.

———— 1996f: And now it's all over was it worth all the effort? *Journal of Geography in Higher Education* 20, 159–65.

JOHNSTON, R. J. and BRACK, E. V. 1983: Appointment and promotion in the academic labour market: a preliminary survey of British University Departments of Geography. *Transactions, Institute of British Geographers* NS8, 100–11.

JOHNSTON, R. J. and CLAVAL, P. (eds) 1984: *Geography since the Second World War: an international survey*. London: Croom Helm.

JOHNSTON R. J. and DOORNKAMP, J. C. (eds) 1982: *The changing geography of the United Kingdom*. London: Methuen.

JOHNSTON, R. J. and GARDINER, V. (eds) 1990: *The changing geography of the United Kingdom* (second edition). London: Routledge.

JOHNSTON, R. J. and GREGORY, S. 1984: The United Kingdom. In R. J. Johnston and P. Claval (eds) *Geography since the Second World War: an inter-national survey*. London: Croom Helm, 107–31.

JOHNSTON, R. J., HAUER, J. and HOEKVELD, G. A. (eds) 1990: *Regional geog-raphy: current developments and future prospects*. London: Routledge.

JOHNSTON, R. J. and HERBERT, D. T. 1978: Introduction. In D. T. Herbert and R. J. Johnston (eds) *Social areas in cities: processes, patterns and problems*. London: John Wiley, 1–33.

JOHNSTON, R. J., JONES, K. and GOULD, M. 1995: Department size and research in English Universities: inter-university variations. *Quality in Higher Education* 1, 41–7.

JOHNSTON, R. J. and PATTIE, C. J. 1990: The regional impact of Thatcherism: attitudes and votes in Great Britain in the 1980s. *Regional Studies* 24, 479–93.

JOHNSTON, R. J. and TAYLOR, P. J. (eds) 1986a: *A world in crisis? Geographical perspectives.* Oxford: Basil Blackwell.

———— 1986b: Political geography: a polities of places within places. *Parliamentary Affairs* 39, 135–49.

———— (eds) 1989: *A world in crisis? Geographical perspectives* (second edition). Oxford: Basil Blackwell.

JOHNSTON, R. J., TAYLOR, P. J. and O'LOUGHLIN, I. 1987: The geography of violence and premature death. In Vayrynen, R. (ed.) *The quest for peace.* London: Sage Publications.

JOHNSTON, R. J., TAYLOR, P. J. and WATTS, M. J. (eds) 1995: *Geographies of global change: remapping the world in the late twentieth century.* Oxford: Blackwell Publishers.

JOHNSTON, R. J. and THRIFT, N. J. 1993: Ringing the changes: the intellectual history of *Environment and Planning A*. *Environment and Planning A*, anniversary issue, 14–21.

JONAS, A. 1988: A new regional geography of localities? *Area* 20, 101–10.

JONES, E. 1956: Cause and effect in human geography. *Annals of the Association of American Geographers* 46, 369–77.

———— 1980: Social geography. In E. H. Brown (ed.) *Geography yesterday and tomorrow.* Oxford: Oxford University Press, 251–62.

JONES, K. 1984: Geographical methods for exploring relationships. In G. Bahrenberg, M. M. Fischer and P. Nijkamp (eds) *Recent developments in spatial data analysis.* Aldershot: Gower, 215–30.

———— 1991: Specifying and estimating multi-level models for geographical research. *Transactions, Institute of British Geographers* NS16, 148–60.

JONES, K., JOHNSTON, R. J. and PATTIE, C. J. 1992: People, places and regions: exploring the use of multi-level modelling in the analysis of electoral data. *British Journal of Political Science* 22, 343–80.

JONES, L. V., LINDSEY, G. and COGGLESHALL, P. E. (eds) 1982: *An assessment of research-doctorate programs in the United States: social and behavioral sciences.* Washington, DC: National Academy Press.

KANSKY, K. J. 1963: *Structure of transportation networks.* Chicago: University of Chicago, Department of Geography, Research Paper 84.

KARIYA, P. 1993: The Department of Indian Affairs and Northern Development: the culture-building process within an institution. In J. S. Duncan and D. Ley (eds) *Place/culture/representation.* London: Routledge, 187–204.

KASPERSON, R. E. 1971: The post-behavioral revolution in geography. *British Columbia Geographical Series* 12, 5–20.

KATES, R. W. 1962: *Hazard and choice perception in flood plain management.* Chicago: University of Chicago, Department of Geography, Research Paper 78.

———— 1972: Review of *Perspectives on resource management. Annals of the Association of American Geographers* 62, 519–20.

———— 1987: The human environment: the road not taken, the road still beckoning. *Annals of the Association of American Geographers* 77, 525–34.

———— 1995: Labnotes from the Jeremiad experiment: hope for a sustainable transition. *Annals of the Association of American Geographers* 85, 623–40.

KATES, R. W. and BURTON, I. (eds) 1986a: *Geography resources and environment* (two volumes). Chicago: University of Chicago Press.

———— 1986b: Introduction. In R. W. Kates and I. Burton (eds) *Geography, resources and environment,* vol. 1: *Selected writings of Gilbert F. White.* Chicago: University of Chicago Press, xi–xiv.

KENNEDY, B. A. 1979: A naughty world. *Transactions, Institute of British Geographers* NS4, 550–8.

KENNEDY, P. 1988: *The rise and fall of the great powers.* New York: Random House.

KING, L. J. 1960: A note on theory and reality. *The Professional Geographer* 12(3), 4–6.

_____ 1961: A multivariate analysis of the spacing of urban settlement in the United States. *Annals of the Association of American Geographers* 51, 222–33.

_____ 1969a: The analysis of spatial form and relationship to geographic theory. *Annals of the Association of American Geographers* 59, 573–95.

_____ 1969b: *Statistical analysis in geography.* Englewood Cliffs, NJ: Prentice-Hall.

_____ 1976: Alternatives to a positive economic geography. *Annals of the Association of American Geographers* 66, 293–308.

_____ 1979a: Areal associations and regressions. *Annals of the Association of American Geographers* 69, 124–8.

_____ 1979b: The seventies: disillusionment and consolidation. *Annals of the Association of American Geographers* 69, 155–7.

_____ 1993: Spatial science and the institutionalization of geography as a social science. *Urban Geography* 14, 538–51.

KING, L. J. and CLARK, G. L. 1978: Government policy and regional development. *Progress in Human Geography* 2, 1–16.

KING, R., CONNELL, J. and WHITE, P. (eds) 1995: *Writing across worlds: literature and migration.* London: Routledge.

KIRK, W. 1951: Historical geography and the concept of the behavioural environment. *Indian Geographical Journal* 25, 152–60.

_____ 1963: Problems of geography. *Geography* 48, 357–71.

_____ 1978: The road from Mandalay: towards a geographical philosophy. *Transactions, Institute of British Geographers* NS3, 381–94.

KISH, G. and WARD, R. 1981: A survival package for geography and other endangered disciplines. *Newsletter Association of American Geographers,* 16, pp. 8,14.

KITCHIN, R. M. 1996: Increasing the integrity of cognitive mapping research: appraising conceptual schemata of environment-behaviour interaction. *Progress in Human Geography* 20, 56–84.

KNOPP, L. and LAURIA, M. 1987: Gender relations and social relations. *Antipode* 19, 48–53.

KNOS, D. S. 1968: The distribution of land values in Topeka, Kansas. In B. J. L. Berry and D. F. Marble (eds) *Spatial analysis.* Englewood Cliffs, NJ: Prentice-Hall, 269–89.

KNOX, P. L. 1975: *Social well-being: a spatial perspective.* London: Oxford University Press.

_____ 1987: The social production of the built environment: architects. architecture and the post-modern city. *Progress in Human Geography* 11, 354–78.

KNOX, P. L. and AGNEW, J. A. 1989: *The geography of the world-economy.* London: Edward Arnold.

KNOX, P. L., BARTELS, E. H., BOHLAND, J. R., HOLCOMB, B. and JOHNSTON, R. J. 1988: *The United States: a contemporary human geography.* London: Longman.

KNOX, P. L. and TAYLOR, P. J. (eds) 1995: *World cities in a world-economy.* Cambridge: Cambridge University Press.

KOBAYASHI, A. and MACKENZIE, S. (eds) 1989: *Remaking human geography*. Boston: Unwin Hyman.

KOFMAN, E. 1988: Is there a cultural geography beyond the fragments? *Area* 20, 85–7.

KOLLMORGEN, W. N. 1979: Kollmorgen as a bureaucrat. *Annals of the Association of American Geographers* 69, 77–89.

KOST, K. 1989: The conception of politics in political geography and geopolitics in Germany 1920–1950. *Political Geography Quarterly* 8, 369–86.

KUHN, T. S. 1962: *The structure of scientific revolutions*. Chicago: University of Chicago Press.

_____ 1969: Comment on the relations of science and art. *Comparative Studies in Society and History* 11, 403–12.

_____ 1970a: *The structure of scientific resolutions* (2nd edn). Chicago: University of Chicago Press.

_____ 1970b: Logic of discovery or psychology of research? In I. Lakatos and A. Musgrave (eds) *Criticism and the growth of knowledge*. Cambridge: Cambridge University Press, 1–23.

_____ 1970c: Reflections on my critics. In I. Lakatos and A. Musgrave (eds) *Criticism and the growth of knowledge*. Cambridge: Cambridge University Press, 231–78.

_____ 1977: Second thoughts on paradigms. In F. Suppe (ed.) *The structure of scientific theories*. Urbana: University of Illinois Press, 459–82, plus discussion 500–17.

LABEDZ, L. 1977: Anarchism. In A. Bullock and O. Stallybrass (eds) *The Fontana dictionary of modern thought*. London: William Collins, 22.

LAKATOS, I. 1978a: Falsification and the methodology of scientific research programmes. In J. Worrall and G. Currie (eds) *The methodology of scientific research programmes, philosophical papers,* vol. 1. Cambridge: Cambridge University Press, 8–101.

_____ 1978b: History of science and its rational reconstructions. In J. Worrall and G. Currie (eds) *The methodology of scientific research programmes, philosophical papers,* vol. 1. Cambridge: Cambridge University Press, 102–38.

LANGTON, J. 1972: Potentialities and problems of adapting a systems approach to the study of change in human geography. In C. Board *et al.* (eds) *Progress in Geography* 4, London: Edward Arnold, 125–79.

_____ 1984: The industrial revolution and the regional geography of England. *Transactions, Institute of British Geographers* NS9, 145–67.

LAPONCE, J. A. 1980: Political science: an import-export analysis of journals and footnotes. *Political Studies* 28, 401–19.

LARKIN, R. P. and PETERS, G. L. 1993: *Biographical dictionary of geography*. Westport, CT: Greenwood Press.

LASH, S. and URRY, J. 1987: *The end of organized capitalism*. Cambridge: Polity Press.

LAVALLE, P., McCONNELL, H. and BROWN, R. G. 1967: Certain aspects of the expansion of quantitative methodology in American geography. *Annals of the Association of American Geographers* 57, 423–36.

LAW, J. 1976: Theories and methods in the sociology of science: an interpretative approach. In G. Lemaine *et al., Perspectives on the emergence of scientific disciplines*. The Hague: Mouton, 221–31.

LEACH, B. 1974: Race, problems and geography. *Transactions, Institute of British Geographers* 63, 41–7.

LEACH, E. R. 1974: *Lévi-Strauss*. London: William Collins.

LEE, R. 1984; Process and region in the A-level syllabus. *Geography* 69, 97–107.

_____ 1985: The future of the region: regional geography as education for trans formation. In R. King (ed.) *Geographical futures*. Sheffield: The Geographical Association, 77–91.

LEE, Y. 1975: A rejoinder to 'The geography of crime: a political critique'. *The Professional Geographer* 27, 284–5.

LEIGHLEY, J. 1937: Some comments on contemporary geographic methods. *Annals of the Association of American Geographers* 27, 125–41.

_____ 1955: What has happened to physical geography? *Annals of the Association of American Geographers* 45, 309–18.

LEMAINE, G., *et al.* 1976: Introduction: problems in the emergence of new disciplines. In G. Lemaine *et al.* (eds) *Perspectives on the emergence of scientific disciplines*. The Hague: Mouton, 1–73.

LEONARD, S. 1982: Urban managerialism: a period of transition. *Progress in Human Geography* 6, 190–215.

LEWIS, G. M. 1966: Regional ideas and reality in the Cis-Rocky Mountain West. *Transactions, Institute of British Geographers* 38, 135–50.

_____ 1968: Levels of living in the Northeastern United States c. 1960: a new approach to regional geography. *Transactions, Institute of British Geographers* 45, 11–37.

LEWIS, J. and TOWNSEND, A. (eds) 1989: *The north-south divide: regional change in Britain in the 1980s*. London: Paul Chapman.

LEWIS, P. W. 1965: Three related problems in the formulation of laws in geography. *The Professional Geographer* 17(5), 24–7.

LEWTHWAITE, G. R. 1966: Environmentalism and determinism: a search for clarification. *Annals of the Association of American Geographers* 56, 1–23.

LEY, D. 1974: *The black inner city as frontier outpost*. Washington, DC: Association of American Geographers.

_____ 1977a: The personality of a geographical fact. *The Professional Geographer* 29, 8–13.

_____ 1977b: Social geography and the taken-for-granted world. *Transactions, Institute of British Geographers* NS2, 498–512.

_____ 1978: Social geography and social action. In D. Ley and M. S. Samuels (eds) *Humanistic geography: problems and prospects*. Chicago: Maaroufa Press, 41–57.

_____ 1980: *Geography without man: a humanistic critique*. Oxford Research Paper 24, School of Geography, Oxford: University of Oxford.

_____ 1981: Behavioral geography and the philosophies of meaning. In K. R. Cox and R. G. Golledge (eds) *Behavioral problems in geography revisited*. London: Methuen, 209–30.

_____ 1983: *A social geography of the city*. New York: Harper & Row.

_____ 1993: Postmodernism, or the cultural logic of advanced industrial capital. *Tijdschrift voor Economische en Sociale Geografie* 84, 171–4.

LEY, D. and DUNCAN, J. S. 1993: Epilogue. In J. S. Duncan and D. Ley (eds) *Place/culture/representation*. London: Routledge, 329–36.

LEY, D. and SAMUELS, M. S. 1978: Introduction: contexts of modern humanism in geography. In D. Ley and M. S. Samuels (eds) *Humanistic geography: prospects and problems*. Chicago: Maaroufa Press, 1–18.

LEYSHON, A., MATLESS, D. and REVILL, G. 1995: The place of music. *Transactions, Institute of British Geographers* NS20, 423–33.

LICHTENBERGER, E. 1984: The German-speaking countries. In R. J. Johnston and P. Claval (eds) *Geography since the Second World War: an international survey*. London: Croom Helm, 156–84.

LITTLE, J., PEAKE, L. and RICHARDSON, P. (eds) 1988: *Women in cities*. London: Macmillan.

LIVINGSTONE, D. N. 1984: Natural theory and neo-Lamarckism: the changing context of nineteenth century geography in the United States and Great Britain. *Annals of the Association of American Geographers* 74, 9–28.

_____ 1992: *The geographical tradition: episodes in the history of a contested enterprise.* Oxford: Blackwell Publishers.

_____ 1995: Geographical traditions. *Transactions, Institute of British Geographers* NS20, 420–2.

LIVINGSTONE, D. N. and HARRISON, R. T. 1981: Immanuel Kant, subjectivism, and human geography: a preliminary investigation. *Transactions, Institute of British Geographers* NS6, 359–74.

LONGLEY, P. A. 1995: GIS and planning for businesses and services. *Environment and Planning B: Planning and Design* 22, 127–9.

LONGLEY, P. A. and CLARKE, G. (eds) 1995: *GIS for business and service planning.* Cambridge: GeoInformation International.

LOSCH, A. 1954: *The economics of location.* New Haven, CN: Yale University Press.

LOVERING, J. 1987: Militarism, capitalism and the nation-state: towards a realist synthesis. *Environment and Planning D: Society and Space* 5, 283–302.

LOWE, M. S. and SHORT, J. R. 1990: Progressive human geography. *Progress in Human Geography* 14, 1–11.

LOWENTHAL, D. 1961: Geography, experience, and imagination: towards a geographical epistemology. *Annals of the Association of American Geographers* 51, 241–60.

_____ (ed.) 1965: *George Perkins Marsh: man and nature.* Cambridge, MA: Harvard University Press.

_____ 1968: The American scene. *The Geographical Review* 48, 61–88.

_____ 1975: Past time, present place: landscape and memory. *The Geographical Review* 65, 1–36.

_____ 1985: *The past is a foreign country.* Cambridge: Cambridge University Press.

LOWENTHAL, D. and BOWDEN, M. J. (eds) 1975: *Geographies of the mind: essays in historical geosophy in honor of John Kirkland Wright.* New York: Oxford University Press.

LOWENTHAL, D. and PRINCE, H. C. 1965: English landscape tastes. *The Geographical Review* 55, 186–222.

LOWENTHAL, D. *et al.* 1973: Report of the AAG Task Force on environmental quality. *The Professional Geographer* 25, 39–46.

LUKERMANN, F. 1958: Towards a more geographic economic geography. *The Professional Geographer* 10(1), 2–10.

_____ 1960a: On explanation, model, and prediction. *The Professional Geographer* 12(1), 1–2.

_____ 1960b: The geography of cement? *The Professional Geographer* 12(4), 1–6.

_____ 1961: The role of theory in geographical inquiry. *The Professional Geographer* 13(2), 1–6.

_____ 1965: Geography; *de facto* or *de jure. Journal of the Minnesota Academy of Science* 32, 189–96.

_____ 1990: *The Nature of Geography:* Post hoc, ergo propter hoc? In J. N. Entrikin and S. D. Brunn (eds) *Reflections on Richard Hartshorne's The Nature of Geography.* Washington: Association of American Geographers, 53–68.

LYNCH, K. 1960: *The image of the city.* Cambridge, MA: MIT Press.

MABOGUNJE, A. K. 1977: In search of spatial order: geography and the new programme of urbanization in Nigeria. In D. R. Deskins *et al.* (eds) *Geographic human-*

ism, analysis and social action: a half century of geography at Michigan. Ann Arbor: Michigan Geography Publications No. 17, 347–76.

MACEACHERN, A. 1995: *How maps work.* New York: Guilford.

MACGILL, S. M. 1981: Liquefied energy gases in the UK: what price public safety? *Environment and Planning A* 13, 339–54.

———— 1983: The Q-controversy: issues and nonissues. *Environment and Planning B: Planning and Design* 10, 371–80.

MACKAY, J. R. 1958: The interactrance hypothesis and boundaries in Canada: a preliminary study. *The Canadian Geographer* 11, 1–8.

MACKENZIE, S. 1989: Restructuring the relations of work and life: women as environmental actors, feminism as geographic analysis. In A. Kobayashi and S. Mackenzie (eds) *Remaking human geography.* Boston: Unwin Hyman, 40–61.

MACMILLAN, B. 1989a: Quantitative theory construction in human geography. In B. Macmillan (ed.) *Remodelling geography.* Oxford: Basil Blackwell, 89–107.

———— 1989b: Modelling through: an afterword to *Remodelling geography.* In B. Macmillan (ed.) *Remodelling geography.* Oxford: Basil Blackwell, 291–313.

MAGEE, B. 1975: *Popper.* London: William Collins.

MAGUIRE, D. J. 1989: The Domesday interactive videodisc system in geography teaching. *Journal of Geography in Higher Education* 13, 55–68.

MAGUIRE, D. J., GOODCHILD, M. F. and RHIND, D. W. (eds) 1991: *Geographical information systems.* London: Longman.

MAIR, A. 1986: Thomas Kuhn and understanding geography. *Progress in Human Geography* 10, 345–70.

MANION, T. and WHITELEGG, J. 1979: Radical geography and Marxism. *Area* 11, 122–4.

MANN, M. 1996: Neither nation-state nor globalism. *Environment and Planning A* 28.

MANNERS, I. R. and MIKESELL, M. W. (eds) 1974: *Perspectives on environment.* Washington: Commission on College Geography, Association of American Geographers.

MARBLE, D. F. and PEUQUET, D. F. 1993: The computer and geography: ten years later. *The Professional Geographer* 45, 446–8.

MARCHAND, B. 1978: A dialectical approach in geography. *Geographical Analysis* 10, 105–19.

MARCUS, M. G. 1979: Coming full circle: physical geography in the twentieth century. *Annals of the Association of American Geographers* 69, 521–32.

MARSHALL, J. U. 1985: Geography as a scientific enterprise. In R. J. Johnston (ed.) *The future of geography.* London: Methuen, 113–28.

MARTIN, A. F. 1951: The necessity for determinism. *Transactions and Papers, Institute of British Geographers* 17, 1–12.

MARTIN, G. J. 1980: *The life and thought of Isaiah Bowman.* Hamden, CT: Archon Books.

———— 1981: Ontography and Davisian physiography. In B. W. Blouet (ed.) *The origins of academic geography in the United States.* Hamden, CT: Archon Books, 279–90.

———— 1990: *The Nature of Geography* and the Schaefer-Hartshorne debate. In J. N. Entrikin and S. D. Brunn (eds) *Reflections on Richard Hartshorne's The Nature of Geography.* Washington: Association of American Geographers, 69–88.

MARTIN, G. J. and JAMES, P. E. 1992: *All possible worlds: a history of geographical ideas* (third edition). New York: John Wiley.

MARTIN, R. L. and OEPPEN, J. 1975: The identification of regional forecasting models using space-time correlation functions. *Transactions, Institute of British Geographers* 66, 95–118.

MASSAM, B. H. 1976: *Location and space in social administration*. London: Edward Arnold.

MASSEY, D. 1975: Behavioral research. *Area* 7, 201–3.

——— 1984a: *Spatial divisions of labour: social structures and the geography of production*. London: Macmillan.

——— 1984b: Introduction: geography matters. In D. Massey and J. Allen (eds) *Geography matters! A reader*. Cambridge: Cambridge University Press, 1–11.

——— 1985: New directions in space. In D. Gregory and J. Urry (eds) *Social relations and spatial structures*. London: Macmillan, 9–19.

——— 1991: Flexible sexism. *Environment and Planning D: Society and Space* 9, 31–58.

——— 1992: Politics and space/time. *New Left Review* 196, 65–84.

MASSEY, D. and MEEGAN, R. A. 1979: The geography of industrial reorganization. *Progress in Planning* 10, 155–237.

——— 1982: *The anatomy of job loss*. London: Methuen.

——— 1985: Introduction: the debate. In D. Massey and R. Meegan (eds) *Politics and method: contrasting studies in industrial geography*. London: Methuen, 1–12.

MASTERMAN, M. 1970: The nature of a paradigm. In I. Lakatos and A. Musgrave (eds) *Criticism and the growth of knowledge*. London: Cambridge University Press, 59–90.

MATHEWSON, K. (ed.) 1993: *Culture, form and place: essays in cultural and historical geography*. Baton Rouge: Louisiana State University, Department of Geography and Anthropology.

MATTHEWS, H. and LIVINGSTONE, I. 1996: Geography and lifelong learning. *Journal of Geography in Higher Education* 20, 5–10.

MAY, J. 1996a: 'A little taste of something more exotic': the imaginative geographies of everyday life. *Geography* 81, 57–64.

——— 1996b: Globalization and the politics of place: place and identity in an Inner London neighbourhood. *Transactions, Institute of British Geographers* NS21, 194–215.

MAY, J. A. 1970: *Kant's concept of geography: and its relation to recent geographical thought*. Toronto: Department of Geography, University of Toronto, Research Publication 4.

——— 1972: A reply to Professor Hartshorne. *The Canadian Geographer* 16, 79–81.

MAYER, H. M. 1954: Urban geography. In P. E. James and C. F. Jones (eds) *American geography: inventory and prospect*. Syracuse: Syracuse University Press, 142–66.

McCARTY, H. H. 1940: *The geographic basis of American economic life*. New York: Harper & Brothers.

——— 1952: *McCarty on McCarthy: the spatial distribution of the McCarthy vote, 1952*. Unpublished Paper, Department of Geography, State University of Iowa, Iowa City.

——— 1953: An approach to a theory of economic geography. *Annals of the Association of American Geographers* 43, 183–4.

——— 1954: An approach to a theory of economic geography. *Economic Geography* 30, 95–101.

——— 1958: Science, measurement, and area analysis. *Economic Geography* 34, facing page 283.

——— 1979: Geography at Iowa. *Annals of the Association of American Geographers* 69, 121–4.

McCARTY, H. H., HOOK, J. C. and KNOS, D. S. 1956: *The measurement of association in industrial geography*. Iowa City: Department of Geography, State University of Iowa.

McCARTY, H. H. and LINDBERG, J. B. 1966: A *preface to economic geography*. Englewood Cliffs, NJ: Prentice-Hall.

McDANIEL, R. and ELIOT HURST, M. E. 1968: A *systems analytic approach to economic geography*. Washington, DC: Commission on College Geography, Publication 8, Association of American Geographers.

McDOWELL, L. 1986a: Feminist geography. In R. J. Johnston, D. Gregory and D. M. Smith (eds) *The dictionary of human geography*. Oxford: Basil Blackwell, 151–2.

—— 1986b: Beyond patriarchy: a class-based explanation of women's subordination. *Antipode* 18, 311–21.

—— 1989: Women, gender and the organisation of space. In D. Gregory and R. Walford (eds) *Horizons in human geography*. London: Macmillan, 136–51.

—— 1991: The baby and the bath water: diversity, deconstruction and feminist theory in geography. *Geoforum* 22, 122–33.

—— 1993a: Space, place and gender relations: Part I. Feminist empiricism and the geography of social relations. *Progress in Human Geography* 17, 157–79.

—— 1993b: Space, place and gender relations: Part II. Identity, difference, feminist geometries and geographies. *Progress in Human Geography* 17, 305–18.

—— 1994: Polyphony and pedagogic authority. *Area* 26, 241–8.

McDOWELL, L. and MASSEY, D. 1984: A woman's place? In D. Massey and J. Allen (eds) *Geography matters!* Cambridge: Cambridge University Press, 128–47.

McKINNEY, W. M. 1968: Carey, Spencer, and modern geography. *The Professional Geographer* 20, 103–6.

McTAGGART, W. D. 1974: Structuralism and universalism in geography: reflections on contributions by H. C. Brookfield. *The Australian Geographer* 12, 510–16.

MEAD, W. R. 1980: Regional geography. In E. H. Brown (ed.) *Geography yesterday and tomorrow*. Oxford: Oxford University Press, 292–302.

MEADOWS, D. H. *et al.* 1972: *The limits to growth*. New York: Universal Books.

MEINIG, D. W. 1972: American wests: preface to a geographical introduction. *Annals of the Association of American Geographers* 62, 159–84.

—— 1978: The continuous shaping of America: a prospectus for geographers and historians. *The American Historical Review* 83, 1186–217.

—— 1983: Geography as an art. *Transactions, Institute of British Geographers* NS8, 314–28.

—— 1989: The historical geography imperative. *Annals of the Association of American Geographers* 79, 79–87.

MERCER, D. C. 1977: *Conflict and consensus in human geography*. Clayton, Victoria, Australia: Monash Publications in Geography No. 17.

—— 1984: Unmasking technocratic geography. In M. Billinge, D. Gregory and R. Martin (eds) *Recollections of a revolution*. London: Macmillan, 153–99.

MERCER, D. C. and POWELL, J. M. 1972: *Phenomenology and related non-positivistic viewpoints in the social sciences*. Clayton, Victoria, Australia: Monash Publications in Geography, No. 1.

MERRIFIELD, A. 1995: Situated knowledge through exploration: reflections on Bunge's 'geographical expeditions'. *Antipode* 27, 49–70.

MEYER, D. R. 1972: Geographical population data: statistical description not statistical inference. *The Professional Geographer* 24, 26–8.

MIDDLETON, N. and THOMAS, D. S. G. 1994: *Desertification: exploding the myth*. Chichester: John Wiley.

MIKESELL, M. W. 1967: Geographical perspectives in anthropology. *Annals of the Association of American Geographers* 57, 617–34.

_____ 1969: The borderlands of geography as a social science. In M. Sherif and C. W. Sherif (eds) *Interdisciplinary relationships in the social sciences.* Chicago: Aldine Publishing Company, 227–48.

_____ (ed.) 1973: *Geographers abroad: essays on the prospects of research in foreign areas.* Chicago: Department of Geography, University of Chicago, Research Paper 152.

_____ 1974: Geography as the study of environment: an assessment of some old and new commitments. In I. R. Manners and M. W. Mikesell (eds), *Perspectives on environment*, Washington, DC: Commission on College Geography, Association of American Geographers, 1–23.

_____ 1978: Tradition and innovation in cultural geography. *Annals of the Association of American Geographers* 68, 1–16.

_____ 1981: Continuity and change. In B. W. Blouet (ed.) *The origins of academic geography in the United States.* Hamden, CT: Archon Books, 1–15.

_____ 1984: North America. In R. J. Johnston and P. Claval (eds) *Geography since the Second World War: an international survey.* London: Croom Helm, 185–213.

MITCHELL, B. and DRAPER, D. 1982: *Relevance and ethics in geography.* London: Longman.

MITCHELL, D. 1995: There's no such thing as culture: towards a reconceptualization of the idea of culture in geography. *Transactions, Institute of British Geographers* NS20, 102–16.

MIYARES, I. M. and McGLADE, M. S. 1994: Specialization in 'Jobs in geography'; 1980–1993. *The Professional Geographer* 46, 170–7.

MOHAN, G. 1994: Destruction of the con: geography and the commodification of knowledge. *Area* 26, 387–90.

MONMONIER, M. S. 1993: What a friend we have in GIS. *The Professional Geographer* 45, 448–50.

MONTEFIORE, A. G. and WILLIAMS, W. M. 1955: Determinism and possibilism. *Geographical Studies* 2, 1–11.

MOODIE, D. W. and LEHR, J. C. 1976: Fact and theory in historical geography. *The Professional Geographer* 28, 132–6.

MOOS, A. I. and DEAR, M. J. 1986: Structuration theory in urban analysis: 1. theoretical exegesis. *Environment and Planning A* 18, 231–52.

MORGAN, M. A. 1967: Hardware models in geography. In R. J. Chorley and P. Haggett (eds) *Models in geography.* London: Methuen, 727–74.

MORGAN, W. B. and MOSS, R. P. 1965: Geography and ecology: the concept of the community and its relationship to environment. *Annals of the Association of American Geographers* 55, 339–50.

MORRILL, R. L. 1965: *Migration and the growth of urban settlement.* Lund Studies in Geography, Series B. 24, Lund: C. W. K. Gleerup.

_____ 1968: Waves of spatial diffusion. *Journal of Regional Science* 8, 1–18.

_____ 1969: Geography and the transformation of society. *Antipode* 1(1), 6–9.

_____ 1970a: *The spatial organization of society.* Belmont, California: Wadsworth, 2nd edn.

_____ 1970b: Geography and the transformation of society: part II. *Antipode* 2(1), 4–10.

_____ 1974: Review of D. Harvey, *Social Justice and the City. Annals of the Association of American Geographers* 64, 475–7.

_____ 1980: Productivity of American PhD-granting Departments of Geography. *The Professional Geographer* 32, 85–9.

_____ 1981: *Political redistricting.* Washington, DC: Resource Publications in Geography, Association of American Geographers.

_____ 1984: Recollections of the 'Quantitative Revolution's' early years: the

University of Washington 1955–65. In M. Billinge, D. Gregory and R. Martin (eds) *Recollections of a revolution*. London: Macmillan, 57–72.

_____ 1985: Some important geographic questions. *The Professional Geographer* 37, 263–70.

_____ 1993: Geography, spatial analysis and social science. *Urban Geography* 14, 442–6.

_____ 1994: Response to Johnston. *Urban Geography*, 15, 296.

MORRILL, R. L. and DORMITZER, J. 1979: *The spatial order: an introduction to modern geography*. North Scituate: Duxbury.

MORRILL, R. L. and GARRISON, W. L. 1960: Projections of interregional patterns of trade in wheat and flour. *Economic Geography* 36, 116–26.

MORRILL, R. L. and WOHLENBERG, E. H. 1971: *The geography of poverty in the United States*. New York: McGraw-Hill.

MOSS, R. P. 1970: Authority and charisma: criteria of validity in geographical method. *South African Geographical Journal* 52, 13–37.

_____ 1977: Deductive strategies in geographical generalization. *Progress in Physical Geography* 1, 23–39.

MOSS, R. P. and MORGAN, W. B. 1967: The concept of the community: some applications in geographical research *Transactions, Institute of British Geographers* 41, 21–32.

MUIR, R. 1975: *Modern political geography*. London: Macmillan.

_____ 1978: Radical geography or a new orthodoxy? *Area* 10, 322–7.

_____ 1979: Radical geography and Marxism. *Area* 11, 126–7.

MULKAY, M. J. 1975: Three models of scientific development. *Sociological Review* 23, 509–26.

_____ 1976: Methodology in the sociology of science: some reflections on the study of radio astronomy. In G . Lemaine *et al.* (eds) *Perspectives in the emergence of scientific disciplines*. The Hague: Mouton, 207–20.

_____ 1978: Consensus in science. *Social Science Information* 17, 107–22.

MULKAY, M. J., GILBERT, G. N. and WOOLGAR, S. 1975: Problem areas and research networks in science. *Sociology* 9, 187–203.

MULLER-WILLE, C. 1978: The forgotten heritage: Christaller's antecedents. In B. J. L. Berry (ed.) *The nature of change in geographical ideas*. de Kalb: Northern Illinois University Press, 37–64.

MUMFORD, L. 1956: Prospect. In W. L. Thomas (ed.) *Man's role in changing the face of the earth*. Chicago: University of Chicago Press, 1141–52.

MURDIE, R. A. 1969: *Factorial ecology of metropolitan Toronto 1951–1961*. Chicago: University of Chicago, Department of Geography, Research Paper 116.

MYRDAL, G. 1957: *Economic theory and underdeveloped regions*. London: Duckworth.

NATIONAL ACADEMY OF SCIENCES-NATIONAL RESEARCH COUNCIL 1965: *The science of geography*. Washington: NAS-NRC.

NEFT, D. 1966: *Statistical analysis for areal distributions*. Philadelphia: Monograph 2, Regional Science Research Institute.

NEWMAN, D. 1996: Writing together separately: critical discourse and the problems of cross-ethnic co-authorship. *Area* 28, 1–12.

NEWMAN, J. L. 1973: The use of the term 'hypothesis' in geography. *Annals of the Association of American Geographers* 63, 22–7.

NYSTUEN, J. D. 1963: Identification of some fundamental spatial concepts. *Papers of the Michigan Academy of Science, Arts, and Letters,* 48, 373–84. Reprinted in B. J. L. Berry and D. F. Marble (eds) *Spatial analysis*. Englewood Cliffs, NJ: Prentice-Hall, 35–41.

_____ 1984: Comment on 'Artificial intelligence and its applicability to geographical problem solving'. *The Professional Geographer* 36, 358–9.

OBERMEYER, N. J. 1994: GIS: a new profession? *The Professional Geographer* **46**, 498–503.

ODUM, H. W. and MOORE, H. E. 1938: *American regionalism – a cultural-historical approach to national integration*. New York: H. Holt & Company.

OLSSON, G. 1965: *Distance and human interaction: a review and bibliography*. Philadelphia: Regional Science Research Institute, Bibliography Series Number 2.

_____ 1969: Inference problems in locational analysis. In K. R. Cox and R. G. Golledge (eds) *Behavioral problems in geography: a symposium*. Evanson: Northwestern University Studies in Geography 17, 14–34.

_____ 1978: Of ambiguity or far cries from a memorializing mamafesta. In D. Ley and M. S. Samuels (eds) *Humanistic geography*. London: Croom Helm, 109–20.

_____ 1979: Social science and human action or on hitting your head against the ceiling of language. In S. Gale and G. Olsson (eds) *Philosophy in geography*. Dordrecht: Reidel, 287–308.

_____ 1980: *Birds in egg/eggs in bird*. London: Pion.

_____ 1982: -/-. In P. R. Gould and G. Olsson (eds) A *search for common ground*. London: Pion, 223–31.

_____ 1992: Lines of power. In T. J. Barnes and J. S. Duncan (eds) *Writing worlds: discourse, text and metaphor in the representation of landscape*. London: Routledge, 86–96.

OPENSHAW, S. 1984a: *The modifiable areal unit problem*. CATMOG 38, Norwich: Geo Books.

_____ 1984b: Ecological fallacies and the analysis of areal census data. *Environment and Planning A* **16**, 17–32.

_____ 1986: *Nuclear power: siting and safety*. London: Routledge & Kegan Paul.

_____ 1989: Computer modelling in human geography. In B. Macmillan (ed.) *Remodelling geography*. Oxford: Basil Blackwell, 70–88.

_____ 1991: A view on the GIS crisis in geography: or, using GIS to put Humpty-Dumpty back together again. *Environment and Planning A* **23**, 621–8.

_____ 1992: Further thoughts on geography and GIS: a reply. *Environment and Planning A* **24**, 463–6.

_____ 1994: Computational human geography: towards a research agenda. *Environment and Planning A* **26**, 499–505.

_____ 1995: Human systems modelling as a new grand challenge area in science: what has happened to the science in social science? *Environment and Planning A* **27**, 159–64.

_____ 1996: Fuzzy logic as a new scientific paradigm for doing geography. *Environment and Planning A* **28**, 761–8.

OPENSHAW, S., CARVER, S. and FERNIE, J. 1989: *Britain's nuclear waste: siting and safety*. London: Belhaven Press.

OPENSHAW, S., CHARLTON, M., CRAFT, A. W. and BIRCH, J. 1988: Investigation of leukaemia clusters by use of a geographical analysis machine. *The Lancet* 6 February, 272–3.

OPENSHAW, S. and GODDARD, J. B. 1987: Some implications of the commodification of information and the emerging information economy for applied geographical analysis in the United Kingdom. *Environment and Planning A* **19**, 1423–40.

OPENSHAW, S. and RAO, L. 1995: Algorithms for reengineering 1991 Census geography. *Environment and Planning A* **27**, 425–46.

OPENSHAW, S., STEADMAN, P. and GREENE, O. 1983: *Doomsday: Britain after nuclear attack*. Oxford: Basil Blackwell.

OPENSHAW, S., WYMER, C. and CHARLTON, M. 1986: A geographical information and mapping system for the BBC Domesday optical discs. *Transactions, Institute of British Geographers* **NS11**, 296–304.

OPENSHAW, S., WYMER, C. and CRAFT, A. W. 1988: A Mark I geographical analysis machine for the automated analysis of point data sets. *International Journal of Geographical Information Systems* 1, 335–58.

O'RIORDAN, T. 1971a: Environmental management. In C. Board *et al.* (eds), *Progress in Geography 3*. London: Edward Arnold, 173–231.

_____ 1971b: *Perspectives in resource management*. London: Pion.

_____ 1976: *Environmentalism*. London: Pion.

_____ 1981: *Environmentalism* (second edition). London: Pion.

O TUATHAIL, G. 1992: Foreign policy and the hyperreal: the Reagan administration and the 'scripting' of South Africa'. In T. J. Barnes and J. S. Duncan (eds) *Writing worlds: discourse, text and metaphor in the representation of landscape*. London: Routledge, 155–75.

_____ 1994: (Dis)placing geopolitics: writing on the maps of global politics. *Environment and Planning D: Society and Space* 12, 525–46.

O TUATHAIL, G. and DALBY, S. 1994: Critical geopolitics: unfolding spaces for thought in geography and global politics. *Environment and Planning D: Society and Space* 12, 513–14.

OWENS, P. L. 1984: Rural leisure and recreation research: a retrospective evaluation. *Progress in Human Geography* 8, 157–88.

PACIONE, M. 1990a: Conceptual issues in applied urban geography. *Tijdschrift voor Economische en Sociale Geografie* 81, 3–13.

_____ 1990b: On the dangers of misinterpretation. *Tijdschrift voor Economische en Sociale Geografie* 81, 26–8.

PAHL, R. E. 1965: Trends in social geography. In R. J. Chorley and P. Haggett (eds) *Frontiers in geographical teaching*. London: Methuen, 81–100.

_____ 1969: Urban social theory and research. *Environment and Planning* 1, 143–54. (Reprinted in R. E. Pahl 1970, *Whose City?* London: Longman, 209–25.)

_____ 1975: *Whose city? And other essays*. Harmondsworth: Penguin Books (2nd edn).

_____ 1979: Socio-political factors in resource allocation. In D. T. Herbert and D. M. Smith (eds) *Social problems and the city: geographical perspectives*. Oxford: Oxford University Press, 33–46.

PAINTER, J. M. and PHILO, C. 1995: Spaces of citizenship: an introduction. *Political Geography* 14, 107–20.

PALM, R. 1979: Financial and real estate institutions in the housing market. In D. T. Herbert and R. J. Johnston (eds) *Geography and the urban environment*, 2. Chichester: John Wiley, 83–124.

PALM, R. and PRED, A. R. 1978: The status of American women: a time-geographic view. In D. Lanegran and R. Palm (eds) *Invitation to geography* (second edition). New York: McGraw-Hill, 99–109.

PAPAGEORGIOU, G. J. 1969: Description of a basis necessary to the analysis of spatial systems. *Geographical Analysis* 1, 213–15.

_____ (ed.) 1976: *Mathematical land use theory*. Lexington, MA: D. C. Heath.

PARKER, G. 1985: *Western geopolitical thought in the twentieth century*. London: Croom Helm.

PARKES, D. N. and THRIFT, N. J. 1980: *Times, spaces and places*. Chichester: John Wiley.

PARRY, M. and DUNCAN, R. 1995: *The economic implications of climate change*. London: Earthscan.

PARSONS, J. J. 1977: Geography as exploration and discovery. *Annals of the Association of American Geographers* 67, 1–16.

PATERSON, J. H. 1974: Writing regional geography. In C. Board et *al.* (eds) *Progress in geography* 6. London: Edward Arnold, 1–26.

PATERSON, J. L. 1985: *David Harvey's geography*. London: Croom Helm.

PATMORE, J. A. 1970: *Land and leisure*. Newton Abbott: David & Charles.

_____ 1983: *Recreation and resources: leisure patterns and leisure places*. Oxford: Basil Blackwell.

PATTERSON, T. C. 1986: The last sixty years: toward a social history of Americanist archaeology in the United States. *American Anthropologist* **88**, 7–26.

PATTIE, C. J. and JOHNSTON, R. J. 1995: 'Its not like that round here': region, economic evaluations and voting at the 1992 British general election. *European Journal of Political Research* **28**, 1–32.

PEACH, C. and SMITH, S. J. 1981: Introduction. In C. Peach, V. Robinson and S. J. Smith (eds) *Ethnic segregation in cities*. London: Croom Helm, 9–24.

PEAKE, L. 1994: 'Proper words in proper places ...' Or, of young Turks and old turkeys. *The Canadian Geographer* **38**, 204–6.

PEET, J. R. 1971: Poor, hungry America. *The Professional Geographer* **23**, 99–104.

_____ 1975a: Inequality and poverty: a marxist-geographic theory. *Annals of the Association of American Geographers* **65**, 564–71.

_____ 1975b: The geography of crime: a political critique. *The Professional Geographer* **27**, 277–80.

_____ 1976a: Further comments on the geography of crime. *The Professional Geographer* **28**, 96–100.

_____ 1976b: Editorial: radical geography in 1976. *Antipode* **8**(3), inside cover.

_____ 1977: The development of radical geography in the United States. *Progress in Human Geography* **1**, 240–63.

_____ (ed.) 1978: *Radical geography*. London: Methuen.

_____ 1979: Societal contradiction and marxist geography. *Annals of the Association of American Geographers* **69**, 164–9.

_____ 1980: The transition from feudalism to capitalism. In J. R. Peet (ed.) *An introduction to marxist theories of underdevelopment*. Canberra: Publication HG/14, Department of Human Geography, Australian National University, 51–74.

_____ 1985a: The social origins of environmental determinism. *Annals of the Association of American Geographers* **75**, 309–33.

_____ 1985b: Radical geography in the United States: a personal history. *Antipode* **17**, 1–7.

_____ 1989: World capitalism and the destruction of regional cultures. In R. J. Johnston and P. J. Taylor (eds) *A World in Crisis?* Oxford: Basil Blackwell, 175–99.

_____ 1991: *Global capitalism: theories of societal development*. London: Routledge.

_____ 1993: Reading Fukuyama: politics at the end of history. *Political Geography* **12**, 64–78.

_____ 1996: A sign taken for history: Daniel Shays' memorial in Peterham, Massachusetts. *Annals of the Association of American Geographers* **86**, 21–43.

PEET, J. R. and LYONS, J. V. 1981: Marxism: dialectical materialism, social formation and the geographic relations. In M . E. Harvey and B. P. Holly (eds) *Themes in geographic thought*. London: Croom Helm, 187–205.

PEET, J. R. and THRIFT, N. J. 1989: Political economy and human geography. In R. Peet and N. J. Thrift (eds) *New models in geography*, vol. 1. London: Unwin Hyman, 3–27.

PELTIER, L. C. 1954: Geomorphology. In P. E. James and C. F. Jones (eds) *American geography: inventory and prospect*. Syracuse: Syracuse University Press, 362–81.

PENNING-ROWSELL, E. C. 1981: Fluctuating fortunes in gauging landscape value. *Progress in Human Geography* 5, 25–41.

PENROSE, J. and JACKSON, P. 1993: Identity and the politics of difference. In P. Jackson and J. Penrose (eds) *Constructions of race, place and nation*. London: UCL Press, 202–10.

PEPPER, D. 1984: *The roots of modern environmentalism*. London: Croom Helm.

―――― 1987: Physical and human integration: an educational perspective. *Progress in Human Geography* 11, 379–404.

―――― 1996: *Modern environmentalism: an introduction*. London: Routledge.

PEPPER, D. and JENKINS, A. 1983: A call to arms: geography and peace studies. *Area* 15, 202–8.

―――― (eds) 1985: *The geography of peace and war*. Oxford: Basil Blackwell.

PEPPER, S. C. 1942: *World hypotheses*. Berkeley, CA: University of California Press.

PERRY, P. J. 1969: H. C. Darby and historical geography: a survey and review. *Geographische Zeitschrift* 57, 161–77.

―――― 1979: Beyond Domesday. *Progress in Human Geography* 3, 407–16.

PETCH, J. R. and HAINES-YOUNG, R. H. 1980: The challenge of critical rationalism for methodology in physical geography. *Progress in Physical Geography* 4, 63–78.

PETER, L. and HULL, R. 1969: *The Peter principle*. London: Bantam Books.

PHILBRICK, A. K. 1957: Principles of areal functional organization in regional human geography. *Economic Geography* 33, 299–366.

PHILLIPS, D. R. and JOSEPH, A. E. 1984: *Accessibility and utilization: perspectives on health care delivery*. London: Harper & Row.

PHILLIPS, M. and UNWIN, T. 1985: British historical geography: places and people. *Area* 17, 155–64.

PHILO, C. (compiler) 1991: *New words, new worlds: reconceptualising social and cultural geography*. Aberystwyth: Cambrian Printers.

―――― 1992: Foucault's geography. *Environment and Planning D: Society and Space* 10, 137–61.

PICKLES, J. 1985: *Phenomenology, science and geography: spatiality and the human sciences*. Cambridge: Cambridge University Press.

―――― 1986: *Geography and humanism*. CATMOG 44, Norwich: Geo Books.

―――― 1988: From fact-world to life-world: the phenomenological method and social science. In J. Eyles and D. M. Smith (eds) *Qualitative methods in human geography*. Cambridge: Polity Press, 233–54.

―――― 1992: Texts, hermeneutics and propaganda. In T. J. Barnes and J. S. Duncan (eds) *Writing worlds: discourse, text and metaphor in the representation of landscape*. London: Routledge, 193–230.

―――― 1993: Discourse on method and the history of discipline: reflections on Dobson's 1983 automated geography. *The Professional Geographer* 45, 451–5.

―――― (ed.) 1995a: *Ground truth: the social implications of geographical information systems*. New York: The Guilford Press.

―――― 1995b: Representations in an electronic age: geography, GIS and democracy. In J. Pickles (ed.) *Ground truth: the social implications of geographical information systems*. New York: The Guilford Press, 1–30.

PILE, S. 1991: Practising interpretative geography. *Transactions, Institute of British Geographers* NS16, 458–69.

―――― 1993: Human agency and human geography revisited: a critique of 'new models' of the self. *Transactions, Institute of British Geographers* NS18, 122–39.

PINCH, S. P. 1985: *Cities and services: the geography of collective consumption*. London: Routledge & Kegan Paul.

PIPKIN, J. S. 1981: Cognitive behavioral geography and repetitive travel. In K. R.

Cox and R. G. Golledge (eds) *Behavioral problems in geography revisited.* London: Methuen, 145–80.

PIRIE, G. H. 1976: Thoughts on revealed and spatial behaviour. *Environment and Planning A* 8, 947–55.

PITTS, F. R. 1965: A graph theoretic approach to historical geography. *The Professional Geographer* 17(5), 15–20.

PLATT, R. H. 1986: Floods and man: a geographer's agenda. In R. W. Kates and I. Burton (eds) *Geography, resources and environment,* vol. 2: *Themes from the work of Gilbert F. White.* Chicago: University of Chicago Press, 28–68.

POCOCK, D. C. D. 1983: The paradox of humanistic geography. *Area* 15, 355–8.

POCOCK, D. C. D. and HUDSON, R. 1978: *Images of the urban environment.* London: Macmillan.

POIKER, T. K. 1983: The shining armor of the white knight. *The Professional Geographer* 35, 348–9.

POOLER, J. A. 1977: The origins of the spatial tradition in geography: an interpretation. *Ontario Geography* 11, 56–83.

POPPER, K. R. 1959: *The logic of scientific discovery.* London: Hutchinson.

_____ 1967: Replies to my critics. In P. A. Schipp (ed.), *The philosophy of Karl Popper,* vol. 2. La Salle, IN: Open Court Publishing Company, 961–97.

_____ 1970: Normal science and its dangers. In I. Lakatos and A. Musgrave (eds), *Criticism and the growth of knowledge.* London: Cambridge University Press, 51–8.

PORTEOUS, J. D. 1977: *Environment and behavior.* Reading, MA: Addison Wesley.

_____ 1985: Literature and humanist geography. *Area* 17, 117–22.

_____ 1986: Bodyscape: the body-landscape metaphor. *The Canadian Geographer* 30, 2–19.

_____ 1988: Topocide: the annihilation of place. In J. Eyles and D. M. Smith (eds) *Qualitative methods in human geography.* Cambridge: Polity Press, 75–93.

PORTER, P. W. 1978: Geography as human ecology. *Human geography: coming of age. American Behavioral Scientist* 22, 15–40.

PORTER, P. W. and LUKERMANN, F. 1975: The geography of utopia. In D. Lowenthal and M. J. Bowden (eds) *Geography of the mind: essays in historical geosophy.* New York: Oxford University Press, 197–224.

POWELL, J. M. 1970: *The public lands of Australia Felix: settlement and land appraisal in Victoria 1834–1891.* Melbourne: Oxford University Press.

_____ 1971: Utopia, millennium and the cooperative ideal: a behavioral matrix in the settlement process. *The Australian Geographer* 11, 606–18.

_____ 1972: *Images of Australia.* Clayton, Victoria, Australia: Monash University Publications in Geography No. 3.

_____ 1977: *Mirrors of the New World: images and image-makers in the settlement process.* Folkestone: Dawson.

_____ 1980a: Thomas Griffith Taylor 1880–1963. In T. W. Freeman and P. Pinchemel (eds) *Geographers: biobibliographical studies,* vol. 5. London: Mansell, 141–54.

_____ 1980b: The haunting of Saloman's house: geography and the limits of science. *Australian Geographer* 14, 327–41.

_____ 1981: Editorial comment: 'professional' geography into the eighties? *Australian Geographical Studies* 19, 228–30.

PRATT, G. 1989: Quantitative techniques and humanistic-historical materialist perspectives. In A. Kobayashi and S. Mackenzie (eds) *Remaking human geography.* Boston: Unwin Hyman, 101–15.

_____ 1992: Spatial metaphors and speaking positions. *Environment and Planning D: Society and Space* 10, 241–4.

PRATT, G. and HANSON, S. 1994: Geography and the construction of difference. *Gender, Place and Culture* 1, 5–29.

PRED, A. R. 1965a: The concentration of high value-added manufacturing. *Economic Geography* 41, 108–32.

———— 1965b: Industrialization, initial advantage, and American metropolitan growth. *Geographical Review* 55, 158–85.

———— 1967: *Behavior and location: foundations for a geographic and dynamic location theory. Part I.* Lund: C. W. K. Gleerup.

———— 1969: *Behavior and location: foundations for a geographic and dynamic location theory. Part II.* Lund: C. W. K. Gleerup.

———— 1973: Urbanization, domestic planning problems and Swedish geographic research. In C. Board *et al.* (eds) *Progress in Geography* 5. London: Edward Arnold, 1–77.

———— 1977a: The choreography of existence: comments on Hagerstrand's time geography and its usefulness. *Economic Geography* 53, 207–21.

———— 1977b: *City-systems in advanced economies.* London: Hutchinson.

———— 1979: The academic past through a time-geographic looking glass. *Annals of the Association of American Geographers* 69, 175–80.

———— 1981a: Production, family, and free-time projects: a time-geographic perspective on the individual and societal change in nineteenth century US cities. *Journal of Historical Geography* 7, 3–36.

———— 1981b: Of paths and projects: individual behavior and its societal context. In K. R. Cox and R. G. Golledge (eds) *Behavioral problems in geography revisited.* London: Methuen, 231–55.

———— 1984a: From here and now to there and then: some notes on diffusions, defusions, and disillusions. In M. Billinge, D. Gregory and R. Martin (eds) *Recollections of a revolution.* London: Macmillan, 86–103.

———— 1984b: Structuration, biography formation, and knowledge: observations on port growth during the late mercantile period. *Environment and Planning D: Society and Space* 2, 251–76.

———— 1984c: Place as historically contingent process: structuration and the time geography of becoming places. *Annals of the Association of American Geographers* 74, 279–97.

———— 1985: The social becomes the spatial and the spatial becomes the social. In D. Gregory and J. Urry (eds) *Social relations and spatial structures.* London: Macmillan, 336–75.

———— 1986: *Becoming places, practice and structure: the emergence and aftermath of enclosures in the plains villages of southwestern Skane.* Oxford: Polity Press.

———— 1988: Lost words as reflections of lost worlds. In R. G. Golledge, H. Couclelis and P. R. Gould (eds) *A ground for common search.* Santa Barbara: The Santa Barbara Geographical Press, 138–47.

———— 1989: The locally spoken word and local struggles. *Environment and Planning D: Society and Space* 7, 211–34.

———— 1990: In other wor(l)ds: fragmented and integrated observations on gendered languages, gendered spaces and local transformation. *Antipode* 22, 33–52.

———— 1992: Straw men build straw houses. *Annals of the Association of American Geographers* 82, 305–8.

———— 1996: Interfusions: consumption, identity and the practices and power relations of everyday life. *Environment and Planning A* 28, 11–24.

PRED, A. R. and KIBEL, B. M. 1970: An application of gaming simulation to a general model of economic locational processes. *Economic Geography* 46, 136–56.

PRED, A. R. and PALM, R. 1978: The status of American women: a time-geographic view. In D. A. Lanegran and R. Palm (eds) *An invitation to geography* (second edition). New York: McGraw-Hill, 99–109.

PRICE, D. G. and BLAIR, A. M. 1989: *The changing geography of the service sector.* London: Belhaven.

PRICE, M. and LEWIS, M. 1993a: The reinvention of cultural geography. *Annals of the Association of American Geographers* 83, 1–17.

_____ 1993b: Reply: on reading cultural geography. *Annals of the Association of American Geographers* 83, 520–2.

PRINCE, H. C. 1961–2: The geographical imagination. *Landscape* 11, 21–5.

_____ 1971a: Real, imagined and abstract worlds of the past. In C. Board *et al.* (eds) *Progress in geography 3.* London: Edward Arnold, 1–86.

_____ 1971b: America! America? Views on a pot melting 1. Questions of social relevance. *Area* 3, 150–3.

_____ 1979: About half Marx for the transition from feudalism to capitalism. *Area* 11, 47–51.

PRUNTY, M. C. 1979: Clark in the early 1940s. *Annals of the Association of America Geographers* 69, 42–5.

PUDUP, M. B. 1988: Arguments within regional geography. *Progress in Human Geography* 12, 369–90.

QUAINI, M. 1982: *Geography and marxism.* Oxford: Blackwell.

RADFORD, J. P. 1981: The social geography of the nineteenth century US city. In D. T. Herbert and R. J. Johnston (eds) *Geography and the urban environment* 4. Chichester: John Wiley, 257–93.

RAVENSTEIN, E. G. 1885: The laws of migration. *Journal of the Royal Statistical Society* 48, 167–235.

RAWSTRON, E. M. 1958: Three principles of industrial location. *Transactions, Institute of British Geographers* 25, 135–42.

RAY, D. M., VILLENEUVE, P. Y. and ROBERGE, R. A. 1974: Functional prerequisites, spatial diffusion, and allometric growth. *Economic Geography* 50, 341–51.

REES, J. 1985: *Natural resources: allocation economics, and policy.* London: Methuen.

REES, P. H. and WILSON, A. G. 1977: *Spatial population analysis.* London: Edward Arnold.

RELPH, E. 1970: An inquiry into the relations between phenomenology and geography. *The Canadian Geographer* 14, 193–201.

_____ 1976: *Place and placelessness.* London: Pion.

_____ 1977: Humanism, phenomenology, and geography. *Annals of the Association of American Geographers* 67, 177–9.

_____ 1981a: Phenomenology. In M. E. Harvey and B. P. Holly (eds) *Themes in geographic thought.* London: Croom Helm, 99–114.

_____ 1981b: *Rational landscapes and humanistic geography.* London: Croom Helm.

RENFREW, A. C. 1981: Space, time and man. *Transactions, Institute of British Geographers* NS6, 257–78.

REYNOLDS, R. B. 1956: Statistical methods in geographical research. *Geographical Review* 46, 129–32.

RHIND, D. W. 1981: Geographical information systems in Britain. In N . Wrigley and R. J. Bennett (eds) *Quantitative geography.* London: Routledge & Kegan Paul, 17–35.

_____ 1986: Remote sensing, digital mapping and GIS: the creation of government policy in the UK. *Environment and Planning C: Government and Policy* 4, 91–100.

_____ 1989: Computing, academic geography, and the world outside. In B. Macmillan (ed.) *Remodelling geography.* Oxford: Basil Blackwell, 177–90.

_____ 1996: Differential research funding – a comment on Smith. *Area* 28, 96–7.

RHIND, D. W. and ADAMS, T. A. 1980: Recent developments in surveying and mapping. In E. H. Brown (ed.) *Geography, yesterday and tomorrow.* Oxford: Oxford University Press, 181–99.

RHIND, D. W. and HUDSON, R. 1981: *Land use* London: Methuen.

RHIND, D. W. and MOUNSEY, H. 1989: The Chorley committee and 'Handling geographical information'. *Environment and Planning A* 21, 571–86.

RICHARDS, K. S. and WRIGLEY, N. 1996: Geography in the United Kingdom 1992–1996. *The Geographical Journal* 162, 41–62.

RICHARDSON, H. W. 1973: *The economics of urban size.* Farnborough: Saxon House.

RIDDELL, J. B. 1970: *The spatial dynamics of modernization in Sierra Leone.* Evanston, IL.: Northwestern University Press.

RIESER, R. 1973: The territorial illusion and the behavioural sink: critical notes on behavioural geography. *Antipode* 5(3), 52–7.

ROBINSON, A. H. 1956: The necessity of weighting values in correlation analysis of area data. *Annals of the Association of American Geographers* 46, 233–6.

_____ 1961: On perks and pokes. *Economic Geography* 37, 181–3.

_____ 1962: Mapping the correspondence of isarithmic maps. *Annals of the Association of American Geographers* 52, 414–25.

ROBINSON, A. H. and BRYSON, R. A. 1957: A method for describing quantitatively the correspondence of geographical distributions. *Annals of the Association of American Geographers* 47, 379–91.

ROBINSON, A. H., LINDBERG, J. B. and BRINKMAN, L. W. 1961: A correlation and regression analysis applied to rural farm densities in the Great Plains. *Annals of the Association of American Geographers* 51, 211–21.

ROBINSON, G. M. 1991: An appreciation of James Wreford Watson with a bibliography of his work. In G. M. Robinson (ed.) *A social geography of Canada.* Toronto: Dundurn Press, 492–506.

ROBINSON, M. E. 1982: Representation, misrepresentation, and uncritical rhetoric. *The Professional Geographer* 34, 224–6.

ROBSON, B. T. 1969: *Urban analysis.* Cambridge: Cambridge University Press.

_____ 1972: The corridors of geography. *Area* 4, 213–14.

_____ 1982: Introduction. In B. T. Robson and J. Rees (eds) *Geographical agenda for a changing world.* London: Social Science Research Council, 1–6.

_____ 1984: A pleasant pain. In M. Billinge, D. Gregory and R. Martin (eds) *Recollections of a revolution.* London: Macmillan, 104–6.

RODER, W. 1961: Attitudes and knowledge on the Topeka flood plain. In G. F. White (ed.) *Papers on flood problems.* Chicago: University of Chicago, Department of Geography, Research Paper 70, 62–83.

RODGERS, A. 1955: Changing locational patterns in the Soviet pulp and paper industries. *Annals of the Association of American Geographers* 45, 85–104.

ROONEY, J. F., ZELINSKY, W. and LOUDER, D. R. (eds) 1982: *This remarkable continent: an atlas of United States and Canadian society and cultures.* College Station: Texas A and M University Press.

ROSE, C. 1980: Human geography as text interpretation. In A. Buttimer and D. Seamon (eds) *The human experience of space and place.* London: Croom Helm, 123–34.

_____ 1981: Wilhelm Dilthey's philosophy of human understanding. In D. R. Stoddart (ed.) *Geography, ideology and social concern.* Oxford: Basil Blackwell, 99–133.

_____ 1987: The problem of reference and geographic structuration. *Environment and Planning D: Society and Space* 5, 93–112.

ROSE, D. 1987: Home ownership, subsistence, and historical change: the mining district of West Cornwall in the late nineteenth century. In N. J. Thrift and P. Williams (eds) *Class and space: the making of urban society*. London: Routledge, 1, 8–53.

ROSE, G. 1993: *Feminism and geography*. Cambridge: Polity Press.

_____ 1994: The cultural politics of place: local representation and oppositional discourse in two films. *Transactions, Institute of British Geographers* NS19, 46–60.

_____ 1995: Tradition and paternity: same difference? *Transactions, Institute of British Geographers* NS20, 414–16.

ROSE J. K. 1936: Corn yield and climate in the Corn Belt. *Geographical Review* 26, 88–102.

ROSS, R. S. J. 1983: Facing Leviathan: public policy and global capitalism. *Economic Geography* 59, 144–60.

ROTHSTEIN, J. 1958: *Communication, organization and science*. Colorado: Falcon's Wing Press.

ROWLES, G. D. 1978: Reflections on experiential field work. In D. Ley and M. S. Samuels (eds), *Humanistic geography: problems and prospects*. London: Croom Helm, 173–93.

ROWNTREE, L., FOOTE, K. E. and DOMOSH, M. 1989: Cultural geography. In G. L. Gaile and C. J. Willmott (eds) *Geography in America*. Columbus, OH: Merrill, 209–17.

RUSHTON, G. 1969: Analysis of spatial behavior by revealed space preference. *Annals of the Association of American Geographers* 59, 391–400.

_____ 1979: On behavioral and perception geography. *Annals of the Association of American Geographers* 69, 463–4.

SAARINEN, T. F. 1979: Commentary: critique of Bunting-Guelke paper. *Annals of the Association of American Geographers* 69, 464–8.

SACK, R. D. 1972: Geography, geometry and explanation. *Annals of the Association of American Geographers* 62, 61–78.

_____ 1973a: Comment in reply. *Annals of the Association of American Geographers* 63, 568–9.

_____ 1973b: A concept of physical space in geography. *Geographical Analysis* 5, 16–34.

_____ 1974a: The spatial separatist theme in geography. *Economic Geography* 50, 1–19.

_____ 1974b: Chorology and spatial analysis. *Annals of the Association of American Geographers* 64, 439–52.

_____ 1981: *Conceptions of space in social thought*. London: Macmillan.

_____ 1983: Human territoriality: a theory. *Annals of the Association of American Geographers* 73, 55–74.

_____ 1986: *Human territoriality: its theory and history*. Cambridge: Cambridge University Press.

SAID, E. 1978: *Orientalism*. New York: Harper.

SAMUELS; M. S. 1978: Existentialism and human geography. In D. Ley and M. S. Samuels (eds) *Humanistic geography: problems and prospects*. Chicago: Maaroufa Press, 22–40.

SANDBACH, F. 1980: *Environment, ideology and policy*. Oxford: Blackwell.

SANTOS, M. 1974: Geography, marxism and underdevelopment. *Antipode* 6(3), 1–9.

SARRE, P. 1987: Realism in practice. *Area* 19, 3–10.

SARRE, P., PHILLIPS, D. and SKELLINGTON, R. 1989: *Ethnic minority housing: explanations and policies*. Aldershot: Avebury.

SAUER, C. O. 1925: The morphology of landscape. *University of California Publications in Geography* 2, 19–54.

_____ 1941: Foreword to historical geography. *Annals of the Association of American Geographers* 31, 1–24.

_____ 1956a: The education of a geographer. *Annals of the Association of American Geographers* 46, 287–99.

_____ 1956b: The agency of man on earth. In W. L. Thomas (ed.) *Man's role in changing the face of the earth.* Chicago: University of Chicago Press, 49–69.

_____ 1956c. Retrospect. In W. L. Thomas (ed.) *Man's role in changing the face of the earth.* Chicago: University of Chicago Press, 1131–5.

SAUNDERS, P. and WILLIAMS, P. R. 1986: The new conservatism: some thoughts on recent and future developments in urban studies. *Environment and Planning D: Society and Space* 4, 393–9.

_____ 1987: For an emancipated social science. *Environment and Planning D: Society and Space* 5, 427–30.

SAYER, A. 1979: Epistemology and conceptions of people and nature in geography. *Geoforum* 10, 19–44.

_____ 1981: Defensible values in geography. In D. T. Herbert and R. J. Johnston (eds) *Geography and the urban environment*, 4. Chichester: John Wiley, 29–56.

_____ 1982: Explanation in economic geography. *Progress in Human Geography* 6, 68–88.

_____ 1983: Notes on geography and the relationship between people and nature. In The London Group of the Union of Socialist Geographers, *Society and nature.* London, 47–57.

_____ 1984: *Method in social science: a realist approach.* London: Hutchinson.

_____ 1985a: Realism and geography. In R. J. Johnston (ed.) *The future of geography.* London: Methuen, 159–73.

_____ 1985b: The difference that space makes. In D. Gregory and J. Urry (eds) *Social relations and spatial structures.* London: Macmillan, 49–66.

_____ 1987: Hard work and its alternatives. *Environment and Planning D: Society and Space* 5, 395–9.

_____ 1989a: The new regional geography and problems of narrative. *Environment and Planning D: Society and Space* 7, 253–76.

_____ 1989b: On the dialogue between humanism and historical materialism in geography. In A. Kobayashi and S. Mackenzie (eds) *Remaking human geography.* Boston: Unwin Hyman, 206–26.

_____ 1992a: *Method in social science: a realist approach* (second edition). London: Routledge.

_____ 1992b: What's left to do? A reply to Hadjimichalis and Smith. *Antipode* 24, 214–17.

_____ 1992c: Radical geography and Marxist political economy: towards a re-evaluation. *Progress in Human Geography* 16, 343–60.

_____ 1994a: Realism and space: a reply to Ron Johnston. *Political Geography* 13, 107–9.

_____ 1994b: Postmodernist thought in geography: a realist view. *Antipode* 25, 320–44.

_____ 1995: *Radical political economy: a critique.* Oxford: Blackwell Publishers.

SAYER, A. and MORGAN, K. 1985: A modern industry in a declining region: links between method, theory and policy. In D. Massey and R. Meegan (eds) *Politics and method: contrasting studies in industrial geography.* London: Methuen, 144–68.

SAYER, A. and WALKER, R. A. 1992: *The new social economy: reworking the division of labour.* Oxford: Blackwell Publishers.

SCHAEFER, F. K. 1953: Exceptionalism in geography: a methodological examination. *Annals of the Association of American Geographers* 43, 226–49.

SCHOENBERGER, E. 1993: On knowing what to know. *Environment and Planning A* 25, 1225–8.

SCOTT, A. J. 1982: The meaning and social origins of discourse on the spatial foundations of society. In P. R. Gould and G. Olsson (eds) A *search for common ground*. London: Pion, 141–56.

_____ 1985: Location processes, urbanization, and territorial development: an exploratory essay. *Environment and Planning A* 17, 479–501.

_____ 1986: Industrialization and urbanization: a geographical agenda. *Annals of the Association of American Geographers* 76, 25–37.

_____ 1988: *Metropolis*. Los Angeles: University of California Press.

SCOTT, A. J. and COOKE, P. N. 1988: The new geography and sociology of production. *Environment and Planning D: Society and Space* 6, 241–4.

SCOTT, A. J. and STORPER, M. (eds) 1985: *Production, work, territory*. Boston: George Allen & Unwin.

SEAMON, D. 1984: Phenomenology and environment-behavior research. In G. T. Moore and E. Zube (eds) *Advances in environment, behavior and design*. New York: Plenum, 3–36.

SEMPLE, E. C. 1911: *Influences of geographical environment*. New York: Henry Holt.

SHANNON, G. W. and DEVER, G. E. A. 1974: *Health care delivery: spatial perspectives*. New York: McGraw-Hill.

SHARP, J. 1993: Publishing American identity: popular geopolitics, myth, and *The Reader's Digest*. *Political Geography* 12, 491–504.

SHEPPARD, E. S. 1979: Gravity parameter estimation. *Geographical Analysis* 11, 120–33.

_____ 1993: Automated geography: what kind of geography for what kind of society? *The Professional Geographer* 45, 457–60.

_____ 1995: Dissenting from spatial analysis. *Urban Geography* 16, 283–303.

SHORT, J. R. 1984: *The urban arena: capital, state and community in contemporary Britain*. London: Macmillan.

SIDDALL, W. R. 1961: Two kinds of geography. *Economic Geography* 36, facing page 189.

SIMMONS, I. G. 1993: *Interpreting nature: cultural constructions of the environment*. London: Routledge.

SIMON, H. A. 1957: *Models of man: social and rational*. New York: John Wiley.

SINCLAIR, J. C. and KISSLING, C. C. 1971: A network analysis approach to fruit distribution planning. *Proceedings, Sixth New Zealand Geography Conference*, vol. 1, Christchurch, 131–6.

SLATER, D. 1973: Geography and underdevelopment – 1. *Antipode* 5(3), 21–33.

_____ 1975: The poverty of modern geographical enquiry. *Pacific Viewpoint* 16, 159–76.

_____ 1977: Geography and underdevelopment – 2. *Antipode* 9, 1–31.

SMAILES, A. E. 1946: The urban mesh of England and Wales. *Transactions and Papers, Institute of British Geographers* 11, 85–101.

SMALLMAN-RAYNOR, M, and CLIFF, A. D. 1990: Aquired Immune Deficiency Syndrome (AIDS): literature, geographical origins and global patterns. *Progress in Human Geography* 14, 157–213.

SMALLMAN-RAYNOR, M, CLIFF, A. D. and HAGGETT, P. 1992: *Atlas of AIDs*. Oxford: Blackwell Reference.

SMITH, C. T. 1965: Historical geography: current trends and prospects. In R. J. Chorley and P. Haggett (eds) *Frontiers in geographical teaching*. London: Methuen, 118–43.

SMITH, D. M. 1971: America! America? Views on a pot melting. 2. Radical geography – the next revolution? *Area* 3, 153–7.

_____ 1973a: Alternative 'relevant' professional roles. *Area* 5, 1–4.

_____ 1973b: *The geography of social well-being in the United States*. New York: McGraw-Hill.

_____ 1977: *Human geography: a welfare approach*. London: Edward Arnold.

_____ 1979: *Where the grass is greener: living in an unequal world*. London: Penguin.

_____ 1981: *Industrial location: an economic geographical analysis* (2nd edn). New York: John Wiley.

_____ 1984: Recollections of a random variable. In M. Billinge, D. Gregory and R. Martin (eds) *Recollections of a revolution*. London: Macmillan, 117–33.

_____ 1985: The 'new blood' scheme and its application to geography. *Area* 17, 237–43.

_____ 1986: UGC research ratings: pass or fail? *Area* 18, 247–9.

_____ 1988a: Towards an interpretative human geography. In J. Eyles and D. M. Smith (eds) *Qualitative methods in human geography*. Cambridge: Polity Press, 255–67.

_____ 1988b: On academic performance. *Area* 20, 3–13.

_____ 1994: *Geography and social justice*. Oxford: Blackwell Publishers.

_____ 1995: Against differential research funding. *Area* 27, 79–83.

_____ 1996: Reply to Rhind – value for money, the continuing debate. *Area* 28, 97–101.

SMITH, J. M. 1996: Geographical rhetoric: modes and tropes of appeal. *Annals of the Association of American Geographers* 86, 1–20.

SMITH, N. 1979: Geography, science and post-positivist modes of explanation. *Progress in Human Geography* 3, 365–83.

_____ 1984: *Uneven development: nature, capital and the production of space*. Oxford: Basil Blackwell.

_____ 1986: On the necessity of uneven development. *International Journal of Urban and Regional Research* 10, 87–104.

_____ 1987a: Danger of the empirical turn: The CURS initiative. *Antipode* 19, 59–68.

_____ 1987b: Rascal concepts, minimalizing discourse, and the polities of geography. *Environment and Planning D: Society and Space* 5, 377–83.

_____ 1990: Geography as museum: private history and conservative idealism in *The Nature of Geography*. In J. N. Entrikin and S. D. Brunn (eds) *Reflections on Richard Hartshorne's The Nature of Geography*. Washington: Association of American Geographers, 89–120.

_____ 1991: What's left? A lot's left. *Antipode* 23, 406–18.

_____ 1992: History and philosophy of geography: real wars, theory wars. *Progress in Human Geography* 16, 257–71.

_____ 1994: Shaking loose the colonies: Isaiah Bowman and the 'de-colonization' of the British Empire. In A. Godlewska and N. Smith (eds) *Geography and empire*. Oxford: Blackwell Publishers, 270–99.

SMITH, N. and WILLIAMS, P. (eds) 1986: *Gentrification of the city*. Boston: Allen and Unwin.

SMITH, S. J. 1984: Practising humanistic geography. *Annals of the Association of American Geographers* 74, 353–74.

_____ 1988: Constructing local knowledge: the analysis of self in everyday life. In J. Eyles and D. M. Smith (eds) *Qualitative methods in human geography*. Cambridge: Polity Press, 17–38.

_____ 1989: Society, space and citizenship: a human geography for the 'new times'? *Transactions, Institute of British Geographers* NS14, 144–56.

SMITH, T. R. 1984: Artificial intelligence and its applicability to geographical problem solving. *The Professional Geographer* 36, 147–58.

SMITH, T. R., CLARK, W. A. V. and COTTON, J. 1984: Deriving and testing production system models of sequential decision-making behavior. *Geographical Analysis* 16, 191–222.

SMITH, W. 1949: *An economic geography of Great Britain.* London: Methuen.

SOJA, E. W. 1968: *The geography of modernization in Kenya.* Syracuse: Syracuse University Press.

––––––– 1980: The socio-spatial dialectic. *Annals of the Association of American Geographers* 70, 207–25.

––––––– 1985: The spatiality of social life: towards a transformative retheorization. In D. Gregory and J. Urry (eds) *Social relations and spatial structures.* London: Macmillan, 90–127.

––––––– 1989: *Postmodern geographies.* London: Verso.

SOJA, E.W. and HOOPER, B. 1993: The spaces that difference makes: some notes on the geographical margins of the new cultural politics. In M. Keith and S. Pile (eds) *Place and the politics of identity.* London: Routledge, 183–205.

SPATE, O. H. K. 1957: How determined is possibilism? *Geographical Studies* 4, 3–12.

––––––– 1960a: Quantity and quality in geography. *Annals of the Association of America Geographers* 50, 477–94.

––––––– 1960b: Lord Kelvin rides again. *Economic Geography* 36, facing page 1.

––––––– 1963: Letter to the editor. *Geography* 48, 206.

––––––– 1989: Foreword. In F. W. Boal and D. N. Livingstone (eds) *The behavioural environment.* London: Routledge, xvii-xx.

SPENCER, C. P. and BLADES, M. 1986: Pattern and process: a review essay on the relationship between geography and environmental psychology. *Progress in Human Geography* 10, 230–48.

SPENCER, H. 1892: A *system of synthetic philosophy,* vol. 1: *First principles* (4th edn). New York: Appleton.

SPENCER, J. E. and THOMAS, W. L. 1973: *Introducing cultural geography.* New York: John Wiley.

STAMP, L. D. 1946: *The land of Britain.* London: Longman.

––––––– 1960: *Applied geography.* Harmondsworth: Penguin Books.

––––––– 1966: Ten years on. *Transactions, Institute of British Geographers* 40, 11–20.

STAMP, L. D. and BEAVER, S. H. 1947: *The British Isles.* London: Longman.

STEEL, R. W. 1974: The Third World: geography in practice. *Geography* 59, 189–207.

––––––– 1982: Regional geography in practice. *Geography* 67, 2–8.

STEGMULLER, W. 1976: *The structure and dynamics of theories.* New York: Springer-Verlag.

STEWART, J. Q. 1945: *Coasts, waves and weather.* Boston: Ginn & Co.

––––––– 1947: Empirical mathematical rules concerning the distribution and equilibrium of population. *Geographical Review* 37, 461–85.

––––––– 1956: The development of social physics. *American Journal of Physics* 18, 239–53.

STEWART, J. Q. and WARNTZ, W. 1958: Macrogeography and social science, *Geographical Review* 48, 167–84.

––––––– 1959: Physics of population distribution. *Journal of Regional Science* 1, 99–123.

STODDART, D. R. 1965: Geography and the ecological approach: the ecosystem as a geographic principle and method. *Geography* 50, 242–51.

––––––– 1966: Darwin's impact on geography. *Annals of the Association of American Geographers* 56, 683–98.

––––––– 1967a: Growth and structure of geography. *Transactions, Institute of British Geographers* 41, 1–19.

––––––– 1967b: Organism and ecosystem as geographic models. In R. J. Chorley and P. Haggett (eds) *Models in geography.* London: Methuen, 511–47.

_____ 1975a: The RGS and the foundations of geography at Cambridge. *The Geographical Journal* 141, 216–39.

_____ 1975b: Kropotkin, Reclus and relevant geography. *Area* 7, 188–90.

_____ 1977: The paradigm concept and the history of geography. Abstract of a paper for the conference of the International Geographical Union Commission on the History of Geographic Thought, Edinburgh.

_____ 1981a: The paradigm concept and the history of geography. In D. R. Stoddart (ed.) *Geography, ideology and social concern.* Oxford: Blackwell, 70–80.

_____ 1981b: Ideas and interpretation in the history of geography. In D. R. Stoddart (ed.) *Geography, ideology and social concern.* Oxford: Blackwell 1–7.

_____ 1986: *On geography: and its history.* Oxford: Basil Blackwell.

_____ 1987: To claim the high ground: geography for the end of the century. *Transactions, Institute of British Geographers* NS12, 327–36.

_____ 1990: Epilogue: homage to Richard Hartshorne. In J. N. Entrikin and S. D. Brunn (eds) *Reflections on Richard Hartshorne's The Nature of Geography.* Washington: Association of American Geographers, 163–6.

_____ 1991: Do we need a feminist historiography of geography – and if we do, what should it be? *Transactions, Institute of British Geographers* NS16, 484–7

STORPER, M. 1987: The post-Enlightenment challenge to Marxist urban studies. *Environment and Planning D: Society and Space* 5, 418–26.

STOUFFER, S. A. 1940: Intervening opportunities: a theory relating mobility and distance. *American Sociological Review* 5, 845–67.

SUI, D. Z. 1994: GIS and urban studies: positivism, post-positivism and beyond. *Urban Geography* 15, 258–78.

SUMMERFIELD, M. A. 1983: Population, samples and statistical inference in geography. *The Professional Geographer* 35, 143–8.

SUPPE, F. 1977a: The search for philosophic understanding of scientific theories. In F. Suppe (ed.) *The structure of scientific theories.* Urbana: University of Illinois Press, 3–233.

_____ 1977b: Exemplars, theories and disciplinary matrices. In F. Suppe (ed.) *The structure of scientific theories.* Urbana: University of Illinois Press, 473–99.

_____ 1977c: Afterword – 1977. In F. Suppe (ed.), *The structure of scientific theories.* Urbana: University of Illinois Press, 617–730.

SVIATLOVSKY, E. E. and EELS, W. C. 1937: The centrographical method and regional analysis. *Geographical Review* 27, 240–54.

SYMANSKI, R. and PICKARD, J. 1996: Rules by which we are judged. *Progress in Human Geography* 20, 175–82.

TAAFFE, E. J. 1970: *Geography.* Englewood Cliffs, NJ: Prentice-Hall.

_____ 1974: The spatial view in context. *Annals of the Association of American Geographers* 64, 1–16.

_____ 1979: In the Chicago area. *Annals of the Association of American Geographers* 69, 133–8.

_____ 1993: Spatial analysis: development and outlook. *Urban Geography* 14, 422–33.

TAAFFE, E. J., MORRILL, R. L. and GOULD, P. R. 1963: Transport expansion in underdeveloped countries: a comparative analysis. *The Geographical Review* 53, 503–29.

TATHAM, G. 1953: Environmentalism and possibilism. In G. Taylor (ed.) *Geography in the twentieth century.* London: Methuen, 128–64.

TAYLOR, E. G. R 1937: Whither geography? a review of some recent geographical texts. *Geographical Review* 27, 129–35.

TAYLOR, P. J. 1971a: Distance decay curves and distance transformations. *Geographical Analysis* 3, 221–38.

_____ 1971b: Distances within shapes: an introduction to a new family of finite frequency distributions. *Geografiska Annaler* 53B, 40–53.

_____ 1976: An interpretation of the quantification debate in British geography. *Transactions, Institute of British Geographers* NS1, 129–42.

_____ 1978: Political geography. *Progress in Human Geography* 2, 53–62.

_____ 1979: 'Difficult-to-let', 'difficult-to-live-in', and sometimes 'difficult-to-get-out-of': an essay on the provision of council housing. *Environment and Planning A* 11, 1305–20.

_____ 1981a: Factor analysis in geographical research. In R. J. Bennett (ed.) *European progress in spatial analysis*. London: Pion, 251–67.

_____ 1981b: Geographical scales within the world-economy approach. *Review* 5(1), 3–11.

_____ 1982: A materialist framework for political geography. *Transactions, Institute of British Geographers* NS7, 15–34.

_____ 1985a: The geography of elections. In M. Pacione (ed.) *Progress in political geography*. London: Croom Helm, 243–72.

_____ 1985b: *Political geography: world-economy, nation-state and community.* London: Longman.

_____ 1985c: The value of a geographical perspective. In R. J. Johnston (ed.) *The future of geography*. London: Methuen, 92–110.

_____ 1986: An exploration into world-systems analysis of political parties. *Political Geography Quarterly* 5, S5–S20.

_____ 1988: History's dialogue: an exemplification from political geography. *Progress in Human Geography* 12, 1–14.

_____ 1989: *Political geography: world-economy, nation-state and community* (second edition). London: Longman.

_____ 1990a: Editorial comment: GKS. *Political Geography Quarterly* 9, 211–12.

_____ 1990b: *Britain and the cold war: 1945 as geopolitical transition*. London: Belhaven Press.

_____ 1990c: Journeyman editor. *The Professional Geographer* 42, 359–60.

_____ 1993: Full circle, or new meaning for the global? In R. J. Johnston (ed.) *The challenge for geography. A changing world: a changing discipline.* Oxford: Blackwell Publishers, 181–97.

_____ 1994: The state as container: territoriality in the modern world-system. *Progress in Human Geography* 18, 151–62.

_____ 1995: Beyond containers: internationality, interstateness, interterritoriality. *Progress in Human Geography* 19, 1–15.

_____ 1996: Embedded statism and the social sciences: opening up to new spaces. *Environment and Planning A.*

TAYLOR, P. J. and GUDGIN, G. 1976: A statistical theory of electoral redistricting. *Environment and Planning A* 8, 43–58.

TAYLOR, P. J. and JOHNSTON, R. J. 1979: *Geography of elections.* Harmondsworth: Penguin Books.

_____ 1985: The geography of the British state. In J. R. Short and A. M. Kirby (eds) *The human geography of contemporary Britain.* London: Macmillan, 23–39.

TAYLOR, P. J. and OVERTON, M. 1991: Further thoughts on geography and GIS. *Environment and Planning A* 23, 1087–94.

TESCH, R. 1990: *Qualitative research: analysis types and research tools.* Brighton: Falmer Press.

THOMAN, R. S. 1965: Some comments on *The Science of Geography*. *The Professional Geographer* 17(6), 8–10.

THOMAS, E. N. 1960: Areal associations between population growth and selected factors in the Chicago urbanized area. *Economic Geography* 36, 158–70,

_____ 1968: Maps of residuals from regression. In B. J. L. Berry and D. F. Marble (eds) *Spatial analysts*. Englewood Cliffs, NJ: Prentice-Hall, 326–52.

THOMAS, E. N. and ANDERSON, D. L. 1965: Additional comments on weighting values in correlation analysis of areal data. *Annals of the Association of American Geographers* 55, 492–505.

THOMAS, R. W. 1982: *Information statistics in geography*. Norwich: Geo Books.

_____ 1992: *Geomedical systems: intervention and control*. Routledge: London.

THOMAS, W. L. (ed.) 1956: *Man's role in changing the face of the earth*. Chicago: University of Chicago Press.

THOMPSON, J. H. *et al*. 1962: Toward a geography of economic health: the case of New York state. *Annals of the Association of American Geographers* 52, 1–20.

THORNE, C. R. (ed.) 1993: University Funding Council Research Selectivity Exercise, 1992: implications for higher education in geography. *Journal of Geography in Higher Education* 17, 167–99.

THORNES, J. B. 1989a: Geomorphology and grass roots models. In B. Macmillan (ed.) *Remodelling geography*. Oxford: Basil Blackwell, 3–21.

_____ 1989b: Environmental systems. In M. J. Clark, K. J. Gregory and A. M. Gurnell (eds) *Horizons in physical geography*. London: Macmillan, 27–46.

THRALL, G. 1. 1985: Scientific geography. *Area* 17, 254.

_____ 1986: Reply to Felix Driver and Christopher Philo. *Area* 18, 162–3.

THRIFT, N. J. 1977: *An introduction to time geography*. CATMOG 13, Norwich: Geo Books.

_____ 1979: Unemployment in the inner city: urban problem or structural imperative? A review of the British experience. In D. T. Herbert and R. J. Johnston (eds) *Geography and urban environment: progress in research and applications*, vol. 2. London: John Wiley, 125–226.

_____ 1981: Behavioural geography. In N. Wrigley and R. J. Bennett (eds) *Quantitative geography*. London: Routledge & Kegan Paul, 352–65.

_____ 1983a: On the determination of social action in space and time. *Environment and Planning D: Society and Space* 1, 23–57.

_____ 1983b: Literature, the production of culture and the politics of place. *Antipode* 15, 12–23.

_____ 1987: No perfect symmetry. *Environment and Planning D: Society and Space* 5, 400–7.

THRIFT, N. J. and OLDS, K. 1996: Refiguring the economic in economic geography. *Progress in Human Geography* 20, 311–17.

THRIFT, N. J. and PRED, A. R. 1981: Time geography: a new beginning. *Progress in Human Geography* 5, 277–86.

TIMMERMANS, H. and GOLLEDGE, R. G. 1990: Applications of behavioural research on spatial problems: II Preference and choice. *Progress in Human Geography* 14, 311–54.

TIMMINS, N. 1995: *The five giants: a biography of the welfare state*. London: HarperCollins.

TIMMS, D. 1965: Quantitative techniques in urban social geography. In R. J. Chorley and P. Haggett (eds) *Frontiers in geographical teaching*. London: Methuen, 239–65.

TOBLER, W. R. 1995: Migration: Ravenstein, Thornthwaite and beyond. *Urban Geography* 16, 327–43.

TOCALIS, T. R. 1978: Changing theoretical foundations of the gravity concept of human interaction. In B. J. L. Berry (ed.) *The nature of change in geographical ideas*. de Kalb: Northern Illinois University Press, 65–124.

TOMLINSON, R. F. 1989: Geographic information systems and geographers in the 1990s. *The Canadian Geographer* 33, 290–8.

TOULMIN, S. E. 1970: Does the distinction between normal and revolutionary science hold water? In I. Lakatos and A. Musgrave (eds), *Criticism and the growth of knowledge.* London: Cambridge University Press, 39–48.

TREWARTHA, G. T. 1973: Comments on geography and public policy. *The Professional Geographer* 25, 78–9.

TRUDGILL, S. T. 1990: *Barriers to a better environment.* London: Belhaven Press.

TUAN, YI-FU 1971: Geography, phenomenology, and the study of human nature. *The Canadian Geographer* 15, 181–92.

_____ 1974: Space and place: humanistic perspectives. In C . Board *et al.* (eds), *Progress in Geography* 6. London: Edward Arnold, 211–52.

_____ 1975a: Images and mental maps. *Annals of the Association of American Geographers* 65, 205–13.

_____ 1975b: Place: an experiential perspective. *The Geographical Review* 65, 151–65.

_____ 1976: Humanistic geography. *Annals of the Association of American Geographers* 66, 266–76.

_____ 1977: *Space and place.* London: Edward Arnold.

_____ 1978: Literature and geography: implications for geographical research. In D. Ley and M. S. Samuels (eds) *Humanistic geography: prospects and problems.* Chicago: Maaroufa Press, 194–206.

_____ 1979: *Landscapes of fear.* Oxford: Basil Blackwell.

_____ 1982: *Segmented worlds and self.* Minneapolis: University of Minnesota Press.

_____ 1984: *Dominance and affection.* New Haven: Yale University Press.

TULLOCK, G. 1976: *The vote motive.* London: Institute of Economic Affairs.

TURNER, B. L. 1989: The specialist-synthesis approach to the revival of geography: the case of cultural ecology. *Annals of the Association of American Geographers* 79, 88–100.

TURNER, B. L., CLARK, W. C., KATES. R. W., RICHARDS, J. F., MATTHEWS, J. T. and MEYER, W. B. (eds) *The earth as transformed by human action: global and regional changes in the biosphere over the past 300 years.* Cambridge: Cambridge University Press.

TURNER, B. L. and MEYER, W. B. 1985: The use of citation indices in comparing geography programs: an exploratory study. *The Professional Geographer* 37, 271–8.

TURNER, B. L. and VARLYGUIN, D. 1995: Foreign-area expertise in U.S. geography: an assessment of capacity based on foreign-area dissertations, 1977–1991. *The Professional Geographer* 47, 308–14.

TURNER, R. K. 1993: Sustainability: principles and practice. In R. K. Turner (ed.) *Sustainable environmental economics and management: principles and practice.* London: Belhaven Press, 3–36.

ULLMAN, E. L. 1941: A theory of location for cities. *American Journal of Sociology* 46, 853–64.

_____ 1953: Human geography and area research. *Annals of the Association of American Geographers* 43, 54–66.

_____ 1956: The role of transportation and the bases for interaction. In W. L. Thomas (ed.) *Man's role in changing the face of the earth.* Chicago: University of Chicago Press, 862–80.

UPTON, G. J. G. and FINGLETON, B. 1985: *Spatial data analysis by example.* vol. 1: *Point pattern and quantitative data.* Chichester: John Wiley.

URLICH, D. U. 1972: Migrations of the North Island Maoris 1800–1840: a systems view of migration. *New Zealand Geographer* 28, 23–35.

URLICH CLOHER, D. 1975: A perspective on Australian urbanization. In J. M. Powell and M . Williams (eds) *Australian space, Australian time: geographical perspectives.* Melbourne: Oxford University Press, 104–59.

URRY, J. 1985: Space, time and the study of the social. In H. Newby *et al.* (eds) *Restructuring capital.* London: Macmillan, 21–40.

—— 1986: Locality research: the case of Lancaster. *Regional Studies* 20, 233–42.

—— 1987: Society, space and locality. *Environment and Planning D: Society and Space* 5, 435–44.

VALENTINE, G. 1993: Negotiating and maintaining multiple sexual identities. *Transactions, Institute of British Geographers* NS18, 237–43.

VAN DEN DAELE, W. and WEINGART, P. 1976: Resistance and receptivity of science to external direction: the emergence of new disciplines under the impact of science policy. In G. Lemaine *et al.* (eds) *Perspectives on the emergence of scientific disciplines.* The Hague: Mouton, 247–75.

VAN DER LAAN, L. and PIERSMA, A. 1982: The image of man: paradigmatic cornerstone in human geography. *Annals of the Association of American Geographers* 73, 411–26.

VAN DER WUSTEN, H. and O'LOUGHLIN, J. 1986: Claiming new territory for a stable peace: how geography can contribute. *The Professional Geographer* 38, 18–27.

VAN PAASSEN, C. 1981: The philosophy of geography: from Vidal to Hagerstrand. In A. Pred and G. Tornquist (eds) *Space and time in geography.* Lund: C. W. K. Gleerup, 17–29.

VANCE, J. E. 1970: *The merchant's world.* Englewood Cliffs, NJ: Prentice-Hall.

—— 1978: Geography and the study of cities. *Human Geography: Coming of Age. American Behavioral Scientist* 22, 131–49.

VON BERTALANFFY, L. 1950: An outline of general systems theory. *British Journal of the Philosophy of Science* 1, 134–65.

WAGNER, P. L. 1976: Reflections on a radical geography. *Antipode* 8(3), 83–5.

WALBY, S. 1986: *Patriarchy at work.* Cambridge: Polity Press.

WALKER, R. A. 1981a: A theory of suburbanization. In M. J. Dear and A. J. Scott (eds), *Urbanization and urban planning in capitalist societies.* London: Methuen, 383–430.

—— 1981b: Left-wing libertarianism, an academic disorder: a reply to David Sibley. *The Professional Geographer* 33, 5–9.

—— 1989a: What's left to do? *Antipode* 21, 133–65.

—— 1989b: Geography from the left. In G. L. Gaile and C. J. Willmott (eds) *Geography in America.* Columbus, OH: Merrill Publishing Company, 619–51.

WALKER, R. A. and STORPER, M. 1981: Capital and industrial location. *Progress in Human Geography* 5, 473–509.

WALLACE, I. 1989: *The global economic system.* London: Unwin Hyman.

WALMSLEY, D. J. 1972: *Systems theory: a framework for human geographical enquiry.* Research School of Pacific Studies. Department of Human Geography Publication HG/7, Canberra: Australian National University.

—— 1974: Positivism and phenomenology in human geography. *The Canadian Geographer* 18, 95–107.

WALMSLEY, D. J. and SORENSEN, A. D. 1980: What marx for the radicals? an Antipodean viewpoint. *Area* 12, 137–41.

WARD, D. 1971: *Cities and immigrants: a geography of change in nineteenth-century America.* New York: Oxford University Press.

WARF, B. 1986: Ideology, everyday life and emancipatory phenomenology. *Antipode* 18, 268–83.

—— 1988: The resurrection of local uniqueness. In R. G. Golledge, H. Couclelis and P. R. Gould (eds) *A ground for common search:* Santa Barbara: The Santa Barbara Geographical Press, 51–62.

—— 1993: Postmodernism and the localities debate: ontological questions and epistemological implications. *Tijdschrift voor Economische en Sociale Geografie* 84, 162–8.

_____ 1995: Separated at birth? Regional science and social theory. *International Regional Science Review* 17, 185–94.

WARNTZ, W. 1959a: *Toward a geography of price.* Philadelphia: University of Pennsylvania Press.

_____ 1959b: Geography at mid-twentieth century. *World Politics* 11, 442–54.

_____ 1959c: Progress in economic geography. In P. E. James (ed.) *New viewpoints in geography.* Washington: National Council for the Social Studies, 54–75.

_____ 1968: Letter to the editor. *The Professional Geographer* 20, 357.

_____ 1984: Trajectories and coordinates. In M. Billinge, D. Gregory and R. Martin (eds) *Recollections of a revolution.* London: Macmillan, 134–52.

WATERMAN, S. 1985: Not just the milk and honey – now a way of life: Israeli human geography since the six-day war. *Progress in Human Geography* 9, 194–234.

WATERMAN, S. and KLIOT, N. 1990: The political impact on writing the geography of Palestine-Israel. *Progress in Human Geography* 14, 237–60.

WATKINS, J. W. N. 1970: Against normal science. In I. Lakatos and A. Musgrave (eds) *Criticism and the growth of knowledge.* London: Cambridge University Press, 25–38.

WATSON, J. D. 1968: *The double helix: a personal account of the discovery of the structure of DNA.* London: Weidenfeld & Nicolson.

WATSON, J. W. 1953: The sociological aspects of geography. In G. Taylor (ed.) *Geography in the twentieth century.* London: Methuen, 453–99.

_____ 1955: Geography: a discipline in distance. *Scottish Geographical Magazine* 71, 1–13.

_____ 1983: The soul of geography. *Transactions, Institute of British Geographers* NS8, 385–99.

WATSON, M. K. 1978: The scale problem in human geography. *Geografiska Annaler* 60B, 36–47.

WATTS, M. 1988: Deconstructing determinism. *Antipode* 20, 142–68.

_____ 1989: The agrarian crisis in Africa: debating the crisis. *Progress in Human Geography* 13, 1–41.

WATTS, S. J. and WATTS, S. J. 1978: The idealist alternative in geography and history. *The Professional Geographer* 30, 123–7.

WEAVER, J. C. 1943: Climatic relations of American barley production. *Geographical Review* 33, 569–88.

_____ 1954: Crop-combination regions in the Middle West. *Geographical Review* 44, 175–200.

WEBBER, M. J. 1972: *Impact of uncertainty on location.* Canberra: Australian National University Press.

_____ 1977: Pedagogy again: what is entropy? *Annals of the Association of American Geographers* 67, 254–66.

WEBBER, M. M. 1964: The urban place and the non-place urban realm. In M. M. Webber *et al. Explorations into urban structure.* Philadelphia: University of Pennsylvania Press, 79–153.

WESCOAT, J. L. 1992: Common themes in the work of Gilbert White and John Dewey: a pragmatic appraisal. *Annals of the Association of American Geographers* 82, 587–607.

WESTERN, J. S. 1978: Knowing one's place: 'the Coloured people' and the Group Areas Act in Cape Town. In D. Ley and M. S. Samuels (eds) *Humanistic geography: problems and prospects.* Chicago: Maaroufa Press, 297–318.

WHEELER, P. B. 1982: Revolutions, research programmes and human geography. *Area* 14, 1–6.

WHITE, G. F. 1945: *Human adjustment to floods*. Chicago: University of Chicago, Department of Geography, Research Paper 29.

———— 1972: Geography and public policy. *The Professional Geographer* 24, 101–4.

———— 1973: Natural hazards research. In R. J. Chorley (ed.) *Directions in geography*. London: Methuen, 193–216.

———— 1985: Geographers in a perilously changing world. *Annals of the Association of American Geographers* 75, 10–16.

WHITE, P. E. 1985: On the use of creative literature in migration study. *Area* 17, 271–83.

———— 1995: Geography, literature and migration. In R. King, J. Connell and P. White (eds) *Writing across worlds: literature and migration*. London: Routledge, 1–19.

WHITEHAND, J. W. R. 1970: Innovation diffusion in an academic discipline: the case of the 'new' geography. *Area* 2(3), 19–30.

———— 1984: The impact of geographical journals: a look at ISI data. *Area* 16, 185–7.

———— 1985: Contributors to the recent development and influence of human geography: what citation analysis suggests. *Transactions, Institute of British Geographers* NS10, 222–3.

WHITEHAND, J. W. R. and PATTEN, J. H. C. (eds) 1977: Change in the town. *Transactions, Institute of British Geographers* NS2(3).

WILBANKS, T. 1995: Employment trends in geography: introduction. *The Professional Geographer* 47, 315–17.

WILBANKS, T. J. and LIBBEE, M. 1979: Avoiding the demise of geography in the United States. *The Professional Geographer* 31, 1–7.

WILLIAMS, P. R. 1978: Urban managerialism: a concept of relevance? *Area* 10, 236–40.

———— 1982: Restructuring urban managerialism: towards a political economy of urban allocation. *Environment and Planning A* 14, 95–106.

WILSON, A. G. 1967: A statistical theory of spatial distribution models. *Transportation Research* 1, 253–69.

———— 1970: *Entropy in urban and regional modelling*. London: Pion.

———— 1974: *Urban and regional models in geography and planning*. London: John Wiley.

———— 1976a: Catastrophe theory and urban modelling: an application to modal choice. *Environment and Planning A* 8, 351–46.

———— 1976b: Retailers' profits and consumers' welfare in a spatial interaction shopping model. In I. Masser (ed.) *Theory and practice in regional science*. London: Pion, 42–57.

———— 1978: Mathematical education for geographers. Department of Geography, University of Leeds, Discussion Paper 211, Leeds

———— 1981a: *Catastrophe theory and bifurcation: applications to urban and regional systems*. London: Croom Helm.

———— 1981b: *Geography and the environment: systems analytical methods*. Chichester: John Wiley.

———— 1984a: Making urban models more realistic: some strategies for future research. *Environment and Planning A* 16, 1419–32.

———— 1984b: One man's quantitative geography: frameworks, evaluations, uses and prospects. In M. Billinge, D. Gregory and R. L Martin (eds) *Recollections of a revolution: geography as spatial science*. London: Macmillan, 200–26.

———— 1989a: Classics, modelling and critical theory: human geography as structured pluralism. In B. Macmillan (ed.) *Remodelling geography*. Oxford: Basil Blackwell, 61–9.

_____ 1989b: Mathematical models and geographic theory. In D. Gregory and R. Walford (eds) *Horizons in human geography*. London: Macmillan, 29–47.

WILSON, A. G. and BENNETT, R. J. 1985: *Mathematical methods in human geography and planning*. Chichester: John Wiley.

WILSON, A. G., REES, P. H. and LEIGH, C. 1977: *Models of cities and regions*. London: John Wiley.

WISE, M. J. 1975: A university teacher of geography. *Transactions, Institute of British Geographers* 66, 1–16.

_____ 1977: On progress and geography. *Progress in Human Geography* 1, 1–11.

WISNER, B. 1970: Introduction: on radical methodology. *Antipode* 2, 1–3.

WOLCH, J. and DEAR, M. J. (eds) 1989: *The power of geography: how territory shapes social life*. Boston: Unwin Hyman.

WOLDENBERG, M. J. and BERRY, B. J. L. 1967: Rivers and central places: analogous systems? *Journal of Regional Science* 7, 129–40.

WOLF, L. C. 1976: Comments on the Harries-Peet controversy. *The Professional Geographer* 28, 196–8.

WOLPERT, J. 1964: The decision process in spatial context. *Annals of the Association of American Geographers* 54, 337–58.

_____ 1965: Behavioral aspects of the decision to migrate. *Papers and Proceedings Regional Science Association* 15, 159–72.

_____ 1967: Distance and directional bias in inter-urban migratory streams. *Annals of the Association of American Geographers* 57, 605–16.

_____ 1970: Departures from the usual environment in locational analysis. *Annals of the Association of American Geographers* 60, 220–9.

WOLPERT, J., DEAR, M. J. and CRAWFORD, R. 1975: Satellite mental health facilities. *Annals of the Association of American Geographers* 65, 24–35.

WOMEN AND GEOGRAPHY STUDY GROUP, 1984: *Geography and gender: an introduction to feminist geography*. London: Hutchinson.

WOOLDRIDGE, S. W. 1956: *The geographer as scientist*. London: Thomas Nelson.

WOOLDRIDGE, S. W. and EAST, W. G. 1958: *The spirit and purpose of geography*. London: Hutchinson.

WOOLF, R. D. and RESNICK, S. A. 1987: *Economics: marxian versus neoclassical*. Baltimore: Johns Hopkins University Press.

WOOLGAR, S. W. 1976: The identification and definition of scientific collectivities. In G. Lemaine *et al.* (eds) *Perspectives on the emergence of scientific disciplines*. The Hague: Mouton, 233–45.

WRIGHT, J. K. 1925: *The geographical lore of the time of the crusades: a study in the history of medieval science and tradition in western Europe*. New York: American Geographical Society.

_____ 1947: *Terrae incognitae*: the place of imagination in geography. *Annals of the Association of American Geographers* 37, 1–15.

WRIGLEY, E. A. 1965: Changes in the philosophy of geography. In R. J. Chorley and P. Haggett (eds) *Frontiers in geographical teaching*. London: Methuen, 3–24.

WRIGLEY, N. 1979: Developments in the statistical analysis of categorical data. *Progress in Human Geography* 3, 315–55.

_____ 1984: Quantitative methods: diagnostics revisited. *Progress in Human Geography* 8, 525–35.

_____ 1985: *Categorical data analysis for geographers and environmental scientists*. London: Longman.

_____ 1995: Revisiting the modifiable areal unit problem and the ecological fallacy. In A. D. Cliff, P. R. Gould, A. G. Hoare and N. J. Thrift (eds) *Diffusing geography: essays for Peter Haggett*. Oxford: Blackwell Publishers, 49–71.

WRIGLEY, N. and BENNETT, R. J. (eds) 1981: *Quantitative geography: a British view*. London: Routledge & Kegan Paul.

WRIGLEY, N. and LONGLEY, P. A. 1984: Discrete choice modelling in urban analysis. In D. T. Herbert and R. J. Johnston (eds) *Geography and the urban environment: progress in research and applications*, vol. 6. Chichester: John Wiley, 45–94.

WRIGLEY, N. and MATTHEWS, S. 1986: Citation classics and citation levels in geography. *Area* 18, 185–94.

YOUNG, I. M. 1990: *Justice and the politics of difference*. Princeton: Princeton University Press.

ZELINSKY, W. 1970: Beyond the exponentials: the role of geography in the great transition. *Economic Geography* 46, 499–535.

_____ 1973a: Women in geography: a brief factual report. *The Professional Geographer* 25, 151–65.

_____ 1973b: *The cultural geography of the United States*. Englewood Cliffs, NJ: Prentice-Hall.

_____ 1974: Selfward bound? Personal preference patterns and the changing map of American society. *Economic Geography* 50, 144–79.

_____ 1975: The demigod's dilemma. *Annals of the Association of American Geographers* 65, 123–43.

_____ 1978: Introduction. *Human Geography: Coming of Age*. *American Behavioral Scientist* 22, 5–13.

ZELINSKY, W., MONK, J. and HANSON, S. 1982: Women and geography: a review and prospectus. *Progress in Human Geography* 6, 317–66.

ZIMMERMAN, E. W. 1972: *World resources and industries*. Cambridge, MT: The Twentieth Century Fund.

ZIPF, G. K. 1949: *Human behavior and the principle of least effort*. New York: Hafner.

General Index

Author Index